Methods in Enzymology

Volume 114

DIFFRACTION METHODS FOR BIOLOGICAL MACROMOLECULES

Part A

METHODS IN ENZYMOLOGY

EDITORS-IN-CHIEF

Sidney P. Colowick   Nathan O. Kaplan

Methods in Enzymology

Volume 114

Diffraction Methods for Biological Macromolecules

Part A

EDITED BY

Harold W. Wyckoff

DEPARTMENT OF MOLECULAR
BIOPHYSICS AND BIOCHEMISTRY
YALE UNIVERSITY
NEW HAVEN, CONNECTICUT

C. H. W. Hirs

DEPARTMENT OF BIOCHEMISTRY,
BIOPHYSICS, AND GENETICS
UNIVERSITY OF COLORADO HEALTH SCIENCES CENTER
DENVER, COLORADO

Serge N. Timasheff

GRADUATE DEPARTMENT OF BIOCHEMISTRY
BRANDEIS UNIVERSITY
WALTHAM, MASSACHUSETTS

1985

ACADEMIC PRESS, INC.

Harcourt Brace Jovanovich, Publishers

Orlando San Diego New York Austin
London Montreal Sydney Tokyo Toronto

ACADEMIC PRESS, INC.
Orlando, Florida 32887

United Kingdom Edition published by
ACADEMIC PRESS INC. (LONDON) LTD.
24–28 Oval Road, London NW1 7DX

LIBRARY OF CONGRESS CATALOG CARD NUMBER: 54-9110

ISBN 0–12–182014–9

PRINTED IN THE UNITED STATES OF AMERICA

85 86 87 88 9 8 7 6 5 4 3 2 1

Table of Contents

Section III. Data Collection

A. Photographic Techniques

B. Diffractometry

Contributors to Volume 114

Article numbers are in parentheses following the names of contributors.
Affiliations listed are current.

TSUTOMU ARAKAWA (3), *Protein Chemistry, Amgen, Thousand Oaks, California 91320*

U. W. ARNDT (29), *Medical Research Council, Laboratory of Molecular Biology, Cambridge, CB2 2QH, England*

P. J. ARTYMIUK (26), *Laboratory of Molecular Biophysics, Department of Zoology, Oxford OX1 3PS, England*

RICHARD DURBIN (19), *Medical Research Council, Laboratory of Molecular Biology, Cambridge, England*

G. EICHELE (9), *Department of Physiology and Biophysics, Harvard University Medical School, Boston, Massachusetts 02115*

MICHAEL ELDER (18), *Science and Engineering Research Council, Daresbury Laboratory, Warrington WA4 4AD, England*

G. FEHER (4), *Department of Physics, University of California San Diego, La Jolla, California 92093*

ROBERT J. FLETTERICK (25), *Department of Biochemistry and Biophysics, School of Medicine, University of California, San Francisco, California 94143*

R. FOURME (21), *Biochimie Moléculaire et Cellulaire, VA 1131 CNRS and LURE (CNRS and Université-Paris Sud), 91405-Orsay, France*

RONALD HAMLIN (27), *Department of Physics, University of California, San Diego, La Jolla, California 92093*

STEPHEN C. HARRISON (19), *Department of Biochemistry and Molecular Biology, Harvard University, Cambridge, Massachusetts 02138*

STEPHEN R. HOLBROOK (15), *Chemical Biodynamics Laboratory and Lawrence Berkeley Laboratory, University of California at Berkeley, Berkeley, California 94720*

A. J. HOWARD (28), *Protein Engineering Division, Genex Corporation, Gaithersburg, Maryland 20877*

J. N. JANSONIUS (9), *Biozentrum der Universität Basel, Abteilung Strukturbiologie, CH-4056 Basel, Switzerland*

R. KAHN (21), *Physicochimie Structurale, Université Paris-Val de Marne, 94010-Créteil, France and LURE (CNRS and Université-Paris Sud), 91405-Orsay, France*

KENNETH KALATA (30), *Rosenstiel Basic Medical Sciences Center, Brandeis University, Waltham, Massachusetts 02254*

Z. KAM (4), *Polymer Department, Weizmann Institute of Science, Rehovot, Israel*

R. KARLSSON (9), *Biozentrum der Universität Basel, Abteilung Strukturbiologie, CH-4056 Basel, Switzerland*

SUNG-HOU KIM (14, 15), *Department of Chemistry, University of California at Berkeley, Berkeley, California 94720*

ANTHONY A. KOSSIAKOFF (32), *Department of Biocatalysis, Genentech, Inc., South San Francisco, California 94080*

B. W. MATTHEWS (16), *Institute of Molecular Biology and Department of Physics, University of Oregon, Eugene, Oregon 97403*

ALEXANDER MCPHERSON (5, 6, 7), *Department of Biochemistry, University of California at Riverside, Riverside, California 92521*

C. NIELSEN (28), *Department of Chemistry, University of California at San Diego, La Jolla, California 92093*

MAX PERUTZ (1), *Medical Research Council, Laboratory of Molecular Biology, Cambridge CB2 2QH, England*

GREGORY A. PETSKO (12, 13), *Department*

of Chemistry, Massachusetts Institute of Technology, Cambridge, Massachusetts 02139

D. C. PHILLIPS (26), *Laboratory of Molecular Biophysics, Department of Zoology, Oxford OX1 3QU, England*

GEORGE N. PHILLIPS, JR. (8), *Department of Physiology and Biophysics and Department of Biochemistry, University of Illinois, Urbana, Illinois 61801*

WALTER C. PHILLIPS (22, 23), *Rosenstiel Basic Medical Sciences Center, Brandeis University, Waltham, Massachusetts 02254*

IVAN RAYMENT (10, 23), *Department of Biochemistry, University of Arizona, Tucson, Arizona 85721*

MICHAEL G. ROSSMANN (20), *Department of Biological Sciences, Purdue University, West Lafayette, Indiana 47907*

F. R. SALEMME (11), *Central Research and Development Department, E. I. Du Pont de Nemours & Co., Wilmington, Delaware 19898*

B. P. SCHOENBORN (31), *Biology Department, Brookhaven National Laboratory, Upton, New York 11973*

CLARENCE E. SCHUTT (19), *Medical Research Council, Laboratory of Molecular Biology, Cambridge CB2 Q2H, England*

WHAN-CHUL SHIN (14), *Department of Chemistry, Seoul National University, Seoul 151, Korea*

STEVEN A. SPENCER (32), *Genentech, Inc., South San Francisco, California 94080*

ROBERT M. SWEET (2), *Biology Department, Brookhaven National Laboratory, Upton, New York 11973*

JURGEN SYGUSCH (25), *Department of Biochemistry, Faculty of Medicine, University of Sherbrooke, Sherbrooke, Québec, J1H 5N4, Canada*

C. THALLER (9), *Department of Physiology and Biophysics, Harvard University Medical School, Boston, Massachusetts 02115*

SERGE N. TIMASHEFF (3), *Graduate Department of Biochemistry, Brandeis University, Waltham, Massachusetts 02254*

R. W. WARRANT (14), *310 Moonlite Drive, Idaho Falls, Idaho 83401*

L. H. WEAVER (9), *Institute of Molecular Biology, University of Oregon, Eugene, Oregon 97403*

EDWIN M. WESTBROOK (17), *Division of Biological and Medical Research, Argonne National Laboratory, Argonne, Illinois 60439*

E. WILSON (9), *Department of Cell Biology, Stanford University School of Medicine, Stanford, California 94305*

FRITZ K. WINKLER (19), *F. Hoffmann-La Roche & Co. A. G., 4002 Basel, Switzerland*

ALEXANDER WLODAWER (33), *Center for Chemical Physics, National Bureau of Standards, Gaithersburg, Maryland 20899*

HAROLD W. WYCKOFF (24), *Department of Molecular Biophysics and Biochemistry, Yale University, New Haven, Connecticut 06511*

NG. H. XUONG (28), *Departments of Physics, Chemistry, and Biology, University of California at San Diego, La Jolla, California 92093*

Preface

The aim of "Methods in Enzymology" volumes is to present as comprehensively as possible current techniques used in biochemistry, encompassing biological mechanisms, chemistry, and structure. In previous volumes, detailed coverage of solution physical–chemical techniques for the study of protein conformations, conformational changes, and interactions has been provided.

The two volumes on Diffraction Methods for Biological Macromolecules, Parts A and B, are devoted to a description of diffraction methods for biological macromolecules and assemblies. Different aspects of the methods involved in solving, presenting, and interpreting structure so that the reader can proceed knowledgeably and productively toward his goals are presented. We believe that an understanding of the fundamentals of each aspect of the overall method is both intellectually satisfying and practically important.

These two volumes have been divided according to the logical sequence of steps in structure determination. Part A is devoted to the experimental aspects of X-ray crystallography, starting from crystal growth and crystal handling, followed by methods of data collection. Part B includes analysis of the data, covering various aspects of phasing and refinement as well as the structures and methods for their analysis.

The goal which we hoped to attain was twofold: to give biochemists an introduction to the field of macromolecular structure determination, offering them guidance to pathways that are available to determine the structure of a protein, and to give practitioners of X-ray crystallography a comprehensive summary of techniques available to them, some of which are at the state-of-the-art level. We wish to acknowledge with pleasure and gratitude the generous cooperation of the contributors. Their suggestions during the planning and preparation stages have been particularly valuable. Academic Press has provided inestimable help in the assembly of this material. We thank them for their many courtesies.

HAROLD W. WYCKOFF
C. H. W. HIRS
SERGE N. TIMASHEFF

METHODS IN ENZYMOLOGY

EDITED BY

Sidney P. Colowick and Nathan O. Kaplan

VANDERBILT UNIVERSITY
SCHOOL OF MEDICINE
NASHVILLE, TENNESSEE

DEPARTMENT OF CHEMISTRY
UNIVERSITY OF CALIFORNIA
AT SAN DIEGO
LA JOLLA, CALIFORNIA

METHODS IN ENZYMOLOGY

EDITORS-IN-CHIEF

Sidney P. Colowick and Nathan O. Kaplan

VOLUME VIII. Complex Carbohydrates
Edited by ELIZABETH F. NEUFELD AND VICTOR GINSBURG

VOLUME IX. Carbohydrate Metabolism
Edited by WILLIS A. WOOD

VOLUME X. Oxidation and Phosphorylation
Edited by RONALD W. ESTABROOK AND MAYNARD E. PULLMAN

VOLUME XI. Enzyme Structure
Edited by C. H. W. HIRS

VOLUME XII. Nucleic Acids (Parts A and B)
Edited by LAWRENCE GROSSMAN AND KIVIE MOLDAVE

VOLUME XIII. Citric Acid Cycle
Edited by J. M. LOWENSTEIN

VOLUME XIV. Lipids
Edited by J. M. LOWENSTEIN

VOLUME XV. Steroids and Terpenoids
Edited by RAYMOND B. CLAYTON

VOLUME XVI. Fast Reactions
Edited by KENNETH KUSTIN

VOLUME XVII. Metabolism of Amino Acids and Amines (Parts A and B)
Edited by HERBERT TABOR AND CELIA WHITE TABOR

VOLUME 73. Immunochemical Techniques (Part B)
Edited by JOHN J. LANGONE AND HELEN VAN VUNAKIS

VOLUME 74. Immunochemical Techniques (Part C)
Edited by JOHN J. LANGONE AND HELEN VAN VUNAKIS

VOLUME 75. Cumulative Subject Index Volumes XXXI, XXXII, and XXXIV–LX
Edited by EDWARD A. DENNIS AND MARTHA G. DENNIS

VOLUME 76. Hemoglobins
Edited by ERALDO ANTONINI, LUIGI ROSSI-BERNARDI, AND EMILIA CHIANCONE

VOLUME 77. Detoxication and Drug Metabolism
Edited by WILLIAM B. JAKOBY

VOLUME 78. Interferons (Part A)
Edited by SIDNEY PESTKA

VOLUME 79. Interferons (Part B)
Edited by SIDNEY PESTKA

VOLUME 80. Proteolytic Enzymes (Part C)
Edited by LASZLO LORAND

VOLUME 81. Biomembranes (Part H: Visual Pigments and Purple Membranes, I)
Edited by LESTER PACKER

VOLUME 82. Structural and Contractile Proteins (Part A: Extracellular Matrix)
Edited by LEON W. CUNNINGHAM AND DIXIE W. FREDERIKSEN

VOLUME 83. Complex Carbohydrates (Part D)
Edited by VICTOR GINSBURG

VOLUME 84. Immunochemical Techniques (Part D: Selected Immunoassays)
Edited by JOHN J. LANGONE AND HELEN VAN VUNAKIS

VOLUME 85. Structural and Contractile Proteins (Part B: The Contractile Apparatus and the Cytoskeleton)
Edited by DIXIE W. FREDERIKSEN AND LEON W. CUNNINGHAM

VOLUME 86. Prostaglandins and Arachidonate Metabolites
Edited by WILLIAM E. M. LANDS AND WILLIAM L. SMITH

VOLUME 87. Enzyme Kinetics and Mechanism (Part C: Intermediates, Stereochemistry, and Rate Studies)
Edited by DANIEL L. PURICH

VOLUME 88. Biomembranes (Part I: Visual Pigments and Purple Membranes, II)
Edited by LESTER PACKER

VOLUME 89. Carbohydrate Metabolism (Part D)
Edited by WILLIS A. WOOD

VOLUME 90. Carbohydrate Metabolism (Part E)
Edited by Willis A. Wood

VOLUME 91. Enzyme Structure (Part I)
Edited by C. H. W. HIRS AND SERGE N. TIMASHEFF

VOLUME 92. Immunochemical Techniques (Part E: Monoclonal Antibodies and General Immunoassay Methods)
Edited by JOHN J. LANGONE AND HELEN VAN VUNAKIS

VOLUME 93. Immunochemical Techniques (Part F: Conventional Antibodies, Fc Receptors, and Cytotoxicity)
Edited by JOHN J. LANGONE AND HELEN VAN VUNAKIS

VOLUME 94. Polyamines
Edited by HERBERT TABOR AND CELIA WHITE TABOR

VOLUME 95. Cumulative Subject Index Volumes 61–74 and 76–80
Edited by EDWARD A. DENNIS AND MARTHA G. DENNIS

VOLUME 96. Biomembranes [Part J: Membrane Biogenesis: Assembly and Targeting (General Methods; Eukaryotes)]
Edited by SIDNEY FLEISCHER AND BECCA FLEISCHER

VOLUME 97. Biomembranes [Part K: Membrane Biogenesis: Assembly and Targeting (Prokaryotes, Mitochondria, and Chloroplasts)]
Edited by SIDNEY FLEISCHER AND BECCA FLEISCHER

VOLUME 98. Biomembranes [Part L: Membrane Biogenesis (Processing and Recycling)]
Edited by SIDNEY FLEISCHER AND BECCA FLEISCHER

VOLUME 99. Hormone Action (Part F: Protein Kinases)
Edited by JACKIE D. CORBIN AND JOEL G. HARDMAN

VOLUME 100. Recombinant DNA (Part B)
Edited by RAY WU, LAWRENCE GROSSMAN, AND KIVIE MOLDAVE

VOLUME 101. Recombinant DNA (Part C)
Edited by RAY WU, LAWRENCE GROSSMAN, AND KIVIE MOLDAVE

VOLUME 102. Hormone Action (Part G: Calmodulin and Calcium-Binding Proteins)
Edited by ANTHONY R. MEANS AND BERT W. O'MALLEY

VOLUME 103. Hormone Action (Part H: Neuroendocrine Peptides)
Edited by P. MICHAEL CONN

VOLUME 104. Enzyme Purification and Related Techniques (Part C)
Edited by WILLIAM B. JAKOBY

VOLUME 105. Oxygen Radicals in Biological Systems
Edited by LESTER PACKER

VOLUME 106. Posttranslational Modifications (Part A)
Edited by FINN WOLD AND KIVIE MOLDAVE

VOLUME 107. Posttranslational Modifications (Part B)
Edited by FINN WOLD AND KIVIE MOLDAVE

VOLUME 108. Immunochemical Techniques (Part G: Separation and Characterization of Lymphoid Cells)
Edited by GIOVANNI DI SABATO, JOHN J. LANGONE, AND HELEN VAN VUNAKIS

VOLUME 109. Hormone Action (Part I: Peptide Hormones)
Edited by LUTZ BIRNBAUMER AND BERT W. O'MALLEY

VOLUME 110. Steroids and Isoprenoids (Part A)
Edited by JOHN H. LAW AND HANS C. RILLING

VOLUME 111. Steroids and Isoprenoids (Part B)
Edited by JOHN H. LAW AND HANS C. RILLING

VOLUME 112. Drug and Enzyme Targeting (Part A)
Edited by KENNETH J. WIDDER AND RALPH GREEN

VOLUME 113. Glutamate, Glutamine, Glutathione, and Related Compounds
Edited by ALTON MEISTER

VOLUME 114. Diffraction Methods for Biological Macromolecules (Part A)
Edited by HAROLD W. WYCKOFF, C. H. W. HIRS, AND SERGE N. TIMASHEFF

VOLUME 115. Diffraction Methods for Biological Macromolecules (Part B) (in preparation)
Edited by HAROLD W. WYCKOFF, C. H. W. HIRS, AND SERGE N. TIMASHEFF

VOLUME 116. Immunochemical Techniques (Part H: Effectors and Mediators of Lymphoid Cell Functions) (in preparation)
Edited by GIOVANNI DI SABATO, JOHN J. LANGONE, AND HELEN VAN VUNAKIS

VOLUME 117. Enzyme Structure (Part J) (in preparation)
Edited by C. H. W. HIRS AND SERGE N. TIMASHEFF

VOLUME 118. Plant Molecular Biology (in preparation)
Edited by ARTHUR WEISSBACH AND HERBERT WEISSBACH

VOLUME 119. Interferons (Part C) (in preparation)
Edited by SIDNEY PESTKA

VOLUME 120. Cumulative Subject Index Volumes 81–94, 96–101

VOLUME 121. Immunochemical Techniques (Part I: Hybridoma Technology and Monoclonal Antibodies) (in preparation)
Edited by JOHN J. LANGONE AND HELEN VAN VUNAKIS

VOLUME 122. Vitamins and Coenzymes (Part G) (in preparation)
Edited by FRANK CHYTIL AND DONALD B. MCCORMICK

VOLUME 123. Vitamins and Coenzymes (Part H) (in preparation)
Edited by FRANK CHYTIL AND DONALD B. MCCORMICK

VOLUME 124. Hormone Action (Part J: Neuroendocrine Peptides) (in preparation)
Edited by MICHAEL CONN

VOLUME 125. Biomembranes (Part M: Transport in Bacteria, Mitochondria, and Chloroplasts: General Approaches and Transport Systems) (in preparation)
Edited by SIDNEY FLEISCHER AND BECCA FLEISCHER

VOLUME 126. Biomembranes (Part N: Transport in Bacteria, Mitochondria, and Chloroplasts: Protonmotive Force) (in preparation) *Edited by* SIDNEY FLEISCHER AND BECCA FLEISCHER

Methods in Enzymology

Volume 114

DIFFRACTION METHODS FOR BIOLOGICAL MACROMOLECULES

Part A

Section I

Introduction

[1] Early Days of Protein Crystallography

By Max Perutz

In the 1930s, when biochemists began to realize that all chemical reactions in living cells are catalyzed by enzymes, the secret of life was widely believed to lie hidden in the structure of proteins. In earlier times these had been regarded as colloids of indefinite structure, and experiments on gelatin, looked upon as a model of a typical protein, had been a favorite sport among physical chemists with a biological bent. The idea that enzymatic catalysis or oxygen transport might be understood in stereochemical terms seemed utopian. This outlook began to change slowly after the first enzyme, urease, had been crystallized by Sumner in 1928, followed soon afterward by Northrop and Herriott's crystallization of pepsin, trypsin, and chymotrypsin. Early X-ray crystallographers did try to take diffraction pictures of dried crystals of hemoglobin and other proteins, but their weak X-ray patterns were covered up by large backstops or obscured by air scattering. The absence of a diffraction pattern seemed to confirm the belief that proteins lacked a definite structure, until, in the spring of 1934, John Desmond Bernal and Dorothy Crowfoot (now Hodgkin), two young crystallographers at the Department of Mineralogy in Cambridge, England, discovered that crystals of pepsin, suspended in their mother liquor rather than dried, give sharp X-ray diffraction patterns extending to spacings of the order of interatomic distances. Announcing their discovery in *Nature*, they wrote,[1]

> From the intensity of the more distant spots, it can be inferred that the arrangement of atoms inside the protein molecule is also of a perfectly definite kind, although without the periodicities characterising the fibrous proteins. The observations are compatible with oblate spheroidal molecules of diameters about 25 Å and 35 Å, arranged in hexagonal nets, which are related to each other by a hexagonal screw-axis. With this model we may imagine degeneration to take place by the linking up of amino acid residues in such molecules to form chains as in the ring-chain polymerisation of polyoxymethylenes. Peptide chains in the ordinary sense may exist only in the more highly condensed or fibrous proteins, while the molecules of the primary soluble proteins may have their constituent parts grouped more symmetrically around a prosthetic nucleus.
>
> At this stage, such ideas are merely speculative, but now that a crystalline protein has been made to give X-ray photographs, it is clear that we have the means of checking them and, by examining the structure of all crystalline proteins, arriving at far more detailed conclusions about protein structure than previous physical or chemical methods have been able to give.

[1] J. D. Bernal and D. Crowfoot, *Nature (London)* **133,** 794 (1934).

Much of this was nonsense, but in years to come the fulfillment of Bernal and Crowfoot's final prophecy was to become a source of deep satisfaction to crystallographers and of irritation to many physical chemists. In 1935, Dorothy Crowfoot returned to Oxford, and Isidor Fankuchen, a young crystallographer from Brooklyn, New York, took her place as Bernal's assistant. In that year W. M. Stanley crystallized the first virus,[2] tobacco mosaic virus, and the same was done independently by F. C. Bawden and N. W. Pirie at the Department of Biochemistry in Cambridge.[3] They handed their preparations to Bernal and Fankuchen who soon discovered that the virus was not truly crystalline, but formed liquid crystalline tactoids or gels.[4] Both were strongly birefingent and could be aligned in capillaries. The X-ray patterns consisted of a series of equatorial reflections from a hexagonal lattice with spacings that varied between 460 and 150 Å, depending on the virus concentration, and a rich fiber diffraction pattern that was independent of virus concentration and extended to beyond 10 Å spacing. Bernal and Fankuchen called this the intramolecular pattern; it implied that each rod-shaped virus particle is a microcrystal with a regularly repeating molecular pattern along its length.[5] This was a sensational discovery. Judged by its ability to reproduce itself the virus was alive, and yet it was a molecule with a regularly repeating atomic pattern, a duality that seemed to blur the hitherto accepted distinctions between living and nonliving matter.[6]

To obtain X-ray diffraction patterns from dilute virus gels Fankuchen developed techniques that were far ahead of his time. Here is his description[5]:

X-Ray Examination

The standard x-ray apparatus used for the study of small crystals requires to be modified in two directions, both in order to measure low angle reflections and to enhance their intensity. To study adequately the large spacings which were observed, it was necessary either to use longer wave-length radiation or to examine the diffraction at smaller angles. The reason many of the phenomena here described were not discovered earlier was probably due to the choice of the first method. In deciding what radiation to use three factors must be borne in mind—the actual dispersion given by the radiation which at small angles is directly proportional to its wave-length, the absorption in all substances which is approximately proportional to the cube of the wave-length used, and the difficulty of producing the radiation which in turn depends on a number of factors but is certainly far greater for long wave-length radiation than for the standard

[2] W. M. Stanley, *Science* **81,** 644 (1935).

[3] F. C. Bawden and N. W. Pirie, *Proc. R. Soc. London, Ser. B* **123,** 274 (1937).

[4] J. D. Bernal and I. Frankuchen, *Nature (London)* **139,** 923 (1937).

[5] J. D. Bernal and I. Fankuchen, *J. Gen. Physiol.* **25,** 111 (1941).

[6] N. W. Pirie, *in* "Perspectives in Biochemistry" (J. Needham and D. E. Green, eds.), p. 11. Cambridge Univ. Press, London and New York, 1937.

copper K_α radiation. The great advantage of using this radiation is that it is possible to work in air without vacuum cameras, and to use commercial x-ray tubes with large output. The necessary dispersion can be obtained most easily merely by increasing the specimen to film distance. It may, in the future, be necessary to work with long wavelength radiation, but for the moment all the preliminary work can most conveniently be done by changing the camera rather than the radiation. In the second place it was found that the reflection from many of the specimens was extremely feeble and means had to be used to increase the intensity of the scattered radiation, and the contrast achieved on the films. This was done by the adoption, where it was geometrically possible, of slits instead of the usual pinholes, which enabled exposure times to be cut down considerably. Finally, particularly in the case of examination of liquids and wet gels, monochromatic radiation was used. The device introduced by one of us [ref. 7] provided a source of monochromatic radiation, little if anything weaker than that provided by the direct radiation from the tube. Two types of cameras were accordingly used. A long camera for the intermolecular reflections was adapted from the normal Pye camera [ref. 8] by the addition of an arm 40 cm. in length, and the corresponding elongation of the slit system. The slits were made by adjusting lead jaws under the microscope to about 0.10 mm. in width. They were placed 15 cm. apart, and the scattering from the second slit was checked by a slit of 0.20 mm near the crystal specimen. In this way it was possible to obtain a beam which, at a distance of 40 cm. was only 0.7 mm. wide, and therefore subtended an angle of 5′ at the crystal. It was possible to measure diffraction angles of 4.5′ corresponding to a spacing of 1200 Å. In some experiments the whole of the apparatus between the specimen and the plate was replaced by a vacuum camera, but without appreciable improvement. Besides this normal type of long distance camera, monochromatic slitless cameras were also constructed and used. The short cameras used for the intramolecular investigations were of the normal pin-hole type, but with specially small pin-holes giving a beam of divergence 20′, thereby enabling plate distances up to 15 cm. to be used. The longest spacing measurable in this way is about 200 Å. Great care was taken to cut down the lead stop to a size allowing the maximum of pattern to be seen without any overlap of the central beam. It was found useful to use extremely thin lead back stops, leaving a faint trace of the central beam on the plate which could be used as a point of reference. With monochromatic radiation the lead stop could be replaced by one of a definite thickness of aluminum which allowed only a known proportion (1 per cent and 0.01 per cent) of the radiation to be transmitted and could therefore be used to estimate the absolute intensity of reflection.

Of the specimens for x-ray examination only the dry gels could be mounted in air. For other preparations thin-walled tubes or special cells were used. It was found possible to make out of a special low absorption borosilicate glass, tubes of wall thickness of the order of 0.02 mm. These tubes were tested for x-ray absorption when empty and only those which absorbed less than a quarter of the incident characteristic copper K_α radiation were used. There is no advantage in pushing the thinness of the capillaries very much further, because the water with which the specimens are enclosed produces more absorption than the glass.

The specimen was usually secured against change in the capillary by sealing the ends in a small flame. It was of great importance to have absolute sealing to avoid any change in the specimen during the long exposures which often proved necessary. [The best picture from an oriented 6% gel needed an exposure of 400 hr.]

[7] I. Fankuchen, *Nature (London)* **139,** 193 (1937).
[8] J. D. Bernal, *J. Sci. Instrum.* **5,** 241 (1928).

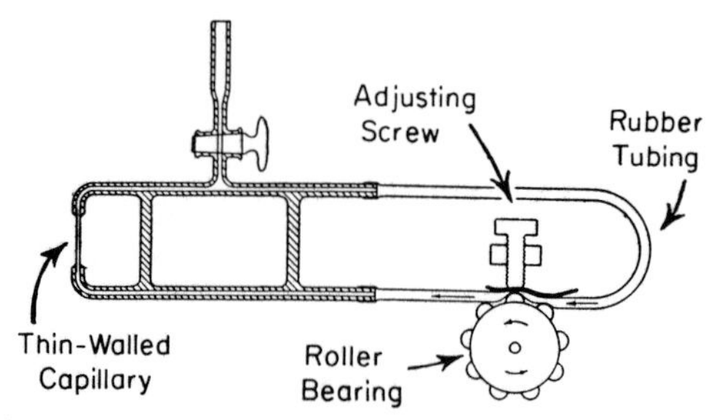

FIG. 1. Peristaltic pump and glass circulatory system for X-ray examination of oriented top layer solution.[5]

These capillaries were used mainly for specimens of wet gel precipitated crystals, and for bottom layer solutions of various strengths. For top layer solution, in order to produce orientation, it was necessary that the liquid should be flowing. A small apparatus was therefore prepared, consisting of a stout capillary tube system, on to which was sealed a piece of thin capillary, the liquid being maintained in motion by means of a peristaltic rubber tube motor [Fig. 1]. In this way both top solution and precipitated crystals were examined in the oriented state.

The disadvantage of capillary tube methods is that specimens which are easily deformable cannot conveniently be introduced into them. For x-ray examinations of such specimens a small brass cell was used. This was prepared as follows. Two brass plates of dimensions $7 \times 7 \times 1$ mm were accurately ground together and a hole of 2 mm. diameter drilled through them. To the ends of this hole were sealed sheets of borosilicate glass blown to bubble thinness. The brass plates were then sealed together with sealing wax. Just before using, the seal was broken with a sharp pointed knife [Fig. 2]. Into this cell the specimen was introduced and the plates were then sealed by closing the fine crack in the sealing wax with a heated knife blade. All these operations were carried out in a low temperature room to avoid evaporation. Such cells showed no appreciable loss of water over periods of many days. It might be thought that the scattering of glass, which though small in volume is of the same order as that of the scattering material to be studied would affect the photographs. This, however, was not found to be so. A certain amount of scattering did occur, but it was in a range equivalent to 4–2 Å, and therefore at angles which were much too high to interfere with any of the significant scattering from the specimens. The same was largely true of the water scattering. It is an interesting but not unexpected fact that prolonged exposures show no appreciable scattering for the water or glass at very low angles. The slight scattering that did occur at low angles was subsequently shown to be largely due to the air in the apparatus, as most of it could be removed by using a vacuum camera.

I remember that vacuum camera. It was a long wooden box which Fankuchen patiently painted over and over again to seal its many leaks. It would have been easier to fill it with hydrogen, but this would have been dangerous since the high tension leads for our X-ray tubes were unprotected wires coming down from the ceiling, and sparks were not uncommon.

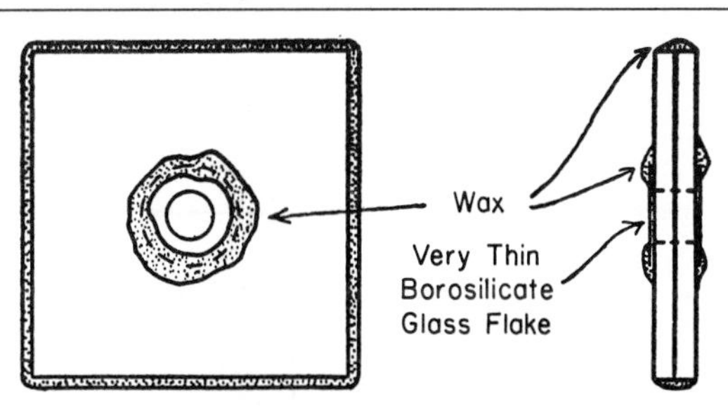

FIG. 2. Small cell for X-ray study of wet gels.[5]

At a lecture at the Royal Institution in London in 1939 Bernal said, "The structure of proteins is the major unsolved problem at the boundary of chemistry and biology today. It is difficult to exaggerate the importance of this study to many branches of science. The protein is the key unit in biochemistry and physiology. . . . All protein molecules that we know now have been made by other protein molecules, and these in turn by others."[9] It does not seem to have occurred to Bernal that the RNA which Bawden and Pirie had found in the tobacco mosaic virus might be concerned with its replication, because nucleic acids were then regarded as undifferentiated molecules, like polysaccharides, which could not fulfill such demanding functions.

I joined Bernal as a graduate student in 1936 and was soon inspired by his visionary faith in the power of X-ray diffraction to solve the structures of molecules as large and complex as enzymes or viruses at a time when the structure of ordinary sugar was still unsolved; he was convinced that knowledge of the structure of proteins would lead us to understand their function. But how was the phase problem to be solved? In his 1939 lecture he suggested "by some physical artifice, such as the introduction of a heavy atom, or the observation of intensity changes on dehydration which have not hitherto been carried out in practice."[9]

I did put the second of these two methods into practice in the years that followed; with immense labor I mapped out the nodes and antinodes of the molecular transform of hemoglobin in the centrosymmetric plane of my monoclinic crystals; helped by W. L. Bragg's minimum wavelength principle,[10] I found many sign relations, but managed to determine the absolute signs of only a few low-order $h0l$ reflections[11] (Fig. 3). I also spent many years collecting and visually measuring the intensities of the reflections needed for a three-dimensional Patterson at 2.8 Å resolution,

[9] J. D. Bernal, *Nature (London)* **143,** 663 (1939).
[10] W. L. Bragg and M. F. Perutz, *Proc. R. Soc. London, Ser. A* **213,** 425 (1952).
[11] M. F. Perutz, *Proc. R. Soc. London, Ser. A* **225,** 264 (1954).

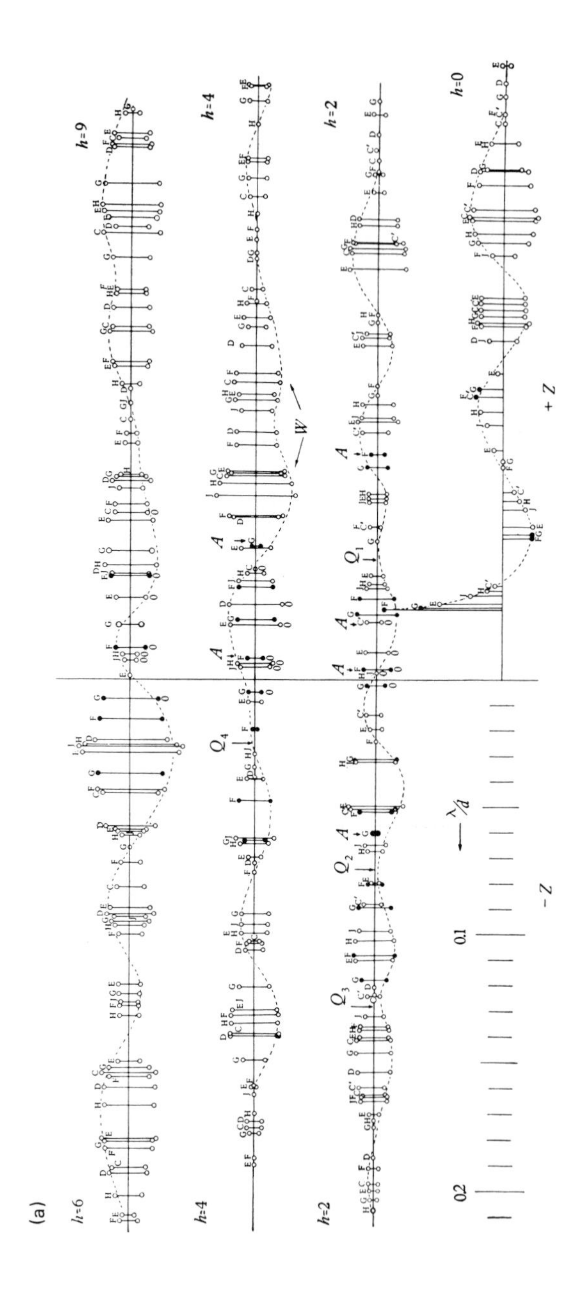
(a)
h=6
h=9
h=4
h=4
h=2
h=2
h=0
W
A
Q_1
Q_2
Q_3
Q_4
+Z
−Z
λ/d
0.1
0.2

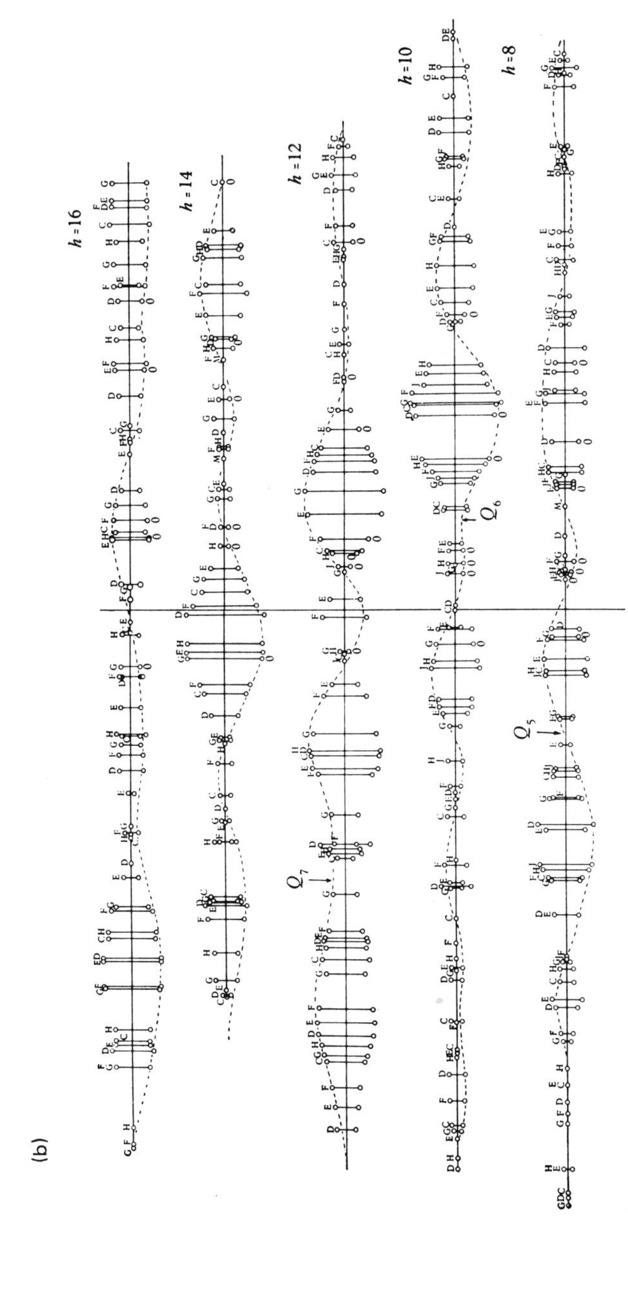

FIG. 3. Salt-free structure amplitudes of *h0l* reflections of horse methemoglobin at different lattice stages. The broken curve gives the correct signs and is derived in combination with the isomorphous replacement method. The layer lines are separated by an arbitrary distance, which is different in (a) and (b). (○) Salt-free observed; (●) salt-free extrapolated; A, anomalous reflection; Q, doubtful node; 0, mark for *h*00 reflection. *F* scale: 1 mm = 200.[11]

hoping that some simplifying regular features of the molecular structure would make that Patterson interpretable.[12] I did in fact fool myself into believing that I had solved the structure of hemoglobin, only to find my solution demolished by my newly arrived research student Francis Crick.

I was desperate that 16 years' toil had led me nowhere when one day the mail brought me reprints from a stranger at Harvard, Austin Riggs, who had compared the SH reactivity of normal and sickle cell hemoglobin. In the course of this work he had found that reaction of hemoglobin with paramercuribenzoate left the sigmoid oxygen equilibrium curve intact.[13] His results implied that 1 mol of hemoglobin could bind 2 mol of paramercuribenzoate without impairment of its function, which suggested that this reaction was unlikely to have disturbed its structure. Had Riggs' papers arrived a few years earlier, I might have doubted that the presence of two mercury atoms would produce significant intensity changes in a diffraction pattern generated by about 5000 atoms of carbon, nitrogen, and oxygen, but in fact I had recently measured the absolute intensity from a hemoglobin crystal and found it to be much lower than I had expected. That made me realize that the vast majority of the scattering contributions of the light atoms cancel out by interference, while the 80 electrons of a mercury atom would scatter in phase. In the paper by D. W. Green, V. M. Ingram, and myself, I outlined the argument[14]:

> Not counting the hydrogens, hemoglobin contains about 5000 atoms. It is not immediately obvious that the replacement of one or two light atoms by heavier ones would allow one to determine many signs. According to statistical arguments [ref. 15] the average intensity per lattice point (i.e. per molecule) should be
>
> $$I = N \times f^{-2} = 5000 \times 7^2 = 250{,}000,$$
>
> whence the r.m.s. structure amplitude $F = 500$. The standard error in measuring F is of the order of 10% so that the difference in intensity to be expected from the attachment of heavy atoms would appear to be hardly larger than the sum of the errors in two separate measurements. In fact, however, the r.m.s. amplitude per molecule is only 300, much smaller than would be expected on statistical grounds. In consequence the contribution of heavy atoms can be detected and measured.
>
> We have prepared crystals of horse oxyhaemoglobin, isomorphous with the normal monoclinic methaemoglobin, in which two mercury atoms are attached to each haemoglobin molecule. Their presence would be expected to change the average structure amplitude by $\pm f_{Hg}\sqrt{2} = \pm 80\sqrt{2} = \pm 114$. This implies a change of F^2 from its normal average of 9×10^4 to either 17×10^4 or 3.5×10^4. It should be possible to measure this

[12] M. F. Perutz, *Proc. R. Soc. London, Ser. A* **195,** 474 (1949).

[13] A. F. Riggs, *J. Gen. Physiol.* **36,** 1 (1952).

[14] D. W. Green, V. M. Ingram, and M. F. Perutz, *Proc. R. Soc. London, Ser. A* **225,** 287 (1954).

[15] A. J. C. Wilson, *Nature (London)* **150,** 152 (1942).

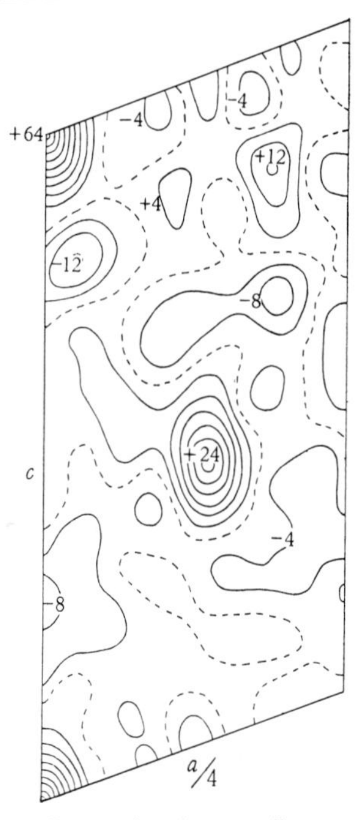

FIG. 4. Difference Patterson of normal and mercuribenzoate hemoglobin, lattice E. The numbers indicate electrons²/Å²; alternate contours around the origin are omitted.[14]

with some accuracy. We have also attached two silver atoms whose mean contribution would be $\pm 47\sqrt{2} = \pm 67$. Their presence would change the average intensity from 9×10^4 to either 13.4×10^4 or 5.4×10^4, a change which could still be measured.

The procedure for finding the signs was as follows. The structure amplitudes of the $h0l$ reflexions were measured from the normal wet lattice E and for the acid-expanded lattice F, both in the presence and absence of mercury. All significant differences between the F's were squared and used to calculate two difference Patterson projections. Each of these showed a clear-cut peak identifiable as the mercury–mercury vector [Fig. 4]. Next, the centre of this vector was found in relation to the centre of the molecule by considering the intensity changes of certain low-order reflexions of known sign. This fixed the mercury positions unambiguously. They were actually found to be slightly different in lattices E and F. Structure factors were then calculated for the mercury atoms in the two lattices, and the signs determined for all reflexions which showed significant intensity changes. The same procedure was followed for the silver compound of lattice E. Finally, all reflexions of definitely known sign, numbered 150 in all, were plotted in reciprocal space and superimposed on the waves of the transform described in part I. It was a triumph to find that there were no inconsistencies [Fig. 5]. The signs determined by the isomorphous replacement method exactly fitted the loops and nodes which had been so laboriously worked out in the course of the previous two years, and confirmed the great majority of the sign relations established by the transform method. The way was now open for the calculation of the Fourier projections.

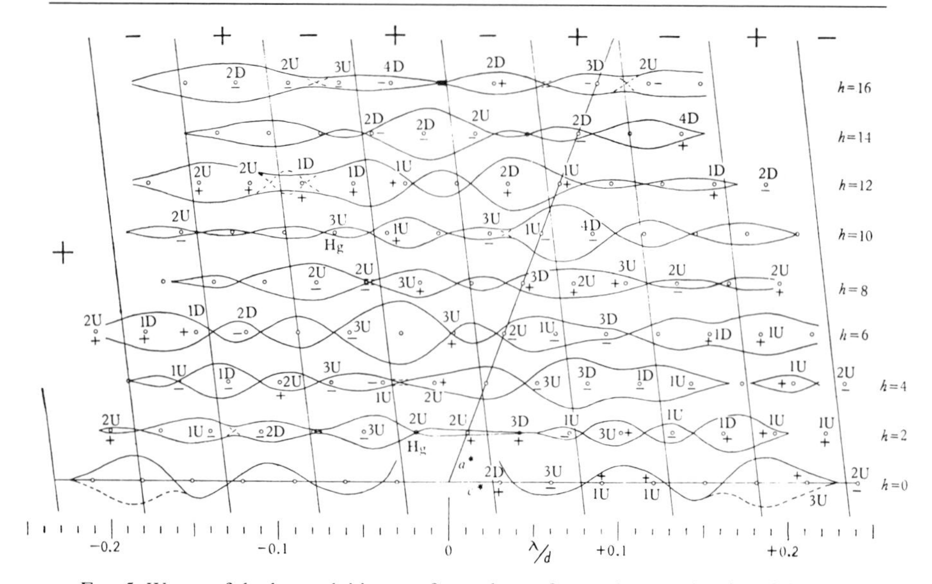

FIG. 5. Waves of the hemoglobin transform, drawn for an electron density of the suspension medium $\rho = 0.39$ electrons/Å^3. Doubtful nodes are marked by broken lines. The points correspond to the reciprocal lattice E of the normal wet crystals. Their markings denote the observed changes in amplitude due to the mercuribenzoate residue, divided by 100 and rounded off; U or D stands for up or down. The oblique straight lines denote the nodes between the fringes due to diffraction by a pair of mercury atoms at $x = 0.068$, $z = 0.290$ and $x = -0.068$, $z = -0.290$. The signs and amplitude of the fringes are indicated at the top of the diagram. Scale of the waves: 1 mm = 400 electrons.[14]

Bragg and I were thrilled to have at last a Fourier projection of the molecule on the centrosymmetric plane, but the overlap of the many features in the 50-Å-thick molecule made it uninterpretable.[16] Not even the hemes could be discerned. Then Dorothy Hodgkin came over from Oxford and drew my attention to the paper by Bokhoven, Schoone, and Bijvoet who had pointed out that the ambiguity left in general phases by a single isomorphous replacement could be resolved by isomorphous replacement with a second heavy atom occupying a position different from that of the first.[17] However, my search for a second heavy-atom derivative proved fruitless until 4 years later Howard Dintzis found a way of reacting the buried SH groups of cysteine 112β with mercuric chloride.[18] From that moment the work leapt ahead. Blow and Crick applied proba-

[16] W. L. Bragg and M. F. Perutz, *Proc. R. Soc. London, Ser. A* **225,** 315 (1954).
[17] C. Bokhoven, J. C. Schoone, and J. B. Bijvoet, *Acta Crystallogr.* **4,** 275 (1951).
[18] A. F. Cullis, H. M. Dintzis, and M. F. Perutz, *N.A.S.-N.R.C. Publ.* **557,** 50 (1958).

FIG. 6. Two different polypeptide chains in the asymmetric unit of hemoglobin compared with myoglobin (left). The heme groups are at the back of the chains.[21]

bility theory to the phase determination by multiple isomorphous replacement[19] and Hilary Muirhead programmed their method of calculating the best or the most probable phase. Michael Rossmann devised a Fourier method of determining the relative Y parameter of the heavy atoms in monoclinic crystals,[20] while Ann Cullis and I did the experimental work. Finally, armed with six different, though not quite independent, heavy-atom derivatives, we obtained phases with a mean figure of merit of 0.78. Thanks to the predominantly α-helical conformation of the polypeptide chain, the tertiary and quaternary structure of the molecule was clearly outlined even at our limited resolution of 5.5 Å.[21] The similarity between the tertiary structures of the α and β hemes and the appearance of four dominant peaks representing the hemes left no doubt that the structure was correct. This conclusion was reinforced by the similarity between the structure of the two hemoglobin chains and that of the single myoglobin chain solved 2 years earlier by Kendrew and collaborators (Fig. 6).

Kendrew began work on myoglobin in 1948, initially from horse muscle,[22] but after having wasted several years trying to grow crystals large enough for a three-dimensional analysis, he and his collaborators discovered that diving mammals and birds are a much better source of myoglobin which can make up as much as 10% of the dry weight of their muscles.[23-25] Myoglobin from the sperm whale formed beautiful monoclinic

[19] D. M. Blow and F. H. C. Crick, *Acta Crystallogr.* **12,** 794 (1959).

[20] M. G. Rossmann, *Acta Crystallogr.* **132,** 221 (1960).

[21] M. F. Perutz, M. G. Rossmann, A. F. Cullis, H. Muirhead, G. Will, and A. C. T. North, *Nature (London)* **185,** 416 (1960).

[22] J. C. Kendrew, *Proc. R. Soc. London, Ser. A* **201,** 62 (1950).

[23] J. C. Kendrew, R. G. Parrish, H. R. Marrack, and E. S. Orlans, *Nature (London)* **174,** 946 (1954).

[24] J. C. Kendrew and P. J. Pauling, *Proc. R. Soc. London, Ser. A* **237,** 255 (1956).

[25] J. C. Kendrew and R. G. Parrish, *Proc. R. Soc. London, Ser. A* **238,** 305 (1956).

crystals that belonged to the space group $P2_1$, and had only two molecules of myoglobin in the unit cell. Kendrew and R. G. Parrish's first paper on sperm whale myoglobin, published in 1956, still deals with Patterson projections,[25] but in that same year Kendrew was joined by two young chemists who began a concentrated effort at attaching heavy atoms to myoglobin. They were Howard Dintzis from the United States and Gerhard Bodo from Austria. Myoglobin contains no cysteine, so that it does not react with *p*-mercuribenzoate. Dintzis had worked with J. T. Edsall at Harvard on the combination of heavy-atom complexes with serum albumin. It occurred to him that the lattices of protein crystals might provide binding sites for such complexes similar to those in single serum albumin molecules. This idea failed in hemoglobin, but proved brilliantly successful in myoglobin. Seven different isomorphous heavy-atom derivatives were prepared; at first a centrosymmetric projection of the electron density was calculated, but even though the myoglobin molecule is only one-quarter the size of hemoglobin, the projection failed to reveal any of its features.[26] However, its small molecular weight and short *b* axis meant that a three-dimensional synthesis at 6 Å resolution needed only 400 terms, 100 of which were real. Today this looks like child's play, but remember that the phase angles had to be determined manually by superposition of circles for each reflection as suggested by Harker.[27] Reflections were measured on precession photographs for the parent compound and six heavy-atom derivatives. Intensities were measured on a new manually operated Joyce Loebel microdensitometer rather than visually as we had done until then; Lorentz and polarization corrections were applied using a desk calculator, and finally about 2000 circles were drawn by hand to determine the phases[28] (Fig. 7). The three-dimensional map was calculated using Cambridge University's homemade electronic digital computer EDSAC I. The first protein molecule beheld by man revealed itself as a long, winding visceral-looking object with a lump like a squashed orange attached to it: that was the heme[29] (Fig. 8). Kendrew concluded his description of it with the words,

> Perhaps the most remarkable features of the molecule are its complexity and its lack of symmetry. The arrangement seems to be almost totally lacking in the kind of regularities which one instinctively anticipates, and it is more complicated than has been

[26] M. M. Bluhm, G. Bodo, H. M. Dintzis, and J. C. Kendrew, *Proc. R. Soc. London, Ser. A* **246,** 369 (1958).

[27] D. Harker, *Acta Crystallogr.* **9,** 1 (1956).

[28] G. Bodo, H. M. Dintzis, J. C. Kendrew, and H. W. Wyckoff, *Proc. R. Soc. London, Ser. A* **253,** 70 (1959).

[29] J. C. Kendrew, G. Bodo, H. M. Dintzis, R. G. Parrish, and H. Wyckoff, *Nature (London)* **181,** 662 (1958).

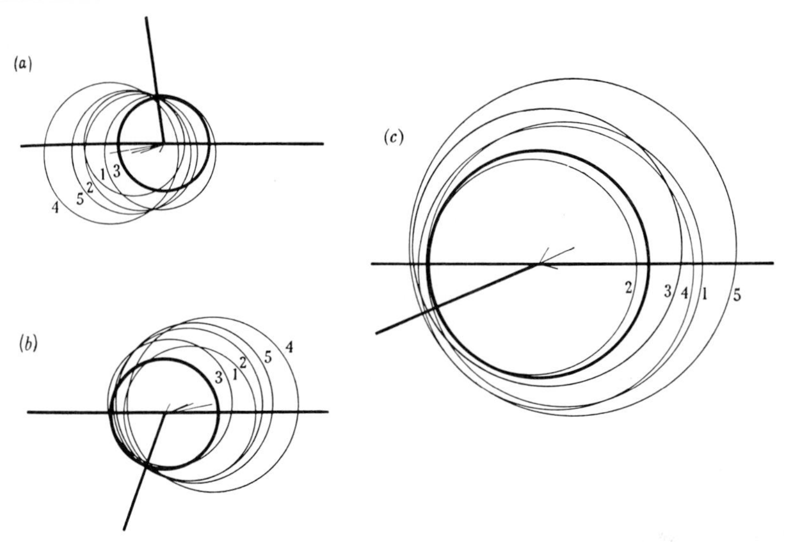

FIG. 7. Examples of phase determination in sperm whale metmyoglobin. The heavy circle represents the amplitude of the reflections from unsubstituted protein, and the light circles those from the derivatives. 1, *p*-chloromercuribenzoate sulfonic acid (PCMBS); 2, $HgAm_2$; 3, Au; 4, PCMBS/$HgAm_2$; 5, PCMBS/Au. The short lines from the centers are the heavy-atom vectors; the heavy line indicates the phase angle eventually selected. (a) (411), (b) (911), (c) (212).[28]

predicated by any theory of protein structure. Though the detailed principles of construction do not yet emerge, we may hope that they will do so at a later stage of the analysis. We are at present engaged in extending the resolution to 3 Å, which should show us something of the secondary structure; we anticipate that still further extensions will later be possible—eventually, perhaps, to the point of revealing even the primary structure.

At this stage Kendrew was joined by two young crystallographers of enormous energy, R. E. Dickerson from the United States and B. Strandberg from Sweden. The three of them decided to raise the resolution of the map, not just to 3 Å, as foreshadowed in Kendrew's 1958 paper in *Nature,*[29] but to 2 Å, requiring the measurement of 9600 reflections for the parent protein and for each of the heavy atom derivatives, an unprecedented undertaking in 1957. It took them no more than 2 years. Here is Kendrew's description of the work[30]:

The data for each derivative were recorded on twenty-two precession photographs; a separate crystal had to be used for each photograph to keep radiation damage within

[30] J. C. Kendrew, R. E. Dickerson, B. E. Strandberg, R. G. Hart, D. R. Davies, D. C. Phillips, and V. C. Shore, *Nature (London)* **185,** 422 (1960).

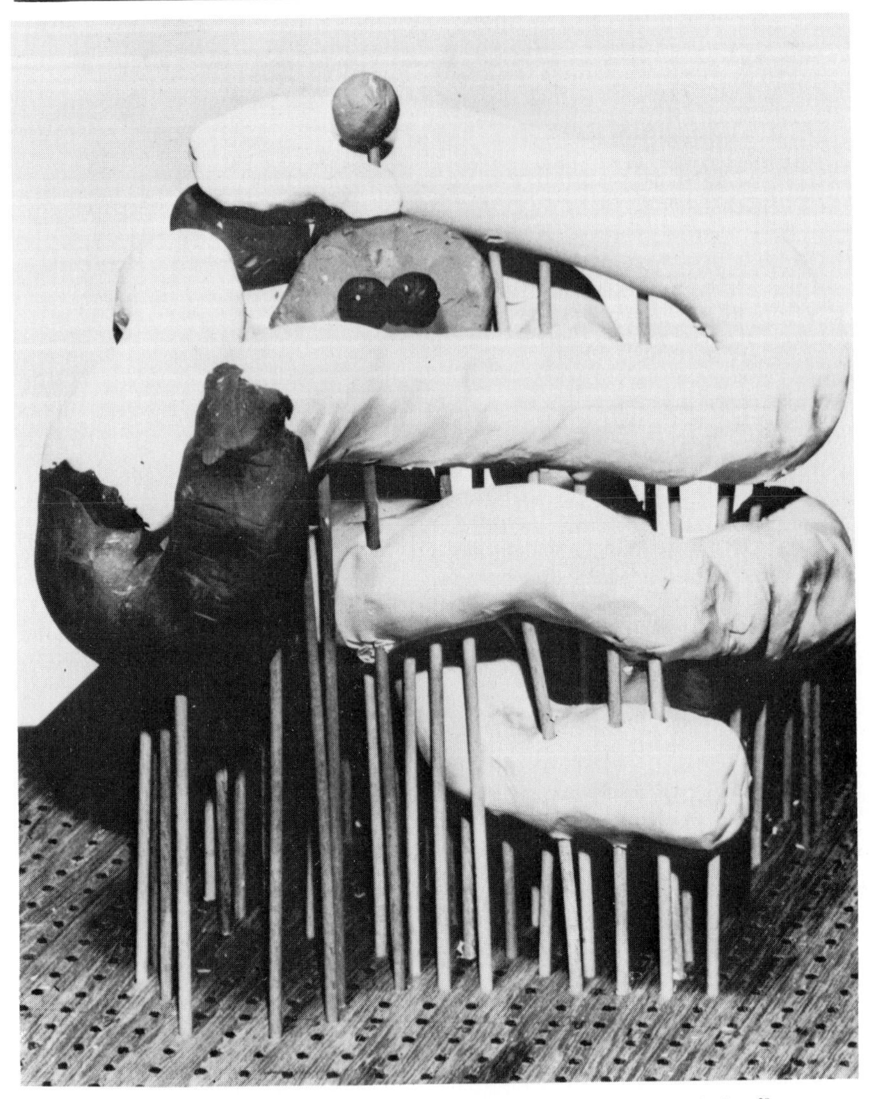

FIG. 8. Plasticine model of sperm whale myoglobin at low resolution.[29]

acceptable limits. The results from the different photographs were scaled together on the computer, the best set of scaling factors being determined by solving an appropriate 22×22 matrix. The degree of isomorphism of each derivative was tested, and found adequate, by means of a computer programme which used the *h*0*l* reflexions to refine the preliminary values of the heavy-atom parameters, temperature factor, etc., and then compared the values of δF_{obs} and δF_{calc} as a function of sin θ. The co-ordinates of the heavy atoms were further refined using correlation functions computed by means of programmes devised by Dr. M. G. Rossmann, and finally refined again during the process of phase determination itself. These phases were determined by essentially the same method as before, but owing to the very large number of reflexions the determination was carried out on the computer rather than graphically. The "best" phases and amplitudes were computed and used in the final Fourier synthesis, to which a moderate degree of sharpening was applied. In all, some hundreds of hours of computer time were required, and the Fourier synthesis itself, which was calculated at intervals of about 2/3 Å., took about 12 hr.

This was the first protein structure to be solved at near-atomic resolution, a fantastic feat at that time since nearly all the methods used had to be developed from scratch as the work progressed[31]; the University of Cambridge's new digital electronic computer EDSAC II, which had only just been completed, was one of the most advanced in the world, and the myoglobin Fourier stretched it to the limit. Kendrew and I were fortunate that important developments in computer technology were taking place not more than 100 yards from our X-ray rooms; without them and the close collaboration with the mathematicians and engineers in the computer laboratory we would not have been able to solve the first protein structures.

Legend has it that they were solved with primitive instruments in a hut next to the Cavendish Laboratory, but in fact the hut merely contained our offices and our biochemical laboratory. The basement of the main building housed what was then the world's best equipment for protein crystallography. Back in 1943 I happened to visit Martin Buerger at MIT, when he showed me his newly invented precession camera and pointed out its great potentialities for recording the diffraction patterns from protein crystals. I also met Buerger's instrument mechanic, Charles Supper, who had built the camera and was setting up in business to make it for sale. Soon after the war, the Rockefeller Foundation bought me his first production model, followed by several others in the years to come. However, a 17° precession picture took days of exposure and contained broad spots on a strong background. In 1949 Kendrew and I engaged a young engineer, D. A. G. Broad, to design a rotating anode tube, and a few years later we also found a first class instrument mechanic, L. G. Hayward. Broad designed a copper tube with a novel type of vacuum seal and an excellent focus,[32] and Hayward possessed the skill to machine its bearings to the small tolerances that were needed to prevent leaks. The Broad tubes reduced our exposure times by an order of magnitude and also gave

[31] R. E. Dickerson, J. C. Kendrew, and B. E. Strandberg, *Acta Crystallogr.* **14,** 1188 (1961).

[32] D. A. G. Broad, U.K. Patent Applications Nos. 5172, 5173, 12761, 38939, 10984 and 13376 (1956).

FIG. 9. J. C. Kendrew building the first atomic model of sperm whale myoglobin.

cleaner and better focused beams than any commercial tubes then available. Without them Kendrew and I could not have collected our X-ray data.

When the electron density of myoglobin map had been plotted on 96 Plexiglas sheets, Kendrew wondered how he could get his hands in to build an atomic model. As this was impossible, he decided to mount his model on a forest of 1/8-in. steel rods on a grid spaced at 2/3 Å intervals in the unit cell, but we had no room large enough to hold it. Fortunately for us, the nuclear physicists dismantled their pre-war cyclotron, and the space vacated there allowed Kendrew to go ahead. He marked electron densities at every 2/3 Å along the rod with Meccano clips painted in an appropriate color code and used 1/8 in. brass spokes screwed together with steel connectors to represent the atoms. Figure 9 shows him constructing the first atomic model of a protein.

[2] Introduction to Crystallography

By ROBERT M. SWEET

Introduction

The modern language of enzymology and molecular biology owes much of its sophistication to the success of X-ray diffraction. By 1965, clever chemists had learned a lot about how enzymes worked. They had found residues that lay close to one another in active sites and had proposed mechanisms of action that were tested by kinetic and model studies. But no one really began to know how it all worked until three-dimensional structures were known for several of those biochemical factories. Today, students of biochemistry take more as dogma than as experimental findings the knowledge of molecular structure that is their heritage. This knowledge has its value, of course. With rare exceptions (Mozart, Einstein) the imagination of man is limited by what he already *knows*. Therefore, we may presume that modern students of enzymology will be able to ask questions we might not have thought of 20 years ago, when we were ignorant of structure.

The winning of the ability to determine these structures was very difficult. The workers who produced most of the results have stood firmly on the shoulders of the scientists whose labors are described in the previous chapter. Figuring out how to determine protein structure required some 40 years. Producing the structural information we have now has been accomplished in 20. The purpose of this chapter is to introduce the general scientific reader to some of the principles that are used in the determination of molecular structure by X-ray diffraction techniques. Our goal is to present the physical and mathematical basis of these techniques and to provide an intuitive approach to understanding them. We refer you to the comprehensive textbooks in the field for a more thorough treatment of this background information. Titles which may be of use are *Protein Crystallography* by Blundell and Johnson and Sherwood's *Crystals, X-Rays and Proteins*. Brief summaries of the field can be found in *Crystal Structure Analysis: A Primer* by Glusker and Trueblood, and in a chapter from Vol. 13 of *Methods of Biochemical Analysis* by Holmes and Blow entitled "The Use of X-Ray Diffraction in the Study of Protein and Nucleic Acid Structure."

We shall introduce you to the field in two steps. The first step is to teach the fundamental principles of diffraction, and we do this in two different ways, in parallel. The first will show how waves constructively and destructively interfere after they are scattered from atoms. The second way will show that X-ray diffraction is mathematically equivalent to the taking of the Fourier transform of the scattering object. The second step is to describe the experimental procedures that a crystallographer must perform in the determination of macromolecular structure. This will be in the form of an approximate chronology of work that might be done in the crystallographic laboratory.

Diffraction

Scattering of X Rays by Atoms. The X rays commonly used for diffraction studies of biological molecules have a wavelength of 0.154 nm and an energy of 8 keV. They are produced when a beam of electrons driven by a potential of roughly 40,000 V strikes a copper anode. These high-energy electrons ionize the copper atoms, removing an inner shell electron. X rays are emitted when a higher energy electron falls to fill the void. The 0.154-nm radiation results when an L shell electron fills a hole in the K shell of a copper atom. This choice of radiation is a compromise: longer wavelength X rays allow investigation of larger molecules, damage the specimen less, and scatter more strongly; shorter wavelength X rays are absorbed less by the specimen and can allow solution of a structure to higher resolution. In some large laboratories, electron synchrotrons or storage rings are used to produce X rays for diffraction studies. In these accelerators electrons travel at nearly the speed of light, their orbits being bent to a circular path by powerful magnets. The radiation is produced in the direction of travel of the electrons, essentially because of the work done on them by the bending magnets. Synchrotron radiation is polychromatic; therefore one may choose precisely the wavelength for use by diffraction of the beam from a monochromator crystal. A major advantage of the synchrotron sources is that they are two to three orders of magnitude more intense than conventional ones.

X rays are electromagnetic radiation. When they pass by the electrons in an atom, the oscillating electric field of the X-ray photon or wave causes the electrons on the atom to oscillate, much like the sloshing of coffee in a cup. These oscillating electrons serve as a new source of X rays which are emitted in almost all directions. (A little thought will show the reader that there are two directions in which no radiation can be emitted when the incident radiation is polarized.) It happens that the scattered radiation is precisely out of phase with the incident radiation, that is, it is phase-shifted by π.

Diffraction of X Rays by Atoms in a Lattice. A crystal has some simple arrangement, or motif, of atoms repeated by reiterated translations in one, two, or three directions. The lattice that describes the translational repetition can be defined by single points chosen at an equivalent point in each motif. The simplest possible crystal is made of two atoms, and its lattice consists of two points at the atomic centers.

When a single X-ray photon passes through a crystal, it diffracts from the entire crystal as if it were a plane wave. This is an example of the wave–particle duality explained by quantum mechanics. Thus, when a photon strikes a lattice of two atoms, as in Fig. 1, it is scattered by both atoms. Because of the constructive and destructive interference of waves, there are special directions in which radiation scattered from these two atoms will form a diffracted beam. One can see in Fig. 1 that the two diffracted waves are in phase because the lower of the two waves travels exactly one wavelength farther than the upper. Notice the phase shift of π in the scattered waves. In general this path difference must be an integral number of wavelengths for diffraction maxima to occur:

$$a \sin \alpha = n\lambda, \qquad n = 1, 2, \ldots \tag{1}$$

The pattern of intensity which would be observed by an X-ray detector is a series of ripples, starting at the direct beam and going off in both directions.

Diffraction and the Fourier Transform. If the elder Bragg was the father of crystallography, then Jean Baptiste Joseph Fourier was surely its godfather. Fourier was a bureaucrat in the government of Napoleon Bonaparte and, among other services to his government, accompanied Napoleon on his visit to Egypt and was an able administrator after their return to France. He was an accomplished mathematician and made contributions that were of great value to crystallographers. He showed that

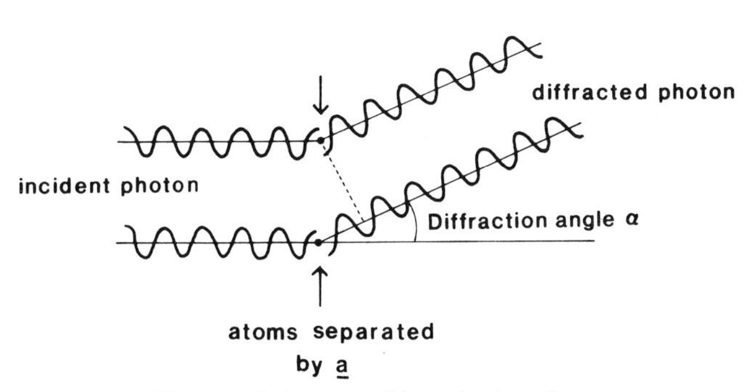

FIG. 1. Diagram of photon striking a lattice of two atoms.

any periodic function can be approximated by sums of sine and cosine functions whose wavelengths were integral fractions of this periodicity. He also devised the technique we now know as the Fourier transform, which again involves sums of trigonometric functions, and which transforms functions between coordinate systems with different dimensionality. The Fourier transform is a precise mathematical description of the physical phenomenon of diffraction, and therefore it is tremendously useful to us.

The one-dimensional Fourier transform of some function $f(x)$ is

$$F(h) = \frac{1}{\sqrt{2\pi}} \int_{-\infty}^{\infty} f(x) e^{ihx}\, dx \tag{2}$$

If $f(x)$ exists in ordinary space, defined by x, then $F(h)$ exists in reciprocal space, defined by the variable h. Because the exponent must be dimensionless, h must have dimensions of reciprocal distance.

The beauty of Fourier's transform is that having gotten there, one can get back again with

$$f(x) = \frac{1}{\sqrt{2\pi}} \int_{-\infty}^{\infty} F(h) e^{-ihx}\, dh \tag{3}$$

Notice several features of these two transformations. First, both contain a complex exponential that yields the trigonometrics mentioned above. Second, the principal difference between them is the sign of the exponent. And third, readers who still remember college calculus can prove that the second follows from the first.

We may now use this transform to calculate an expression for diffraction from a two-atom lattice. (I refer you to Sherwood's book for a thorough treatment.) If we represent each atom by an appropriate δ function, that is, we make the atom a point scatterer of X rays, we get that

$$F(h) = 2 \cos 2ha \tag{4}$$

$F(h)$ is the amplitude (square root of the intensity) of diffraction and a is the spacing between scatterers. A definition of h awaits our discussion of the work of the Braggs. Notice, however, that this function is like the array of ripples that we expected.

Diffraction from a Two-Dimensional Lattice. In our discussion of diffraction from a one-dimensional lattice, we arranged for the incident beam to be perpendicular to the line of atoms. Now we shall show how to calculate conditions for diffraction for any beam that is incident on a two-dimensional lattice of atoms. The situation we shall consider is shown in Fig. 2.

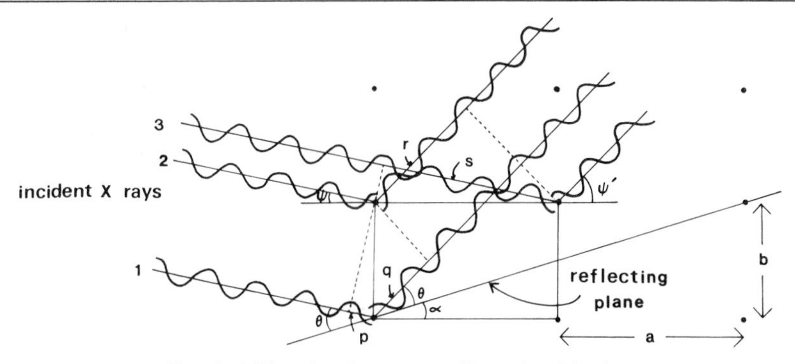

FIG. 2. Diffraction from a two-dimensional lattice.

For simplicity we have chosen a rectangular lattice. The lattice spacing is a in the horizontal and b in the vertical directions. Parallel beams 1, 2, and 3 approach the crystal at angle ψ. We have drawn a number of wave fronts as dotted lines at points where one or another of the waves strikes an atom. We place the same condition on the beams here as in the two-atom case: for diffraction to occur, the path length difference between any two beams must be an integral number of wavelengths. In studying this drawing, note that it has been carefully made, including the phase shift of π on scattering, to show the path length differences. These differences between beams 1 and 2 and between beams 2 and 3 are, respectively,

$$\begin{aligned} p + q &= b \sin \psi + b \sin \psi' = n\lambda \\ s - r &= a \cos \psi - a \cos \psi' = m\lambda \end{aligned} \tag{5}$$

Now we construct a "reflecting" plane which makes angle θ with both incident and diffracted beams. This plane lies at angle α from the horizontal axis. We can now redefine ψ and ψ' in terms of θ and α, as follows:

$$\psi = \theta - \alpha, \qquad \psi' = \theta + \alpha \tag{6}$$

Finally, if we substitute these values into the equations above and rearrange, we are left with a remarkably simple equation which constitutes an important condition for diffraction:

$$\tan \alpha = mb/na \tag{7}$$

In this example $m = 1$ and $n = 2$ and we have drawn the reflecting plane to pass through the appropriate two atoms in the lattice so that this expression holds true. Planes of this sort, that pass through two lattice points (or three in a three-dimensional lattice), or are parallel to planes that do, are known as General Lattice Planes. These planes have useful algebraic properties which we shall discuss later.

But first, we shall place yet another set of constraints on diffraction from a lattice. We have shown that diffraction can be treated as reflection from general lattice planes, but have placed no constraints on the reflection angle θ. The following derivation is similar to that which sent the Braggs on their way and led to the founding of the science of X-ray diffraction crystallography.

Consider two general lattice planes, as in Fig. 3, with interplanar spacing d, and incident and reflected beams of X rays that make angle θ with these planes. As before, we require that the path length difference between the two rays be an integral number of wavelengths, and we arrive at Braggs' Law:

$$n\lambda = 2d \sin \theta \tag{8}$$

Braggs' law does not tell us everything we need to know about diffraction. Coupled with the idea of general lattice planes, it tells us only about the geometry of diffraction, but it says nothing about the relation between diffraction and the way the atoms of the crystal are arranged within its repeating units. The Fourier transform does this. But before we see how the Fourier transform of the crystal is related to Braggs' law, we must first learn a bit more about general lattice planes and the ways they can be represented.

Representation of Lattice Planes. General lattice planes are described by sets of *general indices*. The smallest unit of a crystal that is repeated by translation alone is called the *unit cell*. We often draw unit cells with lattice points at their corners. When general lattice planes are constructed in a unit cell, the indices which describe those planes are the number of segments into which the planes cut each of the unit cell edges. An equivalent definition is that the indices are reciprocals of the fractional lengths into which unit cell edges are cut by the planes. Examples appear in Fig. 4. The sign conventions are as follows. When the plane nearest a lattice point cuts two axes that go both in the positive or both in the negative direction, the two corresponding indices have the same sign. When the

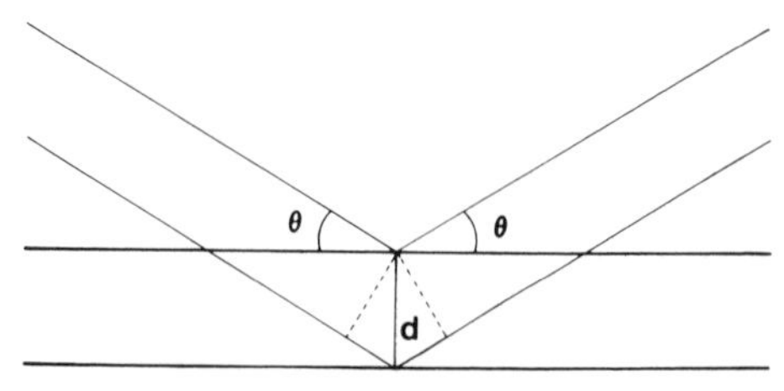

FIG. 3. Diagram of general lattice planes showing interplaner spacing and angles of reflectance and incidence.

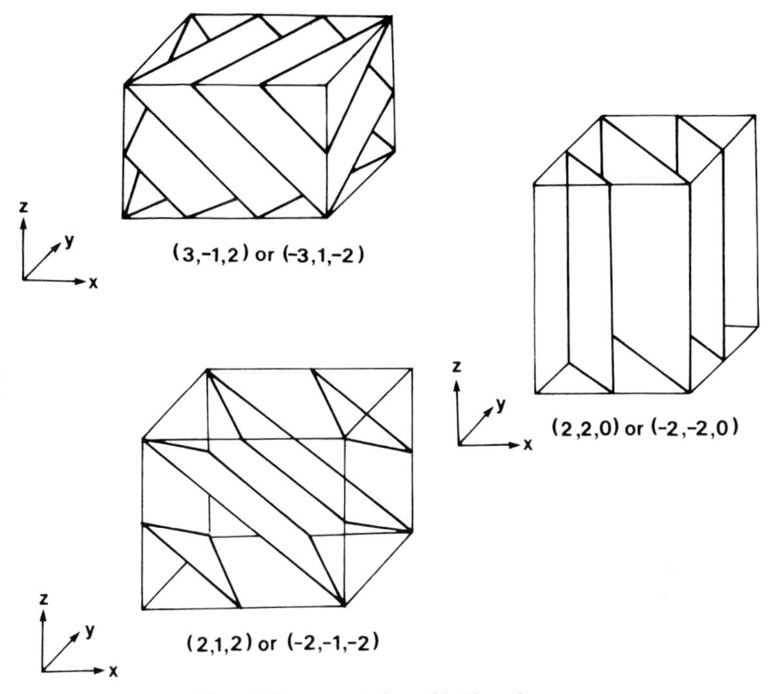

FIG. 4. Representation of lattice planes.

plane cuts axes that go in directions with opposite signs, the indices have opposite signs. A set of indices can be multiplied by -1 and still represent the same set of planes.

The Sphere of Reflection and Reciprocal Space. P. P. Ewald noticed a geometrical simplification of Braggs' law that links this law to the Fourier transform and provides a simplified way of looking at general lattice planes.

Notice first, as in Fig. 5, that if the Bragg reflection angle is θ then the total deflection angle for diffracted X rays must be 2θ. Let us then place this figure, an intersection of beams with a plane, at the center of a sphere, known as the Ewald sphere, that has as its radius the reciprocal of the wavelength of the radiation. Then one can make the construction shown in Fig. 6. One can see immediately that this figure follows the rule

$$\tfrac{1}{2}OA/(1/\lambda) = \sin\theta \qquad \text{or} \qquad \lambda = (2\sin\theta)/OA \tag{9}$$

If we substitute $OA = 1/d$, we obtain

$$\lambda = 2d\sin\theta \tag{10}$$

which is Bragg's law with $n = 1$.

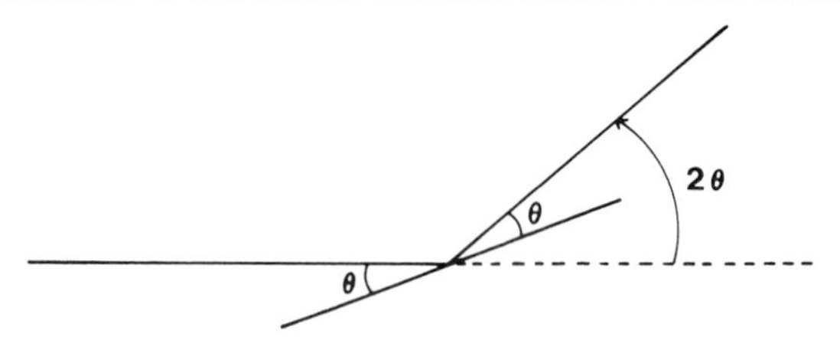

FIG. 5. Diagram of reflection and deflection angles.

Significant features of this construction are first that the chord OA is perpendicular to the reflecting plane, and second that when we define its length to be the reciprocal of the distance between planes which will cause diffraction at angle θ, Braggs' law is obeyed. This suggests that in a crystal each set of general lattice planes *(hkl)*, with spacing d_{hkl} might be represented by a vector $\mathbf{s}_{hkl}$ that is perpendicular to the planes and has length which is $|\mathbf{s}_{hkl}| = 1/d_{hkl}$. The vectors $\mathbf{s}_{hkl}$ define the mathematical space with dimensions of reciprocal distance that we know as *reciprocal space*.

Reciprocal space, the reciprocal lattice (that is, the set of points defined by all of the vectors $\mathbf{s}_{hkl}$), and the Ewald sphere are a remarkably useful heuristic tool for the crystallographer. He can think about diffraction in simple geometric terms as involving the intersection of a point with a sphere rather than having to think about reflecting planes, their spacings, and the angles they make with incident and diffracted rays. In Fig. 6, when point A, which represents vector $\mathbf{s}_{hkl}$ for the plane we have chosen,

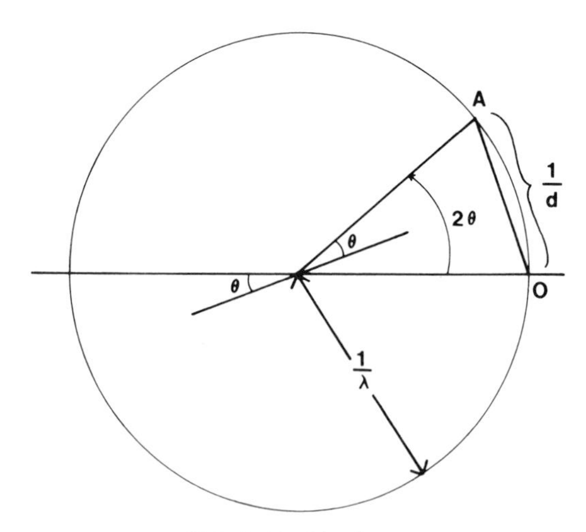

FIG. 6. Ewald sphere.

touches the Ewald sphere, Bragg's law is obeyed and diffraction occurs. Rotating the crystal, so that the angle with the incident beam is no longer θ, will also move A away from the the surface of the sphere since $\mathbf{s}_{hkl}$ is perpendicular to the plane *(hkl)*. Therefore, Braggs' law will not be obeyed and no diffraction will be observed.

It is easy to define the reciprocal lattice vector $\mathbf{s}_{hkl}$ to be perpendicular to the planes (hkl) and to have length $\mathbf{s}_{hkl} = 1/d_{hkl}$:

$$\mathbf{s}_{hkl} = h\mathbf{a}^* + k\mathbf{b}^* + l\mathbf{c}^* \tag{11}$$

The principal reciprocal space vectors $\mathbf{a}^*$, $\mathbf{b}^*$, and $\mathbf{c}^*$ are defined, in terms of the "real space" vectors or unit cell principal axes, as

$$\mathbf{a}^* = \frac{\mathbf{b} \times \mathbf{c}}{\mathbf{a} \times \mathbf{b} \cdot \mathbf{c}}, \quad \mathbf{b}^* = \frac{\mathbf{c} \times \mathbf{a}}{\mathbf{a} \times \mathbf{b} \cdot \mathbf{c}}, \quad \mathbf{c}^* = \frac{\mathbf{a} \times \mathbf{b}}{\mathbf{a} \times \mathbf{b} \cdot \mathbf{c}} \tag{12}$$

One can use the reciprocal lattice vector $\mathbf{s}_{hkl}$ to calculate useful parameters for the crystal. For example, one can easily calculate the spacings for a particular set of lattice planes. For the case where the angle between unit cell edges **a** and **c** is unconstrained (call it β) but both **a** and **c** are perpendicular to **b**, we can readily derive an expression for d_{hkl}:

$$d_{hkl} = (\mathbf{s}_{hkl} \cdot \mathbf{s}_{hkl})^{-1/2} = (h^2a^{*2} + k^2b^{*2} + l^2c^{*2} + hla^*c^* \cos \beta^*)^{-1/2}$$
$$(\beta^* = \pi - \beta)$$

$$a^* = \frac{bc}{abc \cos(\beta - 90°)} = \frac{1}{a \sin \beta}, \quad b^* = \frac{1}{b}, \quad c^* = \frac{1}{c \sin \beta} \tag{13}$$

$$d_{hkl} = \left(\frac{h^2}{a^2 \sin^2 \beta} + \frac{k^2}{b^2} + \frac{l^2}{c^2 \sin^2 \beta} - \frac{hl \cos \beta}{ac \sin^2 \beta}\right)^{-1/2}$$

Therefore, the concept of the reciprocal lattice and Ewald's sphere of reflection are sufficient to provide us with a thorough geometrical description of X-ray diffraction. For a crystal with any particular unit cell dimensions one can define the set of vectors $\mathbf{s}_{hkl}$ that will determine the reciprocal lattice for that crystal. The conditions for diffraction from any particular set of planes are only that the crystal be oriented so that the reciprocal lattice point corresponding to those planes touches the Ewald sphere.

After brief digression to show how the reciprocal lattice and the Ewald sphere play a role in diffraction experiments, we shall return to show how Braggs' law and the diffraction pattern are related to the Fourier transform of a crystal.

The Nature of Diffraction from Crystals. In principle we understand how diffraction might actually occur from a crystal placed in a beam of monochromatic X rays. How does it really work and how is it used?

Notice first that if the crystallographer is to sample diffraction from all possible sets of lattice planes he must bring the reciprocal lattice points for all these planes into contact with the Ewald sphere. To do this, he must *move* the crystal in the beam. Martin Buerger devised an elegant technique for the photography of a diffraction pattern. He developed a camera that moves the crystal in the beam in a precessional motion. This motion rocks single planes of points in the reciprocal lattice through the Ewald sphere. If a metal screen with an annular slit is placed so that it precesses with the crystal and if the film is made to precess about its center point as well, as undistorted image of a single plane of reciprocal space can be recorded on the X-ray film.

In Fig. 7 you can see a diffraction photograph taken from a crystal of the protein phycocyanin with the use of a Buerger precession camera. This is the *image* of a single plane in the reciprocal lattice for this crystal, displayed on a film. Each reciprocal lattice point is represented by a spot of blackened silver grains. The darkness of the spot is proportional to the intensity of the X rays reflected from the set of planes described by that particular reciprocal lattice point.

There are several features one should notice about this photograph. First, as advertised, the Buerger camera produces reflections or diffrac-

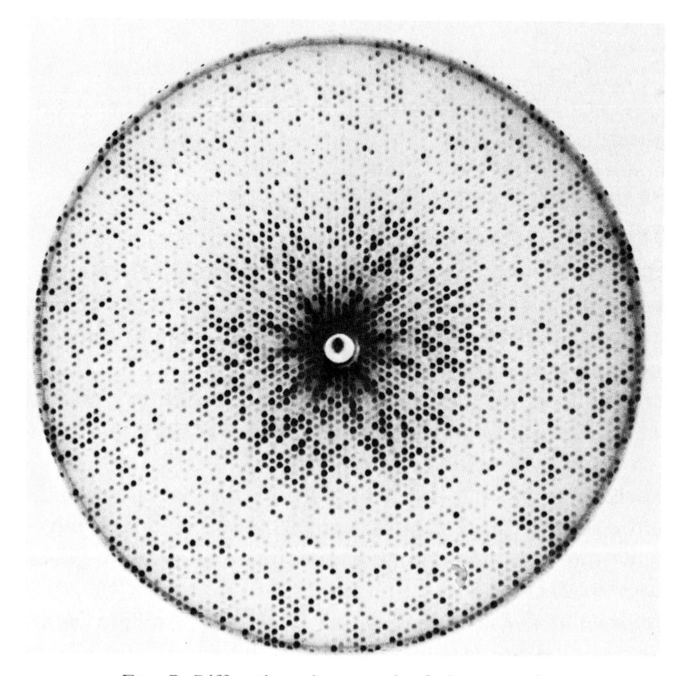

FIG. 7. Diffraction photograph of phycocyanin.

tion maxima that lie on the film along the straight lines of a lattice. The distances between the spots on the lines are proportional to the reciprocal lattice spacings **a***, etc., and these important crystal parameters can be measured directly from a precession film such as this. Second, there is a wide range of intensities in the diffraction maxima that have been recorded. This is because these intensities are determined by the Fourier transform of the contents of the phycocyanin unit cell, sampled at the points that are shown. More directly, while the *arrangement* of spots on the film tells us about the size and shape of the crystal unit cell, only their *intensities* can tell us about the arrangement of atoms in the crystal. The effort to discover how this arrangement of atoms is embedded in these intensities occupies much of the lives of crystallographers and is the subject of several of the chapters that follow. Third, one readily notices that this photograph has striking symmetry. In particular, the pattern of intensities can be put back on itself by a rotation of the photograph by one-sixth of a rotation or by 60°.

The symmetry of this photograph shows much about the total symmetry of intensities in the entire reciprocal lattice for this crystal. One can readily interpret the symmetry of the reciprocal lattice to deduce the symmetry of the molecular arrangement inside the unit cell. This volume will not deal at all with this aspect of crystallography; however the reader can find a comprehensive and comprehensible discussion of crystal symmetry in Martin Buerger's *Elementary Crystallography*.

There are numerous other techniques for the measurement of diffraction intensities, that is, for the sampling of reciprocal space. One of the simplest to understand is use of the single-crystal diffractometer (see Wyckoff [24], this volume, for an exhaustive discussion). On this instrument, a diagram of which appears in Fig. 8, a crystal can be manipulated so that any reciprocal lattice point one chooses can be made to touch the Ewald sphere in the horizontal plane. When the detector is placed at the proper diffraction angle 2θ, it can measure the intensity of diffracted rays. By this technique one can often measure the intensity of one reflection per minute on a computer-controlled instrument. A difficulty with this device is that it measures reflections only one at a time, and geometric constraints make it difficult to use with very finely sampled reciprocal lattices, that is, with very large unit cells.

A technique which suffers neither of these problems is rotation photography, described in some detail in *The Rotation Method* by Arndt and Wonacott and [19], this volume. Here we use the simplest possible geometry: a crystal is mounted in the X-ray beam and a flat piece of X-ray film in a light-tight cassette is placed a short distance away, perpendicular to the beam. The crystal is rotated through an axis perpendicular to the

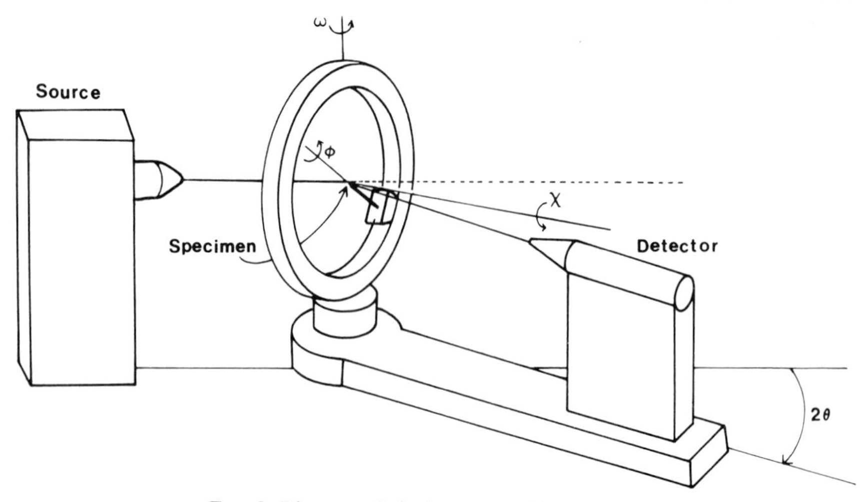

FIG. 8. Diagram of single-crystal diffractometer.

beam through only a small angle, say 1–5°, so that reflections do not superimpose one another on the film. In Fig. 9 appears a "rotation photograph" of another crystal of phycocyanin. Here the crystal was rotated through 3.5° so that regions of several reciprocal lattice planes were swept a short way through the Ewald sphere. The hexagonal arrangement of spots can be seen on the several "lunes" that appear on the photograph. Each of these continuous lunes corresponds to part of a single plane in the reciprocal lattice.

The Sampled Fourier Transform. We now need to understand how the arrangement of molecules in a crystal is manifested in a diffraction pattern, such as those in Figs. 7 and 9. To do this, we return to the idea of the Fourier transform. For simplicity we choose a simple object to represent the pattern of molecules in a crystal's unit cell. We shall use the "top hat" function (Fig. 10), which, also for simplicity, we define in one-dimensional space. The Fourier transform of this function is easily calculated as being

$$F(h) = (2b/h) \sin ah \tag{14}$$

a function that can be plotted as in Fig. 11.

Returning to our analogy of molecular crystals, Fig. 11 shows the transform of the contents of a single unit cell from a crystal. Now we place the object into a crystal, that is, we repeat it many times in Fig. 12, with each repetition being equally spaced from the last. Again we can calculate the transform of this repetitive pattern, and we find that it is a

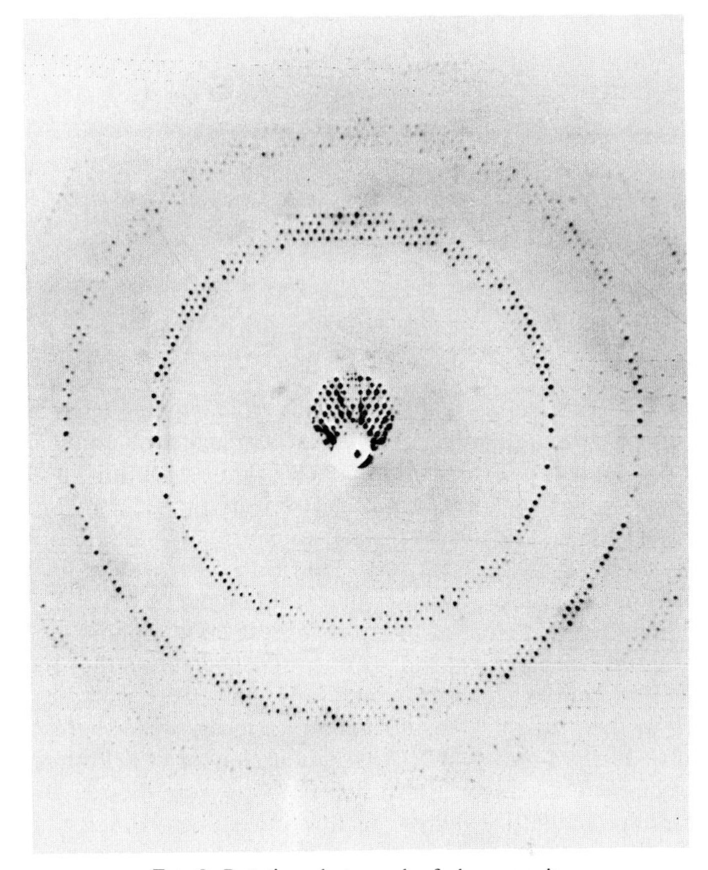

FIG. 9. Rotation photograph of phycocyanin.

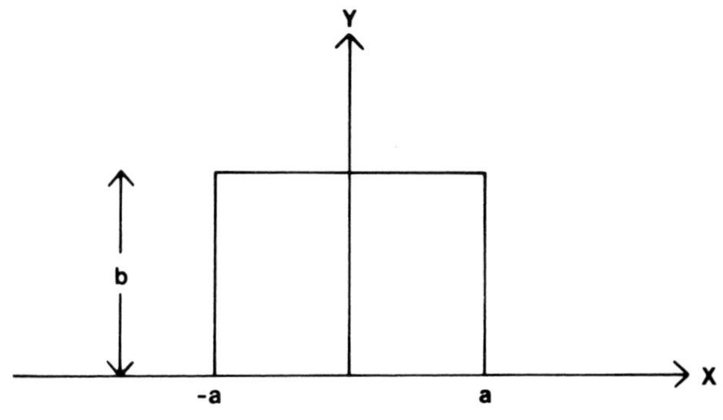

FIG. 10. "Top hat" function.

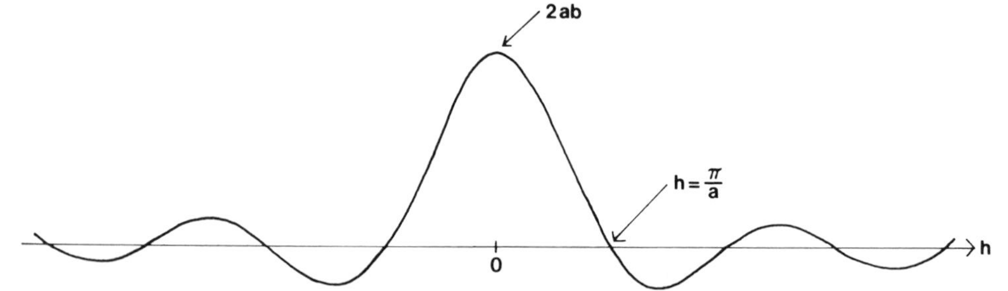

FIG. 11. Plot of Fourier transform of the contents of a single unit cell from a crystal.

series of equally spaced spikes, with spacings that depend upon the reciprocal spacing between the objects in the crystal and with amplitudes that trace out the original transform of the object. This transform is plotted in Fig. 13. We say that the transform of the repeated object (the top hat) is "sampled" at the spikes of the crystal transform. We also can see that when the objects are placed far apart, the transform will be finely sampled; when they are close, the sampling will be coarse.

Notice that the sampling of the transform occurs at points not unlike those in the reciprocal lattice. Indeed it is this observation that completes the connection from Bragg diffraction, the Ewald sphere, and the reciprocal lattice to the idea that the diffraction pattern simply represents the Fourier transform of the crystal. As h, the argument of $F(h)$, has dimensions of reciprocal distance, it also serves to measure the reciprocal lattice. The reciprocal lattice points, each representing a Bragg plane from which reflection occurs, are the same points that have nonzero values for the transform of a repetitive crystal.

Now let us take the next logical step and see how, if we know the structure of the molecules in the unit cell, we can calculate the value of the transform as it is sampled at each reciprocal lattice point.

The Structure Factor and Its Phase. How does one use mathematical notation to represent electromagnetic radiation? We say that the oscillating electric field that accompanies an X-ray photon has a wavelength, an

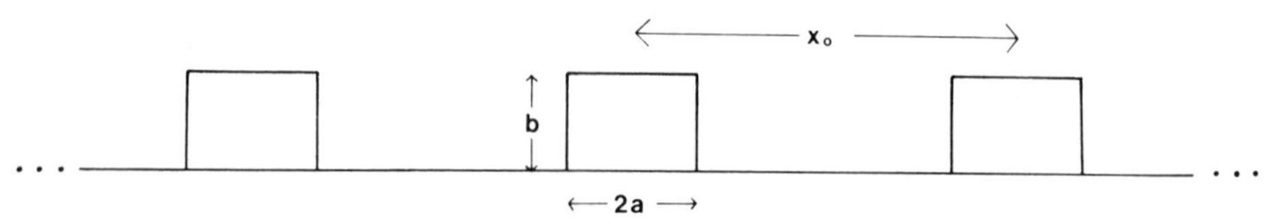

FIG. 12. Repetition of object within a crystal.

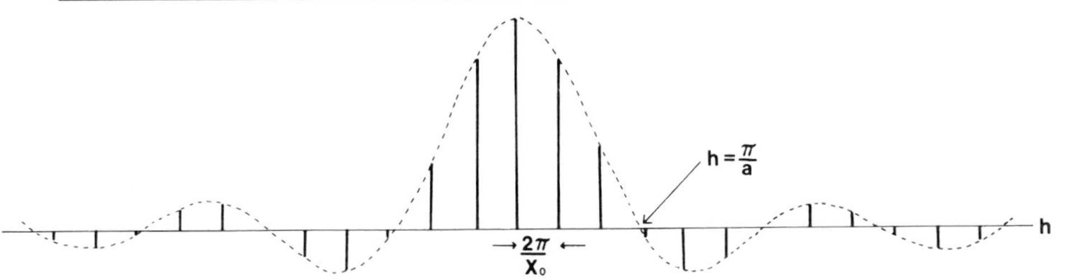

FIG. 13. Plot of Fourier transform of repetitive pattern.

amplitude, and a phase relative to other waves. At any instant the electric field E vaires sinusoidally with distance as in Fig. 14. We call the argument of this sine function the *phase angle*. The origin from which this phase is calculated is really arbitrary, but it must be the same for all waves considered together. This function has the form

$$A(x) = A_0 \cos(2\pi x/\lambda) \tag{15}$$

The phase of this wave, relative to that at the origin, is $2\pi x/\lambda$. How might we represent a wave's amplitude and phase? One could simply use two numbers, the amplitude A_0 and the phase angle ϕ. These two numbers are easily graphed in polar coordinates such as in Fig. 15. Notice that the wave is then easily represented as a point on the graph. Another way to represent a point on a graph is in terms of its horizontal and vertical components A_r and A_i. Yet another is to represent it as a single complex number, having as its real component the horizontal distance to the point and as its imaginary component the vertical distance. This ability to use a single complex number to represent a wave's amplitude and phase is especially useful because there are several interchangeable ways to represent the complex number. The following are all equivalent, with ε being complex and the A values and ϕ being real.

$$\varepsilon = A_r + iA_i = A_0 \exp(i\phi) \tag{16}$$

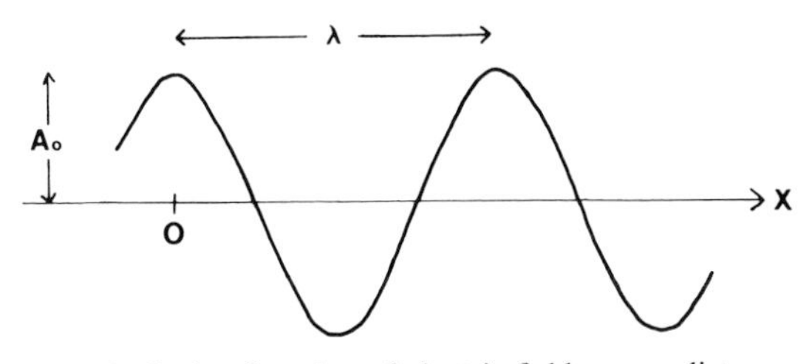

FIG. 14. Cosine function of electric field versus distance.

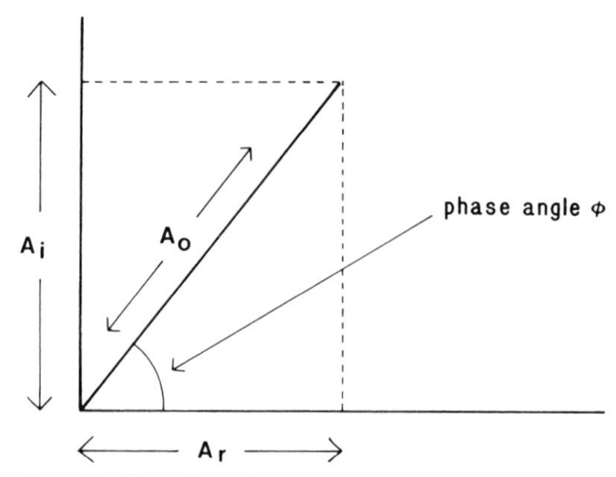

FIG. 15. Plot of amplitude (A) versus phase angle (ϕ).

Another value of this notation is that we can represent the interference among waves as the sum of the complex numbers which describe those waves. As an example, let us evaluate the result of interference between three waves with arbitrary amplitudes and phases. We can see in Fig. 16 that whether we sum the three waves, the vectors, or the complex numbers we arrive at the same result.

We now want to calculate an expression, which we shall call the structure factor, that represents the wave reflected from a single set of Bragg planes in a crystal. We know from the discussion above that this structure factor is the value of the Fourier transform of a single unit cell, evaluated at one of the sampling points that arise from the crystal repetition. These sampling points are reciprocal lattice points.

How do we evaluate the structure factor? Let us extend to three dimensions the expression we defined earlier for the one-dimensional Fourier transform, and at the same time introduce some crystallographic notation. The Fourier transform of a single unit cell is

$$F(\mathbf{s}_{hkl}) = \frac{1}{\sqrt{2\pi}} \int_{v} \rho(\mathbf{r}) \exp(2\pi i \mathbf{s}_{hkl} \cdot \mathbf{r})\, d\mathbf{r} \tag{17}$$

where $d\mathbf{r}$ is volume element. Since we want this transform to represent diffraction of X rays, and since we know that the X rays are scattered by the electrons on the atoms of the structure, we let $\rho(\mathbf{r})$ be the density of electrons at the point represented by the vector $\mathbf{r}$. We define $\mathbf{r}$ as $x\mathbf{a} + y\mathbf{b} + z\mathbf{c}$, where x, y, and z are dimensionless fractional coordinates within the unit cell. We then have that $\mathbf{s}_{hkl} \cdot \mathbf{r} = hx + ky + lz$, a result of the properties of the real and reciprocal lattice vectors. Finally we notice that

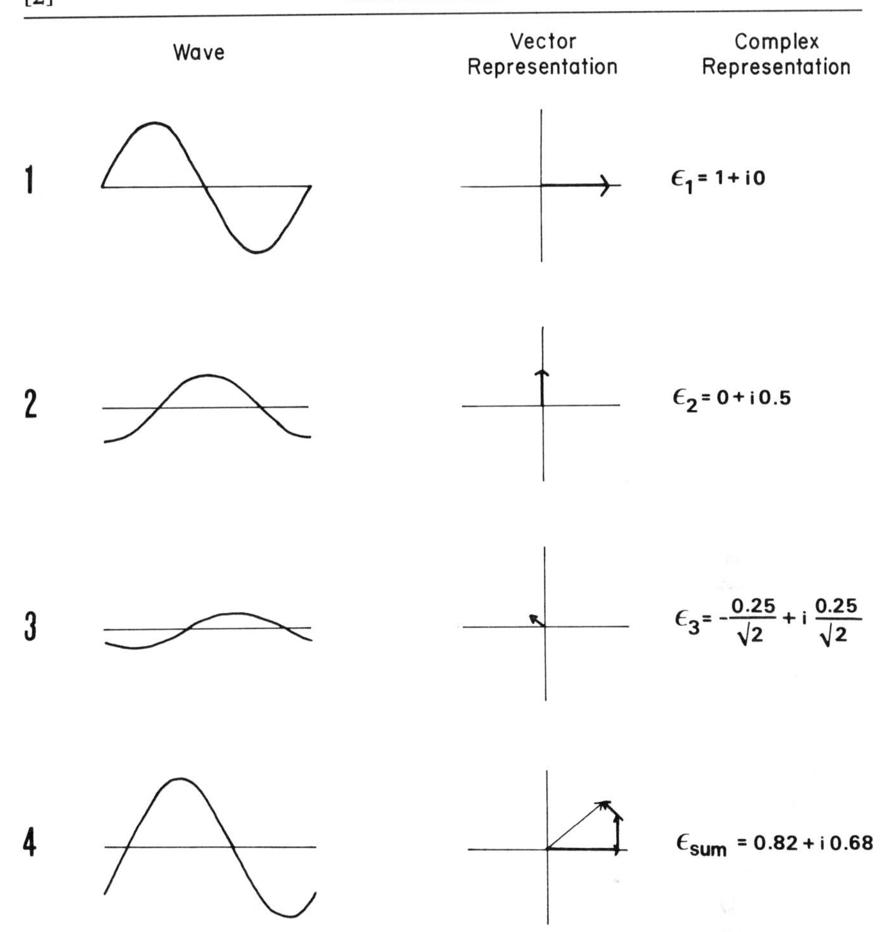

FIG. 16. Representation of result of interference between waves.

we can obtain the integral over the entire unit cell by simply performing a summation over the atoms in that cell.

This gives us the common expression for the structure factor:

$$F_{hkl} = \sum_{\text{atoms}} f_j \exp[2\pi i(hx_j + ky_j + lz_j)] \tag{18}$$

where f_j is the "scattering power" of each atom, x_j, y_j, z_j are its coordinates, and we have dropped the factor $1/\sqrt{2\pi}$.

We can show that this is correct. The structure factor for a single atom is

$$f_{hkl} = f_j \exp[2\pi i(hx_j + ky_j + lz_j)]$$

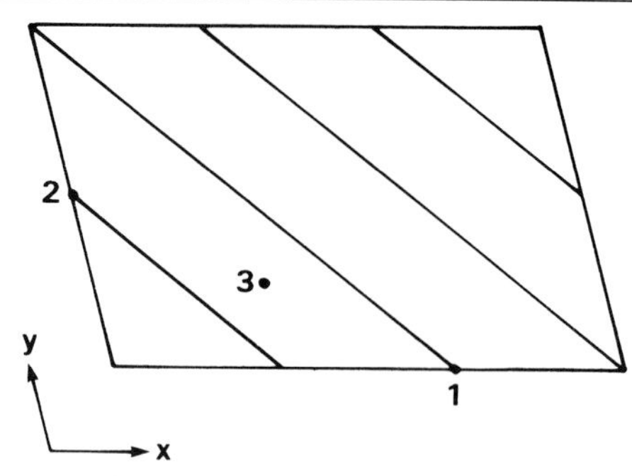

FIG. 17. Two-dimensional unit cell.

where the phase angle is $2\pi(hx_j + ky_j + lz_j)$. Does this makes sense? We can place atoms in a unit cell with the lattice planes *(hkl)* drawn in and see if the phases calculated by this expression are correct. In Fig. 17 we have a two-dimensional unit cell with the (3,2) planes in place and several atomic positions marked. Braggs' law requires that for diffraction to occur, the phases of the waves scattered from any two atoms must be equal when both lie on any one of the lattice planes in question. For example, we calculate that the phase of scattering from atom 1 at (2/3, 0) is $2\pi(3 \cdot 2/3 + 2 \cdot 0) = 4\pi = 0$, and that for atom 2 at (0, 1/2) is $2\pi(3 \cdot 0 + 2 \cdot 1/2) = 2\pi = 0$. We can see that the phase of scattering from atom 3, which lies midway between two planes at (1/3, 1/4), should be π. This turns out to be the case since $2\pi(3 \cdot 1/3 + 2 \cdot 1/4) = 3\pi = \pi$.

Coming to Focus; Regenerating the Image. Recall what we have learned. Starting with very simple physical ideas, the scattering of X rays by atoms and the interference of scattered rays, we have built up a rather complete picture of the way X rays are diffracted from crystals. We understand the geometry of diffraction and can even calculate the amplitude and phase of the diffracted rays. What remains is to see how this information might be used to reconstruct the structure of the crystal.

Let us take our cue from Fourier. He showed that any periodic function can be approximated by a sum of trigonometric functions. For example, Fig. 18 shows a function that is periodic with a repeat distance of **a**. The "Fourier sum" which could approximate this function is

$$g(x) = \sum_{j=0}^{n} \left(A_j \cos 2\pi j \frac{x}{a} + B_j \sin 2\pi \xi \frac{y}{a} \right) \tag{19}$$

The coefficients A_j and B_j are real. This function *g(x)* will come closer and closer to the true shape of *f(x)* as *n*, the number of terms included in the

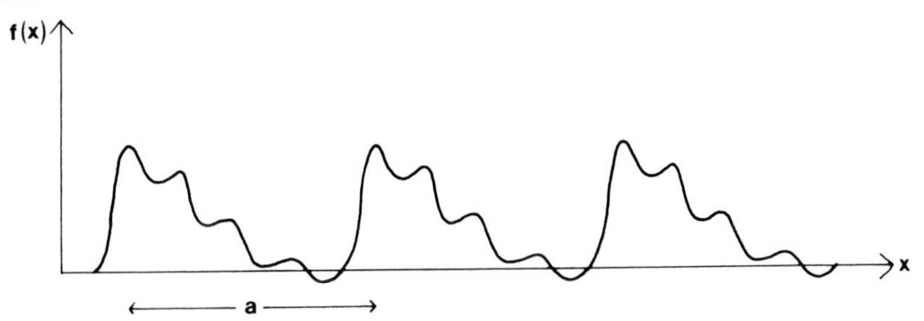

FIG. 18. Periodic function.

sum, gets larger. In Fig. 19 you can see that this is so. The waves represented on the left are shown summed along the right. The more waves used, the more nearly the sum approaches the shape of the periodic function in Fig. 18.

We have a periodic function to represent: the electron density in a crystal. It is periodic in three dimensions over distances that are the lengths of the principal axes of the crystal unit cell. Let us write a Fourier sum that could approximate it. Here we shall use a slightly different form of the Fourier sum. We shall write the function in three dimensions, use a complex exponential to represent the trigonometrics, allow the coefficients

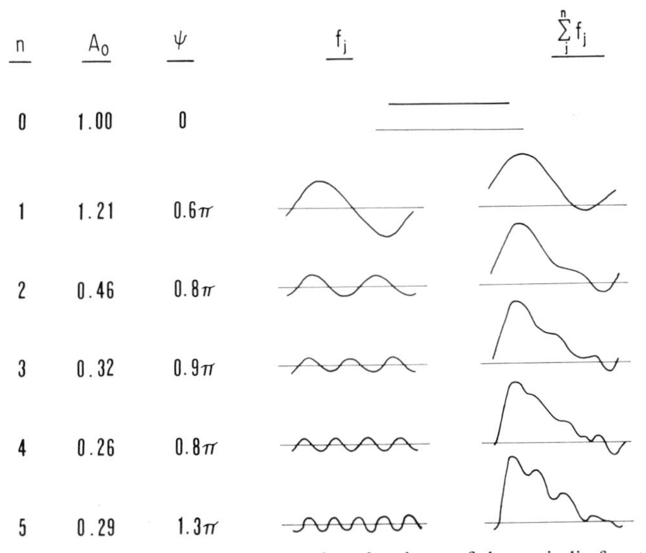

FIG. 19. Summation of waves approaches the shape of the periodic function.

to be complex, and finally sum over all negative and positive values of the indices. The function we want is

$$\rho(x,y,z) = \sum_{h'=-\infty}^{\infty} \sum_{k'} \sum_{l'} C_{h'k'l'} \exp[-2\pi i(h'x + k'y + l'z)] \tag{20}$$

We can evaluate the complex coefficients $C_{h'k'l'}$ by use of a standard mathematical trick. We start by rewriting the structure factor, in a slightly different but recognizable way:

$$F_{hkl} = \int_V \rho(x,y,z) \exp[+2\pi i(hx + ky + lz)]\, dV \tag{21}$$

Now we substitute Eq. (20), the Fourier sum for electron density $\rho(x,y,z,)$, into this expression and simplify the result. What we find is that the complex coefficients $C_{hkl} = F_{hkl}$/(volume of unit cell). This then gives us the correspondence between the complex structure factor and the real electron density in the crystal, the Fourier electron density function:

$$\rho(x,y,z) = \frac{1}{V} \sum_{h=-\infty}^{\infty} \sum_{k} \sum_{l} F_{hkl} \exp[-2\pi i(hx + ky + lz)] \tag{22}$$

One can compare these Eqs. (21) and (22) to Eqs. (2) and (3), written during our first mention of Fourier, and see that the diffracted waves and the structure are Fourier transforms of one another.

Calculation of electron density is a simple computational chore; mathematicians have devised a technique called the ''fast'' Fourier transform in which the summation above is factored in a way that decreases substantially the number of calculations to be made. Something missing from our discussion, however, is a method for determining the phase of the complex structure factor F_{hkl}.

Phase Calculation: The Isomorphous Replacement Method. One of the most important contributions Perutz and his co-workers made to protein structure determination was to develop this method. It depends upon the slight perturbation that is caused by a few very heavy atoms being bound to the protein. When the binding of the heavy atoms causes no substantial changes to the crystal structure, the structure is said to be *isomorphous,* and the crystal with heavy atom bound is an *isomorphous heavy-atom derivative* of the native crystal.

Let us preview the way the method works before we examine the details. First one measures diffraction data for the native protein and for one or more isomorphous heavy-atom derivatives. Then one determines positions of the heavy atoms in the crystal. Doing this depends upon the fact that a hypothetical diffraction pattern from the structure that contains *the heavy atoms alone* is very similar to the *differences* between the diffraction patterns from a native protein crystal and its heavy-atom derivative. As a result, many of the same techniques that are used to solve structures with only a few atoms can be used to locate the heavy atoms in a protein structure; only here the structure factors used are the differences between the two sets of measured structure factors. These techniques usually include the use of the Patterson function, a calculation that results in a knowledge of the *vectors* between atoms in the structure. Determination of the positions of several atoms from knowledge only of the vectors between them is often a problem of exquisite complexity. Other techniques that are sometimes used are the "direct methods" of structure determination. Here the statistical relationships among the amplitudes of structure factors can be used to place constraints on their phases. When the methods are applied to the differences between diffraction patterns, the result is the structure of the constellation of heavy atoms.

After the positions of the heavy atoms have been found, these positions are used in the calculation of structure factors for the heavy atoms alone. As you will see, these structure factors place constraints on the possible values of the phases for the native structure factors.

Recall that the structure factor is a sum of terms, one for each atom in the structure. If the protein atoms in the crystal are not perturbed by the binding of heavy atoms, the structure factor for a heavy-atom derivative of a protein crystal is simply that of the native protein with that for the heavy atoms added on:

$$F_{PH} = F_P + f_H \tag{23}$$

Here F_{PH} is the structure factor for the heavy-atom derivative of the parent protein, F_P is that for the parent, and f_H is that for the heavy atoms alone. This equation involving complex numbers represents a triangle in the complex plane (see Fig. 20). Of course, one does not know the phase of F_P or F_{PH}, but only the amplitudes $|F_P|$ and $|F_{PH}|$ and the phase and amplitude of f_H. As a result, this triangle can be drawn in two possible ways consistant with the data, as in Fig. 21, leaving one with a 2-fold ambiguity for the phase of F_P.

One normally resolves this ambiguity by performing the whole set of measurements and calculations again for at least one more heavy-atom derivative. One hopes that a second derivative will have phase indications close to only one of those from the first. Statistical methods, worked out by Blow and Crick, are used to compare the various indications for the phase and to choose one which will minimize the errors in the final electron density map.

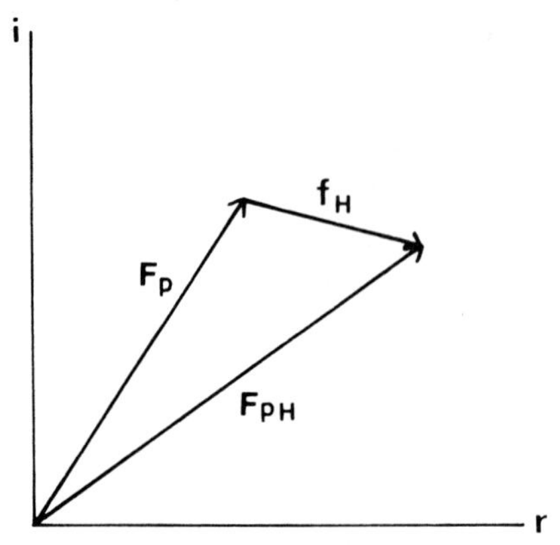

FIG. 20. Plot of the structure factor.

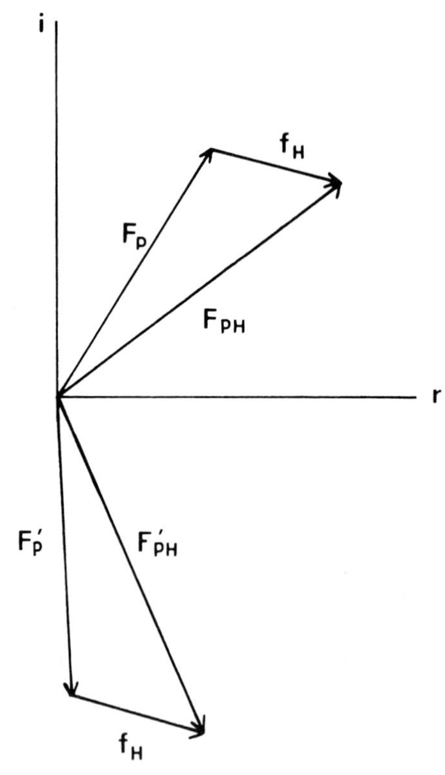

FIG. 21. Plot of 2-fold ambiguity for the phase of F_p.

Practicing Crystallography

From the biochemist's perspective, the crystallographer's chore must seem a long and complicated process. Acknowledging that this is true (which I do) does not make it simpler. It will pay, however, to show that a complete project is a serial sequence of steps whose only dependence upon each other is that they be performed carefully and in the proper order. Very briefly, these steps are

1. Growth of large, perfect crystals and preliminary characterization of the crystals and their diffraction pattern.
2. Preparation of heavy-atom derivatives.
3. Measurement and processing of diffraction data.
4. Calculation of phases for Fourier coefficients.
5. Interpretation of electron density maps and refinement of molecular models.
6. Analysis of the structure.

These items all are treated in the chapters that follow in this volume. Let us preview briefly what we shall find.

Crystal Growth and Characterization. In modern times one has access to a sophisticated arsenal of protein purification techniques, such as ion-exchange and gel-exclusion chromatography plus preparative-scale gel electrophoresis and isoelectric focusing. One forgets that a major purification tool of the early protein chemists, many of whom started life as organic chemists, was crystallization. Even now, some primeval instinct provides a thrill of satisfaction to the modern biochemist when he sees the opalescent sheen of microscopic crystals in a swirled flask of precipitated protein.

The conditions for growth of large crystals for X-ray diffraction are little different from those that produce these "biochemist's" crystals. One need only provide a pure protein and add to it, under the proper conditions, a suitable precipitant and a pinch of patience, and crystals may form. Much of the crystallographer's burden is to find these proper conditions and the proper precipitant. One must choose the correct pH, temperature, and protein concentration. One may select precipitants such as salts, which cause protein to crystallize because of hydrophobic interactions, or organic liquids, such as alcohols, which strengthen electrostatic interactions between protein molecules. In addition, one may need to worry about special conditions, such as the presence of allosteric effectors, that will favor one particular conformation for the molecule.

Section II of this volume discusses many of these considerations. As a testament to the parsimony of many crystallographers, who would rather spend their time thinking about molecular structure than preparing protein, many of the chapters in that section concern methods to grow crystals with the smallest possible quantity of material.

Once crystals have been grown, they require a rather thorough analysis before one can embark upon a full structure determination. Several things must be learned.

1. What is the size of the crystallographic unit cell? What is its symmetry? Both of these are determined by scrutiny and measurement of the diffraction pattern.
2. What form does the molecule adopt in the crystal? How many molecules occupy each unit cell? A number of factors will help to answer these questions. Among them are knowledge of the crystal density, determination of which is discussed in Section III, and knowledge of the size and symmetry of the unit cell and of possible subunit structure in the molecule.
3. How accurately may one hope to learn the structure? Is this a tractable problem? These questions depend upon the quality of the diffraction pattern and upon one's ability to measure the diffraction data in practice. They also depend upon the stability of the crystals in the X-ray beam.

Preparation of Heavy-Atom Derivatives. If crystal growth is an art, this task involves alchemy. Although there are inspired exceptions, variations are usually played on only a few themes. Heavy metals may be bound to the protein before crystals are grown, but more often the heavy-metal compounds are allowed to diffuse into the crystals from solutions in which the crystals are soaked. A number of compounds, for example many containing Pb or Hg, bind tightly to the free thiol on cysteine. Iodine will react with tyrosine. Beyond this, one can find a sizable grab-bag of coordination compounds of Pt, Au, U, etc., that fix themselves to polar or ionic sites on proteins, and even a few, such as dimethyl mercury, that bind in hydrophobic pockets. The process is very tedious, and occasionally a little art, say in the design of a heavy-atom-labeled mimic of a substrate, will save a lot of alchemy.

Measurement of Diffraction Data. As we have mentioned in the first half of this chapter, one may measure diffraction intensities by photographic means, a process that has changed little over decades, or by use of electronic detectors, a field that promises to grow rapidly in the future.

As with all physical measurements, great care must be taken in the collection of X-ray data. One must minimize systematic and random error in the measurements. One must think hard about sources of systematic error—nonuniformity in the detector or X-ray source, absorption of X rays by the specimen, or decay of the specimen in the X-ray beam—and must take steps to eliminate them. Replicate measurements must be made, both to monitor systematic error and to decrease random error. Systematic errors must be corrected if they can be measured. Often this can be accomplished for absorption of X rays and for crystal decay.

The chapters in Section III concern themselves with many aspects of data collection. In them we see both the norm of modern practice and the state of the art as it will stand in the future.

Calculation of Phases for Fourier Coefficients. Before the electron density in a crystal can be calculated, phases must be assigned to the structure factor amplitudes that have been measured. As we mentioned earlier, the most generally successful method available for doing this is *multiple isomorphous replacement.* Here a small number of heavy atoms are bound to each protein molecule in the crystal, perturbing the intensities and phases of the diffracted waves. If the positions of the heavy atoms are known, so that the heavy-atom contribution to the overall diffraction can be calculated, information about the phase can be determined.

Although the method of isomorphous replacement often serves for the initial calculation of phases, one sometimes can use other methods, such as those discussed in Part B, Vol. 115, to improve the initial estimates. This is possible when extra information is available about the structure of the protein in the crystal.

One example of "extra information" is when the molecule possesses symmetry that is not part of the crystal symmetry, that is, when the asymmetric unit of the crystal contains more than one identical piece of protein. In this case, identical but independent portions of the structure can be averaged to produce a new and more accurate electron density map. This is the basis for a powerful technique for phase improvement known as the *molecular replacement method,* described in a collection of papers by that name assembled by Michael Rossmann. It involves several separate steps. First the electron density map is averaged according to the highest possible molecular symmetry. Next, the envelope defining the surface of the molecule is determined by inspection. Third, the regions of density outside this envelope are all set to some average value to represent the interstitial liquid. Finally, this averaged electron density map with smoothed solvent regions is used for calculation of structure factors and a new set of phases is compiled, based on a comparison of the calculated phase and the one from isomorphous replacement. These new phases are used to calculate a new electron density map, and the process is repeated iteratively. The technique is especially powerful when the number of symmetry-related pieces in the asymmetric unit is large, and it has been used with great success in work on the icosahedral viruses and on the coat protein of tobacco mosaic virus.

A variation on this theme is a technique known as *density modification*. Here, in the absence of multiple subunits in the asymmetric unit, one makes whatever improvements seem justified on the electron density and uses this modified density to calculate new phases. The modifications are based on reasonable assumptions about protein crystals. Two assumptions that can be made are first that the electron density in solvent regions of the crystal is fairly smooth, and second that there is an absolute minimum below which the electron density may not go. After the electron density is modified to meet these conditions, structure factors are calculated, calculated phases are combined with those from isomorphous replacement, a new map is calculated, and the process is repeated. Especially in cases where the fraction of the crystal volume occupied by solvent is very high, say with tRNA, the process provides a marked improvement in the structure.

Production of Molecular Models. The *result* of the diffraction experiment is the three-dimensional map of electron density in the crystal. This, map, however, serves merely as *data* to be used to interpret the molecular structure. One must build an atomic model to fit the electron density. Although efforts are being made to automate this process, it often depends greatly upon the chemical intuition of the person doing the work. A fundamental principle of the model building is that one really knows quite a lot about the structure of a protein, particularly if the primary structure is known. In particular, there are fairly strict constraints placed on covalent bond lengths and angles. In addition, although the constraints are not so strong, simple stereochemical arguments make some torsional configurations more likely than others.

One makes use of this extra knowledge as much as possible. Kendrew and Watson invented rigid brass-wire models, which accounted for as many of these covalent constraints as possible, for the model-building work on myoglobin. Modern crystallographers usually use computer graphics to assist in the model-building chore, but again, the known covalent dimensions of amino acid residues are built into the programs that are used.

Because the initial calculation of structure factor phases is inaccurate, the electron density map, hence the model that can be built from it, is also inaccurate. It is always true that more information exists in the original diffraction data than can be represented by a molecular model that fits an electron density map phased by multiple isomorphous replacement. A variety of methods exists to improve this model. These include (1) classical "difference Fourier" techniques, (2) optimization of the model by a least-squares minimization of the difference between observations and the structure factors calculated from the model, and (3) a minimization of the energy of noncovalent interactions in the molecule, a procedure that has nothing to do with the X-ray data. In each case, the only truly unbiased test of the quality of the model is the accuracy with which structure factors calculated from it match the observations.

Analysis of the Structure. There are several senses in which one might "analyze" the structure of a protein molecule. We shall discuss only two.

Crystal structure work on an enzyme is initiated to learn about the structural basis of its biological activity. Knowing the three-dimensional structure of an enzyme is not unlike seeing a lathe or a mill all cleaned up and sitting, silent, on the shop room floor. With a little imagination, one can make pretty good presumptions about how it works. It is not the same, however, as seeing it in action. It is also not the same as seeing the same machine, again silent, with the work and tools in place. To retrieve this metaphor, the crystallographer sees an enzyme in stasis; the active machine is not accessible, although recent work in low-temperature crystal structure analysis shows how the machines can be slowed down a lot. However, it is often possible to catch the enzyme at one end of a process. Specifically, the biochemist can often react the protein in the crystal with cofactors, competitive inhibitors, substrate mimics that bind to an active site, etc. When this has been done, a procedure known as difference electron density synthesis can be used to learn about the complex. Diffraction data are measured from the derivatized crystals, and amplitudes that are the signed difference between the derivative structure factors and the native protein structure factors are used with the native crystal phases to calculate an electron density map. This map closely approximates the difference between the electron density in the derivative and the native crystals. It can often be used to determine the structure of the pseudosubstrate or effector as they are bound to the enzyme and this knowledge often can be used to learn much about the action of the enzyme.

A second way in which one may analyze the structure of a protein molecule is that one can compare it, in a topological sense, with the known structures of other proteins. This is something that has only begun to be possible in the last half-dozen years, after a large number of structures had become known. Some of the ideas that have evolved during this time are reviewed in the following companion volume.

A number of systematic features of protein structure have been noticed. The most often recurring of these is that a rather small number of topologies of folding of the β-pleated sheet structure are observed at a surprisingly high frequency. Another is that at least one rather specific arrangement of peptide chains, the nucleotide-binding fold found in several dehydrogenases, has been found on several occasions to perform a similar function in different proteins.

In the final analysis, the structures themselves, not the X-ray data, are grist for the enzymologist's mill. From these structures we are beginning to understand the way in which enzymes catalyze and control reactions, and to glimpse some of the principles upon which all of molecular, hence cellular, structures are based. A natural consequence of this new knowledge is that our questions become ever more sophisticated and our curiosity about larger and more complicated structures continues to grow. The message in this for the crystallographer is that the success of his methods leads to new demands. Whatever he can discover now will make crucial his ability to discover much more in the future. Although he may not rest on his laurels, he can labor in secure self confidence that the explosion in his capabilities which is occurring now, and which is chronicled in these two volumes, not only will persist, but will continue to be of tremendous value as the future unfolds.

Section II

Crystallization and Treatment of Crystals

[3] Theory of Protein Solubility

By TSUTOMU ARAKAWA and SERGE N. TIMASHEFF

Introduction

The crystallization of proteins is determined both by their solubility, i.e., an equilibrium thermodynamic factor, and by kinetic factors which control nucleation and growth. For crystallization, a protein solution is first brought to a point of supersaturation, a thermodynamically metastable state. Once crystallization starts, the system moves toward equilibrium until it reaches that state. The protein concentration at equilibrium, i.e., protein solubility, is a complex function of a number of factors, such as the physical and chemical natures of the proteins themselves, and environmental parameters (i.e., pH, temperature, the nature of the salt or organic solvent, and its concentration). The size of the crystal depends as well on the degree of supersaturation and on the kinetic pathway of nucleation and growth.

Protein crystals have been grown from a variety of solvent systems (e.g., see Blundell and Johnson[1]). Most prominent among the precipitants used has been $(NH_4)_2SO_4$, because of both its high solubility in water and its strong salting-out properties. Salts, however, affect protein solubility in broadly different ways. Almost all salts, at low concentration, have a general salting-in effect on proteins, and some salts maintain this salting-in property even at high concentrations. Organic solvents, such as ethanol and acetone, have been used for protein crystallization. Most organic solvents, however, have a strong preference for contact with hydrophobic regions and tend to induce protein denaturation. Thus, their use is limited to relatively low concentrations. Recently, two organic compounds, hexylene glycol (2-methyl-2,4-pentanediol, MPD)[2] and polyethylene glycol,[3] have been identified as good protein crystallizing agents. They are very strong protein precipitants[2,4] and do not denature proteins at the

[1] T. L. Blundell and L. N. Johnson, "Protein Crystallography." Academic Press, New York, 1976.

[2] M. V. King, B. S. Magdoff, M. B. Adelman, and D. Harker, *Acta Crystallogr.* **9**, 460 (1956).

[3] A. McPherson, *Methods Biochem. Anal.* **23**, 249 (1976).

[4] I. R. M. Juckes, *Biochim. Biophys. Acta* **229**, 535 (1971).

conditions used for crystallization,[5,6] although they may destabilize the native structure of proteins at extreme conditions.[7,8]

Over the past 50 years, a large number of both theoretical and experimental studies have been devoted to the question of the effects of salts and organic solvents on protein solubility. The general salting-in effect at low salt concentrations has been accounted for reasonably well in terms of the electrostatic interaction between small ions and protein charges or protein dipoles, essentially in terms of the Debye–Hückel theory.[9] Although the electrostatic theory predicts that salts should also have a general salting-out effect, this theory cannot predict a variety of effects at high salt concentrations. In 1959, Kauzmann's classical article brought into perspective the importance of hydrophobic interactions, and gave a new impetus to studies of protein–protein and protein–ligand interactions. Five years later, Sinanoglu and co-workers[10,11] suggested the cavity theory as an important contribution to the hydrophobic effect. This proposal permitted the partial interpretation of salt effects on protein solubility at high salt concentrations in terms of the increase in the surface tension of water induced by salts.[12]

An alternate approach to an understanding of the effects of additives on protein solubility and stability has been through studies of preferential interactions of solvent components with proteins at high concentrations of co-solvents. It has been shown that the preferential interactions with proteins of some salts[13] and sugars,[14,15] MPD,[5] and glycerol[16] can explain their effects on the self-association of proteins and their stability. The results obtained with salts and sugars could be related to their effects on the surface tension of water, whereas MPD and glycerol reflect other poorly understood mechanisms. Thus, a vast amount of information on factors which control solubility has been accumulated. These will be discussed systematically in this chapter.

[5] E. P. Pittz and S. N. Timasheff, *Biochemistry* **17,** 615 (1978).
[6] D. H. Atha and K. C. Ingham, *J. Biol. Chem.* **256,** 12108 (1981).
[7] E. P. Pittz and S. N. Timasheff, unpublished data.
[8] T. Arakawa and S. N. Timasheff, unpublished results.
[9] J. T. Edsall and J. Wyman, "Biophysical Chemistry," Vol. 1, p. 282. Academic Press, New York, 1958.
[10] O. Sinanoglu and S. Abdulnur, *J. Photochem. Photobiol.* **3,** 333 (1964).
[11] O. Sinanoglu and S. Abdulnur, *Fed. Proc., Fed. Am. Soc. Exp. Biol.* **24,** 512 (1965).
[12] W. Melander and C. Horvath, *Arch. Biochem. Biophys.* **183,** 200 (1977).
[13] T. Arakawa and S. N. Timasheff, *Biochemistry* **21,** 6545 (1982).
[14] J. C. Lee and S. N. Timasheff, *J. Biol. Chem.* **256,** 7193 (1981).
[15] T. Arakawa and S. N. Timasheff, *Biochemistry* **21,** 6536 (1982).
[16] K. Gekko and S. N. Timasheff, *Biochemistry* **20,** 4667 (1981).

The Thermodynamic Definition of Protein Solubility

Thermodynamically, the state of a system at equilibrium is described by the temperature, pressure, and composition. The phase rule determines the number of degrees of freedom, f, in terms of state variables:

$$f = c - p + 2 \quad (1)$$

where c is the number of components and p is the number of phases. At fixed temperature and pressure, Eq. (1) reduces to

$$f = c - p \quad (1a)$$

Let us consider a three-component system containing water (component 1), protein (component 2), and additive (component 3) at a defined temperature and pressure. If only one solution phase is present, $f = 2$, i.e., the concentrations of protein and additive can both be varied. At a fixed concentration of additive, $f = 1$ and, as long as there is only one phase, the solution composition is not fixed, i.e., the protein concentration can vary. If there are two phases in the system, $f = 0$ at a fixed concentration of component 3 and, hence, the compositions of both phases must be fixed, meaning that the protein concentration in the solution phase is independent of the total amount of protein in the system. This is the definition of its solubility. Typical solubility curves obtained by Butler[17] for fractions of chymotrypsinogen are shown in Fig. 1. In all cases, the protein concentration in the solution phase remains constant once the solid phase has appeared, in agreement with the phase rule.

When two proteins are present, the system becomes one of four components. In the presence of only one phase, $f = 2$ and the concentrations of both proteins can be varied. If the two proteins do not form a solid solution, the composition of each phase becomes fixed in the presence of one solution phase and two solid phases. In this case, the solubility of each protein can be determined from the total solubility curve unless the proportion of the proteins in solution is equal to the ratio of their solubilities.[18] If the two proteins form a solid solution, the compositions of the phases cannot be fixed and the protein concentration in solution becomes a function of the total amount of proteins.

[17] J. A. V. Butler, *J. Gen. Physiol.* **24,** 189 (1940).

[18] J. F. Taylor, *in* "The Proteins" (H. Neurath and K. Bailey, eds.), Vol. 1, Part A, p. 1. Academic Press, New York, 1953.

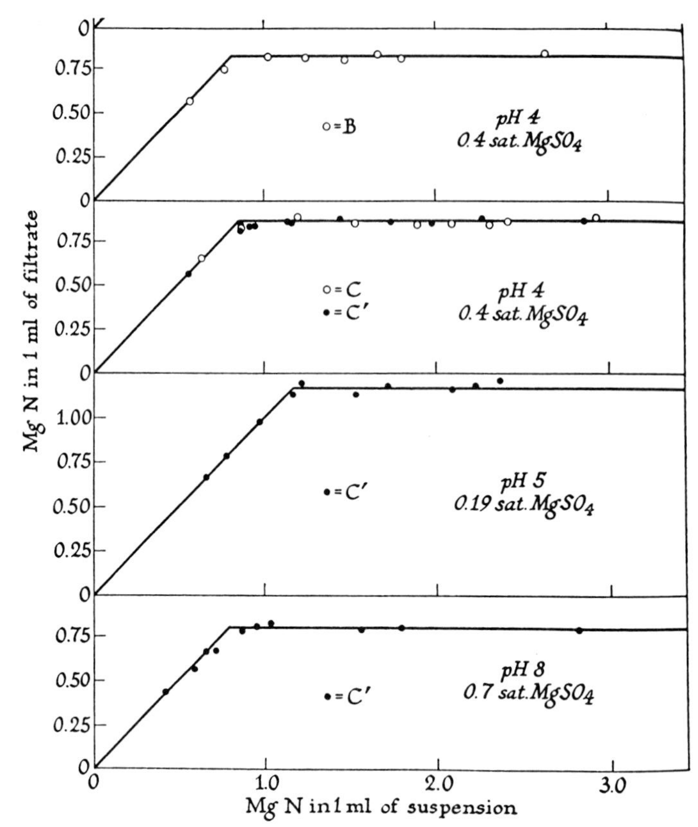

FIG. 1. Solubility curves of chymotrypsinogen fractions in various solvents. B, C, and C′ correspond to the fractions.[17]

Formal application of the phase rule indicates that solubility can be used as a test for protein purity.[18–20] Treece *et al.*[21] have suggested, however, that this method is not a good test of heterogeneity, on the basis of their finding that electrophoretically homogeneous β-lactoglobulins were as heterogeneous by the solubility criterion as natural or artificial mixtures of the two genetic forms, β-lactoglobulins A and B, as shown in Fig. 2.

[19] J. T. Edsall, *in* "Proteins, Amino Acids and Peptides" (E. J. Cohn and J. T. Edsall, eds.), p. 576. Van Nostrand-Reinhold, Princeton, New Jersey, 1943.

[20] R. M. Herriott, V. Desreux, and J. H. Northrop, *J. Gen. Physiol.* **24,** 213 (1940).

[21] J. M. Treece, R. S. Sheinson, and T. L. McMeekin, *Arch. Biochem. Biophys.* **108,** 99 (1964).

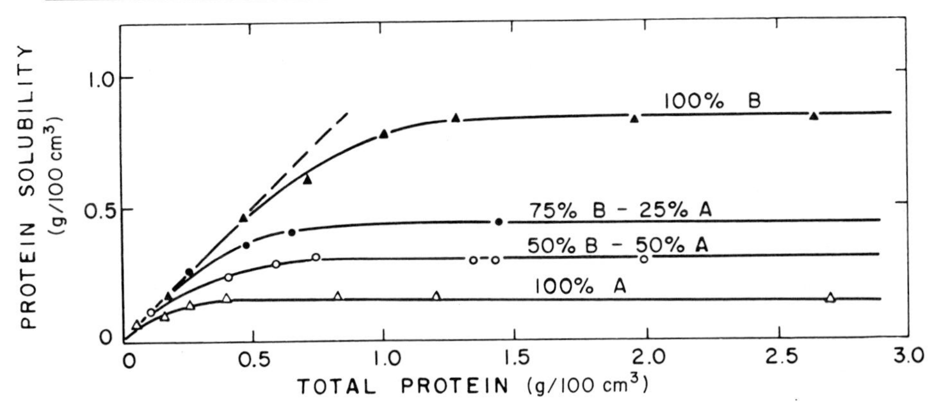

FIG. 2. Solubilities β-lactoglobulins A and B and their mixtures in 0.00625 M NaCl as a function of the total amount of protein present.[21]

The thermodynamic condition of equilibrium between phases is that the chemical potentials of individual components in the phases must be equal. When a protein in the solid phase is in equilibrium with that in the solution phase,

$$\mu_2^{l} = \mu_2^{s} \tag{2}$$

where μ_2 is the chemical potential of the protein and l and s refer to the solution (liquid) and solid phases. If protein solutions in water and in an aqueous solution of component 3 are in equilibrium with the protein in the same solid phase,

$$\mu_2^{s} = \mu_{2,w}^{l} = \mu_2^{l} \tag{3}$$

where w refers to water and μ_2^{l} refers to protein solution in a solvent containing the third component. Since $\mu_2^{(i)} = \mu_2^{\circ(i)} + RT \ln(m_2^{(i)}\gamma_2^{(i)})$,

$$\mu_2^{\circ} - \mu_{2,w}^{\circ} = RT \ln(m_{2,w}/m_2) + RT \ln(\gamma_{2,w}/\gamma_2) \tag{4}$$

where μ_2° is the standard chemical potential of the protein including all the solvent–protein interaction free energies, and γ_2 and m_2 are its activity coefficient and molality in the saturated solution, respectively; γ_2 includes only protein–protein interactions. The left-hand side of Eq. (4) expresses the difference in the free energies associated with protein–solvent interactions in the presence and absence of component 3, i.e., the transfer free energy of the protein from water to the aqueous solution of component 3. Equation (4) indicates that this free energy difference is directly related to the difference between the solubilities of the protein in the two solvent systems, within the assumption that $(\gamma_{2,w}/\gamma_2) = 1$. As stated above, this argument is based on the assumption that the chemical potential of the protein in the solid phase is independent of the composition of solvent in equilibrium with that phase. While this assumption is valid for crystals of small molecules, protein crystals contain significant amounts of both water and small molecules,[22] which may result in a difference between the chemical potentials of the protein in solid phases equilibrated with different aqueous solvents. The chemical potential of a protein in the solid phase is also a function of the nature of this phase, whether it is an amorphous precipitate or a crystal. Results with calf rennin[23] have shown that its solubility is much higher when the solid phase is amorphous than when it is crystalline, suggesting that the crystalline form of this protein has a lower chemical potential than the amorphous precipitate. Unfortunately, it is very difficult to evaluate the chemical potential of a protein in the solid phase.

Theories of Protein Solubility

Electrostatic Theory

Proteins are highly complex polyelectrolytes carrying both positive and negative charges. At low salt concentrations, the protein molecule is surrounded by an ionic atmosphere described by the Debye–Hückel theory, with an excess of ions of charge opposite to its net charge. This screening decreases the electrostatic free energy of the protein, resulting in a decrease in its activity and an increase in its solubility, provided that the chemical potential of the protein in the solid phase remains constant. For compact protein ions the solubility is described by[24]

$$\ln(s_2/s_{2,\mathrm{w}}) = Z^2\varepsilon^2 N\kappa/[2DRT(1 + \kappa a)] \qquad (5)$$

where Z is the net charge of the protein, ε is the electronic charge, N is Avogadro's number, D is the dielectric constant of the medium, R is the universal gas constant, T is the thermodynamic (Kelvin) temperature, a is the sum of the radii of the protein ion and the average of the supporting electrolyte ions in the solution, and κ is given by

$$\kappa = \sqrt{8\pi N\varepsilon^2/1000\,DkT}\,\sqrt{I}$$

$$I = \frac{1}{2}\sum_i C_i Z_i^2$$

[22] J. A. Rupley, *in* "Structure and Stability of Biological Macromolecules" (S. N. Timasheff and G. D. Fasman, eds.), p. 291. Dekker, New York, 1969.

[23] C. W. Bunn, P. C. Moews, and M. E. Baumber, *Proc. R. Soc. London, Ser. B* **178,** 245 (1971).

[24] C. Tanford, "Physical Chemistry of Macromolecules," p. 241. Wiley, New York, 1961.

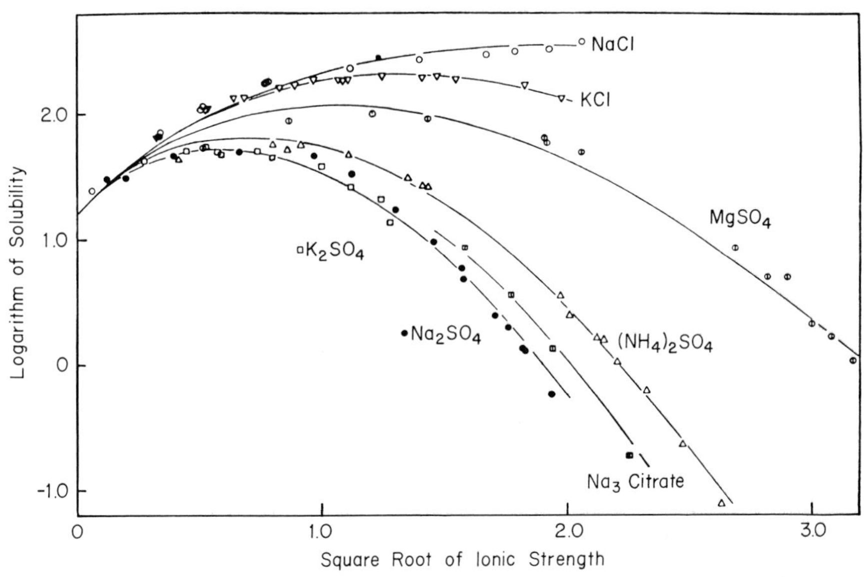

FIG. 3. The solubility of carboxyhemoglobin in various electrolytes at 25°.[25]

where I is the ionic strength and k is the Boltzmann constant. This equation is simply an expression of the Debye–Hückel electrostatic theory for a small ion. According to Eq. (5), protein solubility increases with the square root of the ionic strength. At $Z = 0$, i.e., at the isoelectric point, the protein solubility should also be increased by addition of salt, since, although the average net charge Z is equal to zero, $\bar{Z}^2$ is not zero. As shown in Fig. 3,[25] at low salt concentration all salts have a salting-in effect on proteins, the logarithm of the protein solubility being proportional to $\sqrt{I}$, in the limit as $I \rightarrow 0$. Equation (5) also suggests that protein solubility should have a minimum at the isoelectric point at low ionic strength. This is usually observed.

Kirkwood,[26] assuming simple models for the dipolar ion, has given expressions for the activity coefficient of dipolar ions in dilute salt solutions. All the equations derived have essentially the form:

$$\log f_2 = -KI \tag{6}$$

where f_2 is the activity coefficient of the dipolar ion[27] and K is a constant which depends on the dielectric constant of the medium and on either the

[25] A. A. Green, *J. Biol. Chem.* **95,** 47 (1932).

[26] J. G. Kirkwood, *in* "Proteins, Amino Acids and Peptides" (E. J. Cohn and J. T. Edsall, eds.), p. 276. Van Nostrand-Reinhold, Princeton, New Jersey, 1943.

[27] Note that in this case, f_2 includes the free energy of the solvent–dipolar ion interaction and hence is different in definition from γ_2.

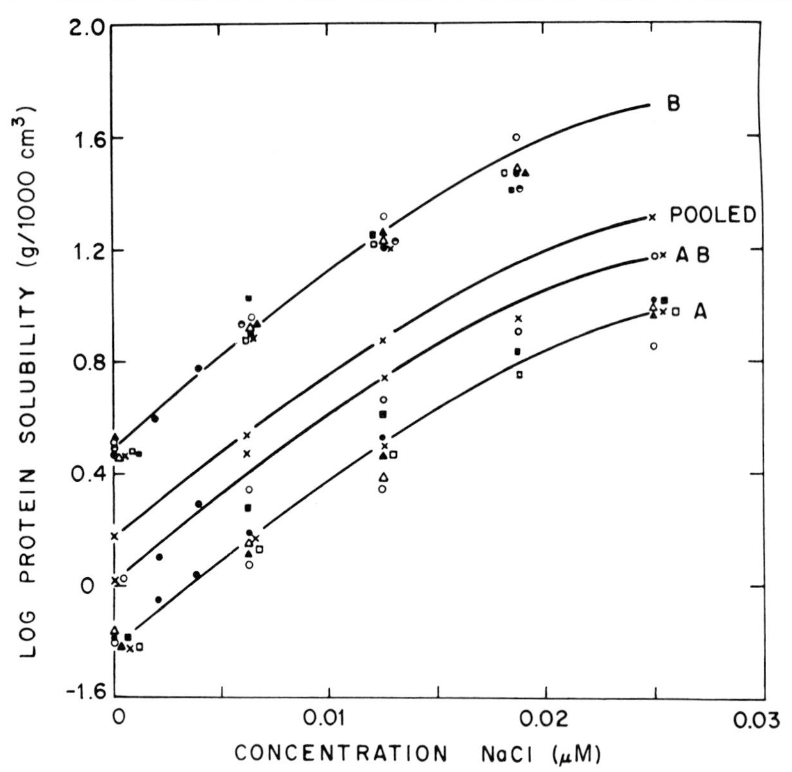

FIG. 4. Solubilities of β-lactoglobulins plotted as a function of ionic strength (μ).[21]

dipole moment or the square of the dipole moment of the dipolar ion, depending on the model used. Again assuming a solvent independent activity of the crystal state, and treating a protein as a dipolar ion, we find

$$\ln(s_2/s_{2,w}) = (K/RT)I \qquad (7)$$

In this case, the logarithm of the protein solubility is proportional to the first power of the ionic strength, rather than its square root. The solubility of β-lactoglobulin, shown in Fig. 4,[21] was found to increase in this manner when measured near its isoelectric point at low NaCl concentrations, a result which possibly reflects the fact that β-lactoglobulin has a large dipole moment.[28]

Small dipolar ions, such as glycine and alanine, should also have a salting-in effect because of electrostatic interactions with proteins. In

[28] E. J. Cohn and J. D. Ferry, *in* "Proteins, Amino Acids and Peptides" (E. J. Cohn and J. T. Edsall, eds.), p. 586. Van Nostrand-Reinhold, Princeton, New Jersey, 1943.

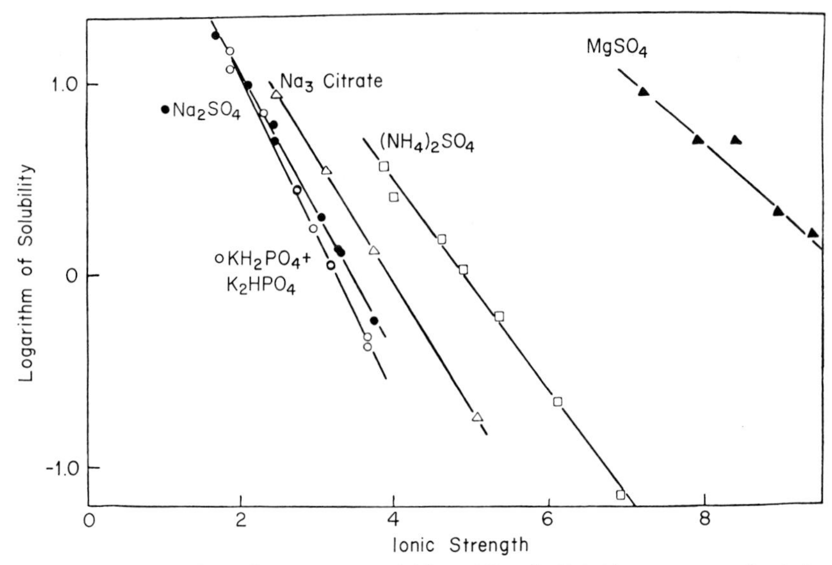

FIG. 5. The solubility of carboxyhemoglobin at 25° and pH 6.6 in concentrated solutions of various electrolytes.[25]

fact, glycine was shown to increase the solubilities of hemoglobin and β-lactoglobulin.[28]

At high salt concentrations, protein solubility is known to decrease, as seen in Figs. 3 and 5. The logarithm of protein solubility decreases almost linearly with salt concentration or ionic strength, as shown in Fig. 5. This salting-out follows the relation:

$$\log S = \beta - K_s m_3 \tag{8}$$

where β and K_s are empirical constants, K_s being the salting-out constant. The parameter K_s is characteristic of the particular salt, the order of salting-out effectiveness being essentially invariant for most proteins.[29] For any given salt it is generally small for smaller proteins, although this rule is not strict. None of the electrostatic theories except that of Kirkwood can predict an extensive salting-out effect and widely different values of K_s at high salt concentrations.

Kirkwood proposed a theory of salting-out in terms of electrostatic interactions. The Kirkwood theory is based on a model in which the polarization of the solvent by the salt ions is reduced as a result of the

[29] P. H. von Hippel and T. Schleich, *in* "Structure and Stability of Biological Macromolecules" (S. N. Timasheff and G. D. Fasman, eds.), p. 417. Dekker, New York, 1969.

presence of cavities represented by the solute (protein) molecules. Salting-out is, then, described by

$$\log f_2 = (2\pi N\varepsilon^2/2303DkT)[b^3\alpha(\rho)/a]I \qquad (9)$$

where b is the radius of the protein molecule, a is the sum of radii of the solute and the salt ions, $\rho = b/a$, $\alpha(\rho)$ is a function of ρ, given by Kirkwood (Table I in ref. 26), and I is the ionic strength. This effect is, therefore, a function only of the geometry of the solute molecule. It arises from the repulsion between the surrounding salt ions and their image charges produced in the cavities of the solute molecule of low dielectric constant. If the dielectric constant of the solute is greater than that of the solvent medium, an attraction will result. Although it is not theoretically valid at high salt concentrations, this equation predicts that the activity coefficient of the solute increases with the ionic strength of the medium, i.e., the interactions between proteins and salt ions are unfavorable and their mixing results in a mutual increase of the activity coefficients.

Effects of Organic Solvents. Assuming a water–organic solvent mixture as a continuum of dielectric constant D, the change in solubility of an ion when transferred from water to the mixture is given by

$$\ln(s_2/s_{2,\mathrm{w}}) = (A/RT)(1/D_0 - 1/D) \qquad (10)$$

where A is a constant and D_0 is the dielectric constant of water. Thus, the solubility of a protein should be decreased by the addition of an organic solvent with a dielectric constant lower than that of water. It is in fact known that organic solvents such as ethanol, acetone, and dimethyl sulfoxide, at relatively high concentrations, can be used for protein precipitation or crystallization.[1] Another factor which should contribute to protein precipitation by organic solvents is the redistribution of water and the organic solvent around the protein molecule, i.e., the preferential interactions of solvent components with the protein. In fact, the water–organic solvent mixture cannot be regarded as a continuous medium in the vicinity of a protein molecule, since the protein surface is a mosaic of regions with different polarities and different affinities for the solvent components. Furthermore, large organic molecules can also be excluded by steric hindrance.

Other Solution Variables. While no detailed discussion of the effects of temperature and pH will be presented, temperature affects protein solubility through its effect on crystal structure and/or on the dielectric constant of the medium.[30] The effect of pH was analyzed in detail by Cohn and Ferry,[28] who based the analysis on the assumptions that the

[30] R. Czok and T. H. Bucher, *Adv. Protein Chem.* **15,** 337 (1960).

solid phase contains only neutral protein molecules in equilibrium with the same forms in the solution phase, which are also in equilibrium with positively or negatively charged protein ions, all protein ions being in solution.

Hydrophobic Interactions

General. In his analysis of the transfer free energy of nonpolar hydrocarbons and nonpolar amino acid side chains from water to organic solvents, as well as of protein–protein and protein–ligand interactions, Kauzmann[31] concluded that the highly unfavorable free energy of interaction between nonpolar groups and water is one of the most important factors involved in the stabilization of protein structure and in protein interactions in aqueous solution. When a protein molecule becomes incorporated from solution into the solid phase, a certain amount of water is removed from the points of contact which contain nonpolar side chains, as well as polar ones. Therefore, hydrophobic interactions should contribute to the difference between the chemical potentials of the protein in aqueous solution and in the solid phase, i.e., to protein solubility.

Cavity Theory. The concept of hydrophobic interactions is based essentially on the uniqueness of water. Sinanoglu and Abdulnur[10,11] have proposed that this uniqueness is expressed by the value of the surface tension of water, which is much larger than that of most organic solvents. Interpreting the enhanced stability of the DNA double helix in water relative to many organic solvents in terms of the solvophobic theory which they had developed, they proposed that a major contribution to the stability stems from ΔG_c, the free energy required to form a cavity in the solvent for accommodating the solute molecule. This parameter is given by the product of the surface area of the cavity, A, and the surface tension of the solvent, σ, i.e., $\Delta G_c = \sigma A$. Although rigorously the surface tension of the solvent should be corrected for curvature,[32] this correction is small and its planar value can be used for comparing solvents.

Most inorganic salts increase the surface tension of water. This suggests that their addition should affect hydrophobic interactions between protein molecules in aqueous solution. Melander and Horvath[12] have extended the cavity theory to protein solubility in aqueous salt solutions. The free energy change ΔG of transfer of a protein from water to aqueous salt solution is

$$\Delta G = \Delta G_e + \Delta G_c \tag{11}$$

[31] W. Kauzmann, *Adv. Protein Chem.* **14,** 1 (1959).

[32] O. Sinanoglu, *J. Chem. Phys.* **75,** 463 (1981).

the electrostatic term ΔG_e being

$$\Delta G_e = -A\sqrt{m_3}/(1 + B\sqrt{m_3}) - C\mu m_3 \qquad (12)$$

where A, B, and C are constants and μ is the dipole moment of the protein. Equation (12) is just a simplified form of the electrostatic theory of the protein–ion and –dipolar models described above. Following Sinanoglu and Abdulnur,[10,11] the second term ΔG_c can be written in a simplified form as

$$\Delta G_c = [NA + 4.8N^{1/3}(\kappa^e - 1)V^{2/3}](\partial\sigma/\partial m_3)m_3 \qquad (13)$$

where A is the surface area of a protein molecule, V is its molar volume, and $(\partial\sigma/\partial m_3)$ is the molal surface tension increment of the salt. κ^e corrects the macroscopic surface tension of the solvent to molecular dimensions. Since protein crystals contain a considerable amount of water, Melander and Horvath[12] assumed that the free energy for cavity formation must be taken into account also for the crystalline protein. Its value should be smaller than that of the dissolved molecules by an amount proportional to the decrease of the protein surface area due to protein–protein contacts. Thus, in the solid phase, the chemical potential of the protein increases with salt concentration due to the increase in surface tension by the salts. Designating the decrease in the protein surface area by ϕ, the solubility change of the protein when transferred from water into the salt solutions is given by

$$\begin{aligned}\ln(S/S_0) = {} & 1/RT[A\sqrt{m_3}/(1 + B\sqrt{m_3}) + C\mu m_3] \\ & - (1/RT)[N\phi + 4.8N^{1/3}(\kappa^e - 1)V^{2/3}](\partial\sigma/\partial m_3)m_3 \qquad (14)\end{aligned}$$

Although the first term is not valid at large values of m_3, it manifests the salting-in effect as described above, while the second term indicates a decrease in solubility with increasing salt concentration. A very important conclusion from this analysis is that the first term does not depend on the nature of the salt, while the second term is strongly dependent on it, since the molal surface tension increments of salts differ greatly. According to Eq. (14), protein solubility should decrease at high salt concentrations at which the second term becomes dominant. Finally, Melander and Horvath[12] concluded that the molal surface tension increment of salts can be used as a measure of salting-out effectiveness, just like the lyotropic number determined empirically. It has to be mentioned, however, that $CaCl_2$, $MgCl_2$, and $BaCl_2$, which have a value of $(\partial\sigma/\partial m_3)$ comparable with or even higher than that of Na_2SO_4, are known not to decrease significantly protein solubility. In addition, this theory cannot be extended readily to other additives such as polyethylene glycol, which is one of the strongest protein precipitants, but which decreases the surface tension of water.[33]

Preferential Protein–Solvent Interactions

Theory

Phase Separation. The preferential interaction pattern of a protein with solvent components is a measure of the activity of the protein and hence its solubility, in a given solvent system. Although the system usually consists of four components, water (component 1), protein (component 2), additive (component 3), and buffer constituents, the last can be neglected usually due to their low concentration, reducing the system to a three-component one. In a three-component system, the preferential interaction of component 3 with the protein can be expressed either as the binding of that component to the protein, $(\partial m_3/\partial m_2)_{T,\mu_1,\mu_3}$, or as the perturbation of the chemical potential of the protein by the co-solvent, $(\partial \mu_2/\partial m_3)_{T,P,m_2}$. The experimental methods for measuring preferential interactions and the theoretical analysis of the data have been presented in detail previously in this series.[34,35] Since the preferential binding and the chemical potential change are related by a negative sign, a positive value of $(\partial \mu_2/\partial m_3)_{T,P,m_2}$, i.e., an increase in the activity of the protein, is accompanied by a preferential exclusion of the co-solvent from the protein surface, i.e., by preferential hydration. When the preferential hydration is large, addition of component 3 may induce a phase separation, resulting in the formation of a crystalline or amorphous protein precipitate. The phase separation can be described by a phase isotherm, similar to that developed by Flory[36] for synthetic polymers to describe changes in their solubility induced by a change in temperature or by addition of a co-solvent, and separation upon heating of aqueous solutions of polyvinyl alcohol–acetate copolymers.[37]

In terms of preferential interaction parameters, the phase isotherm of aqueous protein solutions may be expressed as[5]

$$-(\mu_1 - \mu_1^\circ)/RTV_1 = (C_2/M_2)\{1 + (V_m/2RTM_2)C_2[(\partial \mu_2^{(e)}/\partial m_2)_{T,P,m_3} + (\partial m_3/\partial m_2)_{T,\mu_1,\mu_3}(\partial \mu_2/\partial m_3)_{T,P,m_2}] + O(C_2^2)\} \quad (15)$$

[33] J. C. Lee and L. L. Y. Lee, *Biochemistry* **18,** 5518 (1979).

[34] E. P. Pittz, J. C. Lee, B. Bablouzian, R. Townend, and S. N. Timasheff, this series, Vol. 27, p. 209.

[35] J. C. Lee, K. Gekko, and S. N. Timasheff, this series, Vol. 61, p. 26.

[36] P. J. Flory, "Principles of Polymer Science." Cornell Univ. Press, Ithaca, New York, 1953.

[37] F. F. Nord, M. Bier, and S. N. Timasheff, *J. Am. Chem. Soc.* **73,** 298 (1951).

where μ_1 and μ_1° are the chemical potentials of water in the real solution and in the standard state, respectively. V_1 is the molar volume of water, C_2 is the protein concentration in grams per ml, V_m is the volume (ml) of solution containing 1000 g of water, and $\mu_2^{(e)}$ is the excess chemical potential of the protein. As is evident from Eq. (15), the dependence of μ_1 on C_2 is a function of the preferential protein–solvent interaction parameter as shown by Flory[36] for synthetic polymers. When $(\mu_1 - \mu_1^\circ)$ is negative, the solution is stable, the protein is completely miscible with water, and no phase separation occurs. A positive value means instability of the system, leading to separation into two phases, a solution phase rich in water and a crystalline or amorphous phase rich in protein. This phase isotherm does not give any information on the compositions of the two phases. Since the first interaction term in the parentheses of Eq. (15) is positive, its contribution is to decrease the chemical potential of water, μ_1. The second term is always negative or zero, the signs of $(\partial\mu_2/\partial m_3)_{T,P,m_2}$ and $(\partial m_3/\partial m_2)_{T,\mu_1,\mu_3}$ being always opposite. The condition of phase separation, i.e., $\mu_1 > \mu_1^\circ$, requires that the solvent interaction term be large enough to overcome the contribution of the protein self-interaction term. When the preferential interaction is that of preferential hydration, component 3 is preferentially excluded from the protein and the separated solid phase should contain less component 3 than the solution. On the other hand, for a preferential co-solvent binding system, the solid phase should be enriched with respect to component 3.

Chemical Potential Change. An alternate way of expressing the effect of protein–solvent interactions on the solubility of a protein is through the change in the chemical potential of the protein that is induced by the addition of component 3. Since the preferential interaction measurements are performed on protein solutions, the parameter $(\partial\mu_2/\partial m_3)_{T,P,m_2}$ refers to dissolved protein. Integration of this parameter with regard to m_3 gives the transfer free energy of the protein from water to an aqueous solution of component 3, i.e., $\mu_2 - \mu_{2,w}$[38]

$$\Delta\mu_2 = \mu_2 - \mu_{2,w} = \int_0 (\partial\mu_2/\partial m_3)_{T,P,m_2}\, dm_3 \tag{16}$$

Equation (16) may be decomposed into two terms corresponding to the interactions of components 1 and 3 with the protein, $(\partial\mu_2/\partial m_3)_{T,P,m_2} = -(\partial\mu_3/\partial m_3)_{T,P,m_2}[\nu_3 - (m_3/55.5)\nu_1]$

[38] Since all the preferential interaction measurements are obtained at infinite dilution, the parameters derived from them should correspond to the same conditions; therefore, μ_2 is the standard chemical potential of the protein, μ_2°. For simplicity, however, the superscript will not be used.

and

$$\Delta\mu_2 = -\int_0 \left(\frac{\partial\mu_3}{\partial m_3}\right)_{T,P,m_2} \left(\frac{\partial m_3}{\partial m_2}\right)_{T,\mu_1,\mu_3} dm_3$$
$$= -\int_0 \nu_3 \left(\frac{\partial\mu_3}{\partial m_3}\right)_{T,P,m_2} dm_3 + \frac{1}{55.5}\int_0 \nu_1 m_3 \left(\frac{\partial\mu_3}{\partial m_3}\right)_{T,P,m_2} dm_3 \quad (17)$$

where ν_i is the total binding of component i in moles per mole of protein.[35] If the chemical potential of the protein in the solid phase is independent of the solvent composition, as is true for small molecules, the transfer free energy change of the protein in solution from water to an aqueous solution of component 3 should be directly reflected in the change in its solubility, since

$$\Delta\mu_2 = \mu_2 - \mu_{2,w} = RT \ln(S_{2,w}/S_2) \quad (18)$$

Expressing the chemical potentials of protein in the solid phase by $\mu_{2,w}^s$ and μ_2^s, the protein solubilities in the two systems are

$$-RT \ln S_{2,w} = \mu_{2,w}^l - \mu_{2,w}^s$$
$$-RT \ln S_2 = \mu_2^l - \mu_2^s \quad (19)$$

where l refers to the solution (liquid) phase. The difference in protein solubility is then given by

$$\ln S_2 - \ln S_{2,w} = (\mu_{2,w}^l - \mu_2^l)/RT - (\mu_{2,w}^s - \mu_2^s)/RT \quad (20)$$

The second term represents the variation of the chemical potential of the protein in the solid phase with solvent composition, and represents the transfer free energy of the protein in the solid phase from water to the aqueous solution of component 3. Its determination requires knowledge of the solvent composition and of protein–solvent interactions in the solid phase. This information, however, is very difficult to obtain. As a result, it is usually assumed that there is no change in the chemical potential of the protein in the solid phase, i.e., the second term is set equal to zero. Alternately, one may extrapolate the protein–solvent interaction parameter measured for the dissolved protein into the solid phase in which less surface area is accessible to the solvent. With these approximations, it becomes possible to estimate the solubility change of a protein induced by the addition of component 3 from measurements of the preferential protein–solvent interaction as a function of co-solvent concentration.

Thermodynamic Definition of the Salting-Out Constant. The preferential interactions can be measured at high co-solvent concentrations with a high precision. In the presence of small amounts of electrolytes, there is always a general electrostatic interaction, saturable at low salt concentration, of the small ions with the protein. This interaction may be affected by the addition of organic co-solvents. If we assume that specific interactions between proteins and additives, measured at higher co-solvent concentrations, act in the same manner also at low co-solvent concentrations, the incorporation of the electrostatic interaction term into Eq. (18) gives

$$\ln S_2 - \ln S_{2,w} = (-\Delta\mu_2^{ele}/RT) - (\Delta\mu_2^{spe}/RT)$$
$$= \frac{-\Delta\mu_2^{ele}}{RT} - \frac{1}{RT}\int_0 \left(\frac{\partial\mu_2}{\partial m_3}\right)_{T,P,m_2} dm_3 \tag{21}$$

where $\Delta\mu_2^{ele}$ is the change in the protein chemical potential due to electrostatic interactions and $\Delta\mu_2^{spe}$ is that due to interactions with the specific solvent. Since the salting-out constant K_s is defined as the salting-out effectiveness at high co-solvent concentrations, and $\Delta\mu_2^{ele}$ is significant only at low co-solvent concentrations, K_s may be expressed simply by

$$K_s = -(\partial \log S_2/\partial m_3)_{T,P,m_2} = (1/2.303RT)(\partial\mu_2/\partial m_3)_{T,P,m_2} \tag{22}$$

The actual solubility of the protein in the presence of component 3 and of buffer components is determined by $\Delta\mu_2^{ele}$ and the nature of the dependence of $(\partial\mu_2/\partial m_3)_{T,P,m_2}$ on m_3.

For preferential hydration, [positive value of $(\partial\mu_2/\partial m_3)_{T,P,m_2}$], Eq. (16) shows that the transfer free energy of the protein is positive and, therefore, its solubility should be decreased unless its chemical potential in the solid phase increases sufficiently. On the other hand, for a preferential co-solvent binding system, $\Delta\mu_2^{spe}$ is negative and, therefore, the protein solubility is expected to be increased by the addition of component 3.

From both methods of analysis, it is evident that the sign and magnitude of the preferential protein–solvent interactions can be a strong indication of the effect of an additive on protein solubility in aqueous solution. In fact, a large preferential hydration of proteins has been observed for many substances which are known to induce protein precipitation and crystallization.

Solubility of Systems in Equilibrium. The above thermodynamic treatment may be extended to the case in which a protein exists in equilibrium between two forms, A and B, which differ in solubility. Setting their solubilities as S_A and S_B and the equilibrium constant $K = m_B/m_A$, the solubility curves may be depicted as shown in Fig. 6, where $K < S_B/S_A$. As the total protein concentration m_T increases, the concentrations of both A and B increase. Form A will start to precipitate when m_A becomes equal to S_A, after which a further increase in m_T increases only the amount of the precipitate of A, and the total protein concentration S_T in

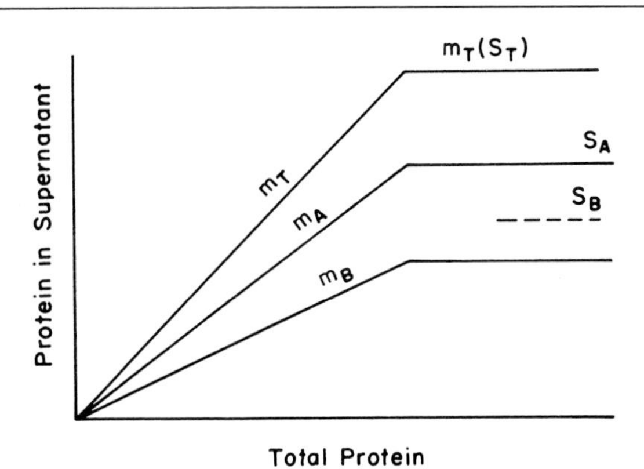

FIG. 6. Model calculation of protein concentration in the supernatant as a function of total protein for a system in equilibrium.

the solution phase can be expressed as $S_A(1 + K)$. Let us assume that the addition of component 3 alters K, S_A, and S_B, and that the transfer free energies of A and B from water to the mixed solvent can be expressed by $\Delta\mu_A^l = am_3$ and $\Delta\mu_A^s = cm_3$, and by $\Delta\mu_B^l = bm_3$ and $\Delta\mu_B^s = dm_3$. Designating the parameters in water by the subscript w, we may write

$$\begin{aligned} S_A = S_{A,w} \exp[(1/RT)(\Delta\mu_A^s - \Delta\mu_A^l)] &= S_{A,w} \exp[(c/RT - a/RT)m_3] \\ &= S_{A,w} \exp[(c' - a')m_3] \end{aligned} \tag{23}$$

and similarly

$$S_B = S_{B,w} \exp[(d' - b')m_3] \tag{23a}$$

$$K/K_w = \exp(-b'm_3)/\exp(-a'm_3) \tag{24}$$

If the chemical potential of the solid phase does not change, i.e., $c' = d' = 0$,

$$(1/K)(S_B/S_A) = (1/K_w)(S_{B,w}/S_{A,w})$$

This relation indicates that, if form A precipitates in the absence of the additive, i.e., if $S_{B,w}/S_{A,w} > K_w$, that form should also be the one to precipitate at any concentration of the aditive. Then, the solubility in the presence of the additive becomes

$$S_T = S_A(1 + K) = S_{A,w} \exp(-a'm_3)\{1 + K_w \exp[(a' - b')m_3]\} \tag{25}$$

When the signs of a' and b' are opposite, S_T may display either a maximum or a minimum.

Let us carry out some sample calculations. Let $S_{A,w} = 10^{-4}$, $K_w = 10^{-8}$, $a' = 2$, and $b' = -2$, and let A be the solid-forming species. The

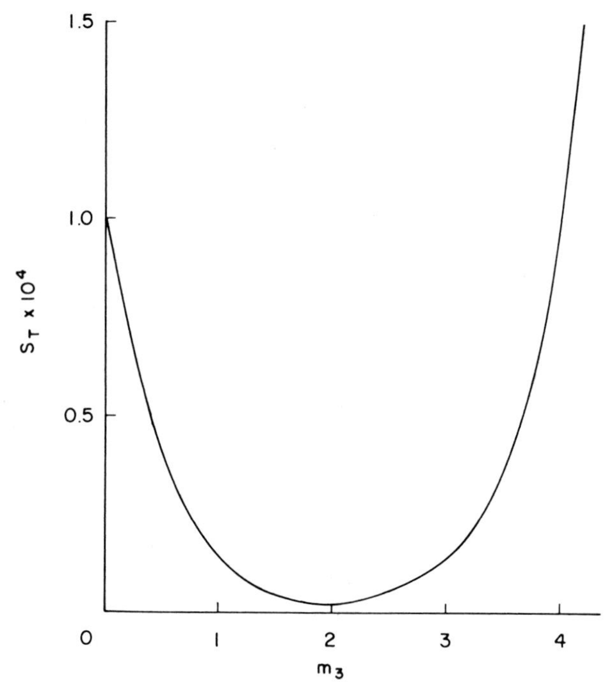

FIG. 7. Model calculation of protein solubility (S_T) as a function of co-solvent concentration for a system in equilibrium. The parameters used are $S_{A,w} = 10^{-4}$, $K_w = 10^{-2}$, $a' = 2$, and $b' = -2$.

resulting dependence of solubility on m_3 is given in Fig. 7. The solubility curve is parabolic, with no inflections and a minimum at a definite value of m_3.

Let us now examine the situation in which the precipitating species changes with solvent composition. This means that there will be a solvent composition, m_3, at which the solubility ratio becomes equal to the equilibrium composition of A and B, i.e., $(S_A/S_B) = (1/K)$. Taking all parameters into account,

$$S_A/S_B = S_{A,w} \exp[(c' - d')m_3] \exp[(b' - a')m_3]/S_{B,w} \tag{26}$$

and setting $S_A/S_B = 1/K$, we obtain

$$\frac{S_{A,w}/S_{B,w}}{1/K_w} = \exp[(d' - c')m_3] \tag{27}$$

For the case where $(S_{A,w}/S_{B,w}) < 1/K_w$, i.e., where A is the solid-forming species in water, the equation has a solution only when $d' < c'$. Let us take an example: let $S_{A,w} = 10^{-4}$, $S_{B,w} = 10^{-7}$, $K_w = e^{-8}$, and $(d' - c') = -0.1$. The solution of Eq. (27) gives $m_3 = 10.9$, for the solvent composition at which both species will coexist in the solid state. Below $m_3 = 10.9$, A is the solid phase-forming species, above this, it is B. Letting S_A be independent of m_3, i.e., setting $a' = c' = 0$, $b' = 1.1$, $d' = -0.1$, both K and S_B increase with greater m_3. The resulting solubility dependence on m_3 is depicted by curve 1 of Fig. 8, where the dotted lines represent hypothetical solubilities of A and B in the regions where the other species precipitates first. There is an inflection at $m_3 = 10.9$, below which S_T increases due to the increase of K and hence the equilibrium concentration of B, and above which the increase in S_T is mainly due to the increase of S_B. Let us take a second example in which all species, both dispersed and solid, interact favorably with component 3. Using the same values of $S_{A,w}$, $S_{B,w}$, and K_w as in the above, and the values of the interaction parameters listed in the legend of Fig. 8, the calculation results in curve 2. The solubility

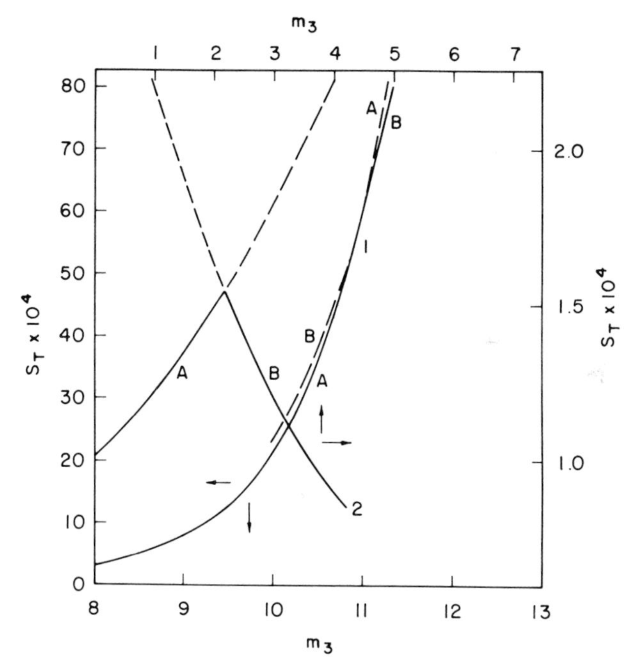

FIG. 8. Model calculations of protein solubility as a function of co-solvent concentration for a system in equilibrium. The curves are calculated as described in the text using the following values of the parameters: solid line 1, $a' = 0$, $c' = 0$, $b' = -1.1$, and $d' = -0.1$; solid line 2, $a' = -0.7$, $c' = -0.5$, $b' = -0.9$, and $d' = -1.0$. Dashed lines are hypothetical solubilities of the more soluble form in the presence of the less soluble form in the solid phase.

TABLE I

PREFERENTIAL INTERACTION PARAMETERS AND CALCULATED SOLUBILITIES OF PROTEINS IN AQUEOUS SALT SYSTEMS[a]

Salt	Concentration (M)	pH	$(\partial g_1/\partial g_2)_{T,\mu_1,\mu_3}$ (g/g)	$(\partial m_3/\partial m_2)_{T,\mu_1,\mu_3}$ (mol/mol)	$(\partial \mu_2/\partial m_3)_{T,P,m_2}$ $\left(\frac{\text{cal/mol protein}}{\text{mol salt}}\right)$	K_s	C_{sat} (g/ml)
			Lysozyme				
Acetate	0.5	4.7	0.684 ± 0.147	−5.14	11,400	8.5	0.30
Acetate	1	4.7	0.478 ± 0.075	−7.55	8,500	6.3	0.25
NaCl	1	4.5	0.424 ± 0.106	−6.20	6,800	5.1	0.45
			Bovine Serum Albumin				
Na_2SO_4	1	5.6	0.524 ± 0.020	−35.4	37,200	27.7	0.06
Acetate[b]	1	5.6	0.312 ± 0.050	−22.4	25,200	18.8	
NaCl	1	5.6	0.243 ± 0.054	−16.8	18,300	13.6	0.28
$MgSO_4$	1	4.5	0.388 ± 0.022	−26.5	20,000	14.9	0.18
NaCl[b]	3	4.5	0.170 ± 0.023	−32.3	16,200	12.1	
KSCN	1	5.6	−0.069 ± 0.046	4.9	−4,900	−3.7	
KSCN[b]	2	5.6	−0.045 ± 0.045	3.3	−3,300	−2.5	
$MgCl_2$[b]	2	4.5	0.015 ± 0.022	−2.1	4,500	3.4	
$CaCl_2$	1	5.6	−0.032 ± 0.021	2.3	−5,200	−3.9	
$BaCl_2$[b]	1	4.8	0.038 ± 0.031	−2.7	5,000	3.7	
$MgCl_2$[b]	1	3.0	0.162 ± 0.027	−11.3	29,600	21.9	0.31
$MgCl_2$[b,c]	1	2.0	0.244 ± 0.013	−9.2	23,900	17.8	0.27

[a] From Arakawa and Timasheff.[13]
[b] From Arakawa and Timasheff.[39]
[c] β-Lactoglobulin.

attains a sharp maximum at a definite value of m_3, here $m_3 = 2.18$. Below this value, S_T increases due to the increase of S_A and above it, the solubility decreases due to decreases of both S_B and the equilibrium concentration of A. Thus, it is evident that equilibrium between two conformers of a protein may give a complex solubility behavior, marked by inflection points, maxima, and minima.

Particular Systems

Salts. Salts such as NaCl, CH_3COONa, and Na_2SO_4 are known as good salting-out agents. The results of preferential interaction measurements given in Table I[13] show that at high concentration they are strongly excluded from the protein. Since the preferential hydration, $(\partial g_1/\partial g_2)_{T,\mu_1,\mu_3}$, is greater than 0.2–0.4 g/g, there is little binding of these salts to the protein. Phase isotherms calculated with Eq. (15) and the assump-

[39] T. Arakawa and S. N. Timasheff, to be published.

tion that the self-interaction term, $(\partial\mu_2^{(e)}/\partial m_2)_{T,P,m_3}$, is due solely to the excluded volume of protein are shown in Fig. 9. The chemical potential of water μ_1 first decreases with increasing protein concentration C_2, then, after passing through a minimum, it starts to increase, finally becoming greater than the value of μ_1°. This leads to phase separation. Below this concentration the protein should be soluble in the aqueous salt solution. Since two phases exist above this concentration, the composition of the system must be fixed according to the phase rule. This means that further addition of protein only increases the amount of the solid phase, while keeping a constant protein concentration in the solution phase. This point

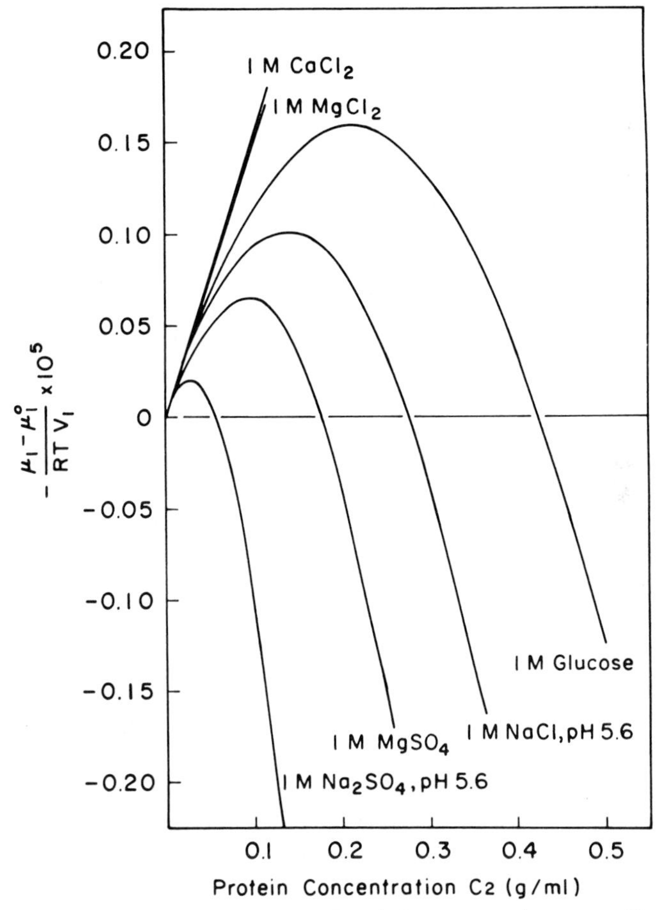

FIG. 9. Phase isotherms for bovine serum albumin in aqueous salt systems. The result for glucose is included for comparison.[15]

is defined as the saturation concentration C_{sat}, i.e., the protein concentration at which $\mu_1 - \mu_1^\circ = 0$ is defined as its solubility in the given solvent system. The values of C_{sat} obtained in this way are listed in Table I. It is evident that the calculated solubility of the protein agrees with the general salting-out effectiveness of salts, a finding which demonstrates the close relation between protein–solvent interactions and protein solubility. Salting-in salts, such as KSCN and $MgCl_2$, are characterized by a small or zero preferential binding to the protein. Therefore, no phase separation could be expected from their phase isotherms. On the other hand, $MgSO_4$, known as a protein precipitant, shows a large preferential hydration. These results demonstrate that the salting-in and salting-out properties of salts at high concentrations are well defined by the preferential protein–solvent interactions.

The dependence of protein solubility on salt concentration can be described also by a combination of the above specific co-solvent interaction parameter with the electrostatic term, i.e.,

$$\log S_2 - \log S_{2,w} = AZ^2\sqrt{I}/(1 + aB\sqrt{I}) - (1/2.303RT)\int_0 (\partial\mu_2/\partial m_3)_{T,P,m_2}\, dm_3 \quad (28)$$

Figure 10 gives the results of calculations for each of the two terms and their summation. The details are given in the figure legend. It is evident that at low salt concentrations, the electrostatic term is predominant, the protein solubility increasing with addition of salt. At high concentrations, the second term becomes dominant, leading to a sharp decrease in the protein solubility. Typical values of K_s calculated with Eq. (22) are given in Table I. Their magnitudes are considerably higher than those determined experimentally. The relative effectiveness is, however, in good agreement with the order of salts as protein precipitants. The high values of K_s could stem from the assumption that the chemical potential of the protein in the solid phase is unchanged by the addition of salt, the complete definition of K_s being

$$K_s = (1/2.303RT)[(\partial\mu_2/\partial m_3)^{l}_{T,P,m_2} - (\partial\mu_2/\partial m_3)^{s}_{T,P,m_2}] \quad (29)$$

Nevertheless the chemical potential change of the protein in solution can predict the relative effectiveness of salts as protein precipitants. The experimental values of K_s are known to be generally, although not strictly, higher for larger proteins, which again agrees with the calculated values.

A change in pH affects both the electrostatic and specific interaction terms, as well as $S_{2,w}$. The results of preferential interaction measurement for $MgCl_2$ and calculated values of K_s at pH 2.0 and 3.0 are given in Table I. These values are considerably higher than those at higher pH, a result

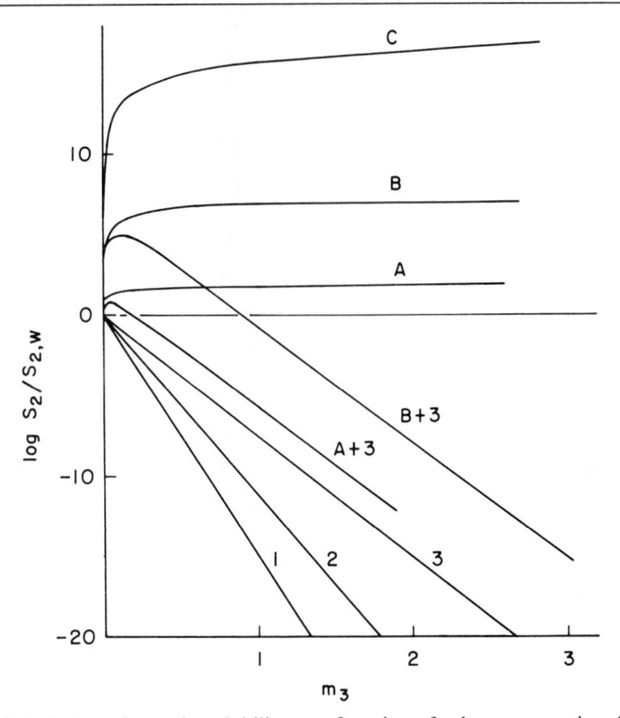

FIG. 10. Calculation of protein solubility as a function of salt concentration. The salting-in term was calculated for a monovalent salt with $Z = 5$ (A), 10 (B), and 15 (C). The salting-out term was calculated assuming $\Delta\mu_2 = m_3(\partial\mu_2/\partial m_3)_{T,P,m_2}$ with $(\partial\mu_2/\partial m_3)_{T,P,m_2} = 20{,}000$ (1), 15,000 (2), and 10,000 (3), at 20°. The summation of the two terms is given for two situations: $(A + 3)$ and $(B + 3)$. This calculation was carried out for a 1:1 salt and $a = 25 \times 10^{-8}$ cm, setting $I = m_3$, $A = 0.645$, $B = 3.31 \times 10^7$ at 20°.

which indicates that $MgCl_2$ should act as a salting-out salt at those conditions. In fact protein solutions in $MgCl_2$ show either turbidity or precipitation when their pH is lowered.

2-Methyl-2,4-pentanediol (MPD). MPD is another solvent known for its protein crystallizing properties. At pH 5.8, it is strongly excluded from ribonuclease, as shown in Table II by the measured preferential interaction parameters.[5] Phase isotherms calculated with these values indicate that the protein solubilities should be close to 50 and 15 mg/ml at 40 and 50% MPD, respectively, agreeing with the observed crystallization of the enzyme from aqueous MPD solutions.[2]

The effect of MPD on solubility may also be described by Eq. (21). In this case, the first term expresses the variation of electrostatic interactions between buffer ions and protein charges and dipoles in the presence of MPD, mainly through the effect of MPD on the dielectric constant of

TABLE II
PREFERENTIAL INTERACTION PARAMETERS AND CALCULATED SOLUBILITIES OF RNASE A IN AQUEOUS MPD SYSTEMS AT pH 5.8[5]

MPD (%)	$(\partial g_1/\partial g_2)_{T,\mu_1,\mu_3}$ (g/g)	$(\partial \mu_2/\partial m_3)_{T,P,m_2}$ $\left(\frac{\text{cal/mol protein}}{\text{mol MPD}}\right)$	K_s	C_{sat} (g/ml)
20	0.196 ± 0.106	1600	1.2	
30	0.555 ± 0.146	4500	3.4	0.7
40	0.810 ± 0.051	6500	4.9	0.049
50	1.031 ± 0.058	8700	6.5	0.015

the medium. If this term is neglected, K_s can be calculated with Eq. (22), just as in salt systems. The resulting values, listed in Table II, are comparable to those of CH_3COONa and NaCl obtained with lysozyme, which has a molecular weight similar to that of RNase.

Preferential interactions may also be expressed formally in terms of binding isotherms. For a system in which the co-solvent binding is very low and its exclusion increases linearly with its concentration, such as aqueous MPD, preferential interaction may be expressed by

$$\left(\frac{\partial m_3}{\partial m_2}\right)_{T,\mu_1,\mu_3} = \frac{nkm_3}{1 + km_3} - \left(\frac{m_3}{55.5}\right) hm_3 \tag{30}$$

where n is the number of co-solvent binding sites, k is the binding constant, and h is the increment of protein hydration due to co-solvent exclusion. By definition, $\nu_1 = hm_3$. Since

$$\begin{aligned}(\partial \mu_2/\partial m_3)_{T,P,m_2} &= -(\partial m_3/\partial m_2)_{T,\mu_1,\mu_3}(\partial \mu_3/\partial m_3)_{T,P,m_2} \\ &= -(RT/m_3)(\partial m_3/\partial m_2)_{T,\mu_1,\mu_3}\end{aligned}$$

the free energy of transfer of the protein from water to the mixed solvent is, by Eq. (16),

$$\Delta\mu_2 = -RT[n \ln(1 + km_3) - hm_3^2/111] \tag{31}$$

Using the preferential interaction data for ribonuclease in aqueous MPD, a good fit to the experimental results was obtained with $n = 40$, $k = 0.03$, and $h = 120$. Furthermore, the above values of the parameters gave the m_3 dependences of $\Delta\mu_2$ and of the protein solubility shown by curve 1 of Fig. 11. It is evident that the calculated solubility change is unreasonably large. This calculation, however, assumes no contact between pro-

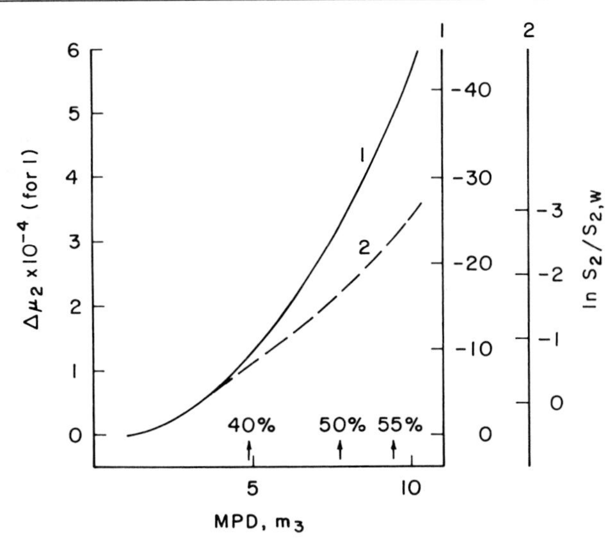

FIG. 11. Calculation of the free energy of transfer and the solubility change of RNase in the MPD system. Sold line (1) was calculated assuming no change in the chemical potential of the solid phase. The dashed line (2) was calculated assuming that the protein in the solid phase is also preferentially hydrated, with protein–solvent contacts reduced by 6.9% from the liquid phase. The solubility scales on the right-hand ordinate refer to curves 1 and 2, respectively.

tein and solvent in the solid phase. Assuming that in the solid phase such contacts exist, but at a level somewhat below that in solution, i.e., keeping the change in the chemical potential of the protein crystal explicit,

$$\log(S_2/S_{2,w}) = (-1/2.303RT)[(\Delta\mu_2)^l - (\Delta\mu_2)^S] \tag{32a}$$

$$= 1/2.303[\Delta n \ln(1 + km_3) - \Delta h m_3^2/111] \tag{32b}$$

where Δn and Δh are the differences in the numbers of MPD binding sites and extents of hydration between proteins in solution and in the solid. Curve 2 of Fig. 11 shows the results of such a calculation for the ribonuclease in aqueous MPD systems for a 6.9% decrease in contacts, i.e., $\Delta n = 2.76$ and $\Delta h = 8.28$. This gives a solubility change, $S_2/S_{2,w}$, within a reasonable range, i.e., 0.025 around 50% MPD. Thus, the protein in the solid phase should contain within its domain 7 mol of MPD (0.06 g/g) and 860 mol of water (1.1 g/g) per mole of protein.

Polyethylene Glycol. The same treatment as for MPD may be used for the aqueous polyethylene glycol (PEG) system in order to estimate its

TABLE III
PREFERENTIAL INTERACTION PARAMETERS OF PROTEINS IN AQUEOUS POLYETHYLENE GLYCOL SYSTEMS

MW of PEG	Protein	Concentration (%)	$(\partial g_1/\partial g_2)_{T,\mu_1,\mu_3}$ (g/g)	$(\partial \mu_2/\partial m_3)_{T,P,m_2}$ $\left(\frac{\text{cal/mol protein}}{\text{mol PEG}}\right)$	K_s
200[a]	β-Lactoglobin	20	0.098 ± 0.098	1100	0.8
200[a]	β-Lactoglobin	40	0.035 ± 0.053	700	0.3
400[b]	Lysozyme	44.8	0.7	6000	4.5
600[a]	β-Lactoglobin	10	0.627 ± 0.125	6700	5.0
1000[a]	β-Lactoglobin	10	1.22 ± 0.18	13100	9.8
1000[b]	Bovine serum albumin	10	0.3	13000	9.7
1000[b]	RNase	10	1.9	15000	11.2
1000[b]	Lysozyme	30	0.8	6000	4.5
4000[b]	Lysozyme	2.5	10.2	85000	63.4

[a] Arakawa and Timasheff.[8]
[b] Lee and Lee.[40]

salting-out effectiveness, as shown in Table III.[8,40] PEG 6000 gave an extraordinarily large value of K_s, much higher than that of Na_2SO_4. This agrees with the observed strong salting-out effectiveness of high molecular weight PEGs, such as 4000 and 6000. Since the value of K_s for PEG ranges from low to high depending on the molecular size, it seems possible to manipulate the protein solubility over a very wide range simply by changing the molecular size and concentration of PEG.

Other Co-Solvents. Significant values of preferential hydrations have been measured for many substances. These encompass some sugars,[14,15] glycerol,[16] some amino acids,[41] organic salts such as guanidinium sulfate and monosodium glutamate, and dimethyl sulfoxide.[8] Since their effects on protein solubility have not been measured, they will not be discussed here. It seems useful to point out, however, that they may be employed for enhancing protein precipitation at favorable conditions.

Mechanisms of Preferential Interaction

Since the protein solubility is closely related to the preferential interactions of solvent components with proteins, knowledge of the mechanism of preferential interactions may shed light on the salting-out mechanism. Several such mechanisms have been proposed.

[40] J. C. Lee and L. L. Y. Lee, *J. Biol. Chem.* **256,** 625 (1981).
[41] T. Arakawa and S. N. Timasheff, *Arch. Biochem. Biophys.* **224,** 169 (1983).

Cohesive Force. Salts, glycerol, sugars, and amino acids have a strong cohesive effect on water through ion–dipole and dipole–dipole interactions. They will, therefore, favor contacts with water over those with the protein surface. This should result in their exclusion from the protein surface. There have been many attempts to interpret the widely different effects of salts on protein solubility in terms of various macroscopic physical properties of aqueous salt solutions. Surface tension seems to be the most useful property, since it is a measure of the ability of a substance to migrate from the water–air interface into bulk water. A similar macroscopic property of aqueous salt solutions is the internal pressure, defined as $(V_s - \bar{V}_s^\circ)/\beta$, where V_s is the molar volume of the pure "liquid" salt, $\bar{V}_s^\circ$ is the partial molar volume of the salt in water at infinite dilution, and β is the compressibility of water. As is evident from this definition, this expresses the ability of salts to bind water and thereby to contract its volume. In fact, this property has been proposed by Robinson and Jencks[42] as a likely explanation of the salting-out effectiveness of salts on the model peptide component ATGEE. Since certain sugars[43] and amino acids[44] increase the surface tension of water, it is possible that the cohesive force is the source of the preferential exclusion not only for salts but also for these substances. This rule, however, is not universal, since for some substances this correlation between the macroscopic property of their aqueous solutions and their preferential interactions and, hence, salting-out effectivenesses does not hold. Typical examples of such exceptions are divalent cation salt, such as $MgCl_2$, $BaCl_2$, and $CaCl_2$. They have large surface tension increments and internal pressures, just as Na_2SO_4. These large values could be expected from their high charge density and/or relatively small ion size. Robinson and Jencks[42] have proposed that divalent cations form hydrates and can bind to amide groups on proteins through hydrogen bonds by the same mechanism as GuHCl and urea. Since this interaction is not electrostatic, the binding constant may be low, requiring high salt concentrations. This may be the reason why these salts exhibit different interactions from those for Na_2SO_4. Indeed, when the affinity of the divalent cations to proteins was lowered by inducing a net positive charge on the protein, preferential hydration was observed in agreement with their large cohesive force (see Table I).

The converse situation is found with glycerol[45] and betaine,[44] which have negative surface tension increments. This is not unexpected, since their polar part would tend to go into water whereas their nonpolar part

[42] D. R. Robinson and W. P. Jencks, *J. Am. Chem. Soc.* **87,** 2470 (1975).

[43] E. Landt, *Ztschr. Ver. Dtsch. Zuckerind. Techn. T. Bd.* **81,** 119 (1931).

[44] J. R. Pappenheimer, M. P. Lepie, and J. Wyman, *J. Am. Chem. Soc.* **58,** 1851 (1936).

[45] "International Critical Table," Vol. 2. McGraw-Hill, New York, 1928.

would favor a hydrophobic environment. When these substances are introduced into aqueous protein solutions, their polar part should favor contact with water, while the nonpolar part should have only weak interactions with the protein surface, since the protein surface is more hydrophilic than the air–water interface. As a consequence of this balance, they may become excluded from the domain of the protein.

Charges on Protein. The surface tension increment of MPD is negative,[46] eliminating cohesive force as a source of preferential hydration. An observation that aqueous MPD systems separated into two phases on the addition of salt has led to the suggestion that charges on the protein surface are the cause of its exclusion from the protein.[5] According to this mechanism, it should be expected that a decrease in the dielectric constant of the medium must increase the MPD exclusion, which is in agreement with the observation that an increase in MPD concentration, which itself decreases the dielectric constant, increases greatly the preferential hydration. A similar mechanism[40] has been suggested for the exclusion of PEG, which is known not to increase the surface tension of water.[33]

Organic solvents such as ethanol or acetone are also used to induce the crystallization of proteins. Such organic solvents generally lower the dielectric constant of the medium and increase the solubility of nonpolar substances or side chain groups of amino acids.[47] This should lead to an exclusion of the organic substances from the native protein due to the presence of charges on its surface, just as the case of MPD, as long as they do not induce denaturation.

Steric Exclusion. This mechanism, proposed by Kauzmann and introduced by Schachman and Lauffer,[48] expresses the exclusion of large molecules from the protein surface simply in terms of steric hindrance. Unless the specific affinity between the protein surface and the additive is significant, i.e., if the additive molecule can move in solution without any perturbation by the protein molecule, excess water must be present in the volume corresponding to the region from the protein surface to the radius of the additive molecule. The large size of PEG and the increasing exclusion for larger PEG suggest that steric hindrance is at least a part of the strong exclusion of PEG from the protein surface.

Complex Formation with Water. If the additive molecule forms a stoichiometric complex with water leading to a decrease in its activity, it should not be able to bind directly to a protein even though it has a strong affinity for the protein. This should result in preferential exclusion. Di-

[46] Y. Kita, T. Arakawa, and S. N. Timasheff, unpublished results.

[47] Y. Nozaki and C. Tanford, *J. Biol. Chem.* **246,** 2211 (1971).

[48] H. K. Schachman and M. A. Lauffer, *J. Am. Chem. Soc.* **71,** 536 (1949).

methyl sulfoxide (DMSO) may be an example of this case, since it binds to the hydrophobic groups of proteins, thereby inducing denaturation at high co-solvent concentrations where free molecules must become available.

Of the salting-out materials, substances such as Na_2SO_4, glycine, and sucrose can be used without any fear of denaturation since they have presumably at most a very low affinity for nonpolar groups. On the other hand, substances such as MPD and DMSO must be used with care. It has been shown, however, that MPD is not an effective denaturant even at 55%.[49]

[49] E. P. Pittz and J. Bello, *Arch. Biochem. Biophys.* **146,** 513 (1971).

[4] Nucleation and Growth of Protein Crystals: General Principles and Assays

By G. FEHER and Z. KAM

If you can look into
the seeds of time
and say which grain will
grow and which will not,
Speak then to me . . .

Macbeth

I. Introduction

Recent advances in the efficiency of gathering and processing crystallographic data are clearly pointing to crystallization as the bottleneck in the determination of macromolecular structures. Although the first protein was crystallized over a hundred years ago, crystallization has largely remained an art rather than become a science. The reason for this is the complex, multicomponent nature of a crystallizing system and the long time required to observe the effect of a change in a parameter by visual inspection. If one utilizes, for instance, 20 independent variables influencing macromolecular crystallization (see Table III of McPherson, this volume [5]) and assumes only 2 possible values for each variable, one will have $2^{20} \simeq 10^6$ independent conditions. Since one usually needs to wait several days for the possible appearance of a visible crystal, a systematic investigation clearly becomes prohibitively time consuming.[1] A possible way to overcome this difficulty is to eliminate the necessity of visual inspection (and, hence, the long time intervals) by monitoring the formation of dimers, trimers, tetramers, . . ., *n*-mers. This is the approach that we have undertaken[2,3] and will discuss in this chapter. As we shall show, a determination of the size distribution of small aggregates makes it possible to approach the problem of nucleation in a systematic way. The nucleation of crystals represents only the first (albeit the most important) step; postnucleation growth and the final size and quality of the crystals also need to be considered. Consequently, we have divided the processes leading to the production of usable crystals into three, temporally distinct phases: nucleation, postnucleation growth, and cessation of growth. We have investigated all three phases of crystallization.[2,3] We chose as a model compound the easily crystallizable protein, hen egg-white lysozyme.[4–6]

The *homogeneous nucleation* of crystals[7] has many similarities with the condensation of droplets from supersaturated vapor first described by

[1] An approach to reduce the number of experiments has been proposed by C. W. Carter, Jr. and C. W. Carter [*J. Biol. Chem.* **254,** 12219 (1979)]. However, even with their method, a complete systematic investigation is still not practical.

[2] Z. Kam, H. B. Shore, and G. Feher, *J. Mol. Biol.* **123,** 539 (1978).

[3] Z. Kam, A. Shaikevitch, H. B. Shore, and G. Feher, *in* "Electron Microscopy at Molecular Dimensions" (W. Baumeister and W. Vogell, eds.), p. 302. Springer-Verlag, Berlin and New York, 1980.

[4] G. Alderton and H. L. Fevold, *J. Biol. Chem.* **164,** 1 (1946).

[5] L. K. Steinrauf, *Acta Crystallogr.* **12,** 77 (1959).

[6] P. Jollès and J. Berthou, *FEBS Lett.* **23,** 21 (1972); Z. Kam and G. Feher, *Fed. Proc., Fed. Am. Soc. Exp. Biol.* **33,** 1310 (1974).

[7] Homogeneous nucleation occurs in the absence of foreign particles as opposed to heterogeneous nucleation which takes place on the surface of foreign bodies. Crystals grown from seeds avoid the nucleation problem altogether.

Gibbs[8] and later elaborated by Volmer and co-workers.[9-11] Since at high enough concentrations of macromolecules, aggregates are formed, the main question connected with the nucleation process is whether an ordered, crystalline entity or a disordered, amorphous precipitate is energetically favored. We have used the technique of quasi-elastic light scattering[12-15] to determine the size distribution of aggregates during the prenucleation phase. This information in conjunction with the theoretical model is capable of answering the above question.

The *growth of crystals* is determined by the rate of transport of protein molecules to the crystal and the probability of attachment of a molecule. The simple approach of measuring the time dependence of the size of a growing crystal encounters quantitative difficulties in interpretation due to the difficulties in knowing the decrease in protein concentration near the surface of a growing crystal. This is due to the slow diffusion of protein molecules which causes the formation of a protein depletion layer. However, we shall show that by measuring the concentration profile in the vicinity of the crystal, the growth rate and the attachment coefficient can be determined.

The *cessation of growth* often results in crystals that are too small for structural analysis. An understanding of this phenomenon is, therefore, of considerable practical importance. Relatively little work has been done on this phase of crystallization and the problem is very poorly understood. We shall discuss our preliminary findings on terminal-sized crystals and advance a working hypothesis for the cessation of growth.

Before proceeding with a description of the work on the crystallization of proteins, it may be useful to delineate the differences between macromolecules and small molecules on which a great deal of work has been done.[16-18] Several characteristics of proteins and other biological macro-

[8] J. W. Gibbs, *Trans. Conn. Acad. Art Sci.* **3,** 108 (1875–76) and p. 343 (1877–1878) (reprinted in "Collected Works of J. Willard Gibbs," Vol. 1, Chapter 3, p. 55. Longmans, Green, New York, 1928).

[9] M. Volmer and A. Weber, *Z. Phys. Chem. (Leipzig)* **119,** 277 (1926).

[10] M. Volmer and H. Flood, *Z. Phys. Chem. (Leipzig)* **170,** 273 (1934).

[11] M. Volmer, *in* "Kinetik der Phasenbildung" (K. F. Bonhoeffer, ed.). Steinkopff, Dresden, 1939.

[12] G. B. Benedek, *in* "Polarisation, Matière et Rayonnement, Livre de Jubilaire en l'honneur d'Alfred Kastler," p. 49. Presses Universitaires de France, Paris, 1969.

[13] H. Z. Cummins and H. L. Swinney, *Prog. Opt.* **8,** 135 (1970).

[14] S. B. Dubin, this series, Vol. 26, Part C, p. 119.

[15] B. Chu, "Laser Light Scattering." Academic Press, New York, 1974.

[16] J. J. Gilman, "The Art and Science of Growing Crystals." Wiley, New York, 1963.

[17] R. F. Strickland-Constable, "Kinetics and Mechanism of Crystallization." Academic Press, New York, 1967.

[18] J. C. Brice, *in* "The Growth of Crystals from Liquids" (E. P. Wohlfarth, ed.), Vol. 12. Am. Elsevier, New York, 1973.

molecules differ sufficiently from those of small molecules that relatively few conclusions can be carried over from this large amount of work. The main differences arise from the size, low symmetry, large solvent content of proteins, and relatively small number of contact points that result in small binding energies per unit volume. The latter causes the protein crystals to be soft and sensitive to small changes in external conditions. Proteins (as distinguished from small molecules) have a large number of potential attachment sites that are energetically almost as favorable as the small number of *specific* sites through which an ordered array (crystal) is formed. Consequently, one often obtains a disordered aggregate instead of a crystal. The slow translational diffusion of proteins together with the low probability of properly orienting large molecules with respect to the lattice results in a slow growth of crystals of macromolecules. This is important from an experimental point of view since it enables us to monitor the size distribution of aggregates even before nucleation has occurred.

II. Nucleation

A. *Qualitative Discussion*

Suppose we have a supersaturated solution of macromolecules in which conditions prevail that energetically favor the formation of crystals. The bulk of the molecules will rapidly establish a distribution of small aggregates reaching a state that is called *quasi-equilibrium*. However, the nucleation process (i.e., the formation of aggregates exceeding a critical size) will proceed more slowly, causing a delay in the appearance of crystals. The reason for this is the energy barrier for nucleation shown in Fig. 1. Its origin lies in the competition of two free energy terms: one positive surface energy term proportional to r^2 and the other a negative (i.e., binding) energy term proportional to r^3. Thus, as the size of the aggregate, r, increases, the volume term becomes more important until a critical size r_c is reached. An aggregate having a critical size is called a nucleus; it can grow spontaneously to form macroscopic crystals.

How does the energetics of amorphous precipitation differ from that of crystallization? An amorphous aggregate can be approximated by a linear chain to which monomers can be added only at the ends. (A small number of weak bonds, in addition to these two bonds per monomer, is expected to have a small effect on the energetics of aggregate formation and will be neglected in our treatment.) Thus the incremental energy difference on adding a monomer is independent of the size of the aggregate. Consequently, the competition between a volume and surface energy term has been eliminated and with it the energy barrier (see Fig. 1). Experimen-

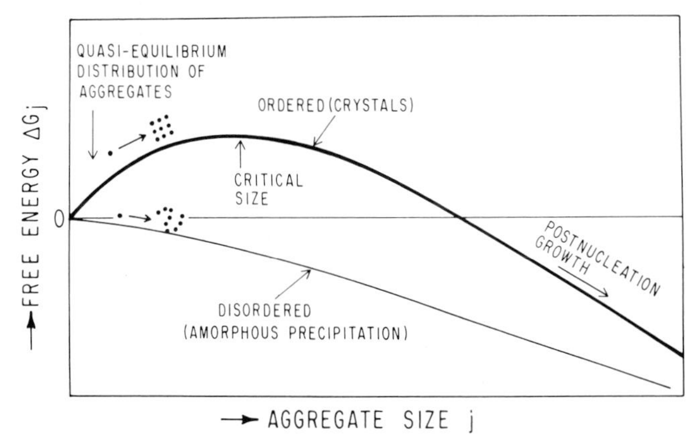

FIG. 1. Schematic representation of the energetics of crystallization (thick line) and of chain-like amorphous precipitation (thin line). (From Kam *et al.*[3])

tally, this exhibits itself in a short delay time and a small degree of supersaturation that can be sustained.

B. Quantitative Theoretical Model

Thermodynamic Considerations. The nucleation process can be formally described by a chain of aggregation reactions

$$A_j^{\alpha} + A_{j'}^{\alpha'} \rightleftharpoons A_{j+j'}^{\alpha''} \qquad (j, j' = 1, 2, \ldots) \tag{1}$$

where A_j^{α} denotes an aggregate of j molecules (j-mer) having a concentration C_j with a configuration (i.e., shape) symbolically denoted by the index α. An important requirement of the theoretical model is to be able to distinguish, in terms of experimentally measurable parameters, aggregation processes that lead to crystallization from those that lead to amorphous precipitation. For this purpose we shall make the simplifying assumptions that for a given j, there are only two extreme structures that need to be considered. The first is a compact structure[19] with bonds in all three dimensions that ultimately becomes a macroscopic crystal; the second is an essentially one-dimensional polymer-like structure with only two bonds per molecule that leads to an amorphous precipitate as discussed in the previous section.

The energetics of the nucleation process are determined by the number of bonds formed in the j-mer. In analogy with the qualitative discus-

[19] Although a j-mer can have many compact configurations, we consider only the ones with the lowest energy. By varying the external conditions, different compact configurations may assume the lowest energy. This gives rise to the observed polymorphism.

sion of the previous section we divide the total free energy into two terms. The first is a negative bulk term that includes all j molecules of the aggregate. In calculating this term we assume that all the molecules have a full complement of bonds. Since this is not true for molecules at the surface, we need to add a positive surface term to account for the loss of free energy due to the open bonds. This surface term depends in reality on the exact surface configuration (e.g., steps, edges). These configuration-dependent effects are assumed to be small for small aggregates and are neglected here.[20] Thus, the change in standard free energy,[21] ΔG_j°, for aggregation of j monomers to form a j-mer is (for a review, see ref. 22):

$$\Delta G_j^\circ = vjG_B + \beta j^\gamma G_S \tag{2}$$

where G_B is the bulk energy/unit volume (as compared to j monomers in solution) and G_S is the energy/unit area of surface.[23] The total volume of the j-mer is vj and the surface area is βj^γ. The values for β and γ depend on the shape of the aggregate. For example, for spherical aggregates $\beta = (36 \pi v^2)^{1/3}$ and $\gamma = 2/3$ (for a linear chain $\gamma \simeq 0$). In Fig. 2a, we show ΔG_j for several values of protein concentration, C, versus j for the spherical case. (The numerical values of Fig. 2 were taken for lysozyme as discussed later.) For linear chains, the surface term is constant and the volume term is proportional to j.

The additional free energy required to form a $(j + 1)$-mer from a j-mer and a monomer is

$$\Delta G_{j+1} - \Delta G_j \simeq d\Delta G_j/dj = vG_B + \gamma\beta j^{\gamma-1} G_S \tag{3}$$

As seen from Fig. 2a, the expression in Eq. (3) is positive for small aggregates and becomes negative for j larger than a critical value j^*. Thus j^* (which depends on concentration) represents an aggregate of critical size, beyond which aggregation by addition of monomers becomes favorable. Aggregates with $j < j^*$ have to fight their way energetically uphill in order to grow. The necessary energy for this "uphill fight" is provided by thermal fluctuations.

[20] They do, however, play an important role in the shape and rate of growth of macroscopic crystals. For instance, the addition of a molecule to an incomplete plane is more favorable than the addition to a new plane. Thus, crystals formed by deposition of monolayers grow in spurts. Certain imperfections, such as screw dislocations, provide a continuous incomplete plane, thereby facilitating crystal growth (see, e.g., M. Ohara and R. C. Reid, "Modeling Crystal Growth Rates from Solution." Prentice-Hall, Englewood Cliffs, New Jersey, 1973.

[21] The "standard" free energy ΔG_j° assumes unit concentration. The actual free energy for a specific concentration C_1 is denoted by $\Delta G_j = \Delta G_j^\circ - jkT \ln C_1$.

[22] A. C. Zettlemoyer, "Nucleation." Dekker, New York, 1969.

[23] Both G_B and G_S contain enthalpy and entropy contributions. These can be separated experimentally by measuring the temperature dependence of ΔG°.

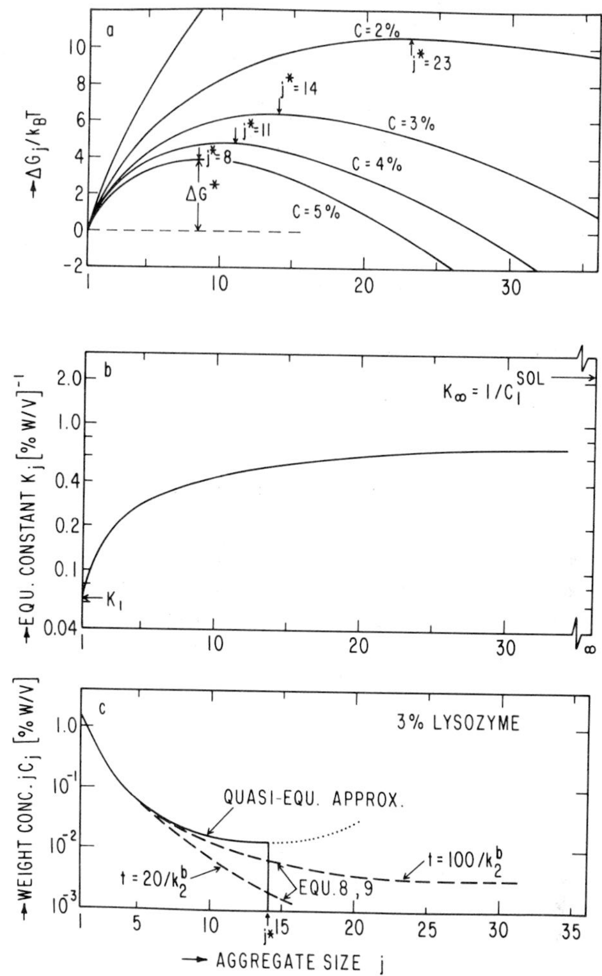

FIG. 2. Parameters that enter into the crystallization process of proteins. The numerical values were chosen for lysozyme at $T = 20°$, pH = 4.2, and 5% NaCl. (a) Free energy change, $\Delta G_j/k_BT$, of forming a j-mer from monomers for different protein concentrations C. The nucleation barrier depends on C and reaches its maximum value ΔG^* at the critical size j^*. (b) The equilibrium constant versus j [Eq. (5)] with $K_\infty = 2.2$ (%, w/v)$^{-1}$ and $K_\infty/K_1 = 35$. (c) The establishment of quasi-equilibrium. The broken lines were obtained by numerically solving Eqs. (8) and (9) for different times. The solid lines represent the quasi-equilibrium approximation. (From Kam *et al.*[2])

From the law of mass action, one obtains a set of equilibrium constants K_j, which are related to the free energy changes by the relation[24]

$$K_j = \frac{C_{j+1}}{C_j C_1} = \exp\left[-\frac{(d\Delta G_j^\circ)}{dj}\bigg/ k_B T\right]$$
$$= \exp\{-[vG_B + \gamma\beta j^{\gamma-1} G_S)/(k_B T)]\} \quad (4)$$

where the C_j terms are equilibrium concentrations of j-mers. For the present discussion we consider a compact three-dimensional spherical structure with a maximum number of bonds. From Eq. (4), with $\gamma = 2/3$ we can express K_j for all j values in terms of K_1 and K_∞:

$$K_j = K_1(K_\infty/K_1)^{1-j^{-1/3}} = K_\infty(K_1/K_\infty)^{j^{-1/3}} \quad (5)$$

Henceforth, we shall regard K_1 and K_∞ as the independent parameters that determine the nucleation process, rather than G_B and G_S from which they were obtained. Equations (4) and (5) can be used to calculate the normalized monomer concentration C_1/C as a function of the dimensionless parameter CK_1, where C is the total protein concentration. This is a useful relation that will occur in subsequent discussions; it is plotted for different ratios of K_∞/K_1 in Fig. 3. Next, we shall show that the ratio K_∞/K_1 is very different for compact crystalline aggregates and for linear amorphous precipitates. Superscripts K^{XTAL} and K^{AMOR} will be used for the two cases.

For a compact configuration, the addition of a monomer to a larger j-mer creates on the average at least two new bonds compared to a single bond for the creation of a dimer from two monomers. Since the energies of these processes enter into the exponent of Eq. (4), it follows that $K_\infty{}^{XTAL}/K_1{}^{XTAL} \gg 1$. A typical value of K_∞/K_1 obtained for lysozyme crystals is ~35 (see the following section). This large ratio, which arises from the three-dimensional character of crystalline aggregates, is an important aspect of the theory. A consequence of this large ratio is the known fact that even if one prepares a solution with sufficient concentration of monomers for large crystals to grow (i.e., $C_1K_\infty{}^{XTAL} > 1$), it is still possible for small aggregates to be unstable and dissolve ($C_1K_1{}^{XTAL} < 1$). Thus, one speaks of a free energy barrier for nucleation of a crystal; this is evident in Fig. 2a. The dependence of K_j on j [Eq. (5)] is illustrated in Fig. 2b.

By contrast, for an amorphous aggregate, which in this work we are approximating by a linear chain, ΔG_j° is proportional to j. Thus, $d\Delta G_j^\circ/dj$

[24] F. Reif, "Fundamentals of Statistical and Thermal Physics." McGraw-Hill, New York, 1965.

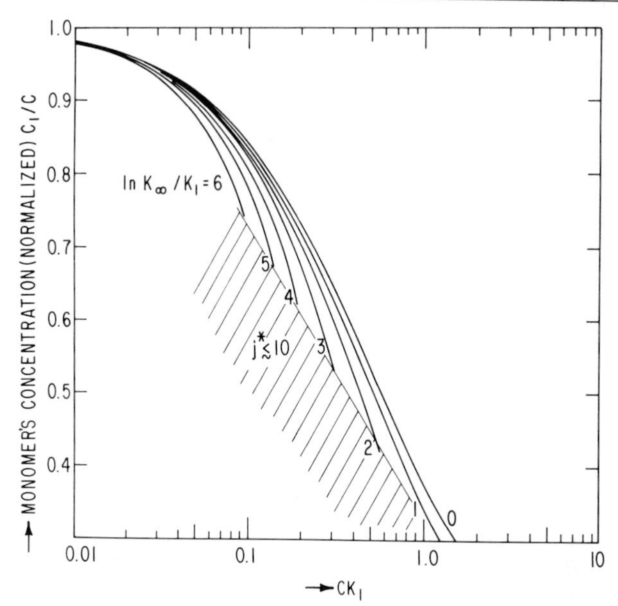

FIG. 3. The normalized concentration of monomers C_1/C as a function of the total concentration C times K_1 for different values of K_∞/K_1. Obtained from Eqs. (4) and (5). The calculation is applicable only for $j < j^*$ (see cutoff in Fig. 2c). The total concentration was calculated by summing C_j for $j < j^*$. For $j^* < 10$ a large error is made in the summation (see cross-hatched area).

and hence K_j is independent of j, i.e., $K_\infty{}^{\mathrm{AMOR}}/K_1{}^{\mathrm{AMOR}} \approx 1$.[25] Thus, if C_1 is larger than the solubility of the linear amorphous precipitate, $C_1 K_\infty{}^{\mathrm{AMOR}} \approx C_1 K_1{}^{\mathrm{AMOR}} > 1$ and small size aggregates will be stable. Thus, no barrier exists for the growth of linear aggregates (see Fig. 1).

Kinetics of Aggregation. In order to study the *kinetics* of aggregation we make the additional assumption that the aggregation occurs via addition of monomers. Experimental evidence supporting this assumption is presented in Section III. With this assumption the aggregation process [see Eq. (1)] can be described by

$$\mathrm{A}_j + \mathrm{A}_1 \underset{k_{j+1}^{\mathrm{b}}}{\overset{k_j^{\mathrm{f}}}{\rightleftharpoons}} \mathrm{A}_{j+1}, \qquad K_j = k_j^{\mathrm{f}}/k_{j+1}^{\mathrm{b}} \tag{6}$$

where k_j^{f} and k_{j+1}^{b} 1 are the forward and backward reaction rates of a monomer with a j-mer.

To complete the model, we need to specify the j dependence of k^{f} and

[25] One can see this formally from Eq. (4) by writing for $K_\infty/K_1 = \exp[(\gamma\beta G_S)/k_B T)]$. For a linear chain $\gamma \simeq 0$ and hence $K_\infty/K_1 = 1$. Weak additional bonds neglected in the simplified theory presented here should cause a small increase in this ratio (see Fig. 9).

k^b so that their ratio obeys Eq. (5). Following Courtney's treatment of the nucleation of water droplets,[26,27] we expect both k^f and k^b to be proportional to the surface area, i.e., to obtain a factor $j^{2/3}$. This factor, of course, cancels out in the ratio k^f/k^b. The additional j dependence of K_j is attributed to the backward (dissolving) reaction rate.[26,27] The reason is as follows. When a molecule attaches to a surface, the rate-determining step is the formation of the *first bond*. However, when a molecule detaches itself, it must break *all the bonds* holding it to the surface. Since the number of bonds depends on the surface-to-volume ratio, the entire j dependence of K arises from the backward rate. In terms of a single rate parameter k_2^b, we can write

$$k_j^f = k_1^f j^{2/3} = k_2^b K_1 j^{2/3}, \qquad k_{j+1}^b = k_2^b (K_1/K_j) j^{2/3} \tag{7}$$

Assuming uniform spatial concentrations, the time development of $C_j(t)$ for $j > 1$ is given by the rate equations:

$$dC_j/dt = k_{j-1}^f C_1 C_{j-1} - (k_j^b + k_j^f C_1) C_j + k_{j+1}^b C_{j+1} \tag{8}$$

The time dependence for $j = 1$ follows from the conservation law:

$$\sum_{j=1}^{\infty} jC_j = C \tag{9}$$

since C, the total concentration, is independent of time.

The time dependence $C_j(t)$ is obtained by solving Eqs. (8) and (9) numerically for an initial set of concentrations $C_j(0)$, equilibrium constants K_1, K_∞, and a scale factor for time, $(k_2^b)^{-1}$. We shall discuss the solution with the initial condition in which at $t = 0$, all $C_j(0) = 0$ except $C_1(0) = C$, and with $K_\infty/K_1 = 35$, as estimated for lysozyme (see Section II,C). A necessary condition for nucleation is that $C_1(0)$ be larger than the solubility limit C_1^{SOL}, defined by $C_1^{SOL} K_\infty^{XTAL} = 1$.

The numerical solutions [i.e., $C_j(t)$] of Eqs. (8) and (9) for $C_1 > 1/K_\infty^{XTAL}$ and $C_1 \ll 1/K_1^{XTAL}$ show that the concentrations C_j stop changing significantly after a relatively short time interval of $\sim(10\text{–}20)/k_2^b$. Thus, a state called *quasi-equilibrium* is attained. This trend is exhibited even for 3% lysozyme (see Fig. 2c, broken lines) for which the condition $C_1 \ll 1/K_1^{XTAL}$ is not well fulfilled ($C_1 K_1^{XTAL} \approx 0.2$). The concentrations C_j at quasi-equilibrium can be adequately approximated by repeated application of the relation $C_{j+1} = C_1 C_j K_j$ [Eq. (4)] subject to the normalizing condition of Eq. (9) and assuming $C_j = 0$ for $j > j^*$. The solid line of Fig. 2c represents the result of these calculations. The dotted line in Fig. 2c was obtained by using Eq. (4) for $j > j^*$ and does not correspond to a physical solution.

[26] W. G. Courtney, *J. Chem. Phys.* **36,** 2009 (1962).

[27] W. G. Courtney, *J. Chem. Phys.* **36,** 2018 (1962).

Eventually, $C_j(t)$ must proceed from quasi-equilibrium to the true equilibrium in which $C_1(t)$ decreases toward C_1^{SOL}. The first step in this process is the formation of a stable nucleus which occurs after a "delay" time τ_D that is much longer than the time τ_{QE} that it takes to establish quasi-equilibrium. A qualitative explanation for this delay time is as follows. For nuclei of size $j \approx j^*$ the forward rate $C_1 k_j^f$ equals the backward rate k_j^b, i.e., there is essentially an equal probability for the two processes $j \rightarrow j + 1$ and $j \rightarrow j - 1$. Thus, a critical nucleus initially grows by a random walk process (which is inherently slow) rather than being driven by a free energy gradient (see zero slope at $j = j^*$ in Fig. 2a).[28–30] A nucleus will be relatively stable when it has traversed the energy barrier and reaches a point (j value) at which the energy difference between successive steps is of the order of kT. Thus, the nucleus has to traverse Δj^* steps (see Fig. 2a) via a random walk, each step taking $\sim(k_{j^*}^b)^{-1}$ sec. The estimated delay time is then given by[22]

$$\tau_D \approx (\Delta j^*)^2 (k_{j^*}^b)^{-1} \tag{10}$$

Since K_j is an increasing function of j (see Fig. 2b), once stable nuclei are formed, they will start growing into macroscopic crystals as discussed in Section III. The strong dependence of the delay time on the degree of supersaturation can be seen from Eq. (10). At low lysozyme concentrations the slope $d(\Delta G_j)/dj$ is small in the vicinity of j^* (see Fig. 2a). Consequently, one has to traverse a large Δj^* region before the energy difference between successive steps is of the order of kT.

Quantitative estimates of the time to achieve true equilibrium were obtained by solving Eqs. (8) and (9) for times much longer than indicated in Fig. 2c.[31] The time course of the monomer concentration $C_1(t)$ for $K_\infty/K_1 = 35$ and initial concentration $C = 1.2\%$ (w/v) is shown in Fig. 4. It shows clearly and dramatically the large difference in times to establish quasi-equilibrium ($\tau_{QE} \approx 10/k_2^b$) and the delay time ($\tau_D \approx 10^5/k_2^b$). Similar calculations were performed for other values of C. It was found, as expected, that τ_D was strongly concentration dependent whereas τ_{QE} depended only weakly on concentration. For $C = 3\%$ (w/v), for example, $\tau_D \approx 500/k_2^b$ and $\tau_{QE} \approx 20/k_2^b$.

[28] J. Zeldovich, *Zh. Eksp. Teor. Fiz.* **12,** 525 (1942).

[29] W. J. Dunning, *in* "Nucleation" (A. C. Zettlemoyer, ed.), Chapter 1, p. 1. Dekker, New York, 1969.

[30] P. R. Andres, *in* "Nucleation" (A. C. Zettlemoyer, ed.), Chapter 2, p. 69. Dekker, New York, 1969.

[31] In order to solve the set of "stiff" Eqs. (8) and (9) for long time intervals (up to $\sim 10^6/k_2^b$), we used the A-stable semi-implicit method of D. A. Calahan [*Proc. IEEE* **56,** 744 (1968)]. We also used a continuum approximation for large j values. The difference in Eq. (8) was converted into a continuum differential equation. Numerical solutions with Δj as an increment, up to $j = 10^{10}$, were obtained.

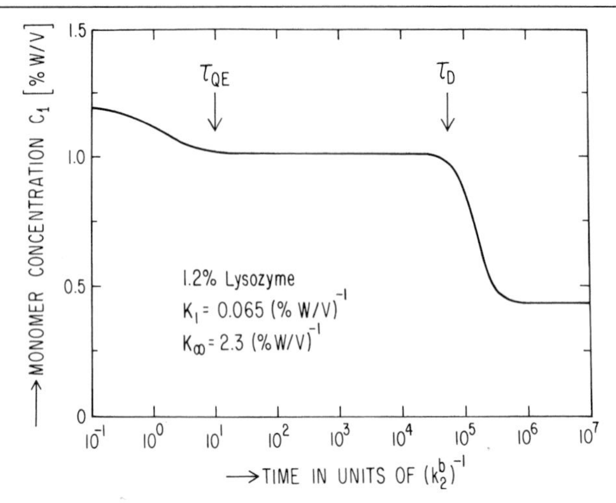

FIG. 4. The time course of monomer concentration for 1.2% (w/v) initial lysozyme concentration, 5% NaCl, $K_\infty/K_1 = 35$, $T = 20°$. Obtained by solving Eqs. (8) and (9) numerically keeping track of $j = 10^{10}$ aggregates (see footnote 31). Note the large difference in time between reaching quasi-equilibrium τ_{QE} and the delay time τ_D. (Calculation performed by H. Shore, private communication.)

The real crystallization times are expected to be longer than the calculated times, since the homogeneous model does not account for the reduced monomer concentration near the surface of a growing aggregate. Furthermore, the smooth j dependence of K_j obtained from the analogy to liquid drops is a crude approximation to the equilibrium between monomers and the surface of a real crystal, where the nucleation of each new surface layer introduces additional delay times.

As shown in the preceding paragraph, the time lag to form crystals at low C values (mild supersaturation) becomes exceedingly long. In contrast, in the case of amorphous aggregates, $K_\infty/K_1 \approx 1$. Thus, even for mild supersaturation, $C_1K_\infty^{AMOR} > 1$, $C_1K_1^{AMOR} > 1$, and the growth of aggregates proceeds in the absence of a barrier without a time lag.

Competition between Crystallization and Amorphous Precipitation. An interesting problem arises when the growth of crystals competes with the growth of an amorphous precipitate in the same solution. As mentioned previously, $K_\infty^{XTAL}/K_1^{XTAL} \gg 1$ while $K_\infty^{AMOR}/K_1^{AMOR} \approx 1$. We want to determine the consequences of having different ratios of K_1^{XTAL} to K_1^{AMOR}. We assume that K_1^{XTAL} and K_1^{AMOR} can be changed[32] by varying the experimental conditions (e.g., pH or salt concentration).

[32] These two quantities are not necessarily equal since different kinds of bonds may be involved.

Three distinct cases can be delineated.

1. If $K_1^{XTAL} > K_1^{AMOR}$, then for any j, $K_j^{XTAL} > K_j^{AMOR}$. Consequently, crystalline aggregates of any size will be energetically more favorable than amorphous aggregates.

2. If $K_1^{XTAL} \ll K_1^{AMOR}$, then it is possible that $K_\infty^{XTAL} < K_\infty^{AMOR}$. In this case the amorphous precipitate is thermodynamically more stable and there is no hope of growing crystals (a crystal placed in such a solution would dissolve to form an amorphous aggregate).

3. If $K_1^{XTAL} < K_1^{AMOR}$ and $K_\infty^{XTAL} > K_\infty^{AMOR}$, crystals are more stable than amorphous aggregates. However, small amorphous aggregates are more stable than small ordered (precrystalline) aggregates. In this case the amorphous aggregation process may interfere with the kinetics of the growth of crystals by depleting the concentration of monomers until the solubility limit $C_1 K_\infty^{AMOR} = 1$ is reached. C_1 may thus be too small to form a crystal.

In concluding this section, let us reiterate the main conclusion obtained from the theoretical model. One can *predict whether the solution is heading toward crystallization or amorphous precipitation by measuring* K_1 *and* K_∞. This can be done by looking at small j-mers at times long before any visible aggregates have appeared in the solution. We shall show how this can be accomplished by using quasi-elastic light scattering.

C. Experimental Results

Classical Batch Crystallization. As mentioned in the introduction, we chose an easily crystallizable protein, hen egg-white lysozyme, as a model system to investigate the crystallization processes and to test the predictions of the theoretical models. Before discussing the results of the more detailed and systematic investigation, let us look first at the gross, qualitative features of the aggregation process as obtained from a classical batch crystallization experiment. Two parameters, the concentrations of lysozyme and salt (NaCl), were picked to prepare a two-dimensional array of crystallizing dishes. Figure 5 shows the time required to form crystals for various values of these parameters. The solubility line is the locus of infinitely long crystallization time and will be discussed more fully in the next section. A convenient parameter is the *degree of supersaturation*, defined as the ratio of the protein concentration to the concentration at the solubility line. A striking feature of the results shown in Fig. 5 is the strong dependence of the crystallization time on the protein (and salt) concentration. The region of crystallization spans a range of values of supersaturation from ~2 to 10 with concomittant crystallization times

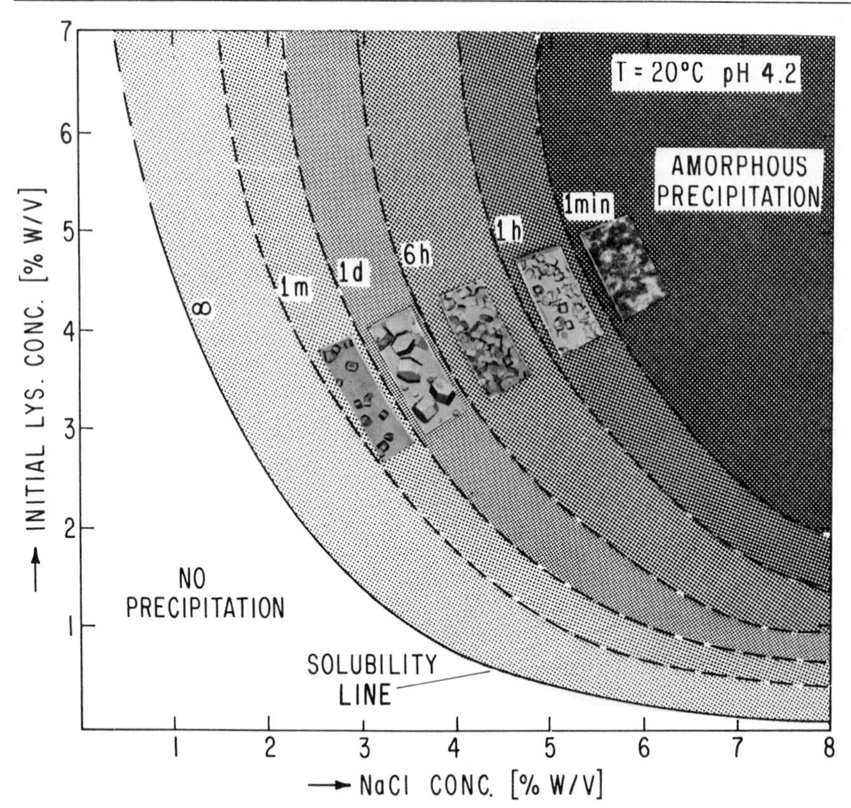

FIG. 5. Schematic map of crystallization kinetics as a function of lysozyme and NaCl concentration obtained from a matrix of dishes. Inserts show photographs of dishes obtained 1 month after preparation of solutions. The range of supersaturation in which crystallization occurs is between ~2 and 10. The crystallization mixtures were prepared as follows. A 10% (w/v) stock solution of three-times crystallized salt-free hen egg-white lysozyme (Sigma Corp.) was prepared in 0.1 *M* sodium acetate–acetic acid buffer pH 4.2. Lysozyme concentrations were checked spectrophotometrically ($A_{281\,\mathrm{nm}}^{1\,\mathrm{cm}}$ = 26.35 for a 1% (w/v) solution) [A. J. Sophianopoulus, C. K. Rhodes, D. N. Holcomb, and K. E. van Holde, *J. Biol. Chem.* **237,** 1107 (1962)]. Equal volumes of lysozyme and NaCl solutions were prepared at twice the desired final concentration. The NaCl solution was heated to 40° and slowly added to the lysozyme solution while stirring. The mixture was passed through a 0.22-μm Millipore filter into carefully cleaned dishes. The dishes were covered with microscope slides, sealed with grease, and placed in temperature-controlled blocks (20 ± 0.5°) on a shock-mounted table. (From Kam *et al.*[3])

varying from months to minutes. Above a supersaturation value of ~10 amorphous precipitation occurs without a significant time lag, as expected from our discussion in connection with Fig. 1.

If we take the example of a 5% NaCl solution we see that an increase of protein concentration from 2 to 4% reduces the crystallization time

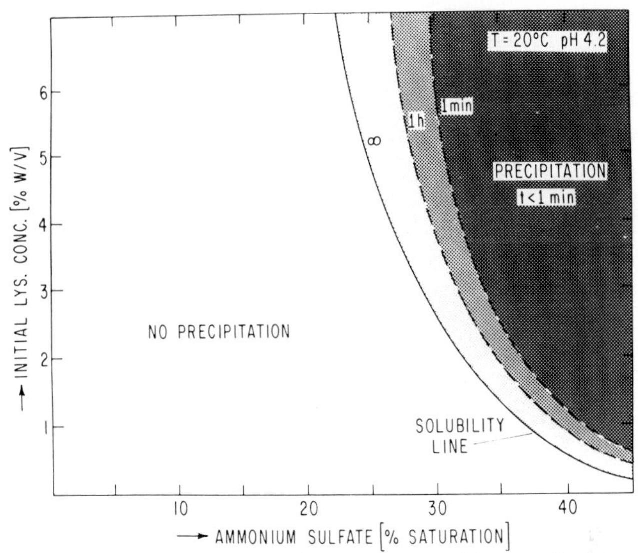

FIG. 6. Schematic map of the kinetics of amorphous precipitation of lysozyme in ammonium sulfate. As compared with crystallization (see Fig. 5), a relatively small degree of supersaturation causes aggregation. The solutions were prepared as described in Fig. 5 (From Kam *et al.*[3])

from about a day to a fraction of an hour, i.e., a reduction by a factor of $\sim 10^2$. This is approximately the value predicted by theory (see previous section) for the ratio of the delay times.

Although at high enough lysozyme concentration an amorphous precipitate was formed in NaCl (see Fig. 5), a more common method of precipitating proteins is with ammonium sulfate as shown in Fig. 6. The characteristic features of amorphous precipitation are the relatively sharp transition from a stable undersaturated mixture to a precipitation at a small value of supersaturation and the lack of a substantial delay in forming the precipitate. Both of these phenomena are a consequence of the lack of an energy barrier as discussed before (see Fig. 1).

Determination of K_∞^{XTAL} from the Protein Solubility. One of the important parameters that appears in the theory discussed in the previous section is K_∞^{XTAL}, the equilibrium constant between monomers and large crystals, i.e.,

$$K_\infty^{\mathrm{XTAL}} = \lim_{j \to \infty} ([C_{j+1}]/[C_j][C_1]) = 1/C_1^{\mathrm{sol}} \tag{11}$$

where C_1^{sol} is the concentration of monomers in equilibrium with the crystal.

The solubility was measured experimentally by determining the concentration of lysozyme in equilibrium with crystals. To ensure that equi-

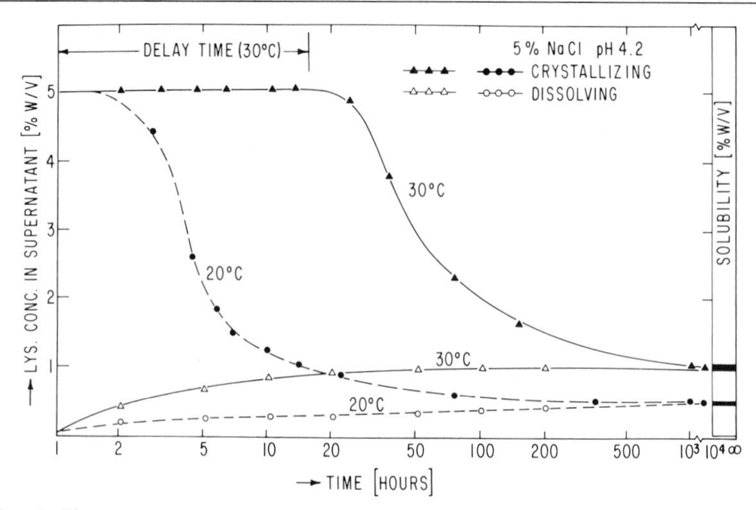

FIG. 7. Time course of lysozyme crystallization and dissolution of crystals as measured by the protein concentration in the supernatant. Both the crystallization and dissolving processes approach asymptotically the solubility concentration. Crystallizing solutions were prepared as described in Fig. 5.

librium was reached, the time course of the concentration was determined for both growing crystals (initial lysozyme concentration 5% w/v) and dissolving crystals (initial lysozyme concentration 0%). The solution was periodically agitated to eliminate concentration gradients of lysozyme near the crystal surfaces and to avoid difficulties with the cessation of growth by nucleating new crystals (see later sections). The experimental results for two temperatures (20 and 30°) are shown in Fig. 7 (5% NaCl, pH 4.2). Both the crystallization and dissolving process approach asymptotically the solubility concentration C^{sol}. The values obtained were

$$\begin{aligned} 20°: &\quad C^{sol} = 0.50 \pm 0.05 \quad (\%\ w/v) \\ 30°: &\quad C^{sol} = 1.10 \pm 0.05 \quad (\%\ w/v) \end{aligned} \tag{12a}$$

To obtain K_∞^{XTAL}, we need C_1^{sol} rather than C^{sol}. The relation between these two quantities is obtained from Fig. 3. Using the values of K_1 = 0.06 $(\%\ w/v)^{-1}$ as determined for lysozyme at 20°, $C^{sol}K_1 = 0.033$, and $\ln(K_\infty/K_1) = 3.6$ (see next section), we obtain $C_1^{sol}/C^{sol} \approx 0.95$ (see Fig. 3) and for K_∞^{XTAL}

$$K_\infty^{XTAL} = 1/(C^{sol})(0.95) = 2.1 \pm 0.2\ (\%\ w/v)^{-1} \tag{12b}$$

The Use of Quasi-Elastic Light Scattering to Determine Whether the Solution is Heading toward Crystallization or Amorphous Precipitation. The theoretical model discussed earlier predicts that the size distribution of aggregates in the quasi-equilibrium state is different in a solution headed for crystallization than in one in which an amorphous precipitate is going to be formed. We have used quasi-elastic light scattering to probe the distribution. It has the advantage over classical light scattering in that it measures not only the intensity but also the spectral distribution (broadening) on the scattered light. The latter arises from the thermal motion of the scatterers and is therefore sensitive to their diffusion coefficient, i.e., their size and shape.[12–15] Thus, the larger the j-mer, the slower it diffuses and consequently the smaller will be the spectral width of the scattered light. Let us look at this process quantitatively.

The intensity of scattered light from an aggregate (j-mer) that is small in comparison to the wavelength of light is proportional to j^2 (since the scattering amplitudes of the j molecules in the aggregate are all in phase). The total scattered intensity from a solution of aggregates is therefore

$$I = \sum_j AC_j j^2 \tag{13}$$

where C_j is the molar concentration of j-mers, and A is the total scattered intensity from 1 mol of monomers.

The frequency (power) spectrum, $S_I(\nu)$, from a solution of j-mers has a Lorentzian shape given by

$$S_I(\nu) = \frac{AC_j j^2}{\pi} \frac{\Delta\nu_j}{(\Delta\nu_j)^2 + (\nu - \nu_0)^2} \tag{14}$$

where ν_0 is the frequency of the incident light and $\Delta\nu_j$ is the width of the spectral distribution given by

$$\Delta\nu = D_j q^2/2\pi \tag{15}$$

D_j is the diffusion coefficient of the j-mer and q is the scattering wave vector. The power spectrum for a distribution of aggregates with concentration C_j and diffusion coefficients D_j is obtained by a summation of Lorentzian curves given by Eq. (14). Experimentally, one usually uses a homodyne system in which one measures the power spectrum, $S_I(\delta\nu)$, of the fluctuating photocurrent around *zero frequency*.[14] Although the summation in this case becomes more involved,[33] the resulting frequency dependence can be numerically approximated with high accuracy by a single Lorentzian with an effective line width $\overline{\Delta\nu}$, which, due to the slower diffusion of larger aggregates, is always smaller than the monomer width $\Delta\nu_1$.

The experimental results obtained from solutions of different lysozyme concentrations are shown in Fig. 8. Light (λ = 4880 Å) from an argon ion laser (Spectra Physics 164) was scattered from the lysozyme

[33] S. B. Dubin, G. Feher, and G. B. Benedek, *Biochemistry* **12,** 714 (1973).

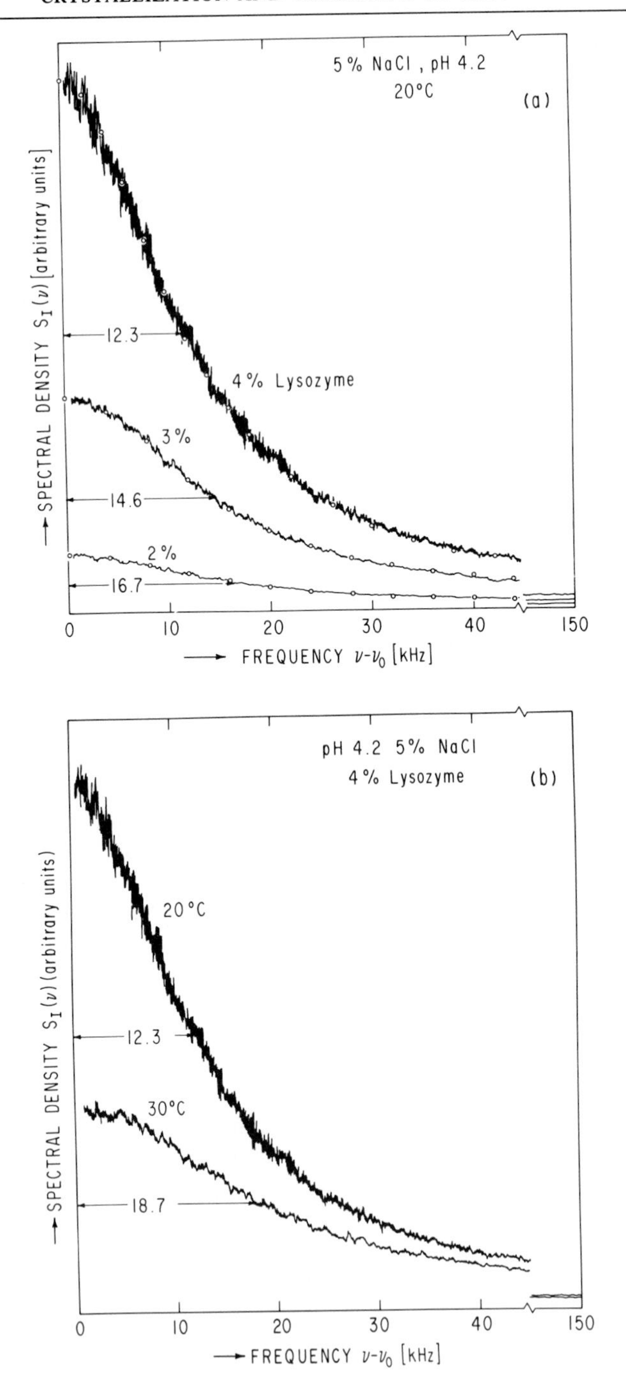
5 % NaCl , pH 4.2
20°C
(a)
4 % Lysozyme
3 %
2 %
12.3
14.6
16.7
SPECTRAL DENSITY $S_I(\nu)$ [arbitrary units]
0
10
20
30
40
150
FREQUENCY $\nu-\nu_0$ [kHz]
pH 4.2 5% NaCl
4 % Lysozyme
(b)
20 °C
30°C
12.3
18.7
SPECTRAL DENSITY $S_I(\nu)$ (arbitrary units)
0
10
20
30
40
150
FREQUENCY $\nu-\nu_0$ [kHz]

solution and detected at 90° by a photomultiplier (RCA 7265) whose output was amplified, fed to a spectrum analyzer[34] (HP 310A, GR 1900, or Fed. Sci. UA 10A), squared, and connected to an x–y recorder. It is evident from Fig. 8 that the linewidth narrows with increasing lysozyme concentration (and decreasing temperature) as would be expected from larger aggregates having a smaller diffusion coefficient.

How do we get the size distribution of the aggregates from a single Lorentzian line shape? We used the theory of the previous section to predict a size distribution at *quasi-equilibrium* (see solid line in Fig. 2c) for different sets of values of C, K_1, and K_∞. By approximating the diffusion coefficients D_j of the various aggregates with Stokes's law for spheres (i.e., $D_j \propto j^{-1/3}$), we calculated a power spectrum S_1. If the calculation for a given K_1 and K_∞ is repeated using a D_j appropriate for ellipsoids rather than spheres, there is only a small change in the calculated S_1 value. In all cases, S_1 closely approximated a Lorentzian curve characterized by a single width $\overline{\Delta\nu}$.

The results of the calculation $\overline{\Delta\nu}/\Delta\nu_1$ versus lysozyme concentration C times K_1 for different ratios of K_∞/K_1 are shown in Fig. 9. The experimental data were processed as follows. The values of $\overline{\Delta\nu}/\Delta\nu_1$ were plotted versus C using the same semilogarithmic scale (see top scale of Fig. 9) as used for the abscissa in Fig. 9. The experimental curve was then shifted horizontally over the theoretical curve until the best fit was obtained. This process is equivalent to choosing the best value for K_1. The result of this procedure is shown by the broken line. For the crystallization conditions (5% NaCl, 20°, pH 4.2), K_1 was found to be 0.065 ± 0.010 $(\%, \text{w/v})^{-1}$ and

[34] More recently, photon correlators (Malvern Instruments, Malvern GB, Langley-Ford, Cambridge, Massachusetts) were introduced. With these one measures the Fourier transform (i.e., autocorrelation) of the spectrum. Since they process data in parallel rather than sequentially, the measuring time is reduced by several orders of magnitude for a given signal-to-noise ratio.

FIG. 8. The power spectrum, S_1, of the photocurrent due to light scattered at 90° from a crystallization mixture of lysozyme after establishment of quasi-equilibrium for different lysozyme concentrations (a) and temperatures (b). The narrowing of the line width with increasing lysozyme concentration and decreasing temperature indicates the aggregation process. (The temperature dependence of the viscosity contributes about 50% to the narrowing of the line between 30 and 20°.) The circles are theoretical fits to a Lorentzian line shape [Eq. (14)] having a width $\overline{\Delta\nu}$, as indicated in the graph. The widths $\overline{\Delta\nu}$ obtained for different lysozyme concentrations are compared with theory in Fig. 9. Crystallizing solutions were prepared as described in Fig. 5. A standard homodyne system was used. Scattering data were taken with an HP 310A spectrum analyzer approximately 5 min after mixing. (Modified from Kam *et al.*[2])

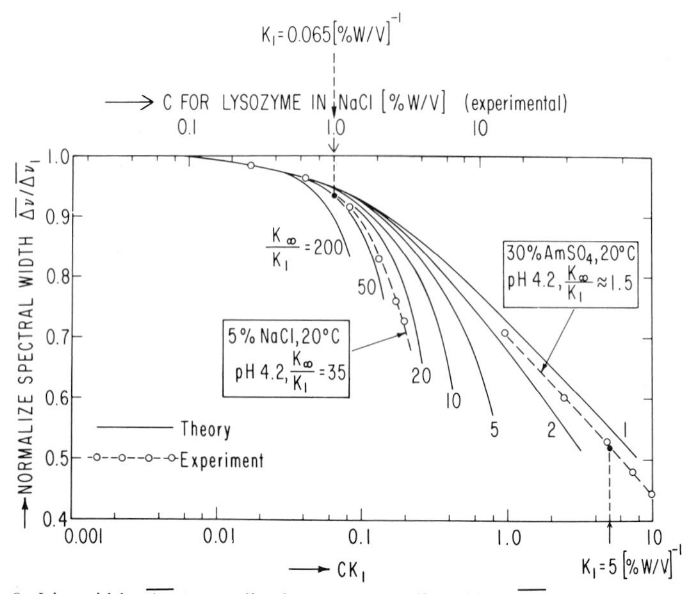

FIG. 9. Linewidths $\overline{\Delta\nu}$ (normalized to monomer linewidths $\overline{\Delta\nu_1}$ obtained at a low lysozyme concentration of ~0.1%) of scattered light versus lysozyme concentration C times K_1. Full lines represent theoretical fits for different values of K_∞/K_1. Broken lines connect experimental points obtained under conditions for which either crystals (in NaCl) or amorphous precipitates [in $(NH_4)_2SO_4$] were formed. Note the large difference in the K_∞/K_1 ratio for the two cases. The top scale represents the experimentally determined lysozyme concentration in NaCl. The value of K_1 can be read off the top scale at the point where the lysozyme concentration C is 1% (w/v) (●). (Modified from Kam *et al.*[2])

$K_\infty = 2.3 \pm 0.5$ (% w/v)$^{-1}$.[35] This value of K_∞^{XTAL} is in good agreement with the value of 2.1 ± 0.2 (% w/v)$^{-1}$ obtained from protein solubility data discussed in the previous section [see Eq. (12b)] and provides evidence for the essential validity of the theoretical model.

A set of quasi-elastic light scattering experiments was also performed under conditions in which lysozyme formed an amorphous precipitate (30% saturated ammonium sulfate dissolved in 0.1 *M* sodium acetate buffer at pH 4.2). For this case, the best fit of the experimental results with theory was obtained for a value of $K_\infty/K_1 = 1.5$, as compared with $K_\infty/K_1 = 35$ for crystallizing solutions. This large difference in K_∞/K_1 for the two cases is just what was predicted by the theoretical model.

[35] The errors quoted represent the precision of the results within the quasi-equilibrium model and are not meant to imply absolute accuracy; in particular at high protein concentrations, the model becomes less adequate.

In principle, we could have used intensity measurements of the scattered light to obtain the same results. However, in view of large deviations from the ideal behavior described by Eq. (13), which occur at high concentrations,[36] absolute intensity measurements are less reliable than the measurements of spectral distribution described above.

Similar results were also obtained on chymotrypsinogen, another water-soluble protein.[37] An attempt to use the method on an integral membrane protein, the reaction center from the photosynthetic bacterium *Rhodopseudomonas sphaeroides*[38] failed; the reaction centers precipitated when illuminated with the laser. The reason presumably is the strong absorption by the chromophores of the reaction center, which causes local heating and denaturation. Reaction centers from two different bacterial species have, however, been recently crystallized[39,40] by the old fashioned "trial and error" method.

Measurement of the Kinetics of Establishing Quasi-Equilibrium. The kinetics of the different temporal phases of the crystallization process were all expressed in the theoretical section in terms of the dissociation constant of a lysozyme dimer, k_2^b. We showed, for instance, that quasi-equilibrium in 3% (w/v) lysozyme was established in $\sim 20/k_2^b$ sec (see Fig. 2c). It was, therefore, of interest to establish the value of the time scale k_2^b by measuring the time to establish quasi-equilibrium. The classical approach of measuring the time course of an equilibration process is by the relaxation method developed by Eigen and DeMaeyer.[41] In this technique, the system at equilibrium is subjected to a sudden perturbation (a "jump" in a parameter), and the subsequent time development of the re-equilibration process is monitored.

Two different perturbations were used to disturb the equilibrium. They were a sudden change in temperature (T-jump)[42] and pressure (P-jump).[43] These experiments were performed in the laboratories of I. Pecht and H. Gutfreund, respectively. The distribution of j-mers was monitored via the intensity of the scattered light. Since in these experiments only the time course is important, *relative* rather than *absolute* light intensities

[36] E. Reisler and H. Eisenberg, *Biochemistry* **10,** 2659 (1971).

[37] A. Shaikevitch and Z. Kam, unpublished results.

[38] For a review, see, for example, G. Feher and M. Y. Okamura, *in* "The Photosynthetic Bacteria" (R.K. Clayton and W.R. Sistrom, eds.), Chapter 9, p. 349. Plenum, New York, 1978.

[39] H. Michel, *J. Mol. Biol.* **158,** 567 (1982).

[40] J. P. Allen and G. Feher, *Proc. Natl. Acad. Sci. U.S.A.* **81,** 4795 (1984).

[41] M. Eigen and L. DeMaeyer, *Tech. Org. Chem.* **8,** 895 (1963).

[42] I. Pecht, D. Haselkorn, and S. Friedman, *FEBS Lett.* **24,** 331 (1972).

[43] J. S. Davis and H. Gutfreund, *FEBS Lett.* **72,** 199 (1976).

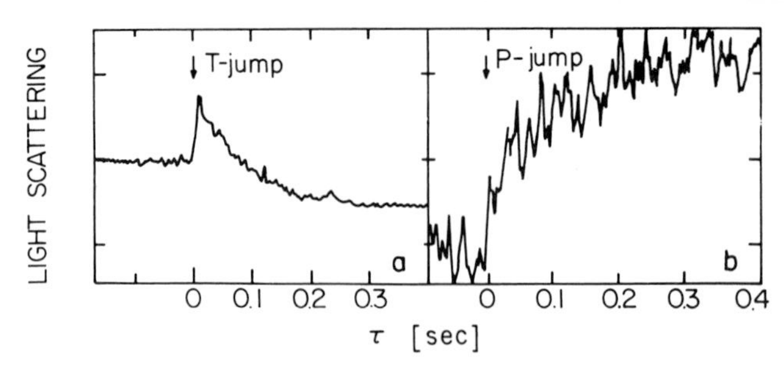

FIG. 10. The intensity of light scattered from a supersaturated lysozyme solution following a perturbation in temperature (a) and pressure (b). The characteristic time (0.1 sec) reflects the establishment of quasi-equilibrium among the different *j*-mers. Conditions: 4% lysozyme, 5% NaCl, $T = 20°$, pH 4.2. (From Kam and Rigler.[45] Reproduced from the *Biophysical Journal* 1982, **39,** pp. 7–13, by copyright permission of the Biophysical Society.)

were of importance, there was no need to employ quasi-elastic light scattering as discussed in the previous section. The results of the two relaxation experiments are shown in Fig. 10. The characteristic time derived from these measurements is 0.1 sec.

An alternate method of obtaining kinetic parameters that does not require an external perturbation is to measure the fluctuations in the concentrations of the reactants.[44] There are basically two ways of analyzing fluctuations. In one, the frequency (Fourier) components are analyzed as was done in Fig. 8; in the other the Fourier transform of the frequency spectrum is taken to obtain the so-called autocorrelation function, which decays with a time constant equal to the characteristic time of the kinetic process.

In the previous section we have shown that the frequency spectrum of the quasi-elastically scattered light is primarily governed by the translational diffusion of the macromolecules. The same will apply for the autocorrelation function. The association–dissociation reactions of the *j*-mers add only an extremely small (and practically unmeasurable) component to the spectra of Fig. 8. Recently, a technique was developed in which the fluctuations due to translational diffusion were eliminated.[45] In this technique one cross-correlates the fluctuating intensities at two different scattering angles. The cross-correlation method was applied to the analysis of the scattering from a supersaturated lysozyme NaCl solution. The results are shown in Fig. 11. The cross-correlation function has a characteristic time of 0.1 sec in agreement with the results obtained with the relaxation method.

[44] See, for example, G. Feher, *Trends Biochem. Sci.* **3** (5) 111 (1978).
[45] Z. Kam and R. Rigler, *Biophys. J.* **39,** 7 (1982).

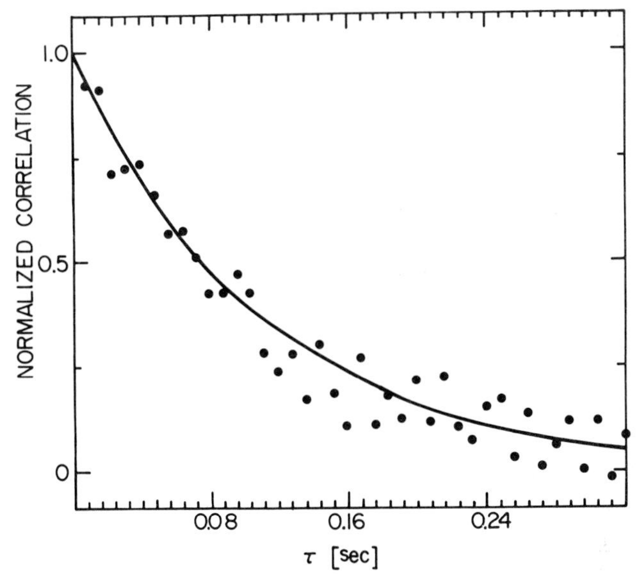

FIG. 11. The cross-correlation of scattered light intensity fluctuations at two scattering angles obtained from a supersaturated lysozyme solution. The characteristic time obtained and conditions used were the same as in Fig. 10. (From Kam and Rigler.[45] Reproduced from the *Biophysical Journal*, 1982, **39**, pp. 7–13, by copyright permission of the Biophysical Society.)

Equating the measured time to reach quasi-equilibrium (0.1 sec) to the theoretically predicted time $\tau_{QE} \approx 10/k_2^b$ (see Fig. 4), one obtains for k_2^b the value

$$k_2^b \approx 100 \text{ sec}^{-1} \tag{16}$$

An independent estimate of this quantity is obtained later.

Assay of the Number of Postcritical Nuclei. So far we have discussed methods to estimate the time to establish quasi-equilibrium, τ_{QE} [see Figs. 10 and 11, and Eq. (16)], to determine the distribution of j-mers (see Fig. 8), and to obtain a rough estimate of the time it takes to grow macroscopic crystals (see Fig. 5). In this section we outline an assay to determine the number of postcritical nuclei and from the time course of this number, obtain a rough experimental estimate of the delay time τ_D (see Fig. 4). A postcritical nucleus is defined as a j-mer that has passed the potential barrier and reached a size $j > j^*$ such that it has a small probability to dissociate below its critical size, $j < j^*$.

The assay is based on the fact that postcritical nuclei will grow even in a mildly supersaturated solution in which there is a negligible likelihood that nuclei will form within the time of the assay. If we place, therefore, aliquots of a highly supersaturated solution containing nuclei into a mildly supersaturated solution, the nuclei will grow into macroscopic crystals that can be counted. In essence this is a classical seeding experiment in which the seed is an (invisible) nucleus rather than a (visible) microcrystal.

The assay was performed as follows. Into one row of a multiwell array were placed 0.05 ml and into the other 0.15-ml aliquots of a 3% lysozyme, 5% NaCl, pH 4.6, crystallizing solution B. After various times, the aliquots were diluted with solutions A, A′, or A″ such that each well contained 0.5 ml of 1% lysozyme, 5% NaCl, pH 4.6. The third row of the array consisted of control wells containing 0.5 ml of 1% lysozyme (solution A″). The array was left undisturbed at 22° for 7 days, and then photographed. The result is shown in Fig. 12. The first two rows show a delay in the appearance of crystals of ~3 min with a subsequent increase in the number of crystals at later times. The second row has approximately three times as many crystals as the first, as expected from the larger aliquots. No crystals were formed in the control wells showing that all

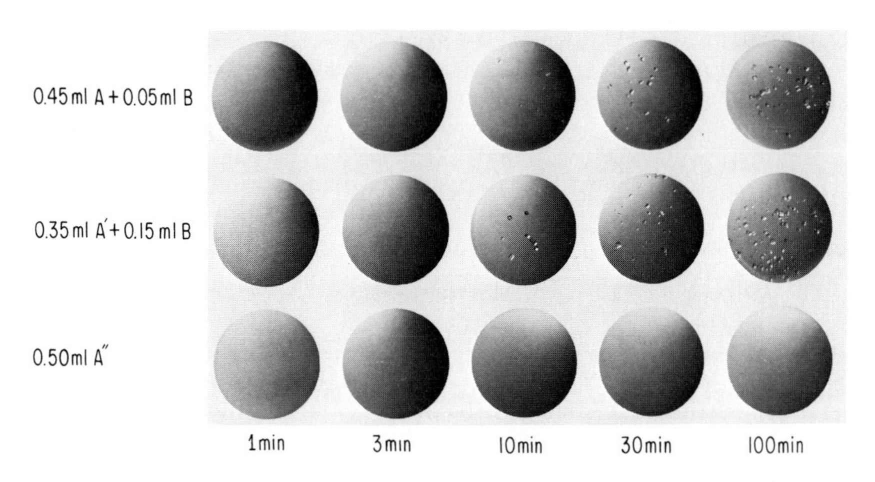

FIG. 12. Assay of the number of postcritical nuclei. Aliquots of a crystallizing solution B (3% lysozyme, 5% NaCl, pH 4.6, $T = 22°$) were diluted after various times with solutions A, A′, or A″ which by themselves do not form nuclei during the time of the experiment (see third row). Solutions A, A′, and A″ were adjusted such that the final composition after mixing was 1% lysozyme, 5% NaCl, pH 4.6. Photograph taken after 7 days of growth (S. D. Durbin, Z. Kam, and G. Feher, unpublished results).

III. Postnucleation Growth

A. *Qualitative Discussion*

After a sufficiently long time interval, a small number of rapidly growing crystals dominates the time development of the system. Once this occurs, the stochastic approach of Eq. (8) with its restriction of spatial homogeneity of monomer concentration is no longer necessary (and in most cases not applicable) and one can focus on the growth of a single crystal.

The growth of crystals from solutions can be visualized as proceeding in two stages: (1) the transport of material to the crystal surface and (2) the association of molecules with the ordered crystal lattice. The observed growth rate may be limited by one or both of these processes. In a quiescent, unstirred crystallization solution the transport will be limited by diffusion, which for macromolecules is a relatively slow process. Lysozyme, for instance, diffuses ~1 cm in 1 week.[33] The slow supply of molecules by diffusion causes a reduction of the protein concentration in the vicinity of a rapidly growing crystal. This has two important practical consequences: the reduced protein concentration prevents the formation of new nuclei and, therefore, eliminates unwanted interferences between crystals. The second consequence is the slowing down of the rate of supply which allows a longer time for thermal motion to properly position the molecules in the lattice. Crystals that grow too fast usually reach a smaller terminal size, presumably because of structural defects that have been incorporated (see Section IV).

The simplest way to measure the growth rate is to monitor the size of a growing crystal as a function of time. However, it is difficult to make a meaningful comparison with theory unless one knows the protein concentration at the surface of the crystal.[46] In this section we describe the determination of lysozyme concentration in the vicinity of a growing crystal and show how this information can be used to obtain values for the growth rate, the average attachment coefficient, and the dimer dissociation constant k_2^b.[2] Since the considerations start out by presupposing the presence of microcrystals, the conclusions reached in this section should

[46] This problem can be avoided by continuously flowing a solution of known protein concentration past the crystal. We have used this technique to obtain crystal growth rates for lysozyme for supersaturation ratios of $1.2 < C/C^{Sol} < 8.6$. Over this range the growth rate varied by ~4 orders of magnitude. (See S. D. Durbin and G. Feher, 1st International Conference on Protein Crystal Growth, Stanford, California, August 1985, to be published in *J. Cryst. Growth.*)

also be applicable to seeding experiments (i.e., heterogeneous nucleation).

B. Determination of Protein Concentration near the Surface of a Growing Crystal

The protein concentration in the solution surrounding the crystal was obtained from the optical absorption at 277 nm (to reduce the absorption the wavelength was shifted from the peak of the band at 281 mm). Crystals were grown in a layer of liquid between two quartz slides 0.18 mm

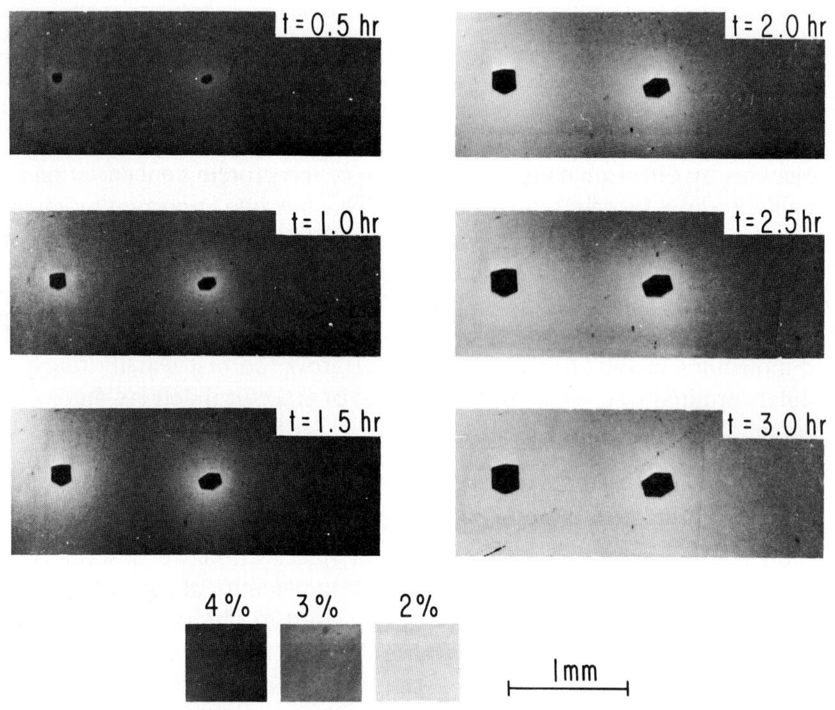

FIG. 13. Ultraviolet light transmission photographs of growing lysozyme crystals. Note the regions around the crystals which have been depleted of protein. Calibration of protein concentration versus film densities is shown at the bottom. Crystallizing mixture: 4% (w/v) lysozyme, 5% (w/v) NaCl mixed at 35°, filtered through 0.22-μm Millipore filters into the space between two quartz slides separated by 0.18 mm. Growth was monitored at $T = 20°$. The ultraviolet light (from a xenon arc lamp) passed through a grating monochromator (Bausch & Lomb 33-86-01) set at 277 nm, 10 nm BW, followed by a filter (Corning Cs7-54). The crystals were imaged with a quartz lens ($f = 39$ mm) on a Polaroid 55P/N film. The negatives were scanned to obtain the concentration profiles of the protein in the vicinity of the crystal (see Fig. 14). (From Kam *et al.*[2])

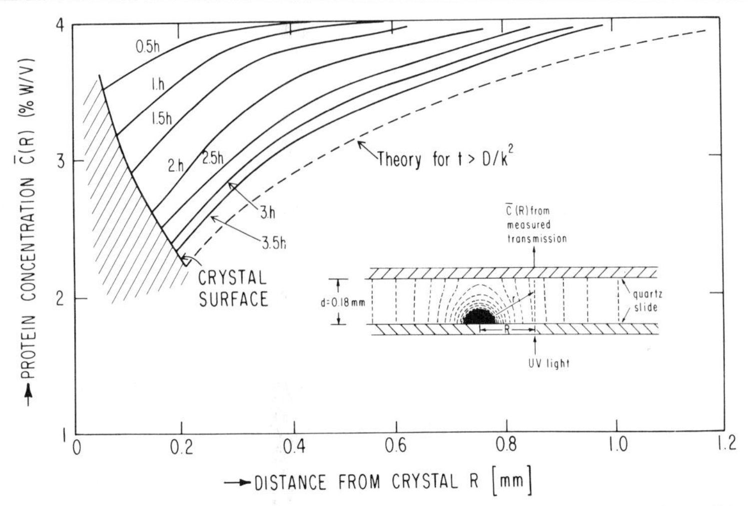

FIG. 14. The concentration profile measured from the center of the crystal as a function of time. The plots were obtained from digital scans of the photograph shown in Fig. 13. The cross-hatched area indicates the growing crystal. The dashed line shows the calculated profile for $t \gg D/k^2$ (i.e., approximately steady state). Insert shows experimental arrangement used to measure $\bar{C}(R)$ and theoretically calculated lines (dashed) along which the protein concentration is equal. For times <1 hr the corrections to $\bar{C}(R)$ is large. Consequently, the curves for 0.5 and 1 hr cannot be reliably used for quantitative analyses.

apart. The crystals were imaged with a quartz lens and photographed every half-hour with an exposure time of 0.5 sec (see Fig. 13). A longer exposure time caused a large number of aggregates to be formed; presumably denatured protein molecules formed nucleation sites. To reduce errors due to fluorescence of the protein, a lens with a long focal length (39 mm) was used. With this configuration, only ~2% of the isotropic fluorescence was collected without essentially reducing the intensity of the collected collimated ultraviolet light. For other details of the experimental arrangements, see legend to Fig. 13.

The results shown in Fig. 13 clearly show the "halos" around the crystals due to the depletion of protein. To obtain the concentration profiles quantitatively, the photographs (negatives) were scanned with a densitometer[47] having a resolution of 0.1 mm, which corresponds to a resolution of 0.013 mm at the sample. The calibration of protein concentration versus optical density of the film was obtained from a set of solutions of known concentrations (see bottom of Fig. 13). The results of the scans are shown in Fig. 14.

[47] N. H. Xuong, *J. Phys. E* **2,** 485 (1969).

The concentration $\bar{C}(R)$ is plotted in Fig. 14 against the distance from the center of the crystal, R, for different times. Unfortunately, $\bar{C}(R)$, the average protein concentration along a path perpendicular to the slide on which the crystal grows (see insert in Fig. 14) is not exactly the quantity that we are interested in. Since the crystal occupies only part of the space between the quartz slides a correction has to be made to $\bar{C}(R)$ in order to obtain the desired three-dimensional concentration distribution $C(r)$ (see insert in Fig. 14). We shall outline here only the essential features of the theoretical model used to calculate this correction. We first obtained the exact solution for a fixed spherical surface in an infinite medium.[48] The solution of a sphere with an absorbing surface and two planes through which there is no flow of material (the quartz slides) was obtained by "imaging" the three-dimensional solution in the boundary planes. The concentration profile near the crystal is then obtained by summing over a large number of images. The effect of images that are far away add only a constant to the solution; the sum can, therefore, be cut off at a finite number of images. The result of this calculation is shown by the dashed lines of equal concentrations in the insert of Fig. 14. We also calculated the average concentration across the slab to simulate our measurement, and obtained the relation between the real and measured concentrations. The results of the calculations show that initially (~0.5 hr) a significant correction (factor of 2) is required but after times longer than 2 hr, the correction becomes negligible (<20%). The calculated concentration profile $C(R)$ for long times (approximately steady state) is shown by the dashed curve in Fig. 14. It is seen to be approximately parallel to the experimental profile obtained after 2.5 hr. Since this curve can be arbitrarily shifted along the vertical axis, good agreement between theory and experiment is obtained.

C. Calculation of the Growth Rate, the Average Attachment Coefficient, and the Dimer Dissociation Constant k_2^b

Having obtained the protein concentration $C|_0$ and the concentration gradient $(dC/dx)|_0$ at the crystal surface ($x = 0$), we can calculate the rate of deposition of protein mass at the crystal surface by Fick's law:

$$dM/dt = \bar{D}\,(dC/dx)|_0 S \qquad (17)$$

where S is the crystal surface area and $\bar{D}$ is the translational diffusion coefficient properly averaged over all D_j values. For lysozyme, $D_1 = 1.06 \times 10^{-6}$ cm²/sec.[33] Assuming the growth rate to be proportional to the

[48] H. S. Carslow and J. C. Jaeger, "Conduction of Heat in Solids," p. 71. Oxford Univ. Press, London and New York, 1959.

number of monomer molecules impinging on the crystal surface, we can define a rate constant k by[49]

$$dM/dt = k(C_1|_0 - C_1^{sol})S \tag{18}$$

Combining Eqs. (17) and (18), we obtain for the rate constant for growth

$$k = \frac{D(dC/dx)|_0}{C_1|_0 - C_1^{sol}} \tag{19}$$

If the growth process were limited by the attachment rate, one would observe a very shallow depletion layer ($dC/dx|_0 \to 0$). If, on the other hand, growth were diffusion limited, as is often the case in inorganic crystals,[17,50,51] the concentration at the crystal would be close to the solubility limit (i.e., $C_1|_0 - C_1^{sol} \approx 0$). In both cases it would be difficult to obtain k. For lysozyme the two rates are comparable and k was determined from Eq. (19). The experimentally determined values of $dC/dx|_0$ and $C|_0$ can be read from the graph of Fig. 14. They are 60% (w/v) cm^{-1} and 2.3% (w/v), respectively (see curve marked 3.5 hr). Using the graph of Fig. 3 to convert $C|_0$ to $C_1|_0$ ($C_1 = 0.95C$), and the previously determined solubility $C_1^{sol} = 0.48\%$ (w/v) [see Eq. (12)] we obtain

$$k = 5 \times 10^{-5} \text{ cm/sec} \tag{20}$$

If our assumption that lysozyme adds as monomers to the crystal surface is correct, the value of k calculated from Eq. (19) should be time independent even though both numerator and denominator vary (decrease) during the time course of the experiment. If the attachment were to proceed via dimers (in equilibrium with monomers), Eq. (19) should be modified to contain the dimer concentration which is proportional to the square of the monomer concentration. Thus, $C_1|_0 - C_1^{sol}$ should be replaced by $(C_1|_0)^2 - (C_1^{sol})^2$. In this case, k obtained from Eq. (19) should vary with time as the concentration changes; a change in k with time by about a factor of 2 is expected. Experimentally we found k, as calculated from Eq. (19), to be approximately time independent. This favors the monomer hypothesis [see Eq. (6)].

It is interesting to interpret the meaning of k on a molecular level. It is the product of the rate of collision of molecules with the crystal per unit

[49] M. Ohara and R. C. Reid, "Modeling Crystal Growth Rates from Solution." Prentice-Hall, Englewood Cliffs, New Jersey, 1973.

[50] S. P. F. Humphreys-Owen, *Proc. R. Soc. London, Ser. A* **197,** 218 (1949).

[51] F. S. A. Sultan, *Philos. Mag.* [7] **43,** 1099 (1952).

concentration and area, Z, times the probability of attachment per collision (attachment coefficient), p, i.e.,

$$k = Zp \tag{21}$$

Substituting for Z the expression $(k_B T/2\pi m)^{1/2}$ obtained from kinetic theory[52] and using the numerical value for k [Eq. (20)] and $m = 14{,}000$ for the molecular weight of lysozyme, we obtain:

$$p = k\sqrt{(2\pi m/k_B T)} \approx 10^{-7} \tag{22}$$

This value is several orders of magnitude smaller than observed for inorganic molecules.[49] This may be a consequence of the weaker binding forces and the more stringent steric requirements imposed by the complexity of the protein structure. However, it should be kept in mind that we are measuring only the *average* attachment coefficient. It will be equal to the probability of attachment per molecule only if all protein molecules attach themselves uniformly over the entire crystal surface S. If, however, the attachment is preferentially along imperfections (e.g., dislocations, steps) then the attachment coefficient at growth sites will be larger than the measured, average p. Preliminary experiments indicate that this is indeed the case, at least for crystals grown under conditions of high supersaturation.[53]

The value of k can also be used to obtain a rough estimate of k_2^b. By converting the difference Eq. (8) into a differential equation, one can show that a given distribution C_j will broaden with time and the mean value of j will increase. From this increase in j (i.e., the term in the differential equation proportional to dC_j/dj), one calculates the average rate of growth of a large aggregate (i.e., $K_j \rightarrow K_\infty$) of size j to be

$$dj/dt = j^{2/3} k_2^b K_1 (C_1 - C_1^{sol}) \tag{23}$$

Comparing Eqs. (23) and (18), and noting that $M = mj$, we obtain

$$k_2^b = \frac{1}{m}\frac{S}{j^{2/3}}\frac{k}{K_1} = \frac{\beta}{m}\frac{k}{K_1} \approx 100 \text{ sec}^{-1} \tag{24}$$

where β for a spherical aggregate of unit density is $(36\pi v^2)^{1/3} \approx 4 \times 10^{-13}$ cm^2. This value of k_2^b agrees with the one obtained from kinetic experiments [see Eq. (16) in Section II,C], although it should be noted that the *exact* agreement is likely to be fortuitous; an order of magnitude agreement is all one can expect from these estimates.

[52] G. W. Castellan, "Physical Chemistry," Chapter 29. Addison-Wesley, Reading, Massachusetts, 1964.

[53] S. Durbin and G. Feher, unpublished results.

D. *Effect of an Acoustic Field on Crystallization*

When a crystallizing dish was placed in an ultrasonic bath, the time for macroscopic crystals to appear was shortened by several orders of magnitude. For example, in a 3% (w/v) lysozyme solution (5% NaCl, pH 4.2, $T = 20°$), the time to observe crystals was reduced from hours to minutes when placed in a standard laboratory (50 W) ultrasonic bath. Furthermore, the number of crystals produced in an ultrasonically treated solution was far in excess of the number observed in an untreated solution. The mechanism of this phenomenon is at present not understood. It is not even clear which of the several times (e.g., τ_{QE}, τ_D, or the postnucleation growth rate) is affected by the ultrasound. However, with the numerous assays available and discussed in this work, this question should be answerable and is now under investigation.

One possible mechanism that in principle could account for the acoustic effect is the following: the sound establishes standing pressure waves in the crystallizing container. A particle exposed to the pressure field will experience a force proportional to the pressure gradient (opposite sides of the particles are subjected to different forces). This will propel the particle to pressure nodes where the gradient is zero. Thus molecules will tend to bunch at the nodes. Since τ_D depends strongly on concentration (see previous section), nucleation may start at an accelerated rate at the nodes.

To test the above hypothesis, we produced a standing wave with a piezoelectric transducer ($\nu = 100$ kHz) attached to a glass tube.[54] When polystyrene spheres of varying sizes (radius 1–10 μm) were introduced, they were visually seen to aggregate in sharp bands at the nodes.[55] The time course of aggregation was measured by monitoring the light absorbed in a beam perpendicular to the axis of the tube. The experimental results could be fitted with the relation

$$\tau^{-1} \approx 10^5 r^2 \tag{25}$$

where τ is the characteristic time and r is the radius of the sphere in centimeters. A theoretical calculation of τ^{-1} is in approximate agreement with Eq. (25).[56]

Unfortunately, when one extrapolates Eq. (25) to macromolecular dimensions, i.e., $r \approx 100$ Å, the times become prohibitively long ($\tau \approx 10^7$ sec). Thus, if the extrapolation of three orders of magnitude is valid, the

[54] Z. Kam, O. Nascimiento, and G. Feher, unpublished results.

[55] This experiment was repeated with red blood cells which quickly separated into sharp bands. This effect may be useful in concentrating particles in lieu of centrifugation.

[56] M. Weissman, Z. Kam, and G. Feber, unpublished results.

simple mechanism described above cannot account for the observed acoustic effect. Another possibility that needs to be investigated is whether the acoustic field breaks up crystals.[56a] This could explain the increased number of crystals but not the accelerated growth.

IV. Cessation of Growth

It is found experimentally that crystals cease to grow after a long time interval even if the protein concentration in the supernatant solution is increased. This means that the mechanism of crystal growth described in the previous section does not apply for very long times, nor can the free energy plot of Figs. 1 and 2 be extrapolated to inifinite j values.

The cessation of growth is of great practical importance since a minimum-size crystal is required to obtain usable diffraction data. We have performed several experiments in an attempt to explain this phenomenon.[2,3] Our preliminary findings on lysozyme can be summarized as follows:

1. Crystals grown under the same set of experimental conditions reached approximately the same terminal size.
2. The faster the crystals grew, the larger their number and the smaller their terminal size. This effect was particularly pronounced when the solution was mechanically agitated or when an external ultrasonic field was applied.
3. When the surfaces of terminal-size crystals were dissolved (in a protein-free buffer), they did not grow beyond their terminal size when subjected to the original growth conditions.
4. When terminal-size crystals (see A and B in Fig. 15) were cut into pieces (A_1, A_2 and B_1, B_2 in Fig. 15) which were spatially separated, each of them grew to approximately the same original terminal size with the regrowth occurring on the freshly exposed surfaces.

What is the mechanism responsible for the cessation of growth? Presumably, the phenomenon is associated with changes of some property of the crystal surface. But what these changes are and what causes them is at present not understood. One possibility is that during crystal growth, errors (either chemical impurities or structural defects) are incorporated until they accumulate to such an extent that further aggregation becomes energetically unfavorable. Consistent with the above hypothesis are the rather remarkable results that pieces broken off from a terminal-size crys-

[56a] We have recently observed with a light microscope that ultrasound breaks up small (5–10 μm) crystals or dislodges them from container walls. The increase in the number of crystals caused by a 30-sec sonication at different times after solution preparation correlated with the number of visible crystals in the unsonicated solutions (S. D. Durbin and G. Feher, unpublished results).

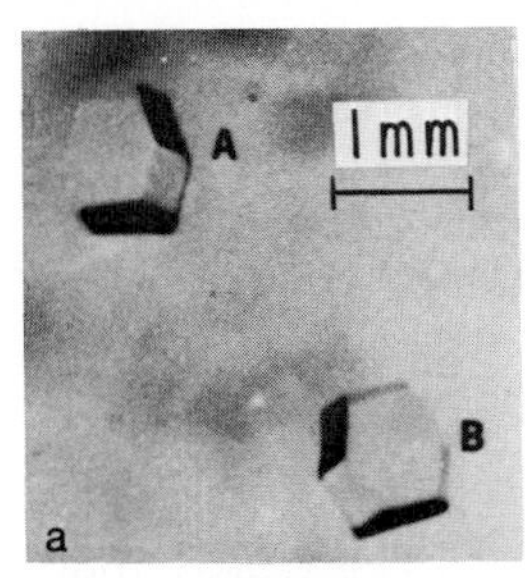

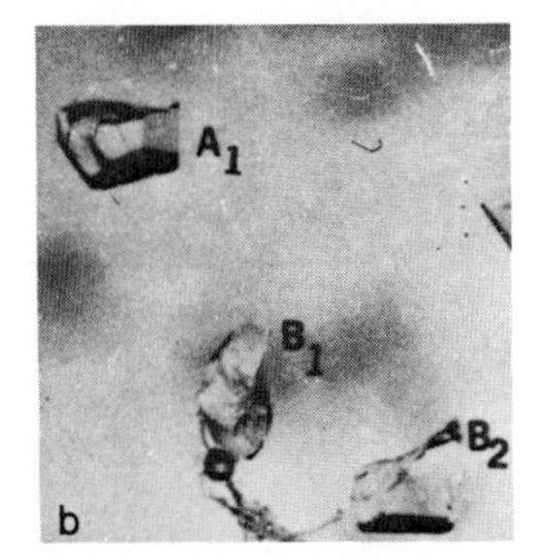

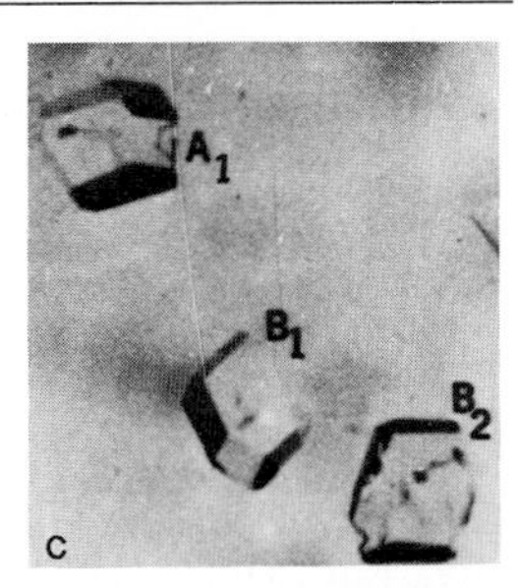

FIG. 15. Regrowth of crystal fragments after cutting terminal-size crystals. Growth conditions: $T = 20°$, 5% (w/v) NaCl, 4.5% (w/v) lysozyme, pH 4.2. After cutting the crystals, they were washed with a freshly made solution of 3.5% (w/v) lysozyme, 5% (w/v) NaCl and regrown in this solution. (a) Terminal size; (b) cleaved; (c) regrown. (From Kam *et al.*[2])

terminal size. Similarly, pure proteins usually yield larger terminal-size crystals than impure ones.

A direct way to test the error hypothesis is to measure the disorder and to correlate it with the size of the crystal. Shaikevitch and Kam[57] studied the long-range order with a specially designed low-angle X-ray diffraction camera. They found the disorder (called mosaic spread) to be very small (0.3 mrad) and independent of the position on the crystal and on the crystal size. Thus, *long-range* order does not seem to be the relevant parameter. It would be instructive, therefore, to study the *short-range* order as function of crystal size. One possible approach is to utilize the unusual electron paramagnetic resonance (EPR) linewidth behavior observed in metmyoglobin, in particular the dependence of the linewidth ΔH on the angle between the external magnetic field and the crystal axis. This dependence has been explained by a misorientation of the individual proteins in the crystal.[58] The linewidth should, therefore, be sensitive to short-range disorder. Another approach to investigate the error hypothesis would be to introduce controlled amounts of errors and correlate them with the terminal size. This could be accomplished by co-crystallizing two proteins or modifying a fraction of the protein molecules by chemical means, denaturation, or association with substrates.

In concluding this section, it is worth noting that the question of terminal size is a recurring theme in biology. Many structures (e.g., virus tails, cilia, tubules) are assembled by an aggregation process until a terminal

[57] A. Shaikevitch and Z. Kam, *Acta Crystallogr., Sect. A* **A37,** 871 (1981).

[58] P. Eisenberger and P. S. Pershan, *J. Chem. Phys.* **47,** 3327 (1967); G. A. Helcke, D. J. E. Ingram, and E. F. Slade, *Proc. R. Soc. London, Ser. B* **169,** 275 (1968); R. Calvo and G. Bemski, *J. Chem. Phys.* **64,** 2264 (1976); A. S. Brill and D. A. Hampton, *Biophys. J.* **25,** 313 (1979).

size is reached. The incorporation of error hypothesis also forms the basis for a model of aging.[59]

V. Summary and Discussion

We have presented a systematic approach to investigate the mechanisms of crystallization of proteins and have developed a simple theoretical framework to guide and interpret the experiments. A protein that can be easily crystallized (i.e., lysozyme) was chosen as a model system. Many of the results, however, should apply not only to proteins but also to other macromolecules (e.g., nucleic acids). From a broader perspective, crystallization may be viewed as a specific molecular recognition event repeated 10^{13}–10^{16} times which results in an ordered crystal lattice. It, therefore, represents a simple and fundamental model for more complicated recognition and assembly processes, such as those that occur in the construction of ribosomes, microtubules, multisubunit enzymes, etc. Thus, knowledge of the mechanisms of crystal growth may not only lead to more effective crystallization procedures, but may also contribute to other fields of biology.

During the growth of the crystals, different, temporally distinct processes become important and were investigated with different experimental techniques. The homogeneous *nucleation process* was investigated with quasi-elastic light scattering. The equilibrium constants of aggregation (K_1 and K_∞) (see Figs. 8 and 9) were determined and used with the aid of a simple model to predict whether one is approaching *crystallization* or the formation of *amorphous precipitates*. These measurements were performed during the prenucleation stage, i.e., at quasi-equilibrium, which is established very rapidly. Thus, the advantage of this approach is that one does not have to wait until visible crystals are formed. The kinetics of establishing quasi-equilibrium were determined by relaxation and correlation techniques (see Figs. 10 and 11).

Stable, *postcritical nuclei* are formed from the pool of aggregates by some aggregates traversing the potential barrier to reach a certain critical size. The time it takes to reach this point can be several orders of magnitude longer than it takes to establish quasi-equilibrium (see Fig. 4). The number of postcritical nuclei was assayed by a straightforward seeding technique (see Fig. 12).

The *growth of the crystals* was obtained from the protein concentration profile in the vicinity of the crystals (see Figs. 13 and 14). An analysis of these data led to an estimate of the dimer dissociation constant k_2^b which forms the basic time scale for the theoretical description. In addi-

[59] L. E. Orgel, *Proc. Natl. Acad. Sci. U.S.A.* **49**, 517 (1963).

tion, the average probability of attachment, p, was obtained and found to be several orders of magnitude smaller than that observed with small molecules. The practical importance of the reduction in protein concentration in the vicinity of growing crystals was discussed.

The mechanisms leading to terminal-size crystals, i.e., the phenomenon of the *cessation of growth*, although of great practical importance to crystallographers, is not understood. A hypothesis was advanced that is based on the incorporation of errors that leads to a "poisoning" of the surface. Several findings (e.g., see Fig. 15) were consistent with this hypothesis. Clearly, more work needs to be done to elucidate this important phase.

We conclude with a discussion of a general strategy for growing protein crystals based on our present understanding of the crystallization processes. One has to distinguish between proteins that have never been successfully crystallized from those that form microcrystals too small to be used for diffraction measurements. In the case of proteins that have resisted crystallization, one needs to focus on the nucleation process. This is accomplished by performing a set of quasi-elastic light scattering experiments and measuring the ratio K_∞/K_1, as described in this chapter. The larger this ratio, the more favorable the nucleation of crystals as compared to forming an amorphous precipitate. Thus, one endeavors to maximize K_∞/K_1 by varying external parameters, e.g., temperature, salt concentration, pH. Note that in the past one concentrated on measuring the solubility of proteins, which is equivalent of measuring only K_∞. But it is *the ratio* of K_∞/K_1 that distinguishes crystallization from amorphous precipitation.

Once nucleation has occurred, it is likely that microcrystals will be formed. This process can be accelerated by agitation or perhaps by exposure to an acoustic field. The conditions for producing large crystals are usually different from those required for nucleation. The crystals should be grown slowly by choosing a low degree of supersaturation. In addition, it is important that the crystals be grown in a vibration-free environment. This eliminates stirring, thereby preventing the growth of "parasitic" crystals in the protein-depleted region around a growing microcrystal (see Fig. 4). Slow growth of crystals also seems to favor the production of large terminal-size crystals. Similar considerations also apply when *seed crystals* are being used.

The ultimate goal is to understand the crystallization processes and to provide a prescription for growing better crystals and for crystallizing proteins that have so far resisted attempts to be crystallized. Although this goal has not been reached yet, a start has been made in the elucidation of some of the mechanisms of crystallization and in proposing assays and strategies for the optimization of crystallization conditions.

Acknowledgments

We thank H. Shore for his guidance and help with the theory and for calculating the curve of Fig. 4 and S. D. Durbin for the results shown in Fig. 12. We are grateful to them as well as A. Shaikevitch and M. Weissman for permission to quote their unpublished results. Helpful discussions with these collaborators as well as with J. Allen, J. Kraut, H. Reiss, L. A. Steiner, N. H. Xuong, and B. Zimm are gratefully acknowledged. Special thanks are due to R. A. Isaacson for his expert technical help. The majority of the work reported here was supported by a grant from the National Institute of Health (GM 13191).

[5] Crystallization of Macromolecules: General Principles

By ALEXANDER MCPHERSON

In principle, the crystallization of a protein, nucleic acid, or virus is little different than the crystallization of conventional small molecules. It requires the gradual creation of a supersaturated solution of the macromolecule and follows the spontaneous formation of crystal growth centers or nuclei. Once growth has begun, the emphasis focuses on the maintenance of essentially invariant conditions so as to sustain the continued, ordered addition of single molecules, or perhaps ordered aggregates, to the surface of the developing crystal. Under these nonequilibrium conditions of supersaturation, the system is driven toward a final state in which the solute is partitioned between a soluble and solid phase. Although the individual macromolecules lose rotation and translational freedom, thereby lowering the entropy of the system, they, at the same time, form many new, stable chemical bonds. This reduces the potential, or free energy, of the system and provides the motivation for the self-ordering process.

The primary differences that arise in the crystallization of macromolecules, in comparison with conventional small molecules, are due to the inherently greater lability or sensitivity of proteins and nucleic acids, and the necessity of maintaining them in an essentially hydrated state at or near physiological pH and temperature. Thus the usual methods of evaporation, dramatic temperature variation, or the addition of strong organic solvents are precluded and must be supplanted with more gentle and restricted techniques.

The most common approach to crystallizing macromolecules is to alter gradually the characteristics of a highly concentrated protein solution and thus achieve a number of objectives. The first is to deprive the protein molecules of an adequate quantity of water molecules to maintain their complete hydration, or to disrupt at least substantially their hydration shell. At the same time, it is desirable to decrease the dielectric constant of the media to reduce the effective electrostatic shielding between the macromolecules. Both of these effects tend to induce phase separation and compel the system to assume a new free energy minimum. This is characterized by the maximization of attractive interactions—change, steric, hydrophobic—and the minimization of dispersive or repulsive interactions. This result can be encouraged as well by modification of the solvent system, as by the addition of polyethylene glycol, for example, in such a way that the protein molecules tend to be excluded from the bulk solvent. A final objective is to alter the system in such a way that specific bonding interactions between macromolecules are promoted or stabilized, and this generally depends on the properties of the specific protein or nucleic acid being crystallized.

The strategy employed to bring about crystallization is to guide the system very slowly toward a state of minimum solubility by modifying the properties of the solvent through equilibration with precipitating agents or by altering some physical property, such as pH, thus achieving a limited degree of supersaturation (see Fig. 1). In extremely concentrated solutions the molecules may aggregate as an amorphous precipitate, a state to be avoided if possible and one indicative that saturation has proceeded too extensively or too rapidly. One must endeavor to approach very slowly the point of inadequate solvation and thereby allow the macromolecules sufficient opportunity to order themselves in a crystalline lattice (see Fig. 2). At the same time, the component variables of the system must be initially set or gradually modified to ensure that the macromolecules will take advantage of the greatest number of favorable interactions with neighbors.

Since a fundamental tenet of crystallization is absolute homogeneity, an important component is the stabilization of the macromolecules into a single population of inflexible and, if possible, compact individuals. This is accomplished by the addition of small molecules that stabilize or interact with the macromolecules and induce intermolecular contacts or by physical and chemical modification of the macromolecule itself. The promotion of crystal formation and growth, therefore, consists of three elements. One must choose the precipitating agent to be employed to modify the system and produce supersaturation, the means by which equilibrium is to be established, and the prevailing conditions at equilibrium.

The behavior of macromolecules in solution is complex and rather unpredictable owing to their varied shapes, polyvalent surface character, and dynamic properties. They can demonstrate a number of distinct solu-

A
B
C
D
E
F
G
H
I
J
K

bility minima. These depend on the nature of the electrolyte, the precipitant, their concentrations, the concentration of the macromolecule, pH, temperature, and a variety of other influences. This multiminima behavior is demonstrated most strikingly by the polymorphism of crystal forms, such as those shown in Fig. 3, exhibited by many proteins and nucleic acids. Many crystalline proteins can be made to grow in at least several different unit cells. For this reason, the methods described in Chapters [6] and [7] should be applied under a large number of sets of conditions in the quest to discover that particular minimum (or minima) which yield crystals. One should, for example, determine the precipitation points of the macromolecule at sequential pH values with a given precipitant, repeat this procedure at different temperatures, and then examine the effects of different precipitating agents.

It might be noted that once a solubility minimum is located, then the approach to supersaturation can be made in basically two ways. In the more common case, the precipitant level is gradually increased until the minimum is reached. A more subtle technique, and one that has been used to grow many crystals, is to increase the precipitant concentration beyond the normal precipitation point of the protein, but at a temperature or pH that maintains the macromolecule soluble. This second variable is then allowed to slowly relax to the desired value and produce supersaturation.

Precipitants fall into three main categories: salts, such as ammonium sulfate; organic solvents, such as ethanol or methylpentanediol; and the polyethylene glycols. For salts, the solubility function for a protein is logarithmic and exponentially decreases as the ionic strength is increased. The rate is a function of the protein and ions involved, and the efficiency of a particular salt compound is proportional to the number of charges carried by its ions that can be involved in the binding of water molecules. Hence it is proportional to its ionic strength $I/2 = \Sigma mz^2$, where m is the concentration of each ion in solution and z its valence. For this reason divalent and trivalent ions, such as sulfate and phosphate, are most commonly used; others are found in Table I.

FIG. 1. Some protein crystals grown by a variety of techniques and using a number of different precipitating agents. They are (A) deer catalase, (B) trigonal form fructose-1,6-diphosphatase from chicken liver, (C) cortisol binding protein from guinea pig sera, (D) concanavalin B from jack beans, (E) beef liver catalase, (F) an unknown protein from pineapples, (G) orthorhombic form of the elongation factor Tu from *Escherichia coli*, (H) hexagonal and cubic crystals of yeast phenylalanine tRNA, (I) monoclinic laths of the gene 5 DNA unwinding protein from bacteriophage fd, (J) chicken muscle glycerol-3-phosphate dehydrogenase, and (K) orthorhombic crystals of canavalin from jack beans.

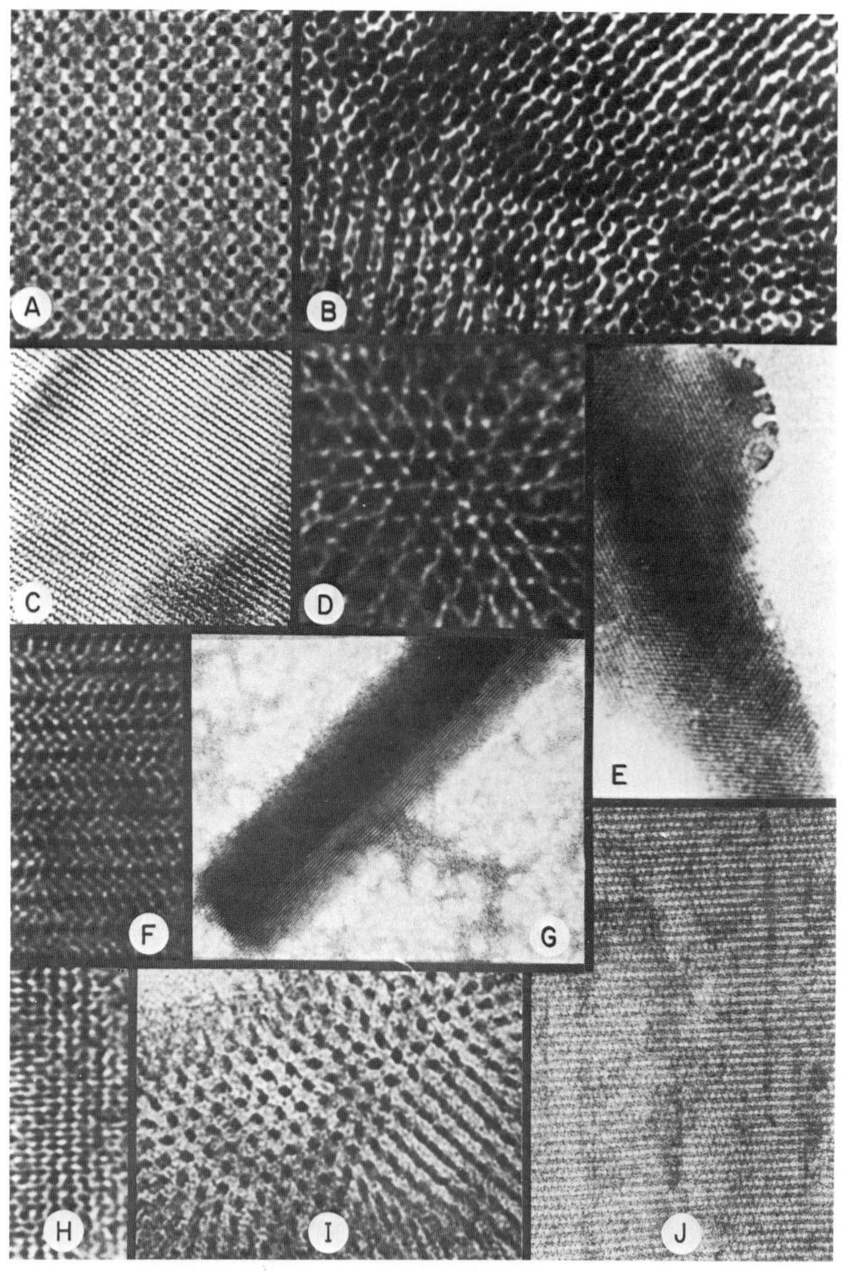

FIG. 2. A collage of electron micrographs of a variety of negatively stained protein microcrystals. They are (A) orthorhombic canavalin from jack bean, (B) the $P2_12_12$ crystal form of pancreatic α-amylase, and (C) the $P2_12_12_1$ form of the same molecule, (D) hexagonal canavalin, (E) the lectin from *Abrus precatorius,* (F) rhombohedral canavalin, (G) *Bacillus subtilis* α-amylase, (H) hexagonal concanavalin B, (I) hexagonal beef liver catalase, and (J) the orthorhombic form of beef liver catalase.

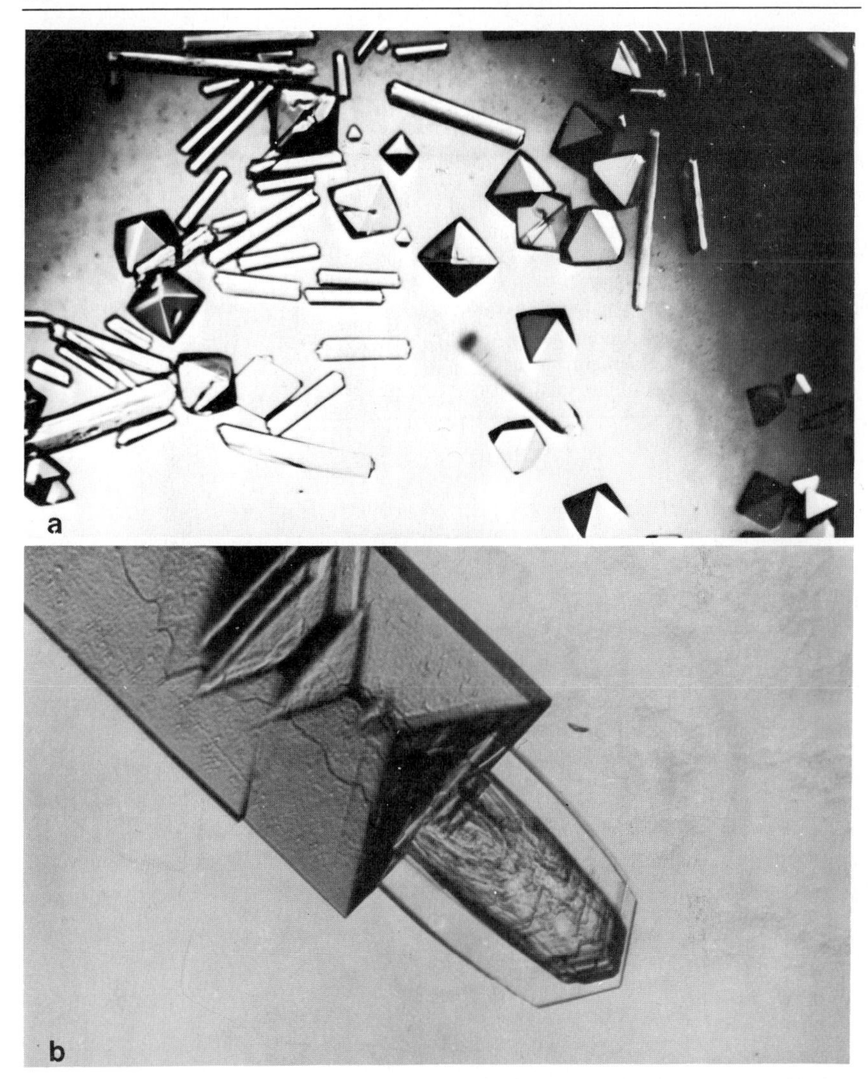

FIG. 3. (a) A single crystallization sample contains both the hexagonal and the cubic form of yeast phenylalanine tRNA crystals; (b) an example of two different orthorhombic forms of pig pancreatic α-amylase growing as twins.

Organic solvents to some extent also bind water molecules and distract them from interacting with protein. They also significantly lower the dielectric constant of the medium. The first of these effects decreases the capacity of the system to fully solvate the macromolecules and the second reduces the effective electrostatic shielding. Thus a reasonably good mea-

TABLE I
COMPOUND USED IN CRYSTALLIZATION

Ammonium or sodium sulfate
Sodium or ammonium citrate
Sodium or potassium or ammonium chloride
Sodium or ammonium acetate
Magnesium sulfate
Cetyltrimethyl ammonium salts
Polyethylene glycol 1000, 4000, 6000, 15,000
Calcium chloride
Ammonium nitrate
Sodium formate
Lithium chloride

sure of the efficiency of a particular solvent for forcing proteins and nucleic acids from solution is the solvent's dielectric constant. Those solvents which have proven of value in crystallizing proteins and nucleic acids are seen in Table II.

The polyethylene glycols which depend primarily on exclusion principles are treated in more detail in the following chapter, [6].

The specific conditions prevailing at supersaturation that best guarantee the macromolecule's integrity and the greatest degree of crystal probability depend almost entirely on the particular protein under investigation. They must, in general, be painstakingly sought by trial and error experiments. Many of the factors that appear to influence crystallization are poorly understood and the bearing of a specific parameter on the problem may only be realized after a great number of trials have been

TABLE II
ORGANIC SOLVENTS USED IN CRYSTALLIZATION

Ethanol
Isopropanol
2-Methyl-2,4-pentanediol (MPD)
Dioxane
Acetone
Butanol
Dimethyl sulfoxide
2,5-Hexanediol
Methanol
1,3-Propanediol
1,3-Butyrolactone

TABLE III
VARIABLES INFLUENCING MACROMOLECULE CRYSTALLIZATION

1. Concentration of precipitant
2. Concentration of macromolecule
3. Temperature
4. pH
5. Pressure
6. Level of reducing agent or oxidant
7. Substrates, coenzymes, and ligands
8. Purity of protein or nucleic acid
9. Preparation and storage of macromolecule
10. Proteolysis and fragmentation
11. Age of macromolecule
12. Degree of denaturation
13. Vibration and sound
14. Volume of crystallization sample
15. Metal ions
16. Seeding
17. Amorphous precipitate
18. Buffers
19. Cleanliness
20. Organism or species from which macromolecule was isolated
21. Gravity, gradients, and convection

completed and all variables carefully evaluated. It is therefore not possible to detail a rational set of rules pertaining to ingredients in the mother liquor or ambient conditions that can ensure, or even give likelihood, of success. The best that can be done at this point is to tabulate those variables for which experience has proved or indicated an influence on crystallization properties, and this is seen in Table III.

Finally, the best guidelines for the crystallization of proteins are still those put forth by the pioneers of the field, Sumner,[1] Kunitz and Northrop,[2] Bailey,[3] and Osborne.[4] Conditions must be chosen that do not denature or otherwise damage the macromolecule. Highly concentrated protein solutions of 10–100 mg/ml must be used and these must be made free of contaminants such as nucleic acids, polysaccharides, and other interfering substances. The protein should be as pure as possible and homogeneous in every way to maximize the probability of success. Some

[1] J. B. Sumner and G. F. Somers, "Chemistry and Methods of Enzymes," 2nd ed. Academic Press, New York, 1947.
[2] J. H. Northrop, M. Kunitz, and R. M. Herriott, "Crystalline Enzymes." Columbia Univ. Press, New York, 1948.
[3] K. Bailey, *Trans. Faraday Soc.* **38,** 186 (1942).
[4] T. B. Osborne, *Am. Chem. J.* **14,** 662 (1892).

TABLE IV
FACTORS CONTRIBUTING TO HETEROGENEITY

1. Presence, absence, or variation in a bound prosthetic group, coenzyme, or metal ion
2. Variation in the length or composition of the carbohydrate moiety on a glycoprotein
3. Proteolytic modification of the protein during the course of isolation
4. Oxidation of sulfhydryl groups during isolation
5. Reaction with heavy metal ions during isolation or storage
6. Presence, absence, or variation in posttranslational side chain modifications such as methylation, amidination, phosphorylation
7. Microheterogeneity in the amino or carboxyl terminus or modification of termini
8. Variation in the aggregation or oligomer state of the protein association/dissociation
9. Conformational instability due to the dynamic nature of the molecule
10. Microheterogeneity due to the contribution of multiple but nonidentical genes to the coding of the protein
11. Partial denaturation of sample
12. Different animals or preparations of enzyme sources

factors often responsible for heterogeneity[5] that are frequently overlooked are found in Table IV. There should be no dust or particulate matter present to provide unwanted nucleii; the solution should be filtered and centrifuged clear. The addition of precipitating agents and other components should proceed very slowly and amorphous precipitate should in general be avoided. Seeding is often successful if microcrystals are available, and the growth of large single crystals can proceed only in the absence of any perturbations or physical disturbances. The macromolecules should be stabilized by the addition of small effector molecules when appropriate, and a broad range of conditions must, in general, be examined.

[5] A. McPherson, "The Preparation and Analysis of Protein Crystals." Wiley, New York, 1982.

[6] Use of Polyethylene Glycol in the Crystallization of Macromolecules

By ALEXANDER MCPHERSON

Polyethylene glycol (PEG) is a polymer produced in various lengths, from several to many hundred units, exhibiting as its most conspicuous feature a regular alteration of ether oxygens and ethylene groups. A consequence is that it significantly perturbs the natural structure of water and replaces it with a more complex network having both water and itself as structural elements. One result of this restructuring of the solvent is that macromolecules, particularly proteins, tend to be excluded and phase separation is promoted. The mechanisms by which PEG induces proteins to crystallize are not fully understood. In addition to its volume exclusion property, it probably shares some characteristics with salts that compete for water and produce dehydration, and with organic solvents, such as methylpentanediol (MPD), which modify the dielectric properties of the medium. A study of the interaction between protein and PEG molecules has appeared that examines the thermodynamic properties of the system.[1] The authors conclude that extensive and generally unfavorable electrostatic interactions result from the introduction of proteins into a PEG solution. The instability reflecting these negative interactions is manifested in phase separation.

Aside from its general applicability and utility in obtaining crystals for diffraction analysis,[2] it also has the advantage that it is most effective at minimal ionic strength and it provides a low electron density medium. The first feature is important because it provides for higher ligand binding affinities than does a high ionic strength medium such as concentrated salt. As a consequence there is greater ease in obtaining isomorphous heavy-atom derivatives and in forming protein–ligand complexes for study by difference Fourier techniques. The second characteristic, a low electron-dense medium, implies a generally lower background or noise level for protein structures derived by X-ray diffraction and presumably, therefore, a more ready interpretation.

It might be noted in passing that a number of protein structures have now been solved using crystals grown from PEG. These along with several studies of a more preliminary nature tend to confirm that the protein molecules are in as native a condition in this medium as in those traditionally used. This is perhaps even more so, since the larger molecular weight PEGs probably do not even enter the crystals and therefore do not directly contact the interior molecules. In addition, it would seem that crystals of a specific protein when grown from PEG are essentially isomorphous with and exhibit the same unit cell symmetry and dimensions as those grown by conventional means.

Polyethylene glycol is produced in a variety of molecular weights. The low molecular weight species are oily liquids while those above 1000, at room temperature, exist as either a waxy solid or a powder. The latter are

[1] J. C. Lee and L. L. Y. Lee, *J. Biol. Chem.* **256,** 625 (1981).

[2] A. McPherson, *J. Biol. Chem.* **251,** 6300 (1976).

preferable for easy dissolution. The size with which it is labeled is the mean molecular weight of the polymeric molecules, and the distribution of weights about that mean may vary appreciably. It is certainly broad for the very high molecular weight species. The most popular sizes currently in use are 1000, 4000, 6000, and 20,000 with 4000 being the author's own personal choice for a first attempt. PEG in its commercial form does contain contaminants; this is particularly true of the high molecular weight forms such as 15,000 or 20,000. These may be removed by simple purification procedures or in the case of PEG 20,000 by dialysis in low-pass dialysis or collodian tubes. Although there have been no reports that repurified PEG has proven more effective, the contaminants could certainly be disadvantageous for some proteins.

All of the PEG sizes from 400 to 20,000 have successfully provided protein crystals, but the most useful are those in the range 2000–6000. There have appeared a number of cases, however, in which a protein could not be easily crystallized using this range but yielded in the presence of 400 or 20,000. The molecular weight sizes are generally not interchangeable for a given protein even within the mid range, some producing the best formed and largest crystals only at, say, 4000 and less perfect examples at other weights. This is a parameter which is best optimized by empirical means along with concentration and temperature. It might be noted that the very low molecular weight PEGs such as 200 and 400 are rather similar in character to MPD and may be substituted for MPD.

A correlation between the molecular weight of a protein under study and that of the PEG used for its crystallization has been suggested. The author, at least, has not found this necessarily to be true. It is certainly the case that the higher molecular weight PEGs have a proportionally greater ability to force proteins from solution, and really obstinate polypeptides might best be approached with PEG 20,000 at the outset. But aside from this, no correlation has been shown.

A very distinct advantage of polyethylene glycol over other agents is that most proteins (but not all) crystallize within a fairly narrow range of PEG concentration; this being from about 4 to 18%. In addition, the exact PEG concentration at which crystals form is rather insensitive and if one is within 2 or 3% (and sometimes much more) of the optimal value some success will be achieved. With most crystallizations from high ionic strength solutions or from organic solvents, one must be within 1 or 2% of an optimum lying anywhere between 15 and 85% saturation. The great advantage of PEG is that when conducting a series of initial trials to determine what conditions will give crystals, one can use a fairly coarse selection of concentrations and over a rather narrow total range. This means fewer trials with a corresponding reduction in the amount of protein expended. Thus it is well suited for particularly precious proteins of very limited availability.

The time required for crystal growth with PEG as the precipitant is also generally much shorter than with ammonium sulfate or MPD but occasionally longer than required by volatile organic solvents such as ethanol. Although equilibration times will depend on the differential between starting and target concentrations, if this is no more than 3 or 4% then crystallization may occur within a few hours or a few days. It seldom requires more than 3 weeks. Thus evaluation of results can be made without undue demands on patience. It should be noted that protein–PEG solutions are excellent media on which to grow microbes, particularly molds, and if crystallization is being attempted at room temperature or over extended periods of time, then some retardant such as azide (commonly 0.1%) must be included in the protein solutions.

Since PEG solutions are not volatile, PEG must be used like salt or MPD and equilibrated with the protein by dialysis, slow mixing, or vapor equilibration. This latter procedure, utilizing either 10-μl hanging drops over 0.5-ml reservoirs or 20-μl drops on multidepression glass plates in a sealed chamber, has proven the most popular. The author has found that when the reservoir concentration is in the range of 5–12%, the protein solution to be equilibrated should be at an initial concentration of about half. That is conveniently obtained by adding 10 μl of the reservoir to 10 μl of the protein solution. When the final PEG concentration to be obtained is much higher than 12%, it is advisable to start the protein equilibrating at no more than 4–5% below the final value. This reduces unnecessary time lags during which the protein might denature.

Crystallization of proteins with PEG has proved more successful when the ionic strength is low. It is quite difficult when ionic strength is high. The author commonly works at 10 to 40 m*M* Tris or cacodylate buffer. If crystallization proceeds too rapidly, addition of some neutral salt may be used to slow growth and better effect crystal form. PEGs are useful over the entire pH range and over a broad temperature range and show no anomalous effects in response to either. PEG appears to be an excellent crystallization agent over the whole spectrum of proteins, although in many specific cases other precipitants may be superior.

A common and useful microprocedure for the growth of protein crystals in association with polyethylene glycol as precipitant is vapor diffusion using hanging drops. It is ideally suited for screening a large number of conditions, particularly final precipitant concentrations, when only a small amount of material is available. It can, however, also be used to grow large, single crystals to be used for diffraction analysis.

With this approach, a microdroplet of mother liquor (as small as 5 μl)

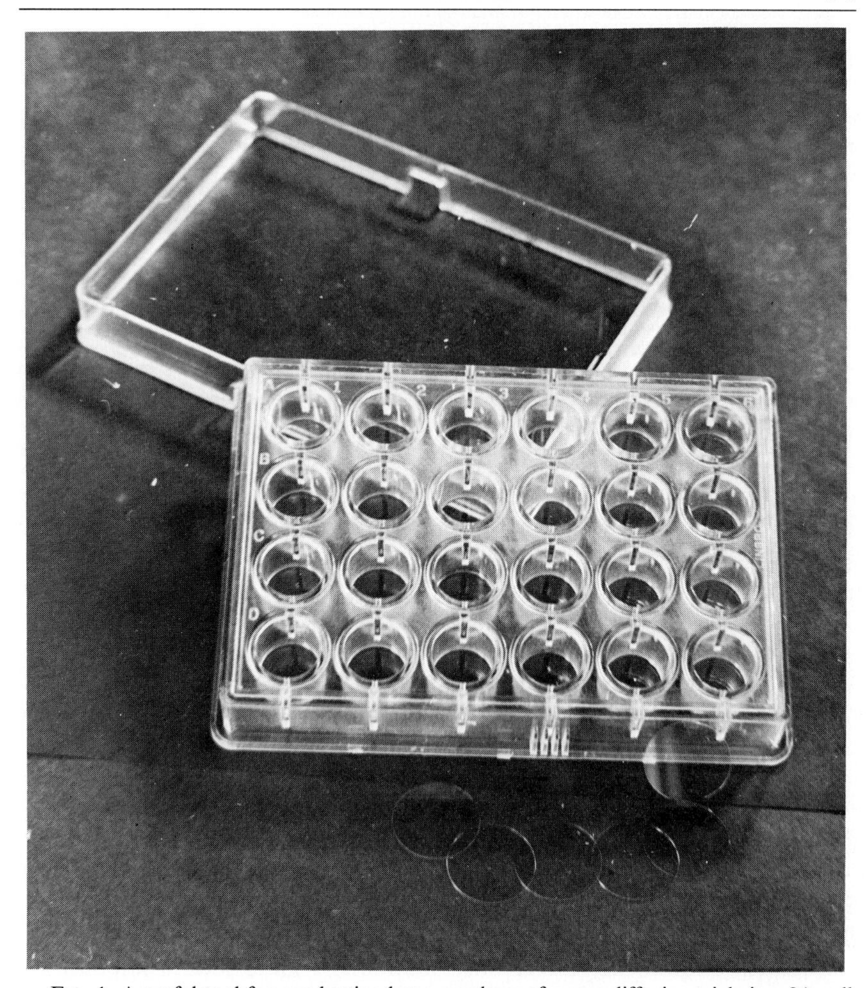

FIG. 1. A useful tool for conducting large numbers of vapor diffusion trials is a 24-well tissue culture tray of clear plastic that provides individual reservoirs over which can be placed microdrops of mother liquor hanging suspended from the undersides of siliconized circular cover slips.

is suspended from the underside of a microscope cover slip, which is then placed over a small well containing 0.5 ml of the precipitating solution. An important point is that the cover slips must be thoroughly and carefully coated with a silicone surface to ensure proper drop formation and prevent spreading. The wells are most conveniently supplied by disposable plastic tissue culture plates (Linbro model FB-16-24-TC) that have 24 wells (1 cm diameter × 2 cm deep) with flat ground rims that permit airtight sealing by application of a single drop of oil at one point on the circumference. These plates provide the further advantages that they can be swiftly and easily examined under a dissecting microscope and allow compact storage. The method utilizing the materials seen in Fig. 1 has been used for the growth of large single crystals. Although it is primarily useful as a means of establishing optimum conditions, it has been successfully applied to the crystallization of tyrosine transfer RNA synthetase[3] and the Tu elongation factor from *Escherichia coli*.[4]

As with other vapor diffusion techniques, the microdroplet comes into equilibrium with the reservoir. Because the latter is several orders of magnitude larger, the final concentration of the precipitant in the drop reaches the essentially constant concentration of the reservoir. If the precipitant is a volatile solvent such as ethanol, then none need additionally be added to the protein sample. In the case of salts or polyethylene glycol, however, the protein sample must contain some initial level, and this is usually about half the desired final concentration. One advantage that the hanging drop method has over other vapor equilibration procedures is that crystals seldom grow from the glass surface, but more often on the liquid surface. Thus they are easier in some cases to retrieve for analysis. Crystals grown in this manner can be mounted in the conventional way if the cover slip containing the crystals suspended in the droplet is simply inverted with a pair of forceps.

[3] B. R. Reid, G. L. E. Koch, Y. Boulanger, B. S. Hartley, and D. M. Blow, *J. Mol. Biol.* **80**, 199 (1973).

[4] F. A. Jurnak, A. McPherson, A. H. J. Wang, and A. Rich, *J. Biol. Chem.* **255**, 6751 (1980).

[7] Crystallization of Proteins by Variation of pH or Temperature

By ALEXANDER MCPHERSON

Because individual proteins display distinct solubilities at varying levels of salt concentration, they can be selectively separated from the solvent as amorphous precipitate or, preferably, as crystals. The precipitation points or solubility minima frequently depend critically on the pH, temperature, and other prevailing conditions. At very low ionic strengths a phenomenon known as "salting-in" occurs in which the solubility of the protein increases as the ionic strength increases. In this range the cations are very important, and the solubility increase occurs because of a decrease in the activity coefficient of the protein.[1] This salting-in effect, when applied in reverse, can be used as a crystallization tool. In practice, one simply dialyzes a protein that is soluble at high ionic strength against distilled water or a low ionic strength buffer to remove ions.

The "salting-out" effect, which Hofmeister[2] suggested to be caused by a reduction of the chemical activity of water by salt and which involves principally the anions, occurs as the ionic strength increases. Since ionic strength is a function of the second power of the valence, multivalent ions are most efficient. In the salting-out region of a protein's solubility profile, the logarithm of the protein solubility as a function of the ionic strength is linear and can be expressed by $\log S = \beta - K_s(I/2)$, where I is the ionic strength in moles per kilogram of water, S is the solubility of the protein in grams per kilogram of water, and β and K_s are constant.[3] K_s serves as a measure of the slope of the solubility extrapolated to zero ionic strength and β is the logarithm of the solubility at that point.

In principle, the precipitation or crystallization of a protein is effected by manipulating the factors that occur in the above equation. One can crystallize a macromolecule by increasing the ionic strength at constant pH and temperature, a K_s fractionation procedure, or it can be brought about at constant ionic strength by manipulating the pH and temperature, which in most cases strongly influence β, a process of β separation. Crystallization of proteins by variations and shifts of the pH or temperature are, therefore, ideal examples of this latter process.

As with other techniques, the objective is to produce a supersaturated protein solution by very gradual and nonperturbing adjustment of conditions. Since both pH and temperature can be accurately and sensitively altered without direct contact with the protein solution, they provide an attractive approach. In practice, a solubility minimum is determined for the protein at some specific pH, temperature, and ionic strength. The protein solution is set initially at the requisite salt concentration for this solubility minimum but either the pH or temperature is made significantly different so as to maintain protein solubility. The temperature is then slowly allowed to relax to a value that minimizes the protein solubility, or the pH is allowed to slowly equilibrate via the liquid or vapor phase to its appropriate value.

[1] M. Dixon and E. C. Webb, *Adv. Protein Chem.* **16,** 197 (1961).

[2] T. Hofmeister, *Naunyn-Schmiedeberg's Arch. Exp. Pathol. Pharmakol.* **24,** 274 (1887).

[3] R. Czok and T. Bücher, *Adv. Protein Chem.* **15,** 315 (1960).

In the case of temperature shifts, the protein is generally raised to a higher temperature as is done, for example, with α-amylase,[4] glucagon,[5] and excelsin or edestin,[6] a process which increases its solubility. The solution is then allowed to cool slowly to a point at which it is no longer soluble. This can be accomplished by placing the warm solution in an insulated or thermally controlled box, or by placing the sample in a Dewar flask. Because of the thermal stability, the approach to supersaturation occurs at an almost imperceptible rate. Exactly the same technique can be applied if the protein is more soluble at cold temperature than warm temperature. Under these circumstances, the protein solution is cooled under a given set of ionic strength conditions and then allowed to warm very slowly. This was used, for example, to grow crystals of DNase by Kunitz[7] and is the basis of Jacobi's sequential extraction procedure using decreasing concentrations of ammonium sulfate.[8]

In the case of pH shift crystallization, the protein, buffered at the final pH and made essentially insoluble by an appropriate level of salt or other precipitant, is raised to a high pH of about 9.0 with ammonium hydroxide to achieve solubilization. Alternatively, it may in some cases be resolubilized by the addition of acetic acid, which lowers the pH to the region of 3.0–4.0. In the first case, NH_3 is then allowed to leave the sample by the vapor phase; in the second case, acetic acid directly evaporates from the sample. Essentially by vapor equilibration, the protein solution returns to neutrality, or whatever final point was chosen by the included buffer. The protein becomes insoluble and, hopefully, crystallizes.

This procedure has been used successfully to grow crystals of a number of proteins including canavalin,[9] concanavalin B,[10] and excelsin, to name only a few. The technique has a number of variations; it can be used in association with direct liquid dialysis or with vapor diffusion methods as well. It can, furthermore, be employed in conjunction with a host of different precipitating agents.

[4] A. McPherson and A. Rich, *Biochim. Biophys. Acta* **285,** 493 (1972).
[5] M. V. King, *J. Mol. Biol.* **11,** 549 (1965).
[6] K. Bailey, *Trans. Faraday Soc.* **38,** 186 (1942).
[7] M. Kunitz, *J. Gen. Physiol.* **33,** 349 and 363 (1950).
[8] W. B. Jacobi, this series, Vol. 11.
[9] A. McPherson and R. Spencer, *Arch. Biochem. Biophys.* **169,** 650 (1975).
[10] A. McPherson, J. Geller, and A. Rich, *Biochem. Biophys. Res. Commun.* **57,** 494 (1974).

[8] Crystallization in Capillary Tubes

By GEORGE N. PHILLIPS, JR.

A variety of techniques involve the use of capillary tubes. These methods have the tremendous advantage of requiring only small amounts of material (5–50 μl per experiment) and yet have been successful in the crystallization of many proteins. Because of the small volume requirements, capillary crystallizations are especially well suited to initial surveys of conditions. The technique involves setting up a gradient of precipitant in the capillary, which slowly equilibrates to yield a supersaturated solution. Either dialysis across a membrane or free interface diffusion can be used to achieve equilibration. The same principles have been applied for growing crystals within thin-walled glass or quartz "X-ray" capillary tubes, so that the crystals can be put in the X-ray beam without transfers or other manipulations.

Microdialysis cells using capillary tubes were first described by Zeppezauer.[1,2] These cells employ a thick-walled glass or plastic capillary 30–50 mm long with a ring of tubing holding a cellulose dialysis membrane over the bottom of the capillaries (Fig. 1a).[3] The protein solution is placed in the capillary, while avoiding the introduction of bubbles. A Parafilm seal is placed over the top end. The cell is then placed in a larger container filled with the appropriate dialysate buffer. "Feet" cut on the ring of tubing on the bottom end of the capillary support the cell, and allow access of the buffer solution to the dialysis membrane. The precipitant concentration, pH, ionic strength, etc. can be incrementally changed without disturbing the cell, so that the technique is very convenient when no prior information is available on crystallizating conditions.

A variation of the Zeppezauer cell that allows continuous increase in precipitant concentration over several weeks has been described by Weber and Goodkin.[4] In this set-up, a long capillary "stem" is placed under the Zeppezauer cell (Fig. 2). The rate of diffusion of precipitant from the reservoir into the protein cell is thereby slowed considerably. This technique reduces the day-to-day tedium of incrementing the precipitant con-

[1] M. Zeppezauer, H. Eklund, and E. S. Zeppezauer, *Arch. Biochem. Biophys.* **126,** 564 (1968).

[2] M. Zeppezauer, this series, Vol. 22, p. 253.

[3] A. McPherson, Jr., *Methods Biochem. Anal.* **23,** 249 (1976).

[4] B. H. Weber and P. E. Goodkin, *Arch Biochem. Biophys.* **141,** 489 (1970).

METHODS IN ENZYMOLOGY, VOL. 114

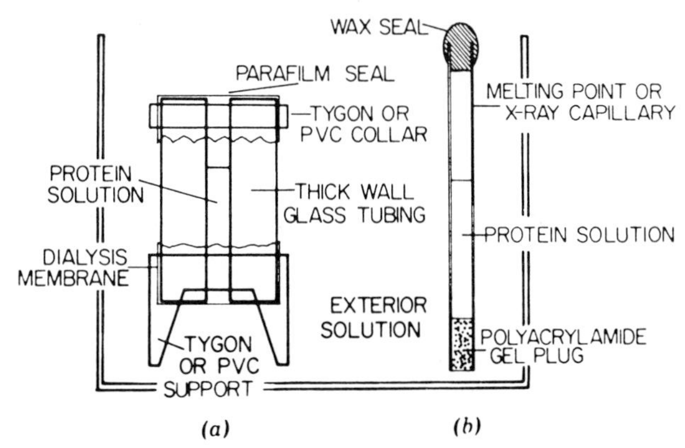

FIG. 1. (a) Thick-walled capillary microdialysis in cell as described by Zeppezauer.[1] (b) Adaptation using a thin-walled capillary. (Drawing reprinted with permission from McPherson.[3] © 1976 John Wiley & Sons, Inc.)

centration, but also results in some uncertainty about the conditions within the cell at any particular time.

A minor difficulty with both the Zeppezauer cell and its modifications is that the refractive properties of the thick-walled capillaries make it

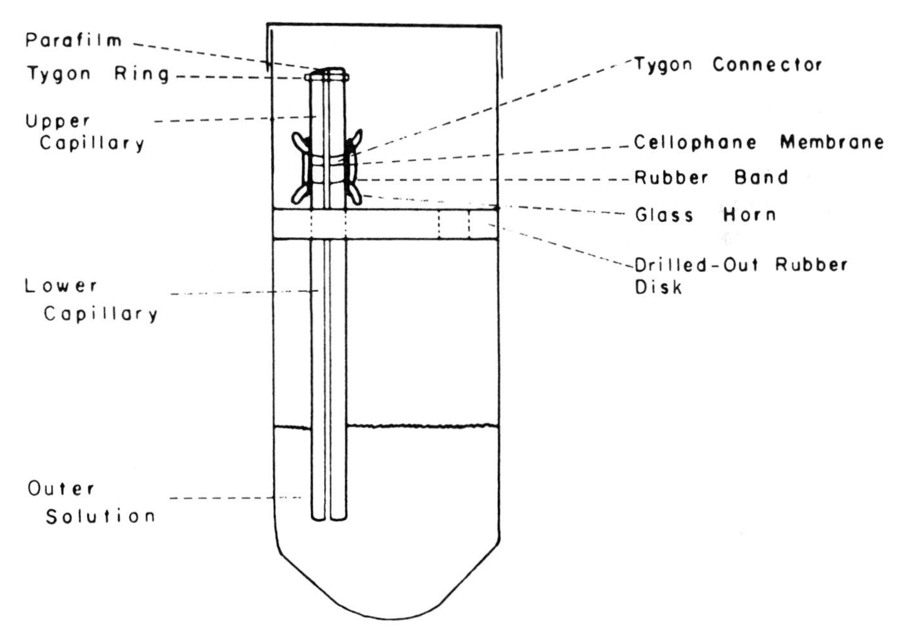

FIG. 2. Diagram of a modified capillary cell with a "stem" to slow the diffusion process. (Drawing reprinted with permission from Weber and Goodkin.[4])

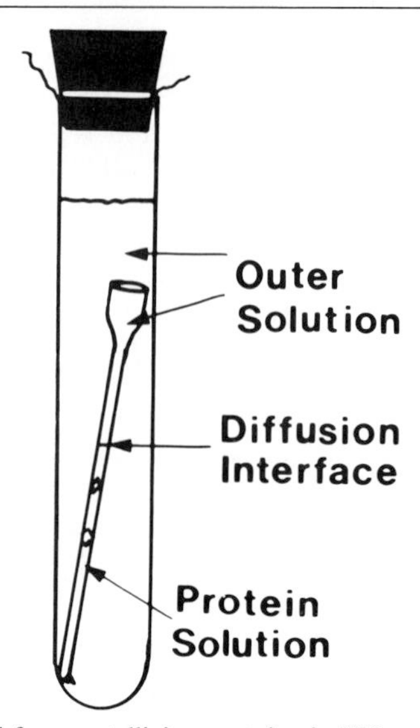

FIG. 3. Method for crystallizing proteins in "X-ray" capillaries.

difficult to recognize microcrystals. Removal of large crystals without damage can also be difficult. Nevertheless, these cells are popular for initial surveys, and have been used to crystallize a number of proteins.

On occasion it has been found to be advantageous to grow crystals directly in thin-walled glass or quartz X-ray capillaries. Such situations can arise if the crystals are so fragile that transfers are difficult, or if problems of crystal slippage during date collection cannot be overcome by other methods (see section on Crystal Mounting in Rayment, this volume [10]). Zeppezauer *et al.*[1] have shown how to adapt equilibrium dialysis to X-ray capillaries. Because membranes cannot be used on the fragile thin-walled capillaries, a short polyacrylamide plug has been substituted as a barrier (Fig. 1b). The composition of the polyacrylamide must be designed to meet the requirements of the particular macromolecule under study and the buffers being used.

A simpler method has been used to grow crystals of tropomyosin in X-ray capillaries.[5] These crystals contain 95% solvent and are mechanically and chemically labile, necessitating their growth in capillaries in order to avoid all manipulations. Quartz, 1-mm capillary tubes (Charles Supper Co., Natick, Massachusetts) are acid washed and sealed by flame on their

[5] D. L. D. Caspar, C. Cohen, and W. Longley, *J. Mol. Biol.* **41**, 87 (1969).

narrow ends. Approximately 10–25 μl of protein solution is placed in the bottom of the capillary using a disposable glass pipet which has been heated and "drawn out" to produce a fine bore. The precipitating solution, in this case one of slightly lower pH than the protein solution, is then layered carefully on top, being careful to avoid mixing. The open capillary is then immersed in a test tube filled with more of the precipitating buffer (Fig. 3). Because the protein solution is more dense than the buffer solution and protein molecules diffuse more slowly than small ions, crystallization takes place rather than diffusion of protein out of the capillaries. Good quality crystals grow in 30–60% of the capillaries within several weeks and are stable in the open capillaries for years. Amazingly, the crystals stick to or wedge themselves in the capillaries and no further treatment is necessary during X-ray diffraction data collection. If the precipitating solution were more dense than the protein solution, the method could be changed by putting the denser buffer in the capillary first, followed by the protein solution. In this arrangement, however, the capillary cannot be in contact with a large reservoir of precipitant buffer, and the final concentration of precipitant will depend on the initial volumes of the protein and precipitant solutions. A similar method of using free interface diffusion within glass pipet tips has also been described by Salemme[6] (see also this volume).

Akervall and Strandberg[7] have grown single crystals of satellite tobacco necrosis virus in X-ray capillaries by introducing a seed crystal, properly oriented, on a flattened place in a capillary. A solution containing the virus was replenished regularly until a crystal had grown sufficiently to wedge itself in place. Problems of transfer and subsequent slippage during data collection were thereby avoided.

For most structure determinations, batch crystallization or bulk dialysis techniques are preferred for growing large numbers of crystals. Individual crystals are then transferred to X-ray capillaries for data collection. The capillary methods described here, however, can be very useful for initial surveys of conditions and for other special requirements.

[6] F. R. Salemme, *Arch. Biochem. Biophys.* **151,** 533 (1972).

[7] K. Akervall and B. Strandberg, *J. Mol. Biol.* **62,** 625 (1971).

[9] Seed Enlargement and Repeated Seeding

By C. THALLER, G. EICHELE, L. H. WEAVER, E. WILSON, R. KARLSSON, and J. N. JANSONIUS

The previous chapters in this section provide a survey of the crystallization techniques presently available in macromolecular crystallography. If none of these methods give rise to sufficiently large single crystals, seeding should be tried. This method has been described[1,2] with reference to submicroscopic seeds. In that form it is sometimes successful, but suffers from two limitations. First of all it is not possible to control the number of seeds that are introduced into an experiment. Second, the crystals resulting from such a seeding experiment may be too small to be of any use. Therefore, seeding with macroscopic seeds is to be preferred, wherever possible. This technique has been used occasionally,[3-9] but only recently has it been suggested as a general technique that can be applied repetitively.[10]

In this chapter we describe how macro-seeding is carried out and when it can be applied.

Origin of the Seed Crystals

There are basically three sources of seed crystals: (1) microscopic crystals of sizes between 1 and 10 μm resulting from, in many instances, the final step of protein purification; (2) small crystals (0.01–0.1 mm) grown by conventional methods which either fail to grow larger or, if they do, result in thin plates or long needles; and (3) pieces of larger but badly intergrown crystals.

[1] T. L. Blundell and L. N. Johnson, "Protein Crystallography." Academic Press, New York, 1976.

[2] A. McPherson, Jr., *Methods Biochem. Anal.* **23,** 249 (1976).

[3] P. D. Martin, A. Tulinsky, and F. G. Walz, Jr., *J. Mol. Biol.* **136,** 95 (1980).

[4] C. M. Metzler, P. H. Rogers, A. Arnone, D. S. Martin, and D. E. Metzler, this series, Vol. 62, p. 551.

[5] W. J. Cook, F. L. Suddath, C. E. Bugg, and G. Goldstein, *J. Mol. Biol.* **130,** 353 (1979).

[6] J. C. Fontecilla-Camps, F. L. Suddath, C. E. Bugg, and D. D. Watt, *J. Mol. Biol.* **123,** 703 (1978).

[7] Yu. N. Chirgadze, M. B. Garber, and S. V. Nikonov, *J. Mol. Biol.* **113,** 443 (1977).

[8] W. Kabsch, H. Kabsch, and D. Eisenberg, *J. Mol. Biol.* **100,** 283 (1976).

[9] K. Åkervall and B. Strandberg, *J. Mol. Biol.* **62,** 625 (1971).

[10] C. Thaller, L. H. Weaver, G. Eichele, E. Wilson, R. Karlsson, and J. N. Jansonius, *J. Mol. Biol.* **147,** 465 (1981).

METHODS IN ENZYMOLOGY, VOL. 114

A fine tungsten needle[11] can be used to isolate single crystalline fragments from composites and bundles. The separation is best carried out under a polarization stereomicroscope since twinning can then be easily detected. Prospective seed crystals from needles or plates should always be trimmed to a compact shape. Experience in the laboratory of the authors has shown that it is advantageous to use as seed a short prism rather than a long thin needle. In the case of mitochondrial aspartate aminotransferase a direct relationship between the shape of the seed and the shape of the resulting crystal was observed.[12]

Seed crystals should preferably be freshly grown and only a few days old. However, sometimes up to 1-year-old crystals have been successfully used.[10]

The Seeding Procedure

The seeds are washed several times in order to remove crystalline debris and denatured molecules on the surface. The washing solution should be less concentrated in precipitant than the solution in which the crystals are normally stored. This washing step usually requires between a few seconds and a minute. The crystals are selected and transferred with the help of a thin glass capillary tube or a micropipet (e.g., Clay-Adams), connected to a length of 3-mm rubber tubing and a mouthpiece. Great care must be taken to transfer the crystals intact as small fragments will also act as nuclei. A washed seed is injected into a solution containing protein and precipitant by means of a fresh capillary tube of diameter just sufficient to accommodate the crystal. The amount of liquid transferred along with the seed should be minimal.

This procedure can be followed with seeds from experiments with the microdialysis cell as described by Zeppezauer,[13] with the hanging drop version of the vapor-phase diffusion technique,[14,15] and with its "sitting" drop counterpart.[16] In the last case about 50 μl of protein–precipitant mixture can be placed in the wells of a microtiter plastic plate. This

[11] A tungsten needle can be made by the following procedure. Place a piece of tungsten wire, 3–5 cm long, halfway in a glass tube and fuse the two together by melting the glass. Dip the end of the wire repeatedly in molten $NaNO_2$ contained in a metal crucible until a sharp tip is formed. Wear safety glasses and use a hood.

[12] C. Thaller, unpublished results (1980).

[13] M. Zeppezauer, this series, Vol. 22, p. 253.

[14] B. R. Reid, G. Koch, Y. Boulanger, B. Hartley, and D. Blow, *J. Mol. Biol.* **80**, 199 (1973).

[15] A. Wlodawer and K. O. Hodgson, *Proc. Natl. Acad. Sci. U.S.A.* **72**, 398 (1975).

[16] S. H. Kim, G. Quigley, F. L. Suddath, and A. Rich, *Proc. Natl. Acad. Sci. U.S.A.* **68**, 841 (1971).

amount is sufficient to seed 4–10 crystals. Thus the protein can be used with considerable economy.

In a successful seeding experiment growth conditions should be such that *ab initio* crystallization cannot occur. To this end, it is best to screen a concentration range of protein and precipitant slightly below the nucleation point. The injected seed crystals also show very different behavior; for some proteins they remain constant in size for days, while for others dramatic growth is observed within a few hours.

The speed of growth can be regulated to some extent by seeding into protein–precipitant mixtures of different composition and by variation of the concentration gradient between hanging drop and reservoir. In a microdialysis cell the growth process of the seed can be controlled continuously by temporarily lowering or increasing the precipitant concentration in the outer solution ("pulsing"[17]).

The Repeated Seeding Technique

A small, carefully washed crystal is used to seed a protein solution. After growth has ceased, the crystal is removed and inserted into a fresh protein solution, which allows it to grow further. This process can be repeated until the crystals have reached the desired size. For example, an increase in crystal volume by three orders of magnitude has been achieved from InGP synthase[18] seeds after a 10-fold repetition of this technique.[10]

Large crystals are more sensitive to changes in the surrounding solution than smaller ones. The conditions for washing between the cycles might have to be adjusted when larger seeds are used. If crystals show a strong tendency for formation of secondary crystals on the surface, the growth process must be monitored frequently. A daily transfer to a new solution was necessary for InGP synthase crystals, for example.

Salt precipitation steps in the protein purification produce very small crystals and repetitive seeding is necessary to grow them to a proper size. In the first seeding step a tungsten needle should be dipped into a drop of microcrystalline suspension and then passed through a 20-μl hanging drop. The concentration in this drop should again be slightly lower than necessary to crystallize *ab initio*. Many small crystals can be obtained in this way. However, in a next step these crystals can be enlarged according to the above-described method.

[17] R. E. Koeppe, II, R. M. Stroud, V. A. Pena, and D. V. Santi, *J. Mol. Biol.* **98,** 155 (1975).
[18] InGP synthase; *N*-(5′-phosphoribosyl) anthranilate isomerase : indoleglycerol phosphate synthase (EC 4.1.1.48).

Further applications for the macro-seeding technique are (1) co-crystallization in the required crystal form of proteins with heavy-atom compounds for formation of isomorphous derivatives and (2) co-crystallization of enzymes with inhibitors or substrate analogs.[10]

Although co-crystallization can lead to nonisomorphism, a native seed crystal used as a nucleus can improve the chances of growing isomorphous crystals. For instance, in order to produce a sodium ethylmercury(II) thiosalicylate derivative of InGP synthase, a native crystal was crushed in a 20-μl drop of storage solution. The fragments were used as seeds in a subsequent hanging drop experiment which contained protein, precipitant, and the heavy-atom compound. Within a few days small, isomorphous derivative crystals grew that were enlarged in a second cycle.[12]

Another potentially useful application of the method is "cross-seeding" in which seed crystals from one protein are used to initiate growth of another (homologous) protein. As an illustration of its success, a seed crystal of mitochondrial aspartate aminotransferase from chicken was inserted into a hanging drop containing the enzyme from its pig counterpart. Large, but badly twinned crystals were produced. In a second cycle, fragments of these crystals were used as seeds which grew into well-diffracting large crystals of the pig enzyme.[19] This recipe might work for other homologous proteins as well.

One of the most encouraging aspects of seeding is the high yield of good-size crystals. A success rate of 6–8 large crystals out of 10 injected seeds is possible after some experience has been obtained.

In summary, the described seeding method is applicable to any of the known micromethods of crystallization. By using an effective crystal growth technique on a microscale, a number of less abundant but important proteins should become available for X-ray crystallographic investigations.

Acknowledgments

We thank Dr. M. G. Vincent for critically reading the manuscript. The original work described here was supported in part by Grant 3.415.78 from the Swiss National Science Foundation.

[19] G. Eichele, G. C. Ford, and J. N. Jansonius, *J. Mol. Biol.* **135,** 513 (1979).

[10] Treatment and Manipulation of Crystals

By IVAN RAYMENT

General Considerations

Throughout a crystal structure analysis, it is necessary to physically and chemically manipulate the crystals being studied. These operations include crystal mounting prior to data collection, preparation of heavy-atom derivatives, and changing the solvent for low-temperature diffraction studies or studying the contrast between proteins, nucleic acid, and solvent in the crystal. The prime objective of all of these manipulations is to preserve the inherent order of the crystals.

Protein crystals are usually grown from an aqueous solution, and when they have finished growing they are in equilibrium with the solution that surrounds them. Any rapid change in the equilibrium may result in damage to the crystals. Thus, the first rule in crystal manipulation is not to make dramatic changes in the crystallizing solution (or mother liquor, as it is generally termed).

The preparation of derivatives often entails placing a crystal in a new solution which contains the heavy-atom compound of interest. It is important to establish a synthetic mother liquor to which the crystals may be transferred without reducing their order. Crystals which are grown in high salt or polyethylene glycol may often be stabilized by slowly increasing the concentration of the precipitant such that they will not dissolve when they are moved. Crystals which are grown from low salt present a greater problem since they are often in equilibrium with a significant protein concentration in the solution. Occasionally they may be stabilized by the addition of a low concentration of polyethylene glycol.

Crystal Mounting

The purpose of crystal mounting is to isolate a single crystal from its growth medium so that its diffraction properties may be studied. It is vital that the manipulation of the crystal introduce as little damage as possible to its three-dimensional order.

Protein crystals usually contain a large proportion of the aqueous medium in which they were grown. This varies from as little as 29% for edestin to as much as 95% for tropomyosin.[1] The solution forms an inte-

[1] B. W. Matthews, *J. Mol. Biol.* **33,** 491 (1968).

gral part of the crystal lattice. Its loss results in physical shrinkage of the crystal and reduction of the three-dimensional order as detected by X-ray diffraction. The most important aspect of crystal mounting is to preserve the crystal in its state of hydration. This may be accomplished by sealing the crystal in a thin-walled (0.001 mm thick) glass or quartz capillary tube.[2,3] The crystal manipulations are most easily viewed using a binocular dissecting microscope and should be performed quickly to minimize the dehydration of crystals. The exact procedures used will depend on the strength and morphology of the crystals and also the vessel in which they are grown. It is generally easier to mount a crystal from a shallow depression slide (routinely used for growing crystals by vapor diffusion) than from a straight-sided vial. The ease of mounting should be considered when growing crystals. The important steps in crystal mounting are illustrated in Fig. 1.

The first step in mounting the crystals is to dislodge them from the surface on which they grew. Robust crystals may be nudged gently with a glass rod or with the open end of the capillary. A fragile or soft crystal may be loosened by squirting a narrow jet of the crystallizing medium at its base. A controlled way of doing this is to use a pipettor set at ~25 μl. Once the crystal is free it may be drawn gently into the capillary using suction from a small volume (0.25 ml) glass syringe or mouth aspirator (Fig. 1a). The end of the capillary should be broken off cleanly; this facilitates drawing up the crystal and reduces the chance of a hairline crack running up the tube which could cause the crystal to dry out. Quartz tubes may be cleaved by scratching them with a diamond marking pencil and bending gently at that point. Glass tubes, which tend to be more fragile, may be broken cleanly by running a thin band of melted wax around the tube and fracturing just beyond the wax. It is good practice to acid-wash the glass capillaries prior to using them in order to remove any alkaline deposits remaining from their manufacture.

Next, the capillary should be inverted to allow the crystal to fall to the inner meniscus. Then any gross excess solution may be returned to the crystallizing droplet without losing the crystal. Conserving the crystallizing solution in this way may allow several crystals to be mounted from a small droplet.

The crystal should be drawn further into the capillary (2–3 cm) and rotated until it lies in roughly the required orientation. Then the surrounding solution may be removed using thin strips of filter paper or by the

[2] M. V. King, *Acta Crystallogr.* **7,** 601 (1954).

[3] K. C. Holmes and D. M. Blow, "The Use of Diffraction in the Study of Protein and Nucleic Acid Structure." Wiley (Interscience), New York, 1966.

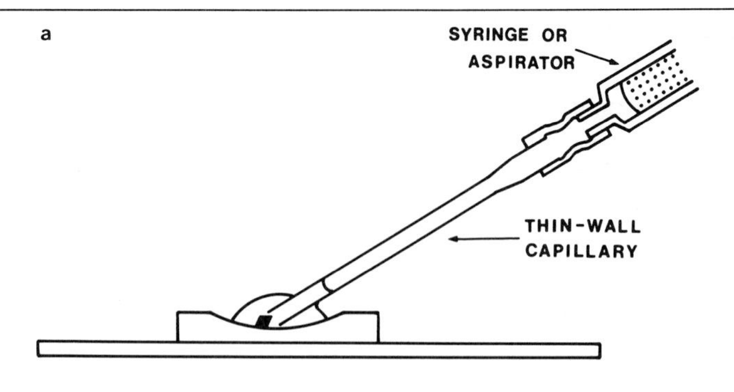

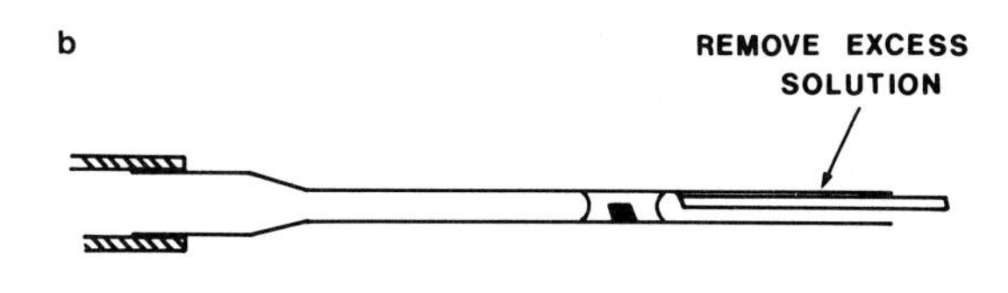

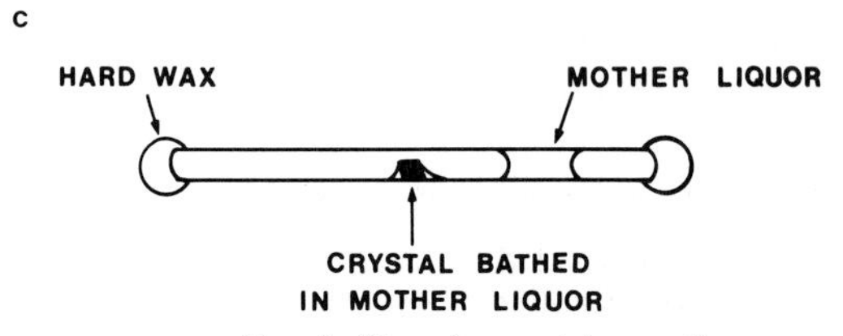

FIG. 1. Steps in crystal mounting.

capillary action of a thin glass rod (Fig. 1b). Care must be taken not to touch the crystal at this stage. The extent to which the crystal should be dried must be determined by experience. Some crystals may be thoroughly dried while others must be left quite wet to preserve their order. Robust crystals may be further oriented by nudging them gently with a glass rod.

The final step in crystal mounting is to place a small volume of mother liquor in the capillary and seal both ends. In principle, if the capillary is totally sealed, there will be sufficient solution directly associated with the crystal to maintain the correct degree of hydration. However, it is common practice to draw a little additional solution into the tube or to place a small piece of filter paper moistened with the crystallizing solution close to the crystal. The presence of this additional solution often proves troublesome if there is any temperature differential across or along the capillary tube.

This situation often arises when collecting diffraction data at reduced temperatures. Migration of water vapor results in the crystal either losing or gaining water, leading to dehydration, slipping, or dissolving. These problems may be overcome by removing the excess mother liquor completely and placing a column of silicone oil on either side of the crystal. Crystals mounted in this way may remain hydrated for several months if the capillary has no holes or cracks. Finally, the open end of the capillary should be sealed thoroughly with melted wax (sealing or high vacuum) and then cleaved 2–3 cm from the crystal and sealed (Fig. 1c) again. If the crystal is not to be used immediately, it may be advantageous to seal the capillary further by coating the wax with epoxy resin. This process strengthens the thin-walled capillary and seals any small cracks which may have been formed when the tube was cut.

Crystal Slippage

Crystal slippage has occasionally been a problem in macromolecular crystallography, particularly if the crystals have to be left very wet. There are several solutions which may work in any given instance.

There is often a tendency for a crystal to slip immediately after mounting. The simplest way to reduce slippage is to allow the crystal to settle for a day or so before use. This is particularly effective if the crystal can be oriented optically on a set of goniometer arcs and allowed to rest in the same orientation that it will be examined. This approach is best suited to photographic data collection in which the angular movement of the crystal is quite small.

An alternate approach is to hold the crystal in place by a thin plastic film.[4] This film is prepared *in situ* with a 0.2% solution of poly(vinyl formal) 15/95 powder (Polyscience Inc., Warrington, Pennsylvania) in 1,2-dichloroethane. This mixture, which is immiscible with water, will spread over an aqueous surface and leave a thin plastic film when the solvent evaporates. After drying the crystal a small volume of this solution may be drawn into the capillary tube until it just touches the crystal. The film will spread instantaneously over the crystal. The excess solution should be expelled and the capillary sealed in the normal way. This proce-

[4] I. Rayment, J. E. Johnson, and D. Suck, *J. Appl. Crystallogr.* **10,** 365 (1977).

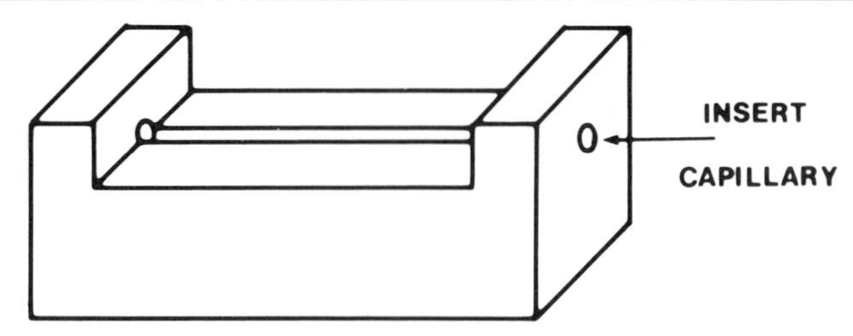

FIG. 2. Block for flattening capillaries.

dure does not increase the background scatter or damage the crystal in any way.

Crystal slippage may also be reduced by mounting the crystal on a portion of a capillary that has been flattened to form a small table. This results in more contact between the crystal and the capillary, thus giving greater adhesion. This approach is particularly advantageous for thin flat crystals which would otherwise crack in order to conform to a round capillary. Capillaries may be flattened using the block shown in Fig. 2. The tube is placed in the trough in the metal block such that the exposed surface may be heated and softened using an oxybutane torch.

[11] Protein Crystallization by Free Interface Diffusion

By F. R. SALEMME

The basic objective in growing crystals for protein crystallography is to create a situation in which the total quantity of protein initially present in solution eventually forms a few large, single crystals. The process of crystallization can be viewed to occur in two steps.[1] The first rate-limiting step involves the ordered association of a relatively small number of protein molecules to form crystalline nucleii. Subsequently, additional molecules can associate with the nucleii to eventually result in macroscopic crystals. The difficulty inherent in producing only a small number of large crystals from solution, under given experimental conditions, stems from the fact that the energetics of nucleation and growth are fundamentally different[1] and, moreover, may vary owing to the solubility characteristics peculiar to a given protein. Consequently, the situation may arise in which the conditions required for nucleation differ substan-

[1] Z. Kam, H. B. Shore, and G. Feher, *J. Mol. Biol.* **123,** 539 (1978).

tially from those associated with optimal (i.e., slow) crystal growth. Difficulties of this sort were encountered in attempts to crystallize several soluble bacterial cytochromes from ammonium sulfate,[2,3] in which batch crystallization attempts lead to either the formation of a protein gel or a shower of microcrystals. This motivated experimental approaches to crystallization[4] which involved creating an initial situation appropriate for nucleation (i.e., high local concentrations of both protein and precipitating salt), followed by rapid relaxation to conditions of lower protein and salt concentration associated with optimal crystal growth from preexisting nucleii. This was experimentally achieved simply by layering a concentrated solution of protein (typically 20–50 mg/ml) over an appropriately concentrated ammonium sulfate solution in either a small-bore glass tube or 5 × 20-mm test tube, according to the crystallization scale desired. In this case, the gradient in both protein and precipitant concentration at the initially formed interface is very high, and so optimal for promoting nucleation. However, subsequent interdiffusion between the two layers relaxes the initial conditions in a few hours. This results in the survival of only a few of the newly created nucleii, which are now in an environment of lower precipitant and protein concentration more optimal for slow crystal growth.

The free interface diffusion method has seen only limited application owing to the efficacy of other methods, whose success presumably reflect a relatively small difference in molecular association free energies for crystal nucleation and growth for most proteins. However, in cases in which experimental evidence suggests the contrary, the free diffusion technique provides an alternative approach.

[2] F. R. Salemme, *Arch. Biochem. Biophys.* **163,** 423 (1974).
[3] P. C. Weber and F. R. Salemme, *J. Mol. Biol.* **117,** 815 (1977).
[4] F. R. Salemme, *Arch. Biochem. Biophys.* **151,** 533 (1972).

[12] Flow Cell Construction and Use

By GREGORY A. PETSKO

The flow cell, a device for immobilization of a protein crystal while mother liquor is continuously flowed over it, was first developed by Wyckoff, Tsernoglou, Richards, and their associates.[1] The basic design

[1] H. W. Wyckoff, M. S. Doscher, D. Tsernoglou, T. Inagami, L. N. Johnson, K. D. Hardman, N. M. Allewell, D. M. Kelley, and F. M. Richards, *J. Mol. Biol.* **127,** 563 (1967).

has been modified since then, but the principles are unchanged. This description is based on techniques taught the author by Prof. Demetrius Tsernoglou, with some refinements by Dr. William A. Gilbert.

There are two types of flow cell, yoked and yokeless. The latter has seen less use, but is, in principle, of equivalent value. It is easier to build a flow cell with a yoke, however, so this operation will be described first.

The function of the yoke is to provide a rigid support for the capillary tube and polyethylene flow line. The yoke is usually made of brass, with a base of a pin 1 cm long by 3 mm in diameter, the size of a mounting pin for a standard goniometer head. To this pin is welded a semicircular brass support, approximately 1 cm in radius, with a 3-mm width and a thickness of 1.2 mm. A set a two holes, each 0.75 mm in diameter, is drilled in perfect registration 0.2 mm from the ends of the semicircle (Fig. 1). A 1-cm length of polyethylene tubing, PE-100 (Intramedic brand, Clay-Adams Division of Becton, Dickinson and Company, Parsippany, New Jersey 07054), 0.86 mm inner diameter, 1.52 mm outer diameter, is tied tangentially to the outer edge of the arc and then epoxied into place.

A quartz capillary tube of 0.7 mm outer diameter is selected. Two criteria are used: the capillary must fit smoothly into the holes drilled in

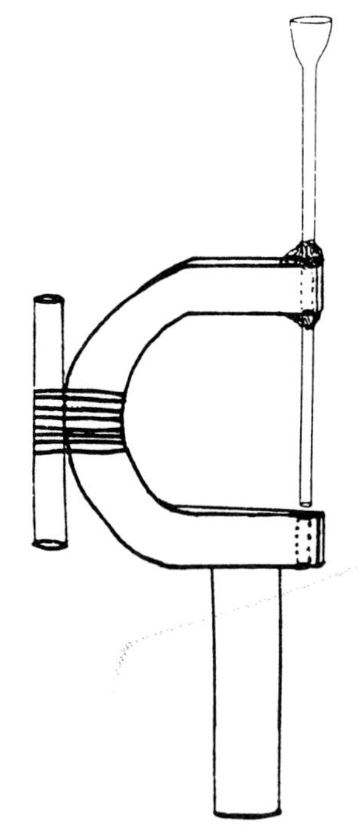

FIG. 1. Diagram of yoke used for a flow cell.

the arc, and size PE-10 polyethylene tubing (0.28 mm inner diameter, 0.61 mm outer diameter) must fit smoothly into the capillary tube. The funnel-shaped end of the capillary will be retained at this time, but the bottom end should be cut cleanly with a diamond-tipped stylus, leaving the tube open at both ends. Mount the yoke on a goniometer head and secure the goniometer to a ring stand with a C clamp.

The capillary is then threaded into the holes in the yoke top so that its open small-diameter end rests just above the lower set of holes (Fig. 1). Five-minute epoxy is applied to the top of the yoke to hold the capillary in place. After the epoxy has dried, polyethylene tubing (PE-10, 2–3 ft in length) is threaded up through the holes at the bottom of the yoke so that 2–4 mm of its length is inside the quartz capillary. Epoxy is then applied to the quartz–polyethylene–brass joint (Fig. 2). Some epoxy will flow up into the quartz tube and make the joint watertight, but the glue must not flow over the end of the PE-10 tubing. The epoxy should be allowed to harden for at least 1 hr before proceeding.

The partially constructed flow cell should now be filled with mother liquor. Insert the free end of the PE-10 tubing into a small test tube containing the mother liquor. Attach the test tube to the ring stand with a

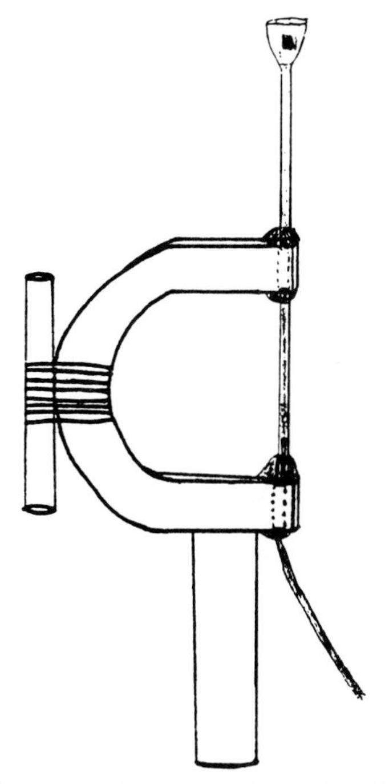

FIG. 2. Application of epoxy to the quartz–polyethylene–brass joint of the yoke.

clamp and raise the tube so that the top of the mother liquor is at least 6–8 in. above the flow cell. If mother liquor does not flow through the polyethylene tubing, release the air pressure by sucking gently on the funnel-shaped end of the capillary. While the flow cell is filling, the epoxy joint should be checked for leaks. Terminate the flow (by lowering the test tube until the mother liquor in it is at the same height as the column of liquid in the quartz tube) when the mother liquor has just reached the bottom of the funnel part of the capillary.

Now, using a 6-in. length of PE-10 as a ramrod, gently push a small amount of pipe cleaner fibers into the funnel and pack them down on top of the end of polyethylene tubing inside the quartz tube (Fig. 2). Only a very small amount of fibers should be used: the height of the bed of fibers above the tubing base, on which the crystal will rest, must not exceed 2 mm or the flow will be restricted.

The height of the liquid in the capillary tube should now be raised until it just reaches the top of the funnel (all adjustments to the height of the mother liquor can be made by either raising the test tube above the flow cell or lowering it below the flow cell). The crystal is then inserted into the capillary tube by means of a Pasteur pipet, and allowed to fall to the fiber

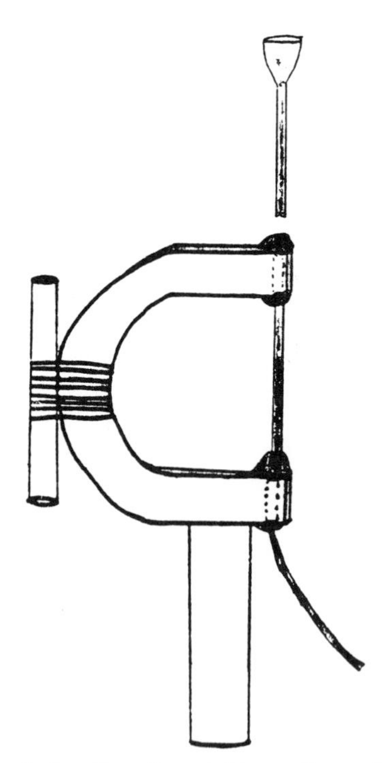

FIG. 3. Loading of crystal in capillary tube.

base by gravity (it should be evident that careful choice of crystal size is necessary to avoid having the crystal jam en route). Lower the liquid level back to the bottom of the funnel and *carefully* pack another 1–2 mm of pipe cleaner fibers on top of the crystal using the PE-10 ramrod (Fig. 3).

At this point the critical top joint must be made and sealed. Adjust the height of the reservoir in the test tube so that it is even with the top of the yoke. No water pressure must be exerted on the joint to be made. Cut the top of the capillary tube with the diamond-tipped stylus and break it off cleanly at the top of the yoke (Fig. 3). Insert one end of a new 2- to 3-ft-long piece of PE-10 tubing into the open top of the flow cell yoke, remove any liquid forced out at the joint by blotting with a tissue, and quickly seal the joint with fresh 5-min epoxy. While the epoxy dries, the test tube must remain at a height such that no liquid flows past the newly created joint but the crystal is kept covered by mother liquor. Also, the top length of polyethylene tubing should be taped or looped carefully on the ring stand so that no torque is applied to the joint while the epoxy is drying. The top joint should be dried for at least 3 hr, and overnight drying is preferable.

Once all joints are completed and dried, the free end of the top length of polyethylene tubing should be threaded through the top end of the attached PE-100 tube (Fig. 4) to make a tight loop at the outer rim of the flow cell arc. A test tube containing mother liquor is attached to the free end and both test tubes are sealed with cork or rubber stoppers containing holes or slits for the polyethylene tubing. The direction and rate of mother liquor flow can now be controlled by adjusting the relative heights of the

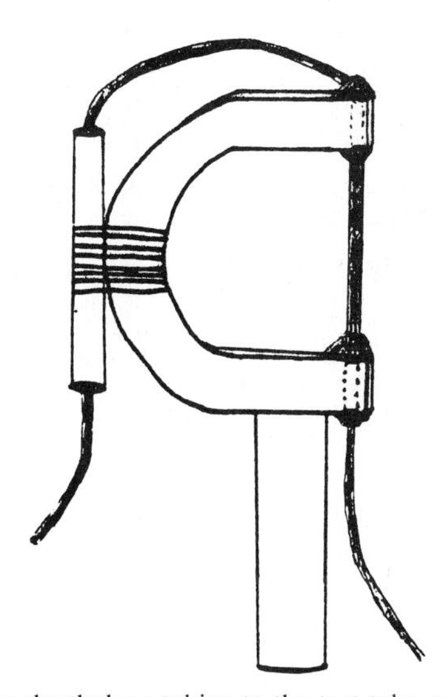

FIG. 4. Attachment of polyethylene tubing to the test tube containing mother liquor.

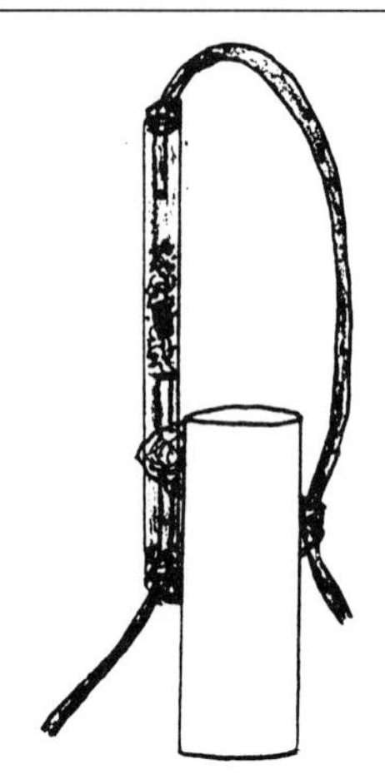

FIG. 5. Placement of tubing for support.

two test tubes. The author prefers, for reasons of crystal stability and ease of temperature control, to flow liquid from the top tubing down over the crystal and out the bottom tubing. A flow rate of 1–2 ml/day is the maximum recommended. Depending on the exact rate, 15–30 min are usually required for a wavefront of new mother liquor to move from the source test tube reservoir to the crystal.

Crystals with the symmetry of high-symmetry space groups are ideal for flow cell work because data collection can be carried out in a region of reciprocal space where the yoke does not interfere with either the incident or the diffracted X-ray beam. Monoclinic and triclinic crystals give rise to a blind region where the yoke obscures a portion of the data. Yokeless flow cells eliminate this difficulty, but at the cost of increased fragility and decreased mechanical stability. In the yokeless design, the capillary tube is initially glued to the side of a crystal mounting pin (a brass cylinder 1 cm long by 3 mm in diameter). From this point on, the assembly and filling of the flow cell are identical to that of the yoked design, but the top joint is even more fragile due to the absence of the supporting brass hemisphere. There is no length of PE-100 tubing to provide a guide for the upper feed tube, so the top tubing must be looped carefully and glued to either the pin or the goniometer head for support (Fig. 5).

There is increased background in the flow cell mount relative to a normal crystal mounting, due to the surrounding liquid and pipe cleaner fibers, but the background is cylindrically symmetrical. The flow cell also essentially eliminates the need for an absorption correction, since there is cylindrical isotropy for the magnitudes of the Bragg reflections as well.

[13] Preparation of Isomorphous Heavy-Atom Derivatives

By GREGORY A. PETSKO

This chapter will concentrate on the chemical logic underlying the preparation of heavy-atom derivatives of proteins. Most previous discussions[1] have tabulated those compounds and reaction conditions that have produced useful derivatives in various crystalline proteins. The present treatment follows the pioneering discussion of Blake,[2] and attempts to set down a systematic approach that can be used on future problems. First, the various amino acid side chains are classified according to their reactivity with different heavy metals. Then, different classes of heavy atoms are discussed, followed by methods for the introduction of reactive functionalities into proteins. The reader is also encouraged to consult the chapter on heavy-atom derivative preparation in Blundell and Johnson.[3]

General Considerations

Whenever a new crystalline protein must be derivatized for phase determination by the method of isomorphous replacement, the crystallographer faces an initial choice: are the derivatives to be prepared by reacting the protein with heavy metal in solution and then crystallizing the complex, or by soaking the crystals in mother liquor containing the heavy-atom compound? Historically, both methods have been employed with success, but this author believes that diffusion into preexisting native crystals should be tried first. Intermolecular contacts in the crystal lattice will reduce the number of potentially reactive groups available for heavy-atom binding, making it more likely that the important first derivative will have a small number of sites. Also, use of the native crystal form eliminates the problem of finding crystallization conditions for a derivatized protein whose solubility may have changed, and greatly reduces the possibility that the derivative crystal will not be isomorphous with the parent form. If the soaking method does not work and the protein seems unreactive, exposure to heavy-atom compounds in solution followed by crystallization should then be tried, as the restrictive effects of lattice contacts will be avoided.

[1] D. Eisenberg, *in* "The Enzymes" (P. D. Boyer, ed.), 3rd ed., vol. 1, p. 1. Academic Press, New York, 1970.

[2] C. C. F. Blake, *Adv. Protein Chem.* **23**, 59 (1968).

[3] T. L. Blundell and L. N. Johnson, "Protein Crystallography." Academic Press, New York, 1976.

Sigler and Blow[4] noted the possible competition of NH_3 derived from $(NH_4)_2SO_4$ for heavy-metal binding to the protein, and it is clear that the preparation of derivatives will depend on the pH, composition of the mother liquor, and temperature. Specific effects will be noted below, but in general the following rules apply.

1. Ammonium sulfate, except at low pH (i.e., below 6), is a poor mother liquor for heavy-atom binding due to the production of the good nucleophile NH_3. When possible, ammonium sulfate should be replaced by sodium or lithium sulfate, or potassium phosphate. However, phosphate is a poor mother liquor for uranium and rare earth binding.
2. Intermediate pH (i.e., 6–8) is better than high pH (>9) because many heavy-atom compounds are alkalai labile or form insoluble hydroxides, but pH values below 6 may cause problems because most reactive groups will be protonated and blocked at low pH.
3. Most heavy-atom binding will be significantly slower at 4° than at room temperature.
4. Tris buffer is a potential competitor with the protein for heavy metals such as platinum, while phosphate buffer is a potential competitor for uranyl and rare earth compounds.

Soaking time and heavy-atom concentration are the other remaining considerations. Published soaking times are as short as several hours or as long as months. If no binding is found in a soak of several days, increasing the soaking time to 1 or more weeks may produce some binding, but cases of this are rare. Generally, a soak of 1–3 days will suffice to screen for binding. The concentration of the heavy-metal compound will depend to some extent on its solubility in the crystal mother liquor, but 1 m*M* seems a useful first value. Concentration is a more useful variable than soaking time if no binding is found, since mass action can be used to produce reasonable occupancy of even weak complexes. Increasing the heavy-atom concentration by an order of magnitude, or more, is worthwhile. If excessive binding is observed, by either complex Patterson maps, huge intensity changes, or crystal cracking, reduction of either soaking time or heavy-atom concentration, or both, is warranted. The author has found that, particularly in cases of covalent binding, both the time and the concentration can be very small. Soaking times as short as 1 hr combined with concentrations of 0.01 m*M* have produced full occupancy mercury derivatives of proteins with reactive sulfhydryl groups.[5]

[4] P. B. Sigler and D. M. Blow, *J. Mol. Biol.* **14,** 640 (1965).

[5] D. Ringe, G. A. Petsko, F. Yamakura, K. Suzuki, and D. Ohmori, *Proc. Natl. Acad. Sci. U.S.A.* **80,** 3879 (1983).

Two final comments concern the use of cross-linking agents and the action of light. If the heavy-metal binding causes severe crystal damage, even at low concentrations with brief incubation, prior cross-linking of the crystal with glutaraldehyde may prevent cracking. Finally, many heavy-atom compounds, particularly platinum and iridium complexes, have a very vigorous photochemistry. Reproducibility of derivative occupancy—and, sometimes, of sites—can best be ensured by carrying out all experiments in the dark.

Reactivities of the Amino Acids

Often the choice of the first heavy-atom compounds to try, and the appropriate conditions of pH and mother liquor composition, can be made from inspection of the amino acid composition of the protein. The following discussion of the reactivities of the 20 commonly occurring amino acids is meant to guide such choices.

Nearly all heavy-atoms react as cations or as complexes by nucleophilic ligand substitution reactions (the chief exception is iodine, which reacts with activated aromatic residues by electrophilic substitution). Therefore, the reactive amino acids can be discussed with reference to their nucleophilicity and their ability to substitute for common heavy-atom ligands.

Unreactive Amino Acids. The side chains of glycine, alanine, valine, leucine, isoleucine, proline, and phenylalanine are not reactive under the chemical conditions possible for crystal soaking experiments.

Cysteine. Cysteine is highly reactive as the thiolate anion, the percentage of which increases rapidly above pH 7. Sulfur in the ionic form is an excellent nucleophile and will react rapidly and essentially irreversibly with mercuric ion or organomercurials. In ligand substitution reactions, the thiolate is a fast-entering attacking group that forms thermodynamically stable complexes with class b metals (copper, silver, iridium, rhodium, platinum, paladium, gold). Reaction with rare earths and uranyl complexes is not expected. The substitution reactions are more sensitive to pH than the covalent attack. At pH values of 6 and below, where cysteine is nearly completely protonated, reaction with mercurials is still rapid but there is little reactivity with $PtCl_4^{2-}$.[6]

Cystine. Surprisingly, disulfide linkages are weakly reactive in ligand substitution reactions. $PtCl_4^{2-}$ has been found to bind to S–S bridges in some crystalline proteins by loss of chloride. Mercurials do not ordinarily

[6] G. A. Petsko, D. C. Phillips, R. J. P. Williams, and I. A. Wilson, *J. Mol. Biol.* **120,** 345 (1978).

insert spontaneously into disulfides but addition of mercury is possible if the cystine is first reduced with, say, dithiothreitol prior to mercuration.[7] Of course, steric restrictions may prevent formation of the S–Hg–S group.

Histidine. Histidine is the most common reactive group in proteins. The protonated imidazolium cation predominates below pH 6 and is not reactive as a nucleophile. The unprotonated imidazole is a good nucleophile and will displace chloride (or hydroxide) from platinum complexes, coordinate to gold complexes, react with copper and silver, and can be mercurated. The best heavy atoms for reaction with histidine are K_2PtCl_4, ethyl mercury phosphate, mersalyl, and $NaAuCl_4$. The imidazole ring is aromatic and electrophilic aromatic substitution reactions, such as iodination, are possible for histidine but the conditions are somewhat severe.

Methionine. The un-ionizable S–CH_3 group is unreactive toward mercurials but its lone pair of electrons makes it a good attacking group in nucleophilic ligand substitution reactions. Methionine will readily displace chloride from all platinum chloride complexes to form a stable bond. Contrary to the earlier hypothesis of Dickerson and associates,[8] this reaction does not involve oxidation of the platinum to the +4 state. It is a simple, S_N2-type ligand exchange reaction.[6] The reaction of platinum compounds with methionine is not pH sensitive in the normal range.

Lysine. The amino group of lysine has a pK_a of about 9 in proteins. At neutral pH or lower, the protonated amine will not participate in substitution reactions but may undergo weak electrostatic complexation with negatively charged heavy-metal compounds. Near and above the pK_a, lysine can react with platinum and gold complexes to displace weaker ligands such as chloride (iodine, bromine, and nitro groups will not be displaced by lysine but may be by histidine and should be by the sulfur-containing amino acids). The particular utility of lysine residues, which are often highly accessible to the solvent, is their use as "handles" to attach other, more reactive, functionalities to proteins (see below).

Arginine. Arginine would seem to be a very attractive group for derivatization as it occurs relatively infrequently in most proteins and, therefore, would provide derivatives with a small number of sites per molecule. Unfortunately, the pK_a of the guanidinium group is >12 in proteins and the cationic form is unreactive in simple substitution reactions. There may be an electrostatic interaction with anionic heavy-atom compounds.

[7] K. R. Ely, R. L. Girling, M. Schiffer, D. E. Cunningham, and A. B. Edmundson, *Biochemistry* **12,** 4233 (1973).

[8] R. E. Dickerson, D. Eisenberg, J. Varnum, and M. L. Kopka, *J. Mol. Biol.* **45,** 77 (1969).

Recently, advantage has been taken of the facile reaction of guanidinium groups with the glyoxal function (phenylglyoxal is a common arginine reagent in protein chemical modification) by using mercury phenylglyoxal as an arginine-specific heavy-atom compound.[9] This reagent is a useful addition to the crystallographic arsenal for protein crystals at pH values of neutral or greater.

Tryptophan. The indole ring is relatively inert to electrophilic aromatic attack by iodine but the ring nitrogen can be mercurated.[10] The pH is not critical for this reaction, but there must be no competing nucleophiles in the mother liquor. Optimal conditions are sodium sulfate or polyethylene glycol as the precipitant with cacodylate or borate buffer. The mercurial reagents that have proven successful are ethylmercury phosphate and mercuric acetate. Tryptophan does not react readily in ligand substitution processes.

Tyrosine. The phenolate oxygen anion is a very good nucleophile and would be expected to bind a variety of heavy atoms if not for its high pK_a (10.5). Only at pH values >10 would there be sufficient concentration of the phenolate form. Therefore, the principal reaction of tyrosine is electrophilic aromatic substitution by iodine. Sigler[11] has given a detailed description of the preparation of the iodination solution and the conditions for its use with crystalline proteins. It is worth remembering that ammonium sulfate mother liquor at high pH should be avoided if iodination is attempted: one may accidently synthesize nitrogen triiodide, which is explosive!

Aspartic and Glutamic Acids. Both of these residues have side chain pK_a values on the order of 3–4. At low pH they will be protonated and unreactive. Above pH 5 they will be anionic and potentially good ligands for heavy metals. Although class b metals prefer sulfur or nitrogen ligands, uranium and various rare earths behave like calcium and display a strong preference for oxygen ligation. Colman *et al.*[12] have shown that lanthanides can replace calcium in thermolysin; the rare earths of higher atomic number, having ionic radii less than or equal to that of calcium, were found to cause less perturbation of the protein structure. In all cases, the primary ligands are carboxylate side chains from aspartic and glutamic acids. Uranyl complexes such as $K_3UO_2F_5$ and $UO_2(NO_3)_2$ are also carboxylate seeking, but these compounds are not very soluble above pH 7 due to the formation of hydroxides. Uranium and rare earth derivati-

[9] I. A. Wilson, J. J. Skeral, and D. C. Wiley, *Nature* (*London*) **289,** 366 (1981).

[10] D. Tsernoglou and G. A. Petsko, *FEBS Lett.* **68,** 1 (1976).

[11] P. B. Sigler, *Biochemistry* **9,** 3609 (1970).

[12] P. M. Colman, L. H. Weaver, and B. W. Matthews, *Biochem. Biophys. Res. Commun.* **46,** 1999 (1972).

zation must not be attempted in phosphate buffer, because the phosphate oxyens will compete with protein carboxylates for the heavy atom, and the phosphate complexes are often insoluble.

Serine and Threonine. The hydroxyl groups of these side chains are fully protonated at all normal pH values and are not reactive nucleophiles. Occasionally, an abnormally reactive serine has been found as in the serine proteases and β-lactamases, and these residues have been derivatized, but such behavior is uncommon.

Asparagine and Glutamine. In model studies with simple metal complexes the free amino acids Asn and Gln will weakly coordinate through their amide nitrogens, but such complexes have not been found frequently in crystalline proteins. At present there is no information that permits a rational approach to the labeling of these amino acids with heavy metals. A systematic study would be worthwhile.

Classes of Heavy Atoms

Platinum Compounds. The most widely successful heavy-atom reagent is the $PtCl_4^{2-}$ ion, which binds to methionine, histidine, and cysteine residues. In the absence of excess chloride ion this compound is rapidly converted to aquo and hydroxide species of neutral or +1 overall charge. Petsko *et al.*[6] gave a detailed account of the chemistry of chloroplatinate in a variety of crystal mother liquors. Briefly, they concluded that ammonium sulfate solutions at high pH will compete with the protein for platinum binding, that pH values greater than 5–6 are needed to unblock histidine so that it will react with chloroplatinate, and that cysteine is a ligand only at pH values near neutrality or above. They concluded further that most other platinum compounds, such as $Pt(NO_2)_4^{2-}$, $Pt(NH_3)_2Cl_2$, or Pt(ethylenediamine)Cl_2 will react in the same way. The exceptions are compounds with ligands of high thermodynamic stability, such as $Pt(CN)_4^{2-}$. The cyanide will not be displaced by any protein side chain, so platinum tetracyanide is inert to nucleophilic ligand substitution. This complex remains dianionic under all conditions, and will bind, if at all, to positively charged protein residues (lysine, arginine, and, at low pH, histidine) by electrostatic attraction.

Mercurials. A wide range of mercury compounds have been observed to bind to crystalline proteins. With only a few exceptions, the protein ligands are either cysteine sulfur or histidine nitrogen. Binding is promoted at pH values of neutral or greater, where these residues are deprotonated. The most commonly used mercurials are *p*-chloromercuribenzoic acid, ethylmercury thiosalicylate, $HgCl_2$, ethylmercury chloride, mercuric acetate, mersalyl, and ethylmercury phosphate. The latter compound is the author's personal favorite. It is commercially available (Chem Service), water soluble (except in the presence of phosphate ions; phosphate buffer should never be used with this reagent), and disproportionates to give the highly reactive, partially hydrophobic $CH_3CH_2Hg^+$ ion.[13] This species will penetrate into proteins to react with buried sulfhydryl or imidazole groups and, being small, usually does not cause lack of isomorphism. If too many sites are found by this reagent, greater steric selectivity may be achieved with a larger, aromatic mercurial. Once even a single site has been substituted with mercury, a second isomorphous derivative often may be prepared by repeating the experiment with a dimercurial. Dimercury acetate is a popular choice; the two mercury atoms are 3.8 Å apart and the compound is not very large. Sprang and Fletterick[14] gave a simple synthesis for this dimercurial. Although competitive binding of NH_3 from the mother liquor is less of a problem for mercurials than it is for platinum compounds, the author's recommendation is still to replace ammonium sulfate with lithium or sodium sulfate if no binding is found.

Rare Earths and Uranyl Salts. As stated above, phosphate buffer will tend to compete with protein carboxylate groups for binding to lanthanides or uranium compounds, so this buffer should be replaced by Tris, cacodylate, or borate (depending on the pH range). Samarium is the preferred lanthanide owing to its large anomalous scattering signal. Of the commonly available uranyl salts, the nitrate is the most reactive but the acetate is slightly more soluble in common mother liquors.

Complex Ions. One species stands out in this category: K_2HgI_4, which disproportionates in aqueous solution to give a mixture of substances: HgI_4^{2-}, HgI_3^-, and HgI_2 as well as I^-. The most important product is the planar, trigonal HgI_3^- group, which will bind electrostatically to cations or penetrate into proteins by virtue of its large flat structure. The dissociation equilibrium for this ion can be driven in the direction of HgI_3^- formation by the addition of excess KI to the K_2HgI_4 soaking solution.

Special Reactions

One potentially reactive nucleophile is present in all proteins, the terminal amino group (the C-terminal carboxylate does not appear to bind heavy metals). This group has been exploited by Drenth and associates,[15] who have reacted the amino terminus of papain with an iodinated Ed-

[13] D. A. Vidusek, M. F. Roberts, and G. Bodenhausen, *J. Am. Chem. Soc.* **104,** 5452 (1982).

[14] S. Sprang and R. J. Fletterick, *J. Mol. Biol.* **131,** 523 (1979).

[15] J. Drenth, J. N. Jansonius, R. Koekoeck, H. M. Swen, and B. G. Wolthers, *Nature (London)* **218,** 929 (1968).

man's reagent. This reaction has been somewhat neglected, in the author's opinion unjustly. Even if an iodine is not heavy enough to serve as a major derivative for a protein, location of the N-terminus would be a valuable aid in chain tracing. There are two complications in the use of Edman-like compounds: protonation of the amino terminus and competition from lysine reaction. As usual, it is the unprotonated amine that is the reactive nucleophile so exposure to the Edman compound should be done above neutral pH. Unfortunately, the unprotonated ε-amino group of lysine will also react with Edman reagents. Control of the reaction may be achieved sometimes by taking advantage of the lower pK_a of the protein N-terminus. Carrying out the reaction between pH 7 and 8 will sometimes give only substitution at the amino terminus. Of course, NH_3 derived from the ammonium ion will rapidly inactivate Edman compounds, so ammonium sulfate mother liquor cannot be used.

Another method of specific chemical labeling of crystalline enzymes is by heavy-atom-substituted substrate analogs or inhibitors. Unfortunately, each protein presents a different case, and no general strategy can be given. However, it is relatively easy to mercurate or iodinate aromatic compounds, so synthesis of suitable analogs is often possible.

Some proteins in their native state bind metal atoms that can be replaced by heavier atoms, thus providing an isomorphous derivative. The metal most often replaced in such studies has been zinc. Removal of zinc is achieved by dialysis against EDTA, 1,10-phenanthroline, histidine, or some combination of these chelators. Then cadmium or mercury or lead (usually as the acetate salt) is added by dialysis or direct soaking. Lanthanides will replace bound calcium atoms in many proteins on simple soaking of the crystals in solutions containing the rare earth.

Insoluble Reagents

Many heavy-atom-containing compounds, such as derivatives of Edman's reagent, are at best only sparingly soluble in water. Addition of these reagents to protein crystals can be achieved by use of a carrier solvent that is less polar. Acetonitrile is the most popular choice. The heavy-atom compound should be dissolved in neat, vacuum-distilled acetonitrile and the organic solvent should then be suspended in the crystal mother liquor at 3–5% by volume.

Introduction of Reactive Groups

When all of the standard, trial-and-error methods for preparation of a heavy-atom derivative have failed, the last resort is to introduce one or more highly reactive functionalities into the protein, and then to react the modified macromolecule with heavy metal. Two strategies are possible: the protein can be modified in solution and then crystallized, or the chemical reactions can be carried out in the crystalline state. The author prefers to start with the latter approach, as the crystal contacts should limit the number of modification sites and reduce the chances for nonisomorphism.

Because of their high degree of reactivity to mercurials, sulfhydryl groups are the logical choice for introduction into proteins. The usual target group for the modification reaction is the amino group of lysine (or the protein N-terminus). Shall and Barnard reacted ribonuclease with *N*-acetylhomocysteine thiolactone and produced two separable products, each containing a new SH group in a different place.[16] This reaction was run in solution, but it should be possible to carry it out in the crystalline state at neutral pH. Gallwitz *et al.*[17] reacted tobacco mosaic virus with 4-sulfophenylisothiocyanate, a modified Edman-like reagent, which derivatized both the N-terminus and lysine 53 of the coat protein. The reaction yielded phenylthiocarbamoyl derivatives that tautomerized to give reactive SH groups. The sulfhydryls reacted with methylmercury nitrate to give a stable thiomercurial that did not suffer replacement of the sulfur by oxygen in air. Gallwitz *et al.*[17] gave instructions for the synthesis of the modification reagent. Reaction of this compound with proteins must be carried out in the dark anaerobically at pH 8.5.

Mowbray and Petsko[18] have described the reaction of amino groups in the galactose chemotaxis receptor protein with carbon disulfide under anaerobic conditions in the crystalline state. They described a simple apparatus for carrying out the modification, in phosphate buffer at pH 8. Subsequent reaction with mercurials and dimercurials produced several different isomorphous derivatives for a crystalline protein that had not been successfully derivatized by conventional methods. The carbon disulfide reaction appears to form a dithiocarbamate with a pK_a of 3.5. The modified amino group is only stable above pH 6, so the mercuration must be carried out at neutral pH or higher.

Finally, metal-chelating groups can be attached to protein amino groups by similar chemistry. Benisek and Richards[19] reacted lysozyme with picolinimidate compounds, with up to seven such groups being covalently attached to the protein, depending on conditions. These groups chelated platinum and gold compounds very strongly.

[16] S. Shall and E. A. Barnard, *J. Mol. Biol.* **41,** 237 (1969).

[17] U. Gallwitz, L. King, and R. N. Perham, *J. Mol. Biol.* **87,** 257 (1974).

[18] S. L. Mowbray and G. A. Petsko, *J. Biol. Chem.* **258,** 5634 (1983).

[19] W. Benisek and F. M. Richards, *J. Biol. Chem.* **243,** 4267 (1968).

Recommendations

Consideration of the specific amino acid composition, presence of endogenous metals, cysteine content, etc. is always desirable; nevertheless, some general rules can be formulated.

1. Whenever possible, avoid using ammonium sulfate unless the pH is below 6.
2. Heavy-metal binding is more likely at pH 7 and above than at pH values below 6.
3. All soaking should be done in the dark.
4. For first screening of possible derivatives, the author recommends (a) K_2PtCl_4 at pH 6 or higher in ammonia-free media; (b) ethylmercury phosphate in phosphate-free media, pH 6 or higher; (c) samarium acetate or uranyl acetate in phosphate-free media; (d) K_2HgI_4 with excess KI; or (e) $K_2Pt(CN)_4$.

It would be surprising (but, alas, not unprecedented) if none of these five trials produced significant intensity changes. In the author's opinion, to produce *some* binding to any given protein, the reagent of choice would be K_2PtCl_4 or the corresponding nitrite. To produce a *suitable* isomorphous derivative of high occupancy with a reasonable number of sites, the author would suggest ethyl mercury phosphate.

Cautionary Postscript

All heavy-atom compounds should be regarded as highly toxic. Mouth pipetting of heavy-atom-containing solutions should never be attempted. Preparation of such solutions must be done in a fume hood. Gloves and protective clothing and eyeguards should be worn at all times.

[14] Heavy Metal Ion–Nucleic Acid Interaction

By SUNG-HOU KIM, WHAN-CHUL SHIN, and R. W. WARRANT

Metal ions play an important role in nucleic acids in two ways: through nonspecific interaction and specific interaction. Since nucleic acids are polyanionic molecules, they interact with any positively charged ion such as various metal ions and amines. This type of interaction is primarily to neutralize the negative charges, and does not require positive ions to be bound site specifically. It is likely that positive ions are statistically bound to the phosphates of the molecule by nonspecific interaction.

The second class of positive ions includes those which bind to nucleic acids site specifically. These ions are usually important for stabilizing the tertiary structure of nucleic acids[1–4] and/or for their biological functions. This type of site-specific binding is of great interest to nucleic acid crystallographers because such binding of metal ions can provide a convenient way of obtaining heavy-atom derivatives.

Due to the relatively short history of nucleic acid crystallography, there have not been many structures determined so far. Among those structures which have been determined one can further investigate the environment of the heavy-atom binding sites. Such studies have been done primarily with difference Fourier methods using heavy-atom derivative data and native phases. Most of these studies have been done on transfer RNA structures. Here we summarize those heavy-metal binding environments which are relatively well defined by crystallographic methods and hope that this knowledge will be helpful for those who will be attempting to make heavy atom derivatives of nucleic acids crystals.

Samarium

Samarium has been introduced into yeast phenylalanine tRNA crystals as samarium(III) acetate.[2,4,5] In the orthorhombic crystal form of yeast phenylalanine tRNA there are three strong binding sites. The strongest site was found at the tight corner of the backbone between residues 8 and 12. Based on distance criteria, it appears that the samarium ion is directly coordinated to the phosphate oxygens of residues 8 and 9 (Fig. 1a). The remaining four coordination sites are occupied by water which in turn forms hydrogen bonds to the O-3′ of residue 8 and the O-3′ of residue 7. The coordination of this samarium ion is compatible with octahedral geometry.

The second samarium binding site has been located at another sharp corner, in the dihydroU-loop between residues 20 and 21. This samarium ion appears again to be directly coordinated to the phosphate oxygens of residues 20 and 21 (Fig. 1b), and can also be interpreted as having octahedral geometry with the remaining four coordination sites occupied by

[1] S. R. Holbrook, J. L. Sussman, R. W. Warrant, G. M. Church, and S. H. Kim, *Nucleic Acids Res.* **4,** 2811 (1977).

[2] A. Jack, J. E. Ladner, D. Rhodes, R. S. Brown, and A. Klug, *J. Mol. Biol.* **111,** 315 (1977).

[3] G. Quigley, M. M. Teeter, and A. Rich, *Proc. Natl. Acad. Sci. U.S.A.* **75,** 64 (1978).

[4] C. D. Stout, H. Mizuno, S. T. Rao, P. Swaminathan, J. Rubin, T. Brennan, and M. Sundaralingam, *Acta Crystallogr., Sect. B* **B34,** 1529 (1978).

[5] F. L. Suddath, G. J. Quigley, A. McPherson, D. Sneden, J. J. Kim, S. H. Kim, and A. Rich, *Nature* (*London*) **248,** 20 (1974).

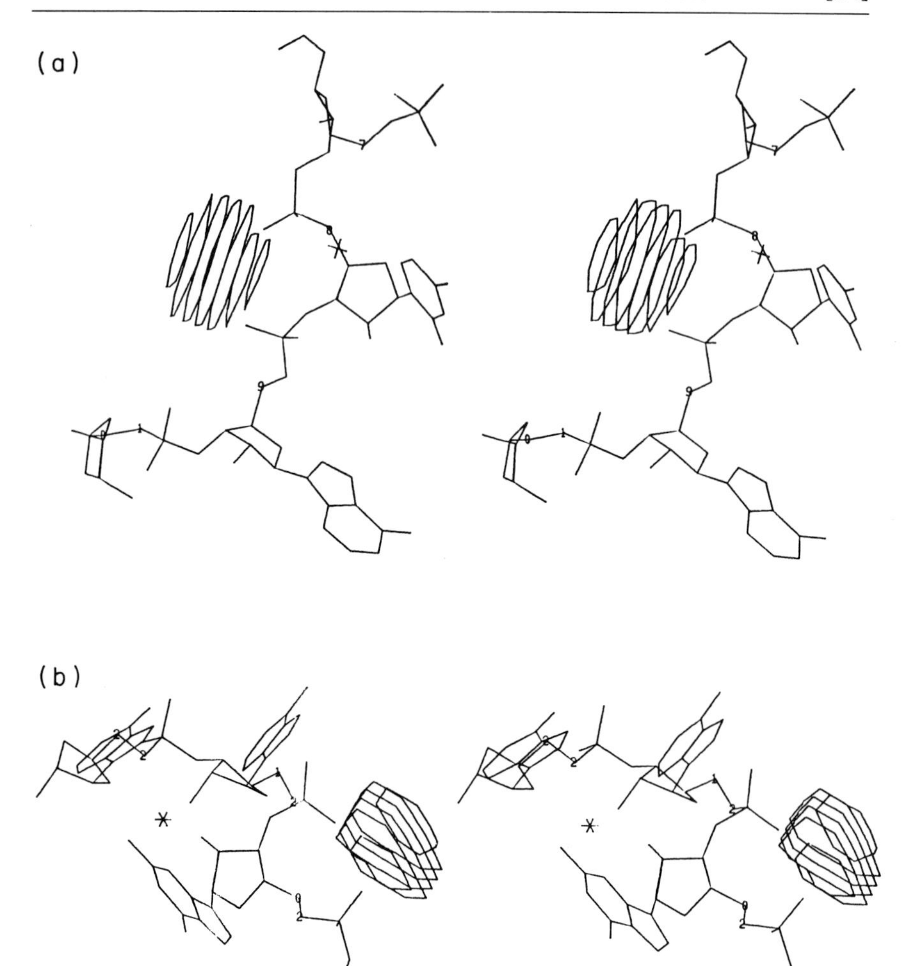

FIG. 1. Difference electron density calculated using Sm acetate derivative data and native phases of yeast tRNA^{Phe} crystals in an orthorhombic lattice: (a) Sm found between phosphates of residues 8 and 9; (b) Sm found between phosphates of residues 20 and 21; (c) Sm found between phosphate of residue 14 of one tRNA and that of residue 56 of a neighboring molecule; (d) coordination geometries of Sm(III) ion.

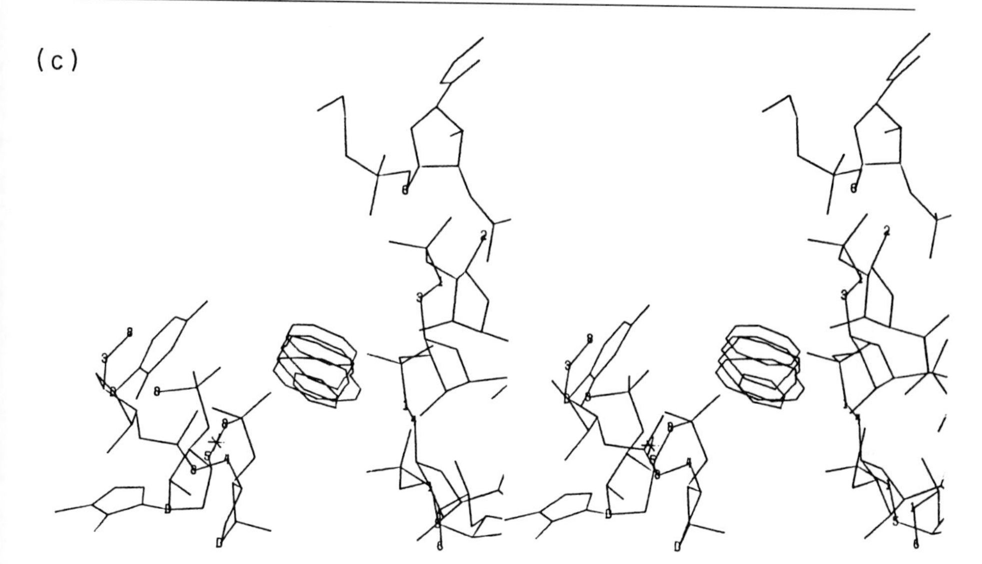

(d)

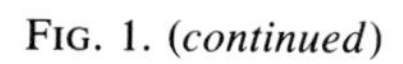

FIG. 1. (*continued*)

water molecules. These waters in turn form hydrogen bonds to the nearby oxygens. The third samarium site has been found between two tRNA molecules and the ion appears to be directly coordinated to a phosphate oxygen of residue 14 of one molecule and to a phosphate oxygen of residue 56 of another molecule (Fig. 1c). This samarium ion, as the other two samarium ions, has octahedral geometry.

In all three samarium binding sites, the samarium makes one or two direct coordinations to phosphate oxygens, and the remainder of the coordination shell is occupied by water molecules which in turn make several hydrogen bonds to nearby atoms. Thus, it appears that samarium does not form direct coordination complexes to bases but exclusively to backbone phosphate oxygens. Our observation agrees with that of others.[2,4,6] Other lanthanide metal ions, such as praseodymium, europium, gadolinium, terbium, and lutetium, have been found to bind to the same sites as samarium.

Platinum

The platinum(IV) ion has been introduced into tRNA crystals as a potassium salt of hexachloroplatinate[5] or a *trans*-dichlorodiammine-Pt(IV) complex.[2,4] For the former there are three tightly bound platinum ions in orthorhombic crystals of yeast phenylalanine tRNA. The first platinum ion has been found to be bound to the bases of residues 62 and 63 (Fig. 2a). Based on distance criteria, it appears that the platinum is directly coordinated to N-4 of C63 (C base of residue 63) and N-7 of A62. The coordination of this platinum is consistent with octahedral geometry for the remaining four coordination sites occupied by water or chloride, which may in turn form hydrogen bonds to nearby atoms. The second platinum has been located near the bases of residues 73 and 74 (Fig. 2b). Again, from distances, it appears that the platinum ion is directly coordinated to N-4 of C74 and N-7 of A73. The coordination geometry of this platinum also appears to be octahedral, with the remaining four coordination sites occupied by water or chloride ions making additional hydrogen bonds with nearby atoms. The interesting aspect of these two sites is that the platinum ions are found to be directly coordinated to two adjacent bases in a specific base sequence, that is the A–C sequence. Throughout the entire nucleotide sequence of yeast phenylalanine tRNA, there are only two A–C sequences, and these two sites are where the platinums have been located.

[6] M. M. Teeter, G. J. Quigley, and A. Rich, "Nucleic Acid-Metal Ion Interaction" (T. G. Spiro, ed.), p. 145. Wiley (Interscience), New York, 1980.

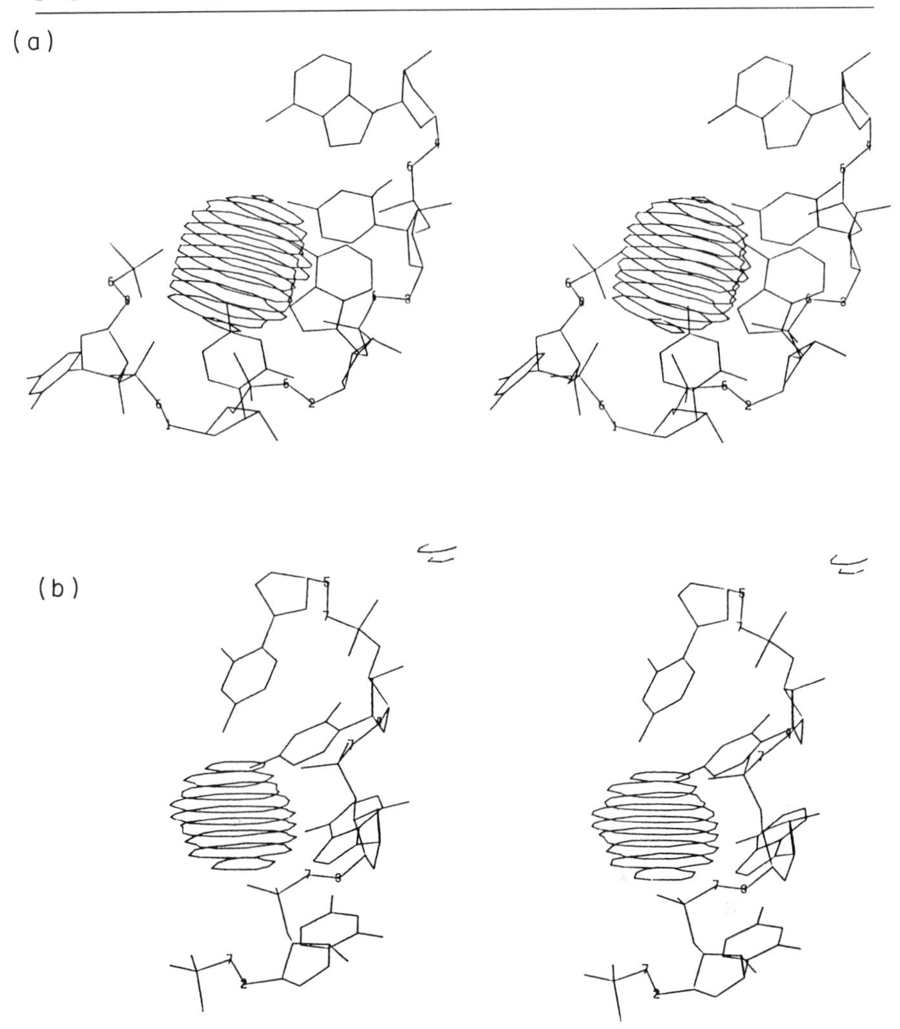

FIG. 2. Difference electron density calculated using the K_2PtCl_6 derivative data and native phases of yeast $tRNA^{Phe}$ crystals in an orthorhombic lattice. (a) Pt found between adenine of residue 62 and cytosine of residue 63; (b) Pt found between adenine of residue 73 and cytosine of residue 74; (c) Pt found between ribose of 3′-terminal residue 76 and base of residue 36 of a neighboring molecule; (d) coordination geometries of Pt(IV) ion. (e) An interpretation of the platinum site from a difference Fourier map of the *trans*-$Pt(NH_3)_2Cl_2$ derivative of yeast $tRNA^{Phe}$ in a monoclinic lattice.[2]

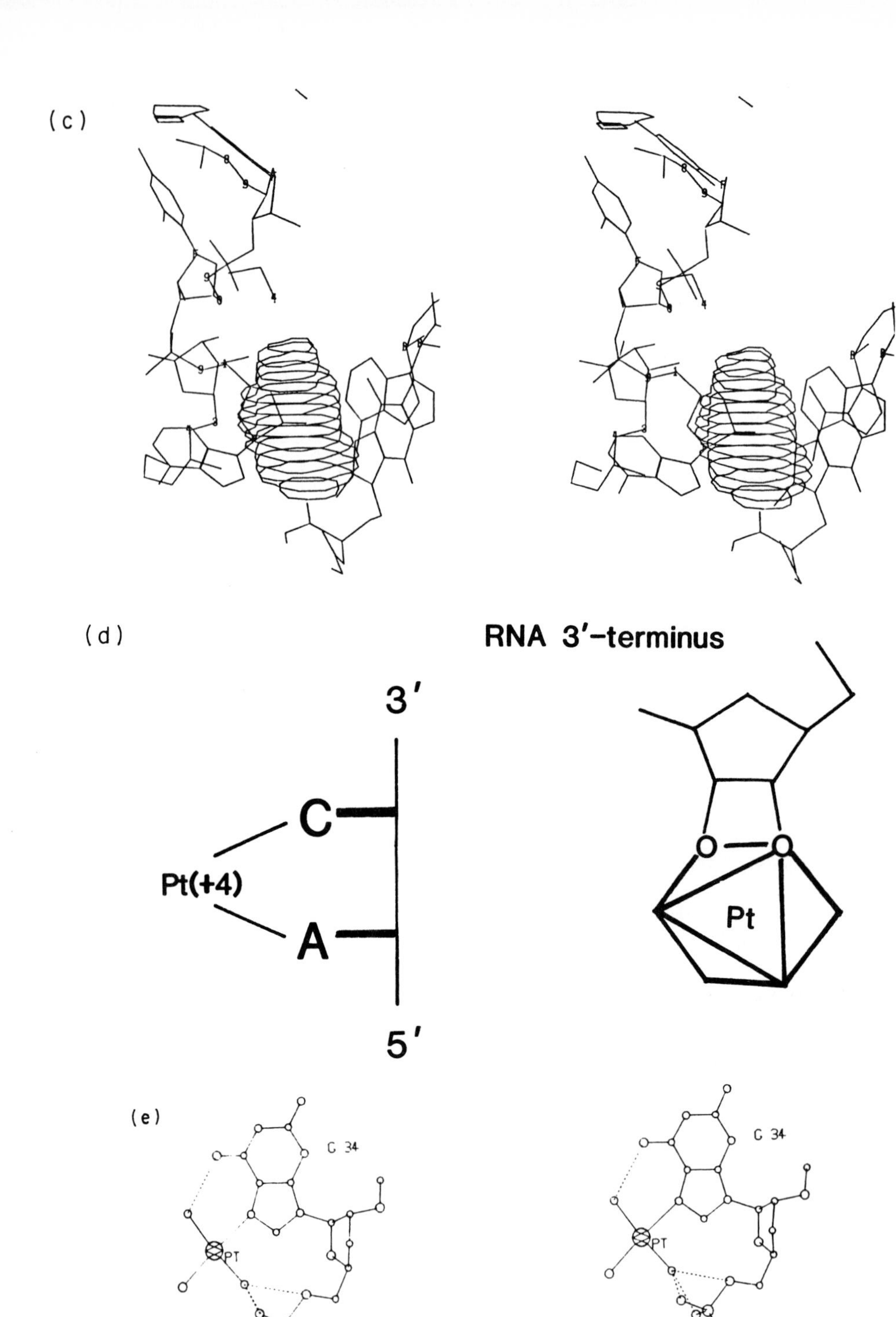

FIG. 2. (*continued*)

The third platinum site has been tentatively located between the 3′-terminal adenosine residue (residue 76) and the adenine of residue 36 of the next molecule (Fig. 2c). Although the electron density for residue 76 is weak, the platinum peak is near the O-2′ and O-3′ hydroxyl groups of the 3′-terminal ribose. The *trans*-dichlorodiammine-platinum(II) has a square planar geometry. The platinum directly coordinates to N-7 of guanine and the remaining coordination ligands make several hydrogen bonds to nearby oxygens of the same nucleotide.[2] An interpretation of this site is shown in Fig. 2e.

The anticancer drug *cis*-dichlorodiammine-platinum(II) provided a low-resolution derivative crystal of a DNA fragment 12 base-pairs long. The platinum atoms of this complex are located near N-7 and O-6 of two out of four G residues.[7]

In summary the potassium salt of hexachloroplatinate coordinates directly only to bases but not to the backbone. Furthermore, this platinum compound appears to be specific for a particular base sequence of A–C. In addition it coordinates directly to the 3′ terminus of the RNA. Similarly, *trans*- as well as *cis*-platinum coordinate to G bases using the additional hydrogen bonding environment. In all cases, platinum forms direct coordination bonds to the bases (except at the 3′ terminus) and thus provides an alternative choice of heavy atoms to the samarium or other lanthanide atoms that specifically coordinate to the backbone of nucleic acids.

Osmium

Osmium ion has been introduced into yeast phenylalanine tRNA crystals as an osmium trioxide bipyridine complex with ATP.[2,4,5] From electron density maps it appears that there is only one major binding site of this osmium complex per tRNA. The shape of the electron density of this complex in the difference Fourier maps of the orthorhombic form of the tRNA crystal is spherical, suggesting that the bound osmium complex no longer has the geometry of the ternary complex. Due to the weak electron density of residue 76 in the native Fourier map, and the shift of the residue in the derivative crystal, it is difficult to precisely assign the coordination geometry of the osmium complexed to residue 76. However, the location of the electron density is very close to the cis-diol groups of the ribose of residue 76 (Fig. 3a). From this observation and the crystal structure of an osmium trioxide bipyridine complex with adenosine,[8] we interpret that in

[7] R. Wing, H. Drew, T. Takano, C. Broka, S. Tanaka, K. Itakura, and R. E. Dickerson, *Nature (London)* **287,** 755 (1980).

[8] J. F. Conn, J. J. Kim, F. L. Suddath, P. Blattman, and A. Rich, *J. Am. Chem. Soc.* **96,** 7152 (1974).

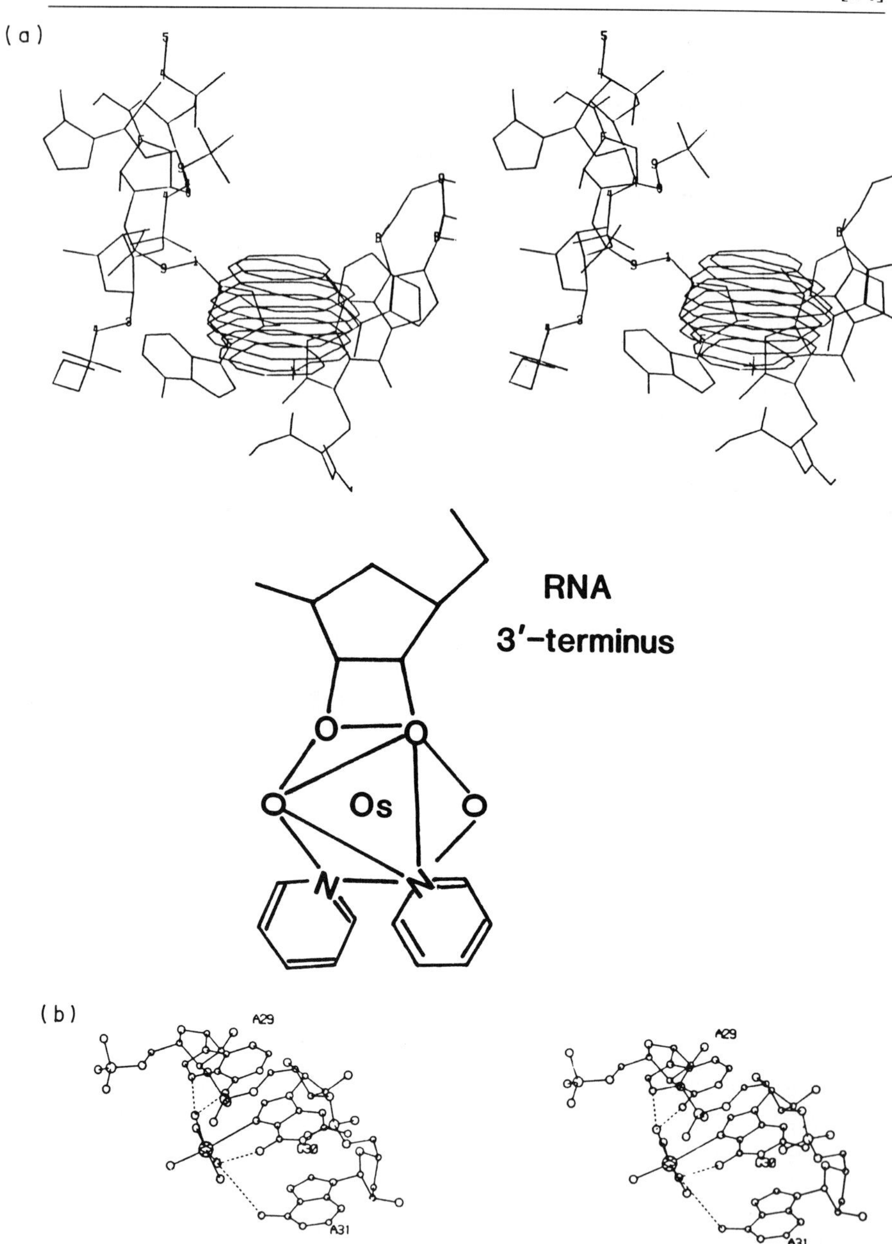

FIG. 3. (a) Difference electron density of $OsO_3 \cdot (Py)_2$ derivative data. The peak is located between the ribose of 3′-terminal residue 76 and the base of residue 36. Assumed coordination geometry is shown. (b) An interpretation of an Os site from a difference Fourier map of the $OsO_3 \cdot (Py)_2$ derivative of yeast $tRNA^{Phe}$ in a monoclinic lattice.[2]

this tRNA crystal osmium forms two direct coordination bonds to the O-2′ and O-3′ hydroxyl oxygens of the ribose of residue 76, and the remaining four coordination sites are occupied by two pyridines and two oxygens. These two oxygens are at a distance close enough to make hydrogen bonds to the bases of residues 36 and 37 of the neighboring molecule.

The same osmium complex was interpreted to coordinate to N-7 of a G residue in monoclinic crystals of this tRNA[2] as shown in Fig. 3b.

In summary, this osmium complex may be a useful derivative to label either the 3′ terminus of RNA or G residues with a large open space near the N-7 position.

Cobalt and Manganese

The chloride salts of cobalt(II) and manganese(II) have not provided good heavy-atom derivative crystals of tRNA. However, difference

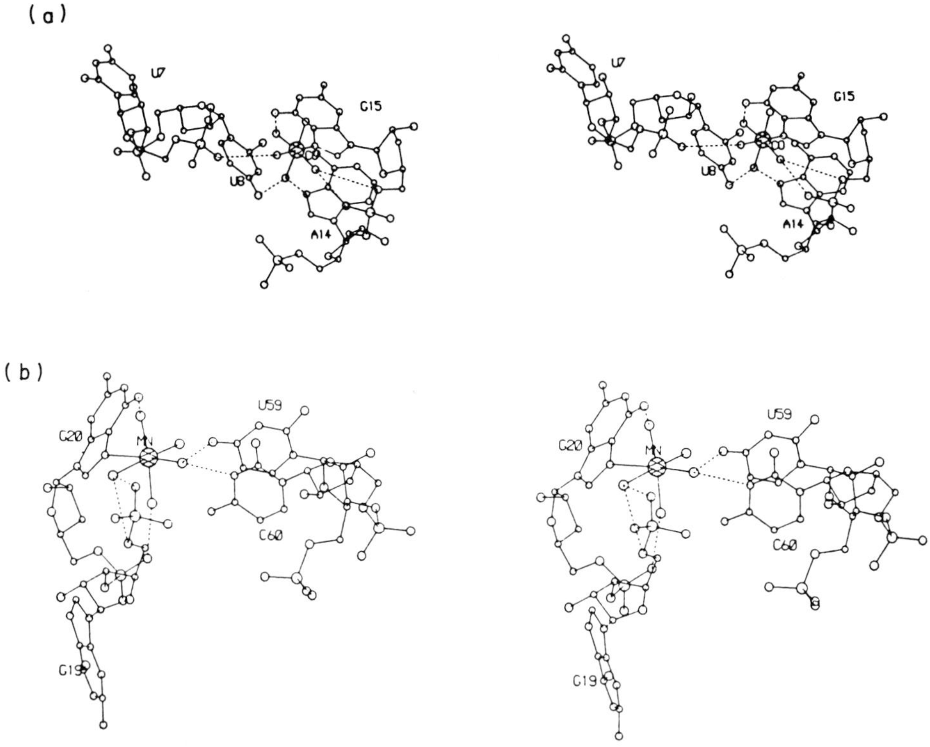

(a)

U7 C15 U8 A14

(b)

U59 C20 MN C60 C19

FIG. 4. (a) Co binding site; (b) Mn binding site interpreted from difference Fourier maps of the respective derivatives of yeast tRNA^{Phe} in a monoclinic lattice.[2]

Fourier maps made after structure refinement show both metal ions making direct coordination bonds to the N-7 atom of a G residue in the tRNA[2] as shown in Fig. 4.

Mercury

Mercury ions introduced as hydroxymercuryhydroquinone-O-O-diacetate (HMHD) bind to O-4 of a uracil in the monoclinic form of the tRNA.[2]

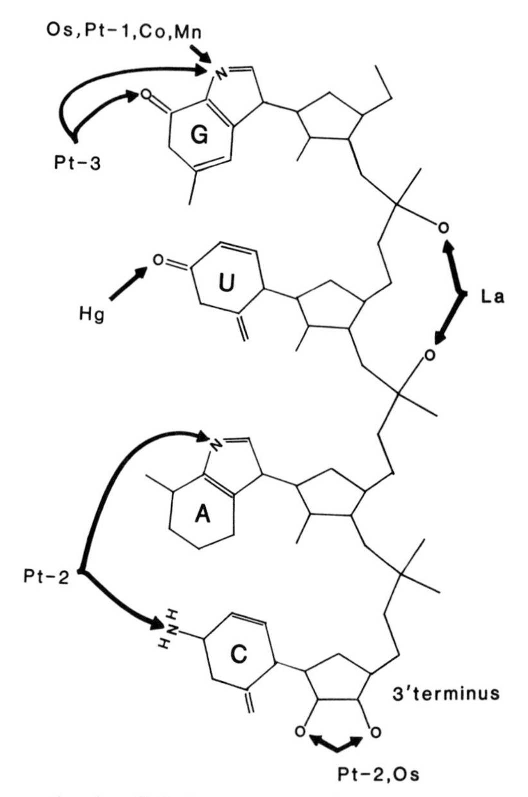

FIG. 5. A diagram showing all the heavy atoms and their presumed binding sites. A single arrow indicates formation of a single coordination bond and a double arrow for two simultaneous coordination bonds. Only those atoms in nucleic acids that form coordination bonds to the heavy atoms are labeled. Heavy atom symbols are Os from $OsO_3 \cdot (Py)_2$; Pt-1 from *trans*-$Pt(NH_3)_2Cl_2$; Pt-2 from K_2PtCl_6; Pt-3 from *cis*-$Pt(NH_3)_2Cl_2$; Co from $CoCl_2$; Mn from $MnCl_2$; Hg from hydroxymercuryhydroquinone-O-O-diacetate; La from acetate or chloride salts of lanthanide atoms.

Summary

All the heavy atoms that have so far been found to provide good derivative crystals do so by forming direct coordination bonds to either the backbone or the bases of nucleic acids in an environment where the coordination shell can be further stabilized by several hydrogen bonds. A summary of coordination sites is shown in Fig. 5 and listed below:

1. Lanthanide ions such as Sm(III), Lu(III), Pr(III), Eu(III), Tb(III), Dy(III), Gd(III) form coordination bonds to oxygen atoms of two adjacent phosphates or to phosphates from different parts of the chain.
2. The N-7 position of guanine is the most common site for heavy atoms. N-7 can become a ligand to many metal ions such as Os(VI) from $OsO_3 \cdot (Py)_2$, Pt(II) from square-planar *cis*- or *trans*-dichlorodiammine complexes, Co(II), and Mn(II).
3. The O-4 position of uracil can be a binding site for the Hg atom of hydroxymercuryhydroquinone-O-O-diacetate.
4. The N-7 of adenine and the N-4 of cytosine in the base sequence A–C can be a binding site for an octahedral platinum(IV) from K_2PtCl_6.

Acknowledgment

Research cited in this chapter has been supported by NIH and NSF.

[15] Crystallization and Heavy-Atom Derivatives of Polynucleotides

By Stephen R. Holbrook and Sung-Hou Kim

Experience in the crystallography of nucleic acids is quite limited when compared with that of proteins. Recently, the number of oligo- and polynucleotides for which crystallization and structure determination have been reported has greatly increased. We now have a large enough data base that a compilation of the published information may be helpful in future work in this rapidly expanding field. We present here summaries of the crystallization conditions which have been successful in producing crystals of sufficient quality to allow structure determination and summaries of the heavy-atom derivatives which were actually used in application of the multiple isomorphous replacement (MIR) method to solve the phase problem.

Crystallization

Due to the difference in chemical composition between nucleic acids and proteins, it was expected that the crystallization of nucleic acids would require quite different conditions than had been used for proteins. The backbone of proteins is nonionic with 20 possible side chains, only a few of which are aromatic. The backbone of nucleic acids is polyanionic and there are only four possible side chains, all of which are aromatic. Therefore, the surface of a nucleic acid has a higher charge density than that of a protein. Consequently, the intermolecular contacts which are essential for crystal formation will be quite different for nucleic acids than for proteins.

Although a wide variety of crystallization conditions have been used to obtain protein crystals suitable for structure determinations,[1] a standard protocol has been proposed which results in usable crystals for many proteins.[2] In 1968, several laboratories succeeded in obtaining single crystals of transfer RNA.[3,4] Since that time many more tRNAs and oligonucleotides have been crystallized in a form suitable for X-ray diffraction studies. The various experimental procedures which have been used to obtain single crystals suitable for X-ray diffraction studies and which have resulted in determination of structures are collected in Table I.[5-21] The crystallization conditions are as published by the authors or obtained by personal communication. Although a variety of conditions are observed to form usable crystals, we are now in a position to make generalizations about the factors which may be important in crystallization of polynucleotides.

[1] A. McPherson, *Methods Biochem. Anal.* **23,** 249 (1976).

[2] A. McPherson, *J. Biol. Chem.* **251,** 6300 (1976).

[3] S.-H. Kim and A. Rich, *Science* **162,** 1381 (1968).

[4] A. Hampel, M. Labanauskas, P. G. Connors, L. Kirkegard, U. L. Rajbhandary, P. B. Sigler, and R. M. Bock, *Science* **162,** 1384 (1968).

[5] S.-H. Kim, G. J. Quigley, F. L. Suddath, and A. Rich, *Proc. Natl. Acad. Sci. U.S.A.* **68,** 841 (1971).

[6] J. Ladner, J. Finch, A. Klug, and B. Clark, *J. Mol. Biol.* **72,** 99 (1972).

[7] T. Ichikawa and M. Sundaralingam, *Nature (London), New Biol.* **236,** 174 (1972).

[8] S.-H. Kim, G. J. Quigley, F. L. Suddath, A. McPherson, D. Sneden, J. J. Kim, J. Weinzierl, and A. Rich, *J. Mol. Biol.* **51,** 523 (1970).

[9] R. Giege, D. Moras, and J. C. Thierry, *J. Mol. Biol.* **115,** 91 (1977).

[10] C. D. Johnson, K. Adolph, J. J. Rosa, M. D. Hall, and P. B. Sigler, *Nature (London)* **226,** 1246 (1970).

[11] N. H. Woo, B. A. Roe, and A. Rich, *Nature (London)* **286,** 346 (1980).

[12] R. Wing, H. Drew, T. Takano, C. Broka, S. Tanaka, K. Itakura, and R. E. Dickerson, *Nature (London)* **287,** 755 (1980); personal communication.

[13] Z. Shakked, D. Rabinovich, W. B. T. Cruse, E. Egert, O. Kennard, G. Sala, S. A. Salisbury, and M. A. Viswamitra, *Proc. R. Soc (London) Ser. B* **213,** 479 (1981); personal communication.

TABLE I

CRYSTALLIZATION CONDITIONS FOR NUCLEIC ACIDS AND OLIGONUCLEOTIDES

Molecule	Concentration	Buffer and ions	Precipitant	T (°C)	Time	Resolution (Å)	Ref.
Yeast $tRNA^{Phe}$, orthorhombic	3–5 mg/ml	10 m*M* Na cacodylate, 10 m*M* $MgCl_2$, 1 m*M* spermine · 4 HCl, ph 6.0	Isopropanol 10% (v/v), vapor diffusion	4	1 day–2 weeks	2.5	5
Yeast $tRNA^{Phe}$, monoclinic	0.2 m*M* (vapor diffusion), 3 mg/ml (dialysis)	10 m*M* K cacodylate, 10–15 m*M* $MgCl_2$, 1.5–2 m*M* spermine, pH 7.0	Dioxane 10–20% (vapor diffusion), 6% (dialysis)	4	10 hr–3 days	2.5	6
Yeast $tRNA^{Phe}$, monoclinic	3.4 mg/ml	10 m*M* Na cacodylate, 10 m*M* $MgCl_2$, 1 m*M* spermine · 4 HCl, 7–11% 2-methyl-2,4-pentanediol, pH 6.0	2-Methyl-2,4-pentanediol 10–30% (v/v), vapor diffusion	7	2 weeks	2.5	7
Yeast $tRNA^{Phe}$, cubic	15 mg/ml	40 m*M* Na cacodylate, 40 m*M* $MgCl_2$ 4 m*M* spermine · 4 HCl, 40 m*M* EDTA, pH 6.0	Isopropanol 9% (v/v), vapor diffusion	4	2 days to months	3.0	8
Yeast $tRNA^{Asp}$	6 mg/ml	15 m*M* Na cacodylate, 15 m*M* $MgCl_2$, 3 m*M* spermine, 40% $(NH_4)_2SO_4$, pH 6.8	$(NH_4)_2SO_4$ 62% vapor diffusion	20	2–3 weeks	2.8	9

(*continued*)

TABLE I (continued)

Molecule	Concentration	Buffer and ions	Precipitant	T (°C)	Time	Resolution (Å)	Ref.
Yeast $tRNA^{fMet}$	4–6 mg/ml	5 m*M* Na cacodylate, 5 m*M* $MgCl_2$, 2 m*M* spermine · 4 HCl, 1.9 *M* $(NH_4)_2SO_4$, pH 6.0	2.7 *M* $(NH_4)_2SO_4$, vapor diffusion	4	4 weeks	4.0	10
Escherichia coli $tRNA^{fMet}$	10–14 mg/ml	50 m*M* $(NH_4)_2SO_4$, 10 m*M* $MgCl_2$, 8 m*M* spermine, 2 m*M* $BaCl_2$	Isopropanol 1–2%, vapor diffusion	4	months	2.8	11
$d(CGCGAATTCGCG)_2$	5 mg/ml	1.25 m*M* $Mg(Ac)_2$, 1 m*M* spermine · 4 HCl, pH 7.5	2-Methyl-2,4-pentanediol 20–30%, vapor diffusion	20	weeks	1.9	12
$d(GGTATACC)_2$	2 m*M*	25 m*M* Na cacodylate, 4 m*M* $BaCl_2$, 2 m*M* spermine · 4 HCl, pH 6.9	2-Methyl-2,4-pentanediol 15%, vapor diffusion	—	—	2.25	13
$d(GG5BrUA5BrUACC)_2$	2 m*M*	1.4 m*M* $Mg(Ac)_2$, 2 m*M* spermine · 4 HCl, pH 6.0	2-Methyl-2,4-pentanediol 15%, vapor diffusion	—	—	2.25	13
$d(CGCGCG)_2$	2 m*M*	30 m*M* Na cacodylate 15 m*M* $MgCl_2$ or 7 m*M* $BaCl_2$, 10 m*M* spermine · 4 HCl, pH 7.0	Isopropanol 5%, vapor diffusion	20	3 weeks	0.9	14

d(CGTACG-daunomycin)$_2$	2 mM	30 mM Na cacodylate, 15 mM $MgCl_2$, 10 mM spermine · 4 HCl, pH 6.5	2-Methyl-2,4-pentanediol 30% (v/v) vapor diffusion	20	1 week	1.5	15
d(CTTTTG): r(CAAAAG)	2 mM	10 mM Na cacodylate, 10 mM $MgCl_2$, 10 mM spermine · 4 HCl, 0.3 mM EDTA, 10% 2-methyl-2,4-pentanediol, pH 7.0	2-Methyl-2,4-pentanediol 30% (v/v), vapor diffusion	4	4 weeks	—	16
d(CGCG)$_2$	2 mM	30 mM Na cacodylate, 15 mM $MgCl_2$, ±10 mM spermine, pH 7.0	Isopropanol 5%, vapor diffusion	20	days	1.3	17
d(CGCG)$_2$ (high salt)	0.5 mg/ml	0.2 M $MgCl_2$, pH 6.3–6.5	2-Methyl-2,4-pentanediol, 10%, vapor diffusion	20	1 month	1.5	18
d(I-CCGG)$_2$	14.4 mg/ml	30 mM Na cacodylate, 0.4 mM spermine · 4 HCl, pH 7.5	Isopropanol 40–80%, vapor diffusion	2	1 month	2.1	19
d(ATAT)$_2$	—	Tris–HCl, pH 7.5	Acetone liquid diffusion	20	2 weeks	1.0	20
r($A^-A^+A^+$)	—	H_2O/acids, pH 2.8–3.5	Evaporation	—	—	0.95	21

Cations. Since nucleic acids are polyanionic molecules, the presence of cations to neutralize the negative charge is essential. Depending on the choice of cation, the precipitation or crystallization conditions are significantly affected. The general rule is that divalent cations such as magnesium enhance precipitation or crystallization. That is, the presence of a divalent cation lowers the amount of precipitating agent required to either precipitate or crystallize the nucleic acid. This is presumably due to more effective shielding of negative charges by divalent cations than by monovalent cations. The divalent cations so far used in successful nucleic acid crystallization are magnesium and barium, the latter being effective at lower concentrations.

Polyamines. The presence of polyamines such as spermine or spermidine also reduces the amount of precipitating agent required for crystallization. Divalent cations and polyamines influence the crystallization in similar ways. This is because polyamines are positively charged near neutral pH and thus behave as polycations. To date spermine has been the only polyamine found useful in crystallization, but it is not clear if this is because it has been the most widely used or because it is the most effective.

pH. Generally, the lower the pH, the faster the precipitation or crystallization of nucleic acids. Of course one has to maintain the pH range within that where the molecule is functional. The p*K* values of various functional groups on nucleic acids are far from neutral pH. For example, the first protonation of adenine occurs at the N-1 position around pH 3. Therefore, one can experiment with a reasonably wide range of crystallization conditions by using pH differences of up to 1.0 on either side of

[14] A. H.-J. Wang, G. J. Quigley, F. J. Kolpak, J. L. Crawford, J. H. van Boom, G. van der Marel, and A. Rich, *Nature (London)* **282,** 680 (1979); personal communication.

[15] G. J. Quigley, A. H.-J. Wang, G. Ughetto, G. van der Marel, J. H. van Boom, and A. Rich, *Proc. Natl. Acad. Sci. U.S.A.* **77,** 7204 (1980); personal communication.

[16] S. R. Holbrook, F. H. Martin, I. Tinoco, T.-S. Young, and S.-H. Kim, *J. Mol. Biol.* **153,** 837 (1981).

[17] J. L. Crawford, F. J. Kolpak, A. H.-J. Wang, G. J. Quigley, J. H. van Boom, G. van der Marel, and A. Rich, *Proc. Natl. Acad. Sci. U.S.A.* **77,** 4016 (1980); personal communication.

[18] H. Drew, R. E. Dickerson, and K. Itakura, *J. Mol. Biol.* **125,** 535 (1978); personal communication.

[19] B. N. Conner, T. Takano, S. Tanaka, K. Itakura, and R. E. Dickerson, *Nature (London)* **295,** 294 (1982); personal communication.

[20] M. A. Viswamitra, O. Kennard, P. G. Jones, G. N. Sheldrick, S. Salisbury, L. Falvello, and Z. Shakked, *Nature (London)* **273,** 687 (1978).

[21] D. Suck, P. C. Manor, G. Germain, C. H. Schwalbe, G. Weimann, and W. Saenger, *Nature (London), New Biol.* **246,** 161 (1973).

pH 7. All crystallizations in Table I were done between pH 6.0 and 7.5 except r(ApApA) which was crystallized at pH 2.8–3.5 in the dipolar ion form which may be less biologically relevant.

Precipitant. Finally, the precipitants of choice to grow crystals of nucleic acids appear to be the simple alcohols isopropanol and 2-methyl-2,4 pentanediol (MPD) using the vapor diffusion technique.[22] Isopropanol diffuses directly into the dip from the reservoir and MPD is added directly to the dip at a low concentration and then equilibrated with a reservoir of higher concentration. This removes water until the dip concentration effects precipitation. The use of simple alcohols as nucleic acid-precipitating agents is in contrast to the broad success of ammonium sulfate and polyethylene glycol as precipitants for crystallization of proteins.[1]

Heavy-Atom Derivatives

The heavy-atom compounds used to obtain heavy-atom derivative crystals of various nucleic acids are listed in Table II.[23–28] Three basic approaches have been taken to obtain heavy-atom derivatives from nucleic acids.

Soaking. In the conventional soaking method, heavy-atom-containing solutions are introduced into the drop containing the crystals. The modes of interaction of some of these are fairly well understood and are discussed in the next chapter. In general, one can classify these heavy-atom derivatives into two groups: those in which heavy atoms form coordination bonds to the backbone of the nucleic acid and those in which the coordination bonds are formed primarily to the bases.

Chemical Modification. The second class of heavy-atom derivative crystals is obtained by chemical modification of the polynucleotide to be crystallized. One example is the iodination of transfer RNA, in which the iodine is specifically introduced at a particular base and the chemically modified nucleic acid then crystallized.

[22] S.-H. Kim and G. J. Quigley, this series, Vol. 59, p. 3.

[23] F. L. Suddath, G. J. Quigley, A. McPherson, D. Sneden, J. J. Kim, S.-H. Kim, and A. Rich, *Nature (London)* **248,** 20 (1974); unpublished results.

[24] A. Jack, J. E. Ladner, and A. Klug, *J. Mol. Biol.* **108,** 619 (1976).

[25] C. D. Stout, H. Mizuno, S. T. Rao, P. Swaminathan, J. Rubin, T. Brennan, and M. Sundaralingam, *Acta Crystallogr., Sect. B* **B34,** 1529 (1978).

[26] D. Moras, M. B. Comarmond, J. Fischer, R. Weiss, J. C. Thierry, J. P. Ebel, and R. Giege, *Nature (London)* **288,** 669 (1980).

[27] R. W. Schevitz, A. D. Podjarny, N. Krishnamachari, J. J. Hughes, and P. B. Sigler, *Nature (London)* **278,** 188 (1979); personal communication.

[28] H. Drew, T. Takano, S. Tanaka, K. Itakura, and R. E. Dickerson, *Nature (London)* **286,** 567 (1980).

TABLE II
HEAVY-ATOM DERIVATIVES USED FOR MIR PHASING IN POLYNUCLEOTIDES

Polynucleotide	Compound	Concentration (m*M*)	Soak time	Maximum resolution (Å)	Figure of merit (%)	Ref.
Yeast $tRNA^{Phe}$, orthorhombic	K_2PtCl_4	1	2 days	4.0	66	23
	$K_2Pt(CN)_4$	1	1 week	4.0		
	$NaAu(CN)_2$	1	1 week			
	$Sm(CH_3COO)_3$	1	2–7 days	3.0		
	$Lu(CH_3COO)_3$	1	2–7 days			
	$Pr(NO_3)_3$	1	2–7 days	4.0–5.0		
	Eu, Tb, Dy, Gd (acetate)	1	2–7 days			
	K_2OsO_4	1	2 weeks	4.0		
	bis(Pyridine) osmate: ATP, CTP, UTP, AMP, GMP	Solid	2 weeks	4.0		
Yeast $tRNA^{Phe}$, monoclinic	*trans*-$Pt(NH_3)_2Cl_2$	0.2	1–2 days	2.5	67	24
	Hydroxymercurihydroquinone-*O*,*O*-diacetate	1	1–2 days	2.5		
	$Sm(CH_3COO)_3$	0.2–0.5	2–3 days	2.5		
	$LuCl_3$	0.2–0.5	2–3 days	2.5		
	Osmium–ATP complex	0.6	5 days	2.5		

Yeast tRNA^{Phe}, monoclinic	$Sm(CH_3COO)_3$	0.3	days	3.0	63	25
	$Gd_2(SO_4)_2$	0.3	days	3.0		
	trans-$Pt(NH_3)_2Cl_2$	0.35	3 days	3.0		
	$OsO_3(pyridine)_2$	3.4	2 weeks	4.0		
	$NaAuCl_2$	1.35	7 weeks	6.0		
Yeast tRNA^{Asp}	$Gd_2(SO_4)_2$	5	5 days	3.0	62	26
	$Au(en)_2Cl_3$	2	1–2 weeks	3.5		
Yeast tRNA^{fMet}	Gadolinium	—	—	4.5	49	27
	Pyridylmercury acetate	—	—	4.5		
$d(CGCGAATTCGCG)_2$	*cis*-$Cl_2(NH_3)_2Pt$	[a]	1–2 weeks	4.0	57	12
	5-Bromocytosine (C3)	Covalent, *de novo* synthesis		2.7		
$d(CGCGCG)_2$	$BaCl_2$	Co-crystallization 7 m*M* $BaCl_2$ replacing $MgCl_2$		2.0	76	14
	$CoCl_2$	[b]	2–3 days			
	$CuCl_2$	[b]	2–3 days			
$d(CGCG)_2$ (high salt)	5-Bromocytosine (C3)	Covalent, *de novo* synthesis		2.7	42	28

[a] Crystals transferred to a new drop containing 35% (sat.) *cis*-$Pt(NH_3)_2Cl_2$ and 60% 2-methyl-2,4-pentanediol (MPD). After soaking in this drop for 1–2 weeks the crystals are transferred to a 60% MPD solution to allow loosely bound heavy atoms to diffuse away.

[b] Crystals were transferred to a drop containing mother liquor plus a high precipitant concentration. From the dimensions of the crystal and estimated contents of oligonucleotide a sufficient amount of $CoCl_2$ or $CuCl_2$ solution was added to displace 2–3 Mg^{2+} ions per molecule of the oligonucleotide.

De Novo Synthesis Using Base Analogs. Heavy atoms can also be incorporated into polynucleotides by chemical syntheses from nucleosides containing a heavy atom on the base. Synthesis of polynucleotides using 5-bromocytosine has been a useful technique for introducing heavy atoms to solve both derivative and isomorphous native structures. These covalent heavy-atom derivatives can be used with Patterson and Fourier methods to directly determine structure.[19]

Acknowledgment

Our results cited in this section have been supported by grants from NIH and NSF.

[16] Determination of Protein Molecular Weight, Hydration, and Packing from Crystal Density

By B. W. MATTHEWS

Introduction

Measurements of the density of protein crystals have long been used to provide information concerning the molecular weight, hydration, partial specific volume, and packing of protein molecules in the crystal phase. Values of the molecular weight can be quite accurate (typically $\pm 4\%$), and the crystallographic method has the advantage that the results are not influenced by the dissociation or aggregation of the protein.

In this chapter we review the information to be obtained from measurements of protein crystal density and give examples of the use of the methods. The treatment closely follows that given previously.[1]

Theory

For a protein, the determination of molecular weight from the crystallographic data is complicated by the fact that the crystal contains not only the protein molecules, but also the intervening solvent.[2–5] In a survey of

[1] B. W. Matthews, *J. Mol. Biol.* **103,** 659 (1974).
[2] M. F. Perutz, *Trans. Faraday Soc.* **42B,** 187 (1946).
[3] B. W. Low and F. M. Richards, *J. Am. Chem. Soc.* **74,** 1660 (1952).
[4] B. W. Low and F. M. Richards, *Nature (London)* **170,** 412 (1952).
[5] B. W. Matthews, *J. Mol. Biol.* **33,** 491 (1968).

226 crystal forms of globular proteins[6] it was found that the nonprotein fraction of the unit cell volume varied from 30 to 78%, although for proteins of similar molecular weight the range was somewhat less. Furthermore, the nonprotein part of the unit cell is not homogeneous, but, as shown by Adair and Adair[7] and by Perutz,[2] consists of two parts: a "free solvent" region accessible to external ions and in equilibrium with the external supernatant, and a "bound water" region, not accessible to salts, which can be visualized as a hydration shell surrounding the protein.

The density D_c of the protein crystal can be written as the sum of three parts:

$$D_c = D_p(V_p/V) + D_w(V_w/V) + D_s(V_s/V) \tag{1}$$

where D_p, D_w, and D_s are the densities of the protein, bound water, and free solvent; (V_p/V), (V_w/V), and (V_s/V) are the fractional volumes occupied by these three components; and V is the unit cell volume.

If the ratio of the bound water volume to the protein volume is denoted f, i.e.,

$$f = V_w/V_p \tag{2}$$

then

$$V_s = V - V_p - V_w = V - V_p - fV_p \tag{3}$$

Substituting Eq. (3) into Eq. (1) we have

$$D_c = D_s - (V_p/V)[(1 + f)D_s - fD_w - D_p] \tag{4}$$

Hence M_p, the molecular weight of protein per asymmetric unit, is given by

$$M_p = \frac{NV_pD_p}{n} = \frac{NVD_p(D_c - D_s)}{n[D_p - D_s - f(D_s - D_w)]} \tag{5}$$

where N is Avogadro's number and n is the number of asymmetric units per unit cell.

Making the simplifying assumption that the density of the protein is equal to the reciprocal of its partial specific volume $\bar{v}_p$, and also using the relation that w, the weight fraction of bound water to protein, is given by

$$w = f\bar{v}_pD_w \tag{6}$$

[6] B. W. Matthews, *in* "The Proteins" (H. Neurath and R. L. Hill, eds.), 3rd ed., p. 403. Academic Press, New York, 1977.

[7] G. S. Adair and M. E. Adair, *Proc. R. Soc. London, Ser B* **120,** 422 (1936).

then from Eq. (5) the protein molecular weight per asymmetric unit is given by

$$M_p = \frac{NV(D_c - D_s)}{n\{1 - \bar{v}_p D_s - w[(D_s - D_w)/D_w]\}} \tag{7}$$

This is the general expression for the determination of molecular weight from protein crystals. The unit cell volume V is determined from the X-ray diffraction measurements, and the crystal density D_c is usually measured in a calibrated density gradient.[3] The partial specific volume of the protein is most easily estimated from its amino acid composition, if known, or measured pycnometrically in dilute solution.[8,9] The protein molecular weight per asymmetric unit, M_p, is the sum of the molecular weights of all polypeptides within that volume. Usually an approximate molecular weight of one polypeptide is available and this, divided into M_p, gives the approximate number of polypeptides per asymmetric unit. The true number is taken to be the closest integer. This integral value is then substituted back to obtain the crystallographic estimate of the protein molecular weight.

In cases in which the density of the free solvent or supernatant is close to that of water, Eq. (7) reduces to the simpler form

$$M_p = \frac{NV(D_c - D_w)}{[n(1 - \bar{v}_p D_w)]} \tag{8}$$

This is actually the simplest and most favorable case for the accurate determination of molecular weight. Note, in particular, that as the supernatant density D_s increases relative to the density of water D_w, the denominator in Eq. (7) becomes smaller, and therefore less accurate. When the supernatant density increases to the point that the crystals float, Eq. (7) becomes indeterminate (see the section below on Allowance for Bound Water). To obtain an accurate value for the molecular weight, one should try to have the crystal density and the mother liquor density far apart. For "high-salt" crystals, it may be advantageous to supplement the mother liquor with a dense salt such as cesium chloride to achieve a more favorable value of D_s than is provided by the "standard mother liquor."

The derivation of Eq. (8) from Eq. (7) and the assumption made above that protein density is reciprocally related to partial specific volume both imply that the density of the bound water is equal to that of bulk water. In other words, it is assumed that electrostriction[8] and the influence of

[8] E. J. Cohn and J. T. Edsall, (1965). "Proteins, Amino Acids and Peptides." Hafner, New York (Originally published by Van Nostrand-Reinhold, Princeton, New Jersey, 1943).

[9] T. L. McMeekin and K. Marshall, *Science* **116,** 142 (1952).

hydrophobic groups[10] have a negligible effect on the density of the bound water. This assumption has some justification in the available crystallographic literature. If, for example, D_w were equal to 1.05 g/cm^3 [10] rather than 0.998 g/cm^3, i.e., the density of bulk water, then for a typical protein crystal in water the apparent molecular weight calculated using Eq. (8) would be overestimated by about 5%. A survey of such calculations, using the density data available for proteins such as hemoglobin, α-chymotrypsin, γ-chymotrypsin, and thermolysin, suggests no such systematic error. Furthermore, Ten Eyck[11] determined, by sedimentation equilibrium experiments, the effect of pressure on the hydration of chymotrypsinogen, and concluded that in this instance the partial molar volume of water changed on hydration by no more than 2.9% in absolute value, the change probably being an increase of about 1%. The survey and Ten Eyck's results also tend to confirm the validity of using partial specific volumes determined pycnometrically or from amino acid composition for crystallographic molecular weight determination, or, conversely, for using crystallographic measurements to determine the partial specific volume of proteins of known molecular weight.

Density Measurement

The method we have found most suitable for crystal density measurement is as follows.

A density gradient is made in a graduated measuring cylinder by combining appropriate ratios of water-saturated xylene and carbon tetrachloride.[3,4] The gradient is calibrated by adding small droplets of various salt solutions of known density. It is advisable to check, before introducing the crystal, that its expected density and the mother liquor density are within the range of the gradient!

The crystal is then picked up in its mother liquor with a Pasteur pipet, and introduced, in a droplet of the mother liquor, to the density gradient. Using a Hamilton microsyringe, previously rinsed with the same mother liquor, the excess liquid can be drawn off from the protein crystal and the crystal shaken off the syringe tip and allowed to settle in the gradient. The prior rinsing of the microsyringe causes the droplet of mother liquor with the protein crystal to adhere to the hydrophilic point of the microsyringe, so that the removal of the excess mother liquor is straightforward. Removal of a small crystal from the microsyringe may be more difficult. If necessary, try a "flick of the tip."

[10] D. L. D. Caspar, *in* "Principles of Biomolecular Organization" (G. E. W. Wolstenholme and M. O'Connor, eds.), p. 7. Little, Brown, Boston, Massachusetts, 1966.

[11] L. F. Ten Eyck, Ph.D. Dissertation, Princeton University, Princeton, New Jersey (1970).

The above method has been found to give much more reproducible results than the method of drying the excess liquor from the crystal before introducing it into the column. It avoids partial air-drying of the crystal and is suitable for relatively small (to 0.25 mm^3) and fragile crystals. If desirable, measurements can be repeated by using a microsyringe to re-wet and redry the same crystal.

Density Extrapolation

Following the early studies of Perutz,[2] Colman and Matthews[12] pointed out that it is possible to determine the molecular weights of protein crystals, even in the presence of concentrated salt solutions, by measuring the crystal density as a function of solvent density. From Eq. (1), the relation between the two densities is linear, and a plot of D_c against D_s will yield a straight line of slope (V_s/V) and of intercept D_0, the density which the crystal would have if all the solvent could be replaced with water. Applications of this method to γ-chymotrypsin and to β-amylase[12] are illustrated in Fig. 1. Additional data for hemoglobin, adapted from the results of Perutz,[2] have been included. In each case, the observed points lie on straight lines which may be extrapolated to find the hypothetical crystal density D_0 of the crystal in water. Substituting D_0 for D_c in Eq. (8) the protein molecular weight may be obtained with an accuracy of about $\pm 4\%$ in a typical case. For hemoglobin, Perutz[2] was able to obtain the same crystal form from water and from concentrated salt solutions, allowing direct verification of the density extrapolation.

The density plots (Fig. 1) also provide additional information about the unit cell components. The slope gives the volume fraction of free solvent in the unit cell, and the ratio of the volume of bound water to the volume of protein is given by

$$f = D_p w / D_w = (D_p - D_e)/(D_e - D_w) \tag{9}$$

where D_e is the equilibrium density at which $D_c = D_s$.[7,12]

Note that the crystal density measurements do not determine $\bar{v}_p$, the partial specific volume of the protein, which is assumed in the equations given here to be equal to $1/D_p$.

Allowance for Bound Water

Often, crystal densities are measured at only one solvent density, typically that of the supernatant from which the crystals were grown.

[12] P. M. Colman and B. W. Matthews, *J. Mol. Biol.* **60,** 163 (1971).

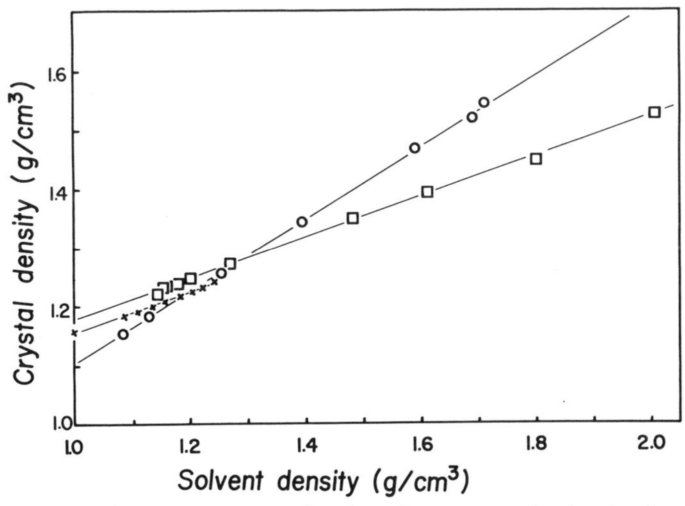

FIG. 1. Plots of crystal density as a function of supernatant density showing the linear relation between them. (□) γ-Chymotrypsin; (○) β-amylase[12]; (×) hemoglobin. (Hemoglobin adapted from the data of Perutz.[2])

Consideration of Eq. (7) shows that the molecular weight could be found from such measurements if the value of w were known.

In theory, w may be obtained very easily by increasing the supernatant density until the crystals remain just suspended [Eq. (9)], although in practice this is not often done. Note that as the crystal density tends toward the solvent density, then from Eq. (9) both the numerator and denominator of Eq. (7) tend to zero, and the molecular weight becomes undefined. In practice, crystal densities measured only for dense supernatant solutions such as saturated ammonium sulfate will tend to give inaccurate molecular weights, because of the increased experimental error in $D_c - D_s$ and the increased uncertainty introduced by the $w(D_s - D_w)$ term.

The supernatant density at which the denominator in Eq. (7) becomes equal to zero is given by

$$D_s = D_c = D_w(1 + w)/(w + \bar{v}_p D_w) \tag{10}$$

which will equal about 1.25 g/cm^3 in a typical case. As this condition is approached, small errors in either $\bar{v}_p$ or w will introduce large errors in the calculated molecular weight, in addition to the experimental uncertainties described above.

It has been shown in studies of several crystals[2,7,12] and by a variety of other techniques[13,14] that for most proteins the value of w lies in the range 0.15–0.35 g water/g protein. Thus, in estimating molecular weight from a pair of crystal and solvent densities, where w is unknown, it is preferable to use Eq. (7), substituting a reasonable value such as 0.25 for w rather than ignoring the bound water, as is sometimes done.

The consequences of ignoring bound water in the estimation of molecular weights are illustrated in the table,[15–17] in which Eq. (7) was used to calculate molecular weights from published density data for several protein crystals in a variety of solvents. In one case the bound water was ignored (i.e., w was put equal to zero) and in the other case w was put equal to 0.25 g water/g protein. No example is included in the table for which the solvent density is close to unity, as in such cases the bound water correction is superfluous.

In each case the omission of the bound water term causes a substantial difference in the molecular weight, the effect becoming more important for higher solvent densities. With the exception of glyceraldehyde phosphate dehydrogenase, the use of $w = 0.25$ g water/g protein results in a reasonably accurate value for the calculated molecular weight, a finding which suggests that this is a suitable value to use in practice (although clearly not in preference to a value determined experimentally, as described above). The solvent density for the crystals of glyceraldehyde phosphate dehydrogenase approaches the critical value given in Eq. (10), so that the apparent error in the calculated molecular weight might be due to relatively small errors either in the assumed partial specific volume or in the fraction of bound water.

Cross-Linking

Because of the simplifications in molecular weight determination which occur for "low-salt" or "salt-free" protein crystals [in which case Eq. (8) can be used instead of Eq. (7)], different methods have been proposed to measure crystal densities under salt-free conditions, even if the crystals are not normally stable in a low-salt environment. One method, in which cross-linking with glutaraldehyde is used to stabilize the

[13] H. Fisher, *Biochim. Biophys. Acta* **109,** 544 (1965).

[14] I. D. Kuntz, Jr., T. S. Brassfield, G. D. Law, and G. V. Purcell, *Science* **163,** 1329 (1969).

[15] B. W. Matthews, *J. Mol. Biol.* **33,** 499 (1968).

[16] L. J. Banaszak, D. Tsernoglou, and M. Sade, *in* "Probes of Structure and Function of Macromolecules and Membranes" (B. Chance, T. Yonetani, and A. S. Mildvan, eds.), Vol. 2, p. 71. Academic Press, New York, 1971.

[17] H. C. Watson and L. J. Banaszak, *Nature (London)* **204,** 918 (1964).

EXAMPLES OF THE CALCULATION OF MOLECULAR WEIGHT

Protein	Molecular weight	$\bar{v}_p$ (cm^3/g)	Crystal supernatant	D_s (g/cm^3)	D_c (g/cm^3)	Calculated[a] molecular weight ($w = 0$)	Calculated[b] molecular weight ($w = 0.25$)	Reference
γ-Chymotrypsin	25,260	0.736	50% saturated $(NH_4)_2SO_4$	1.155	1.232	18,300	24,700	12
γ-Chymotrypsin	25,260	0.736	Cs_2SO_4	2.007	1.525	36,000	23,500	12
Chymotrypsinogen B								
Type B	24,850	0.733	Phosphate	1.092	1.166	19,000	21,300	15
Type C	24,850	0.733	Phosphate, alcohol	1.111	1.226	21,200	25,000	15
Hemoglobin	66,700	0.749	1.6 *M* $(NH_4)_2SO_4$	1.115	1.194	50,300	63,900	2
Malate dehydrogenase	74,000	0.742	65% saturated $(NH_4)_2SO_4$	1.167	1.227	48,000	70,000	16
Glyceraldehyde phosphate dehydrogenase	140,000	0.74	3.0 *M* $(NH_4)_2SO_4$	1.192	1.243	109,000	184,000	17
β-Amylase	197,000	0.74	30% saturated $(NH_4)_2SO_4$	1.085	1.154	184,000	206,000	12
β-Amylase	197,000	0.74	CS_2SO_4	1.710	1.536	344,000	206,000	12

[a] Calculated molecular weight, ignoring bound water.

[b] Calculated molecular weight, assuming $w = 0.25$ g water/g protein.

crystals, will be discussed here; a second method, employing Ficoll gradients,[18] will be discussed in the following chapter.

The use of glutaraldehyde to stabilize protein crystals is well known.[19] In early unpublished experiments Matthews and Colman measured the densities of cross-linked crystals of γ-chymotrypsin and found that the glutaraldehyde caused a significant increase in the apparent molecular weight of the protein. However, Cornick *et al.*[20] independently tested the same approach with crystals of α-chymotrypsin and found that in this case the molecular weight of the protein could be measured with an accuracy of 3%. This apparent discrepancy was investigated by Matthews.[1]

It was found that crystals of γ-chymotrypsin required overnight soaking in a 2% glutaraldehyde solution in 55% saturated ammonium sulfate, pH 5.6, to prevent dissolution on transfer to water. Such crystals cracked when placed into water, but gave diffraction patterns similar, at least to low resolution, to those of the native protein crystals.[6] Over several hours the diffraction pattern gradually deteriorated, being essentially lost after 2 days. Crystals of γ-chymotrypsin soaked in 1% glutaraldehyde solutions exhibited more severe cracking, and partly dissolved when placed in water.

The densities of crystals cross-linked with 2% glutaraldehyde, plotted as a function of supernatant density, are shown in Fig. 2. As with the non-cross-linked native crystals,[12] the density relation is linear. When extrapolated to a solvent density of unity, the crystal density is in good agreement with that observed for the cross-linked crystals placed in water.

Very similar results were obtained for crystals of β-amylase, also cross-linked overnight in a 2% solution of glutaraldehyde in 40% saturated ammonium sulfate, pH 4.0. It is clear from the density plots shown in Fig. 2 that at least under the conditions used for these experiments, cross-linking caused a significant increase in crystal density and, therefore, in apparent protein molecular weight, amounting to 14% for γ-chymotrypsin and 25% for β-amylase. This is obviously much more than would be expected for simple cross-linking of lysine residues, and suggests a substantial amount of polymerization of the glutaraldehyde, or that more than one glutaraldehyde molecule reacts with each lysine ε-amino group.[19,21,22] For both density plots the slope decreases slightly, from 34 to 31% for γ-chymotrypsin, and from 61 to 54% for β-amylase, and presumably reflects

[18] E. M. Westbrook, *J. Mol. Biol.* **103,** 659 (1976).

[19] F. A. Quiocho and F. M. Richards, *Proc. Natl. Acad. Sci. U.S.A.* **52,** 833 (1964).

[20] G. Cornick, P. B. Sigler, and H. S. Ginsberg, *J. Mol. Biol.* **73,** 533 (1973).

[21] F. M. Richards and J. R. Knowles, *J. Mol. Biol.* **37,** 231 (1968).

[22] A. H. Korn, S. H. Feairheller, and E. M. Filachione, *J. Mol. Biol.* **65,** 525 (1972).

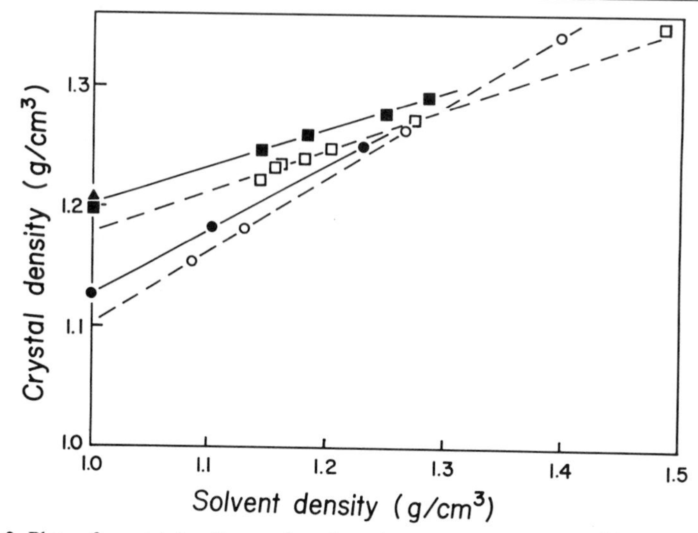

FIG. 2. Plots of crystal density as a function of supernatant density showing the effect of cross-linking with glutaraldehyde. (□, ■), γ-Chymotrypsin; (○, ●), β-amylase. The open symbols are for the normal crystals (data from Colman and Matthews[12]), and the solid symbols are for the cross-linked crystals. (▲), The density of cross-linked γ-chymotrypsin crystals in water, uncorrected for the expansion of the crystal lattice.

a corresponding decrease in the free solvent volume due to displacement by the glutaraldehyde.

In the case of β-amylase it was possible to stabilize the crystals with overnight cross-linking in 1 and 0.5% solutions of glutaraldehyde, although not with 0.25%. The densities of such crystals, after transfer to water were 1.112 and 1.101 g/cm^3, respectively, i.e., less than the more heavily cross-linked crystals (1.127 g/cm^3). The density of the least cross-linked crystals is, within experimental error, equal to that of the extrapolated density for the untreated crystals. Taken together with the results of Cornick *et al.*,[20] these measurements suggest that if protein crystals can be stabilized by soaking in, say, 0.25–0.5% glutaraldehyde solutions, then the error in molecular weight will probably be small (i.e., at most a few percent), but that molecular weights determined from more heavily cross-linked crystals must be regarded with caution. In practice, an estimate of the approximate increase in crystal density due to the cross-linking agent could be obtained by determining the difference in density between the native and cross-linked crystals under solvent conditions in which both can be measured, for example, under the conditions of crystal growth.

During experiments with the cross-linked γ-chymotrypsin crystals the density in water was observed to decrease slowly with time. This could have been due either to a gradual expansion of the crystal lattice, or to slow diffusion of residual salts out of the crystal.

It can be shown that the half-time for diffusion of salts into or out of a typical protein crystal is about 2 min[1,23] so that 30 min is ample time to allow for complete solvent exchange. In the case of γ-chymotrypsin, cross-linked crystals change their volume by about 6% during 1 day, but such changes do not occur for all protein crystals.[1] In any event, it seems desirable to check the cell dimensions of the cross-linked crystals under the same conditions and at the same time that the density measurements are made.

Other Methods

In addition to the techniques described above, a number of alternative methods of determining molecular weight and related parameters have been described. These have not had widespread acceptance, and will be reviewed only briefly here (see Matthews,[1] for additional details).

Measurement of the loss of mass of protein crystals on drying determines the fraction of solvent in the crystals and hence the molecular weight of the protein.[1,24] The method can be moderately accurate in favorable cases but is limited by the uncertainty in the amount of water that remains in the crystals after drying and by the difficulty in determining precisely the fractional loss of mass on drying the crystals.

For large, well-formed crystals, volume can be estimated by direct measurement of the crystal dimensions and the protein content determined from the optical density of a dissolved crystal or by the method of Lowry *et al.*[25] Hence, the molecular weight can be calculated.[26,27] This method has the great advantage that it requires no knowledge of crystal or solvent density, of crystal water content, or of the partial specific volume of the protein. On the other hand, accurate estimation of the volumes of individual crystals may not be feasible.

A final method[1] combines the measured density of the crystals with the mass of the crystals to obtain, in effect, the crystal volume, which is then used to estimate molecular weight by Love's method, as described

[23] H. W. Wyckoff, M. Doscher, D. Tsernoglou, T. Inagami, L. N. Johnson, K. D. Hardman, N. M. Allewell, D. M. Kelly, and F. M. Richards, *J. Mol. Biol.* **27,** 563 (1967).

[24] A. C. T. North, *Acta Crystallogr.* **12,** 512 (1959).

[25] O. H. Lowry, N. J. Rosebrough, A. L. Farr, and R. J. Randall, *J. Biol. Chem.* **193,** 265 (1951).

[26] W. E. Love, *Biochim. Biophys. Acta* **23,** 465 (1957).

[27] E. G. Heidner, B. H. Weber, and D. Eisenberg, *Science* **171,** 677 (1971).

above. This method would be suitable for relatively small or irregular crystals, and in principle could be used with a mass of microcrystals.[1]

Acknowledgments

Preparation of this chapter was supported in part by grants from the National Science Foundation (PCM8014311), the National Institutes of Health (GM20066; GM21967), and the M. J. Murdock Charitable Trust.

[17] Crystal Density Measurements Using Aqueous Ficoll Solutions

By EDWIN M. WESTBROOK

Introduction

The density of a protein crystal must be accurately measured to determine the fraction of its volume which is occupied by protein. This volume fraction can then be used to define the number of macromolecular protomers within the unit cell (n). This number n can yield information on the symmetry of the molecule, and in some cases help to identify the space group. Knowing n can also aid at later stages of the analysis of molecular structure.

Typically, protein occupies 40–50% of the crystal volume, but values between 5 and 73% have been observed for various crystals.[1,2] The remaining volume is occupied by liquid solvent. Although not an exact description, it has been convenient to model the solvent compartment as two distinct volumes: that which has been bound to the anydrous protein during solvation, and that which is free to diffuse without apparent interaction.[3–5] The reality of the "bound solvent" has been manifest in many refined crystal structures, in which significant fractions of the solvent volume contain well-defined water binding sites, associated tightly with protein surfaces.[6] Bound solvent has been shown by many studies to be

[1] B. W. Matthews, *J. Mol. Biol.* **33,** 491 (1968).
[2] C. Cohen, D. L. D. Caspar, D. A. D. Parry, and R. M. Lucas, *Cold Spring Harbor Symp. Quant. Biol.* **36,** 205 (1971).
[3] M. F. Perutz, *Trans. Faraday Soc.* **42B,** 187 (1946).
[4] F. H. C. Crick, this series, Vol. 4, p. 127.
[5] G. S. Adair and M. E. Adair, *Proc. R. Soc. London, Ser. B* **120,** 422 (1936).
[6] M. N. G. James, A. R. Sielecki, G. D. Brayer, and L. T. J. Delbaere, *J. Mol. Biol.* **144,** 43 (1980).

essentially pure water: it is inaccessable to solutes in the external milieu,[3-5,7] although bound ions are sometimes seen. Free solvent, on the other hand, is indistinguishable in composition from the external solvent surrounding the crystal. This has been firmly established by showing for many crystals that density varies as a linear function of the density of the external solvent.[7-9] By measuring the slope of this linear function, it has been possible in many cases to accurately determine the volume fraction of the crystal occupied by free solvent.[7-9] Protein crystals can therefore be modeled as a stable lattice of solvated protein, consisting of protein and bound water, interpenetrated by channels of free solvent, which are extensions of the external solution into the crystal.

The experimental determination of the protein volume fraction ϕ_p requires measurements of crystal density, mean interstitial solvent density, and protein partial specific volume. The following relationship can be readily derived from simple principles:[1,7,8]

$$\phi_p = \frac{\rho_c - \rho_s}{1/\bar{v}_p - \rho_s} = \frac{n\bar{v}_p M}{NV} \quad (1)$$

where ϕ_p is the volume fraction of the crystal occupied by protein; ρ_c, the crystal density (g/cm^3); ρ_s, the density of the total solvent compartment, free plus bound; $\bar{v}_p$, partial specific volume of the unsolvated protein (cm^3/g); n, number of protomers per unit cell; M, molar weight of one protomer (g/mol); N, Avogadro's number (6.023×10^{23} mol^{-1}); V, volume of the unit cell (cm^3).

Since n is an integer, and must be exactly divisible by the number of asymmetric units in the unit cell, it can usually be determined without ambiguity by even rather crude measurements, if good estimates for the protein partial specific volume and molar weight are available. However, if ϕ_p can be accurately measured, the determination of n is simplified, and ϕ_p can be used in the accurate determinations of M or $\bar{v}_p$, using Eq. (1).

The density of a protein crystal has in the past been measured in a gradient of organic solvents, such as bromobenzene and xylene, calibrated with small drops of aqueous salt solutions whose densities have been determined pycnometrically.[10,11] There are several problems with this traditional method. First, the crystal must be handled a great deal when introducing it into the gradient; usually it must be dried on filter paper before being gently scraped or tapped it into the organic solvents.

[7] W. J. Scanlon and D. Eisenberg, *J. Mol. Biol.* **98,** 485 (1975).
[8] P. M. Colman and B. W. Matthews, *J. Mol. Biol.* **60,** 163 (1971).
[9] G. Cornick. P. B. Sigler, and H. S. Ginsberg, *J. Mol. Biol.* **73,** 533 (1973).
[10] B. W. Low and F. M. Richards, *J. Am. Chem. Soc.* **74,** 1660 (1952).
[11] B. W. Low and F. M. Richards, *J. Am. Chem. Soc.* **76,** 2511 (1954).

With too little drying the crystal remains in a surrounding droplet of aqueous solution, but too much drying leaves the crystal crushed and broken. Second, proteins can interact with organic solvents in diverse ways: some crystals are grown from solutions containing alcohols or other miscible solvents[12]; exposed hydrophobic residues of proteins can adsorb organic solvents, while some components of the protein may be soluble in the organic solvent milieu.[13] Third, there is a small but finite capacity of organic solvents to desiccate protein crystals, altering their measured densities.[11]

A second problem with the traditional method is that, even if crystal density can be accurately measured, determining ϕ_p still requires that the mean solvent density ρ_s be accurately known. This parameter is the volume-averaged density of the two solvent regions within the crystal. The volume fractions of free and bound solvent can be determined, permitting ρ_s to be calculated, but this process is cumbersome.[8,9] However the precise value of ρ_s could also be known if the free and bound solvents were compositionally identical, i.e., both were pure water.

For these reasons, a procedure for measuring crystal densities in aqueous density gradients was devised, using Ficoll as the only solute.[14] Ficoll is a large polymer (M_r 400,000) of sucrose, cross-linked with epichlorhydrin. It is very hydrophilic and is extremely soluble, with chemical properties similar to sucrose. Being highly cross-linked, each Ficoll molecule tends to be globular, and it is so large that it effectively is excluded from protein crystals. Highly concentrated Ficoll solutions effectively precipitate proteins from solution in a manner resembling that of polyethylene glycol, and can prevent protein crystals from dissolving even in the absence of other solutes.

Ficoll is manufactured by Pharmacia specifically for making density gradients used in the separation of intracellular organelles or intact cells. Figure 1 shows the density of aqueous Ficoll solutions as a function of Ficoll concentration. Since the polymer is so large, gradients tend to be very stable against diffusion, and the solutions are so viscous that these gradients are not easily disturbed. For a given density, Ficoll solutions have far less osmotic pressure than sucrose solutions. And although these solutions are rather viscous, they are much less viscous than equivalent dextran solutions, because of the compact nature of Ficoll polymers. Indeed, dextran gradients were assessed 30 years ago for this same purpose, and although the method showed promise, it was rejected because of the extreme viscosity of the dextran solutions.[11]

[12] A. McPherson, *Methods Biochem. Anal.* **23,** 249 (1975).

[13] M. L. Collins, Ph.D. Thesis, Washington University, St. Louis (1976).

[14] E. M. Westbrook, *J. Mol. Biol.* **103,** 659 (1976).

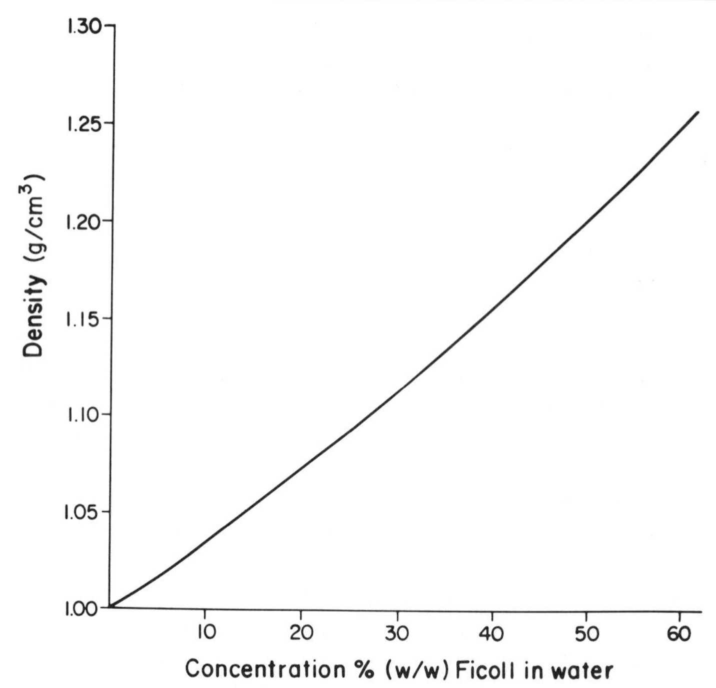

FIG. 1. Density of aqueous Ficoll solutions as a function of Ficoll concentration (at 20°). The maximum density that can be obtained with these solutions is about 1.28 g/cm^3.

Materials and Methods

Ficoll was purchased from Pharmacia, and carbon tetrachloride and toluene from Eastman. Crystals of the following proteins were grown, or furnished as indicated:

Monoclinic steroid isomerase[15]
Hexagonal steroid isomerase[14]
Monoclinic cholera toxin crystals[16] furnished by Dr. P. B. Sigler
Tetragonal hen egg-white lysozyme[17]
Tetragonal γ-chymotrypsin[18]
Orthorhombic microcrystals of frog egg phosvitin–lipovitellin complex[19] furnished by Dr. M. L. Collins

[15] E. M. Westbrook, P. B. Sigler, H. Berman, J. P. Glusker, G. Bunick, A. Benson, and P. Talalay, *J. Mol. Biol.* **103,** 665 (1976).

[16] P. B. Sigler, M. E. Druyan, H. C. Kiefer, and R. A. Finkelstein, *Science* **197,** 1277 (1977).

[17] G. Alderton and H. L. Fevold, *J. Biol. Chem.* **164,** 1 (1946).

[18] P. B. Sigler and H. C. W. Skinner, *Biochem. Biophys. Res. Commun.* **13,** 236 (1963).

[19] D. H. Ohlendorf, M. L. Collins, E. O. Puronen, L. J. Banaszak, and S. C. Harrison, *J. Mol. Biol.* **99,** 153 (1975).

Phosvitin–lipovitellin was a microcrystalline suspension; all others were single crystals.

A 60% (w/w) Ficoll solution was made by adding 75 g of Ficoll to 50 ml of distilled, deionized water at 50° with gentle stirring by hand. The lumpy, white, viscous suspension was covered, kept warm, and allowed to stand overnight to yield a clear, slightly yellow solution having the viscosity of corn syrup and a density of about 1.28 g/ml. A 30% (w/w) Ficoll solution, having a density of about 1.1 g/ml, was made by mixing equal volumes of water and 60% Ficoll solution.

Density gradients were made in 5-ml nitrocellulose ultracentrifuge tubes, which have overall lengths of about 4.0 cm. Gradients were formed with a specially designed gradient maker, which allowed mechanical stirring of these viscous solutions in the mixing chamber (magnetic stirring failed). The solutions were forced from their chambers by piston plungers pressing on a 2-cm column of air above the solutions; the piston for the mixing chamber required a hole to allow passage of the thin metal mixing rod. The mixing rod was driven by an electric motor set to turn at about 60 rpm.

Density gradients were calibrated with 0.005-ml droplets of organic solvents (previously saturated with water), introduced into the gradients with Pasteur pipets whose tips had been flame-narrowed. Calibration droplets were aliquots of solutions made by mixing carbon tetrachloride (density 1.594 g/ml) and toluene (density 0.8669 g/ml) in ratios necessary to give the desired densities. The densities of these calibration solutions were accurately determined pycnometrically prior to use.

Single crystals were introduced into the gradient by capillary, and care was taken to transfer the minimum mother liquor volume. It was not necessary to dry the crystal or otherwise manipulate it. Microcrystals were transferred to the gradient in 10 μl of a suspension. Although the calibration droplets and large crystals could reach their isopycnic positions overnight, it was more satisfactory to use gentle centrifugation (10,000 *g* in a swinging-bucket rotor) for 10–30 min to quickly bring these gradients to equilibrium. Recentrifugation was never noted to alter the relative locations of the calibration droplets and the sample crystal. It was found that crystal densities were determined more reproducibly if the measurements were made quickly, within a short time after the introduction into the gradients of the calibration droplets and the sample crystal.

Positions of the crystal and calibration droplets were measured with a metric ruler to the nearest 0.5 mm. Crystal density was determined by interpolation between bracketing calibration droplets. An attempt was always made to use calibration droplets with densities very close to predicted crystal densities.

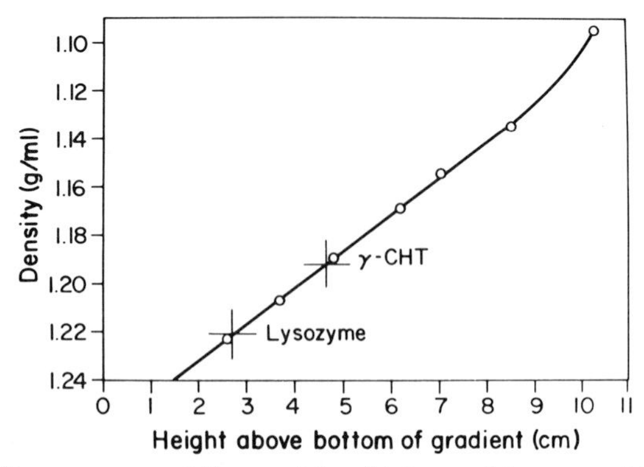

FIG. 2. Density measurement for crystals of tetragonal hen egg-white lysozyme and tetragonal γ-chymotrypsin. Gradient had been centrifuged at 10,000 *g* for 10 min. (○) Locations of calibration droplets; size of circles represents experimental error limits. Vertical lines represent observed locations of crystals, which have experimental errors commensurate with those of the calibration droplets.

Whenever possible, partial specific volumes of the proteins were measured by the method of Kratky *et al.*,[20] using a DMA 601 Density Measuring Cell and DMA 60 Density Meter, manufactured by Mettler/Parr. This simple procedure greatly improved the accuracy of all derived quantities, as assessed by the self-consistency of Eq. (1).

All single crystals whose densities had been determined in these Ficoll gradients were subjected to X-ray diffraction study, to determine their unit cell dimensions in the Ficoll solutions. This was done by mounting the crystals directly from their Ficoll gradients into thin-walled capillaries. If the unit cell volume in Ficoll differed from that of the native form, the value used in Eq. (1) was that found in Ficoll.

Results

Figure 2 shows a plot of the results of one gradient in which both lysozyme and chymotrypsin crystal densities were measured. Figure 3 shows a diagram of the results of steroid isomerase crystal density measurements. The table documents the density measurements of all six crystal forms discussed here, and the determined values of n.

As shown in Figure 2, it is quite possible to reproducibly obtain a linear gradient using the gradient maker described here. Quite linear gradients can also be achieved by hand mixing and gravity feeding the Ficoll solutions, if care is exercised, but this procedure is tedious. However if

[20] O. Kratky, H. Leopold, and H. Stabinger, this series, Vol. 27, p. 98.

Physical Parameters Measured from Six Different Crystals

Type of crystal	Crystal density (g/cm^3)	Molecular weight	Unit cell volume ($Å^3$)	Partial specific volume (cm^3/g)	Protein volume fraction (%)	n (protomers in unit cell)	Protomers per asymmetric unit
Steroid isomerase (monoclinic)	1.163	13,400	862,700	0.737[a]	45.7	24.04	12.02
Steroid isomerase (hexagonal)	1.152	13,400	1,866,000	0.737[a]	42.6	48.47	4.04
Cholera toxin (monoclinic)	1.180	87,500	428,000[b] 435,700[d]	0.733[c]	49.4	2.02	1.01
Lysozyme (tetragonal)	1.221	14,300	237,100[b] 228,200[d]	0.741[a]	63.2	8.20	1.02
γ-Chymotrypsin (tetragonal)	1.192	25,200	474,600	0.734[a]	53.0	8.180	1.02
Phosvitin–lipovitellin (orthorhombic)	1.193	199,000	3,090,000	0.775[a]	66.5	8.02	2.00

[a] Measured value, using Kratky method.[20]
[b] Value of native form.
[c] Measured hydrodynamically.[21]
[d] Value determined in Ficoll solution.

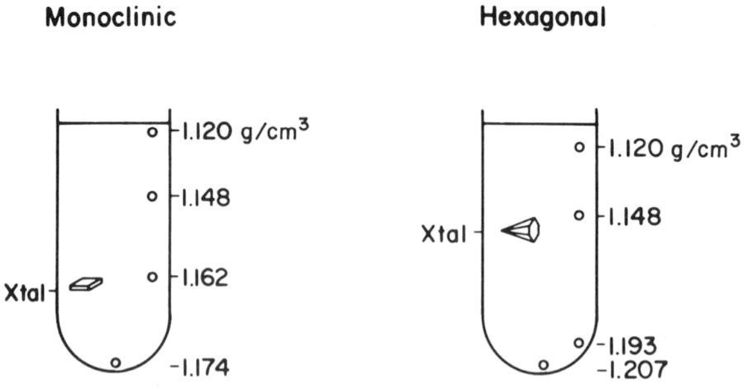

Monoclinic

Xtal

1.120 g/cm^3
1.148
1.162
1.174

Hexagonal

Xtal

1.120 g/cm^3
1.148
1.193
1.207

No. of molecules per A.U.	Expected crystal density, ρ_c (g/cm^3) in Ficoll
10	1.136
11	1.150
12	1.163
13	1.177
14	1.191

No. of molecules per A.U.	Expected crystal density, ρ_c (g/cm^3) in Ficoll
2	1.077
3	1.116
4	1.154
5	1.193

$$\rho_c = 1 + \phi(1 + 1/\bar{v})$$

FIG. 3. Top: Diagrammatic representation of experimental results for two crystal forms of steroid isomerase. Diagrams are roughly to scale, but crystal sizes have been exaggerated. Calibration droplets have been represented by circles on the right sides of each gradient; droplets in the center at the bottom had sunk during centrifugation to the gradient bottom. Bottom: Predicted crystal densities of the two crystal forms of steroid isomerase, for given number of molecules per asymmetric unit (A.U.). In both experiments, the results unambiguously determined this number. Partial specific volume for this determination was 0.737 cm^3/g.

sufficient numbers of calibration droplets are present near the crystal's density, gradient linearity is not essential for high accuracy.

Figures 1 and 2 are quite typical of the results obtained by this method. Calibration droplets reached their isopycnic positions quickly, often without centrifugation. Crystals with volumes greater than 0.03 μl usually reached their isopycnic positions in 30 min without centrifugation. However the most reliable density measurements were obtained after gradients were centrifuged for at least 10 min. The microcrystalline suspension of phosvitin–lipovitellin formed a sharp boundary at its (bottom) high-density side, corresponding to the correct crystal density, while the (upper) low-density side trailed off indistinctly.

The unit cell volume of tetragonal lysozyme shrank 3.4% in its Ficoll solution, as determined by X-ray diffraction. While the diffraction intensities changed greatly, the crystal remained very well ordered and remained in the same space group. The unit cell volume of monoclinic cholera toxin increased 1.8% in the Ficoll solution, and the crystal appeared to have deteriorated as judged by the quality of its diffraction pattern. The unit cell dimensions of all other single crystals remained unchanged in their Ficoll solutions, and the crystals did not appear to deteriorate as a result of this treatment.

Protein volume fractions determined for all six crystals discussed here lie well within the "typical" range for protein crystals.[1] The values for n determined for these crystals are all within 2% of the correct integer value.

Discussion

The Ficoll method is clearly a convenient, accurate means for determining the protein content of a crystal, and hence n. Accuracies of 2% or better can be achieved, if all other pertinent parameters in Eq. (1) have also been well measured. While such accuracy in the determined value of n is seldom important per se, it can be very useful in the determination of unknown or poorly measured molecular weights. It can also shed light on protein quaternary structures. For example, the value given here for ϕ_p in monoclinic cholera toxin crystals defined the molecular weight of the toxin A_1B_m oligomer, where the value of m had not been firmly established. This molecular weight clearly implied $m = 5$, a result later confirmed by other studies.[16,22]

Ficoll is compatible with a wide variety of buffers, and there is no reason that accurate crystal density measurements cannot be made in the presence of high concentrations of buffers or salts along with Ficoll. However if the Ficoll solutions remain essentially solute free, the free solvent regions of the protein crystal contain only water, so that the entire solvent compartment of the crystal may be treated as a uniform region of pure water. This makes $\rho_s = 1.0\ \mathrm{g/cm^3}$ in Eq. (1), and reduces the error in the determined value of the protein volume fraction ϕ_p. The concept of converting the entire solvent compartment to water is the basis for other methods to determine accurately protein volume fractions in crystals.[8,9]

Ficoll density gradients were particularly useful for the density measurement of crystals of phosvitin–lipovitellin complex, a large fraction of which is lipid. The lipid component of this complex dissolved in organic solvent gradients, preventing any proper crystal density measurement by

[21] S. Van Heyningen, *J. Infect. Dis.* **133,** S5 (1976).

[22] D. M. Gill, *Biochemistry* **15,** 1242 (1976).

that method.[13] In contrast, it was not possible to measure accurately the densities of two different tRNA crystals, because their densities lay outside the range that can be achieved with aqueous Ficoll solutions.

Densities measured in Ficoll density gradients are more accurate if one performs the measurement soon after introduction of the crystal and calibration droplets into the gradient. Failure to act quickly may lead to erroneously high values for crystal densities, a problem which is even worse for organic solvent gradients. For example, densities of β-lactoglobulin crystals being measured in bromobenzene–xylene gradients changed measurably in only 10 min,[11] whereas crystal densities in Ficoll gradients generally remain stable for several hours or may never change.

Acknowledgments

The author is most grateful to Dr. Paul Sigler (University of Chicago) and Dr. David Eisenberg (UCLA) for encouragement and suggestions. This work was performed at Chicago in the laboratory of Dr. Sigler, and was supported by Public Health Service Training Grant No. T05 GM01939 (MSTP) and NIH Grant GM22324.

Section III

Data Collection

A. Photographic Techniques
Articles 18 through 23

B. Diffractometry
Articles 24 through 33

[18] Photographic Science and Microdensitometry in X-Ray Diffraction Data Collection

By MICHAEL ELDER

Introduction

The trend toward automation in the collection of X-ray diffraction intensities has favored the use of the counter techniques associated with the modern X-ray single-crystal diffractometer, and militated against the use of film as a method of data capture. This is natural, since film techniques suffer from a number of drawbacks. One is the need for subsequent densitometry, which introduces both a logistical problem and a second potential source of error into intensity measurements. A second is that film provides an inherently less accurate recording method than counter techniques: there is a buildup of X-ray background on a film during exposure, whereas the background associated with a reflection measured with a counter is purely local.

However, the increasing amount of work being done with crystals of large unit cell has tended to reverse this trend. The number of diffraction maxima which have to be measured for a high-resolution study of a crystal of large unit cell means that a technique which makes use of the simultaneous recording of many reflections has greatly increased efficiency over single counter techniques. An increase in data-recording efficiency means a reduction in crystal exposure time, an important factor in the study of crystals of biological macromolecules, which usually deteriorate rapidly in the X-ray beam.

X-ray film provides an area detector which has constant sensitivity over its full area, and good positional accuracy. Film responds linearly to X-radiation from the lowest levels of exposure, which is an essential factor in its use for the accurate measurement of diffraction intensities. For light of longer wavelength, the process of latent image formation in the photographic process requires that several quanta must strike a silver halide grain in order to sensitize it, leading to nonlinear response at low levels of intensity. For X rays, on the other hand, the energy associated with a single quantum is sufficient to sensitize a silver halide grain to subsequent development, so that film responds linearly until it reaches levels of exposure such that there is a significant probability of an incident quantum striking a grain which has already been sensitized.

The minimization of crystal exposure time and the full utilization of these area detector properties of film require a data collection technique which records as many diffraction maxima as possible upon the one film. The screened techniques in use with Weissenberg[1] and precession[2] cameras separate out the reflections from a single reciprocal lattice level and make indexing problems much easier, but are wasteful in terms of crystal exposure to X rays since the majority of the reflections which occur during the exposure of a given photograph are not recorded on the film. Screenless techniques were developed for the study of large unit cell crystals in order to overcome this deficiency. Historically, the first to be developed was the screenless precession technique[3] in which a small precession angle minimizes the amount of reflection overlap on the photographs. There followed the small-angle screenless oscillation technique[4] which, with the commercial development of the Arndt–Wonacott oscillation camera, now dominates the use of film in macromolecular crystallography.

The indexing complications associated with screenless techniques and the quantity of data which need to be measured pose problems which are only made tractable by the use of fast, computer-controlled sampling microdensitometers. Indeed, it was the earlier development of such machines, the first rotating drum scanners, by Abrahamsson[5] and Xuong[6] which made possible the routine use of screenless precession and oscillation methods for film data collection.

The techniques themselves are described in subsequent chapters; the properties and potential of rotating drum microdensitometers and X-ray film are further described here.

Microdensitometry: General Principles

Measurements of Density

In densitometry the optical density (D) of a developed film is measured by determining the ratio of the intensity of incident light (I_0) to that of light transmitted by the film (I_t) over the sample area. The relationship is logarithmic:

$$D = \log(I_0/I_t) \quad (1)$$

[1] K. Weissenberg, *Z. Phys.* **23,** 229 (1924).

[2] M. J. Buerger, "X-Ray Crystallography." Wiley, New York, 1942.

[3] N. H. Xuong, J. Kraut, O. Seely, S. T. Freer, and C. S. Wright, *Acta Crystallogr., Sect. B* **B24,** 289 (1968).

[4] U. W. Arndt, J. N. Champness, R. P. Phizackerley, and A. J. Wonacott, *J. Appl. Crystallogr.* **6,** 457 (1973).

[5] S. Abrahamsson, *J. Sci. Instrum.* **43,** 931 (1966).

[6] N. H. Xuong, *J. Phys. E* **2,** 485 (1969).

from which it follows that a device capable of measuring optical density over the range 0–3 must be capable of measuring light intensities which vary by a factor of 10^3. A microdensitometer is simply a densitometer capable of measuring optical density over a small area of film.

Inherent in the measurement of film optical density with a microdensitometer as a means of measuring diffraction spot intensities is the assumption that density relates linearly to the weight of silver per unit area in the developed emulsion, and hence to the level of exposure to X-ray quanta. The extent to which this assumption is justified for X-ray film is discussed later; it is also important to be aware of any limitations imposed by the densitometer configuration.

Microdensitometer Optical System

The traditional Köhler illumination system of a single-beam microdensitometer such as is of interest in crystallography is shown in diagrammatic form in Fig. 1. The sample plane (the film being scanned) is illuminated from the influx side by the geometric projection of the source aperture. The intensity of the light transmitted by the film is recorded on the efflux side of the sample plane by a photodetector which views the (hypothetical) back-projection of the sampling aperture upon this plane.

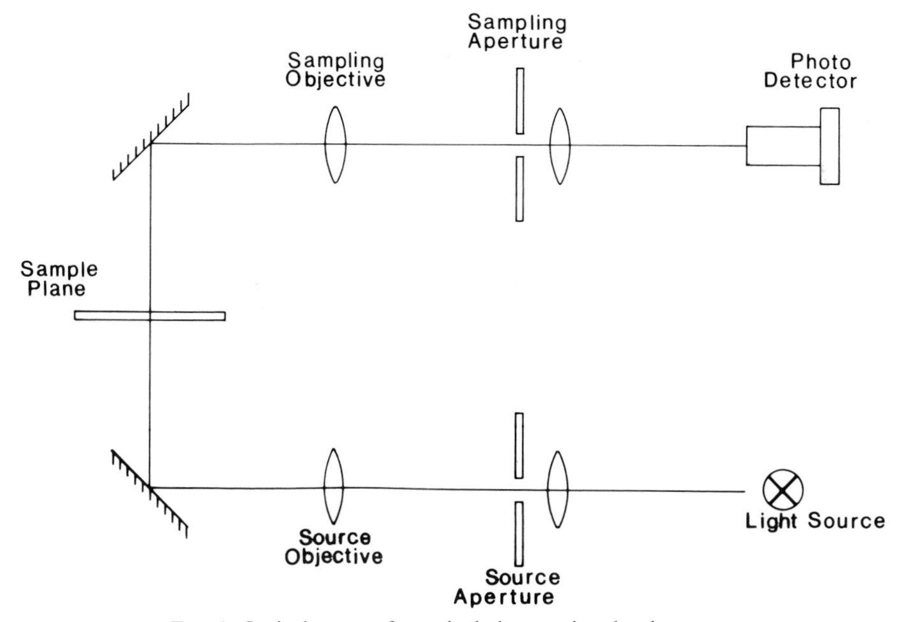

FIG. 1. Optical system for a single-beam microdensitometer.

The sample is conventionally overilluminated to ensure uniformity of illumination, which means that the projection of the source aperture upon the sample exceeds the back-projection of the sampling aperture by a small factor. The light source should be effectively incoherent in order to avoid the nonlinearity of densitometric response associated with an effective point source. There is also the risk of increased light scatter and instrumental flare associated with overillumination to be taken into account. A quantitative description of the configuration of a microdensitometer system designed to balance these effects is given by Swing,[7] and a useful survey of the principles of optical microdensitometry is given in the book by Dainty and Shaw.[8]

The conventional analysis of the conditions that must be met for effective incoherence in a microdensitometer produces the inequalities

$$NA_c/NA_0 \geq 1 + \omega_{max}/\omega_0 \quad (2)$$

where NA_c and NA_0 are the numerical apertures of the condensing (source) objective and the sampling objective, respectively; ω_{max} is the maximum spatial frequency to be measured; and ω_0 is the maximum spatial frequency of the sampling objective; and

$$W \geq 4\lambda/NA_c \quad (3)$$

where W is the width of the source aperture in the sample plane and λ the wavelength of the light used. For a sampling microdensitometer in which intensity measurement is of greater importance than the resolution of fine detail in the image, these requirements impose no great stringency. The cutoff spatial frequency for an aberration-free lens is given by

$$\omega_0 = 2NA/\lambda \quad (4)$$

so for $\lambda = 500$ nm and $NA_0 = 0.25$, ω_0 is 1000 cycles/mm. However, a sampling aperture of 50 μm in a standard rotating drum scanner implies $\omega_{max} = 20$ cycles/mm so that the numerical aperture of the source objective need be little if any greater than that of the sampling objective to satisfy the first inequality, and a source aperture of more than 50 μm easily satisfies the second inequality, which requires $W \geq 8$ μm for $NA_c = 0.25$.

Diffuse and Specular Density

There are two extremes in the method by which density can be measured. A diffuse density measurement involves the use of a wide-angle

[7] R. E. Swing, *Opt. Eng.* **15,** 559 (1976).

[8] J. C. Dainty and R. Shaw, "Image Science." Academic Press, New York, 1974.

condensing lens to collect all the light which passes through the film, including light scattered by film grain. At the other extreme a specular density measurement involves measuring only light transmitted by the film in a direction normal to its surface. Specular density is always higher than diffuse density for the same sample, since less light reaches the photodetector for a given intensity of illumination. The Köhler illumination system is at the specular end of this range as most of the film grain scattered light does not reach the photodetector. However, because the sample is slightly overilluminated some light scatter will occur from regions of film outside the back-projection of the sampling aperture, thus increasing the effective sampling area. This scatter of light by grains in the sample emulsion is known as the Callier effect. It has two effects upon microdensitometer characteristics: by increasing the effective sampling area it causes a reduction in the measured density fluctuation in the sample with increasing overillumination[9]; and it affects the transfer characteristics of the device in much the same way as instrumental flare, discussed next.

Flare

Instrumental flare is another factor in the design of a microdensitometer optical system. Extra light reflected off instrument surfaces and reaching the photodetector will tend to increase the transmittance and hence decrease the measured density, and also to affect the transfer function of the instrument by broadening sharp edges in the film density. This latter effect is more important for resolving peaks in line spectra than for making density measurements from smoothly varying reflection profiles. Flare can be reduced by including light baffles in the instrument and by painting internal surfaces matt black. It can be minimized by reducing the source aperture size as far as possible, subject to the overriding need for uniform sample illumination and effective source incoherence.

Rotating Drum Scanners

The rotating drum microdensitometers in general use by crystallographers for measuring diffraction spot intensities sacrifice resolution and high densitometric accuracy in the interests of scanning speed. The film is mounted on a cylindrical drum which rotates about its axis at speeds of up to 12 rev/sec. The light source and detector are mounted in such a way that they are stationary while density measurements are taken during one

[9] W. Kraus, *Photogr. Sci. Eng.* **12,** 217 (1968).

drum revolution, but can be stepped along the drum axis between successive revolutions in order to sample the film on a square raster. Readings taken every 100 μm from a drum of circumference approximately 40 cm lead to a data transfer rate of about 50 kHz, although a standard film only occupies part of the drum circumference so that the effective data transfer rate is lower than this.

The two instruments most commonly used are the Photoscan manufactured by Optronics, Inc. in the United States and the Scandig by Joyce-Loebl, Ltd. in the United Kingdom. These instruments use, respectively, tungsten and tungsten halogen light sources, and a photomultiplier and a photodiode as light detectors. The voltage output from the photodetectors is amplified logarithmically and this signal is passed through an analog-to-digital converter (ADC) to produce an 8-bit density value. There is some variation in the optical density range that may be covered, but it is usual to select a 0–2 or 0–3D range, discarding values much above 2D as being uncorrectable for the effects of nonlinearity of film response.

Film mounting is achieved by clamping the film over an aperture in the drum on the Optronics instrument, and taping the film to the clear acrylic plastic sleeve that forms the drum on the Scandig. Both instruments work by the single-beam, recurrent auto-zero method. During each drum revolution a zero density value is recorded, through an air gap in the drum on the Optronics or through the plastic on the Scandig, and the inverse of this logarithmically amplified voltage is added to the signal that is presented to the ADC during the remainder of the drum revolution. Thus density values are recorded relative to zero as equivalent to 100% transmittance and are corrected for drift effects.

Both instruments employ Köhler illumination systems and thus measure effectively specular density. The Optronics uses fiber optics to deliver light from the source to the illuminating aperture; in the Scandig the source light passes through a heat filter to remove the infrared component and is the reflected onto the illuminating aperture by a mirror. Figure 2 shows a plot of measured density on the Scandig device for a calibrated step tablet with diffuse densities ranging from 0.05 to 2.05D above clear film as zero. The density response shows little deviation from linearity over the full range of the instrument. Note that the near-specular densities measured by the Scandig exceed the diffuse densities of the calibrated tablet by a factor of 1.3.

X-Ray Film

The choice of a film type for recording X-ray diffraction data is dictated by a number of considerations, paramount among which is the need

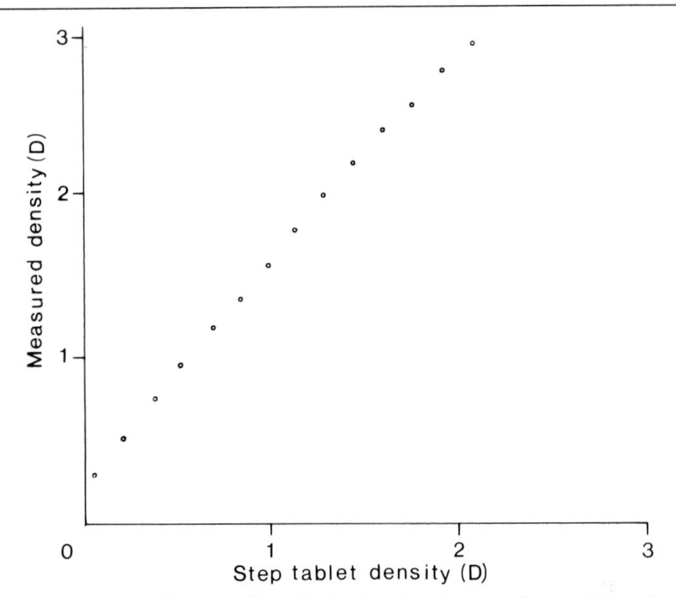

FIG. 2. Measured densities on a Scandig-3 microdensitometer for a calibrated step tablet. The calibrated densities are diffuse, and relative to clear film as zero. The Scandig densities are near specular and relative to an unobstructed light path as zero.

for speed and densitometric accuracy. The intensity of the crystal-diffracted X-ray beam is extremely weak compared with the incident intensity, so the faster the film used to record it the shorter is the exposure time required for a measurable image and the lower is the risk of radiation damage to the crystal. Film speed depends upon the weight of silver in the emulsion and the distribution of this silver in terms of silver halide grain size. The penetrating power of X radiation means that it is practicable to use film bases which are coated with emulsion on both sides, and films are available with a silver content of up to 4 mg cm^{-2}. A coarse-grained emulsion leads to faster film response by maximizing the amount of blackening associated with the absorption of each quantum of radiation, and the crystallographer is not usually concerned with the measurement of the detail which necessitates the use of fine-grained emulsion in many industrial or medical radiological applications.

Emulsion Properties

The table lists some relevant properties for the fastest, relatively coarse grained emulsions which are commercially available.[10] The film

[10] S. C. Dawson, O. S. Mills, and M. Elder, *Acta Crystallogr., Sect. A* **A37**, Suppl., C–311 (1981).

SOME X-RAY FILM EMULSION PROPERTIES MEASURED WITH NI-FILTERED, FLUORESCENT CU K_α RADIATION

Film	Manufacturer	Speed	w_{Ag} (gm cm^{-2})	Film factor	Background (D)	G_0	D_{max}
No screen	Eastman–Kodak (United States)	1.50	4.02	3.75	0.54	2.4	3.7
Reflex 25	CEA (Sweden)	1.15	2.89	2.62	0.22	1.6	4.7
Osray M3	Agfa–Gevaert (Belgium)	0.77	1.87	1.98	0.30	1.8	4.3
Singul X	CEA (Sweden)	0.72	1.24	1.69	0.31	1.6	2.9
Orwo XR1	Orwo (East Germany)	0.65	2.86	2.84	0.21	1.3	5.3

speeds are relative, but are approximately on the same scale as those of a previous survey[11] of films, most of which are no longer available. Film speed correlates with the weight of silver (w_{Ag}) in the emulsion, but the relevance of grain size to film speed can be seen in the difference in speed for two films with essentially the same silver content: Reflex 25 and the finer grained industrial X-ray emulsion, Orwo XR1.

The film factor measures the ratio between the responses of two films placed one behind the other in the same film pack. Since it is the silver halide in the emulsions that dominates the absorption of X rays by the film there is a strong correlation between silver content and film factor. A high film factor is an advantage in a diffraction experiment since it enables a wide range of reflection intensities to be measured using packs of only two or three films. To this end it is common practice to use a two-film pack consisting of a fast film in front with a slower film behind in order further to increase the dynamic range of measurable intensities.

Background and Granularity

The background density represents the optical density measured for an unexposed film developed in the standard way, and thus includes contributions from the film base and from the emulsion fog. Film bases are usually clear or blue-tinted polyester, with an optical density of about 0.1, while the fog varies considerably with the type of film, the age of the film, and the condition in which it has been stored. To minimize fogging, X-ray film should be used as soon after manufacture as possible, and should be stored at low temperature (refrigerator or freezer) and away from areas of

[11] H. Morimoto and R. Uyeda, *Acta Crystallogr.* **16,** 1107 (1963).

high background radiation. A high film background does not in itself reduce the accuracy of measurements, but since all density measurements are made superimposed upon this background it is obvious that a high value does reduce the effective dynamic range of a film. Of greater importance is the noise associated with film background, which is conveniently measured as the quantity termed granularity. As defined by Selwyn, the granularity for a film at optical density D above background is given by

$$G_D = \sigma_D(2S)^{1/2} \tag{5}$$

where σ_D is the standard deviation of a large number of density readings from a constantly exposed area of film, and S is the area of the film scanner measurement aperture. The smaller the value of G, the smaller the variation in density measurement caused by fluctuations in grain size and distribution within the emulsion.

The detective quantum efficiency (DQE) of film as a recording medium is defined by

$$\mathrm{DQE} = \sigma_{\mathrm{p}}^2/\sigma_{\mathrm{out}}^2 \tag{6}$$

where σ_{p}^2 is the photon noise associated with the signal being measured and σ_{out}^2 is the output noise from the measuring process. For the ideal photon detector σ_{out}^2 is no larger than σ_{p}^2 and DQE = 1. Background fog affects σ_{out}^2 in two ways: incident quanta striking grains which have already been activated are wasted, so that the input photon noise is effectively amplified by the fog; and the fogged grains also contribute an additive noise to the output signal. Thus

$$\mathrm{DQE} = \sigma_{\mathrm{p}}^2/(k\sigma_{\mathrm{p}}^2 + \sigma_{\mathrm{fog}}^2 + \sigma_{\mathrm{other}}^2) \tag{7}$$

where the amplification factor k contains a contribution from the fog, and the term $\sigma_{\mathrm{other}}^2$ combines other additive noise effects such as film scanner noise, emulsion thickness variations, and differential absorption effects in the black paper used to protect film packs from light during exposure.

Linearity of Response

The final column in the table, D_{max}, measures the linearity of response of the films. Once some film grains have been sensitized, there is obviously a finite probability that subsequent diffracted quanta striking the film will be absorbed by grains which have already been sensitized and thus make no contribution to the latent image. This leads to a nonlinearity of response usually expressed by an equation of the form

$$D = D_{\mathrm{max}}[1 - \exp(-AE)] \tag{8}$$

where D is the above background density for incident exposure E, and A is a constant.[11] The limiting value for D is D_{max} (usually well beyond the measurable range for a microdensitometer), at which point all silver halide grains have been sensitized and increased exposure has no effect.

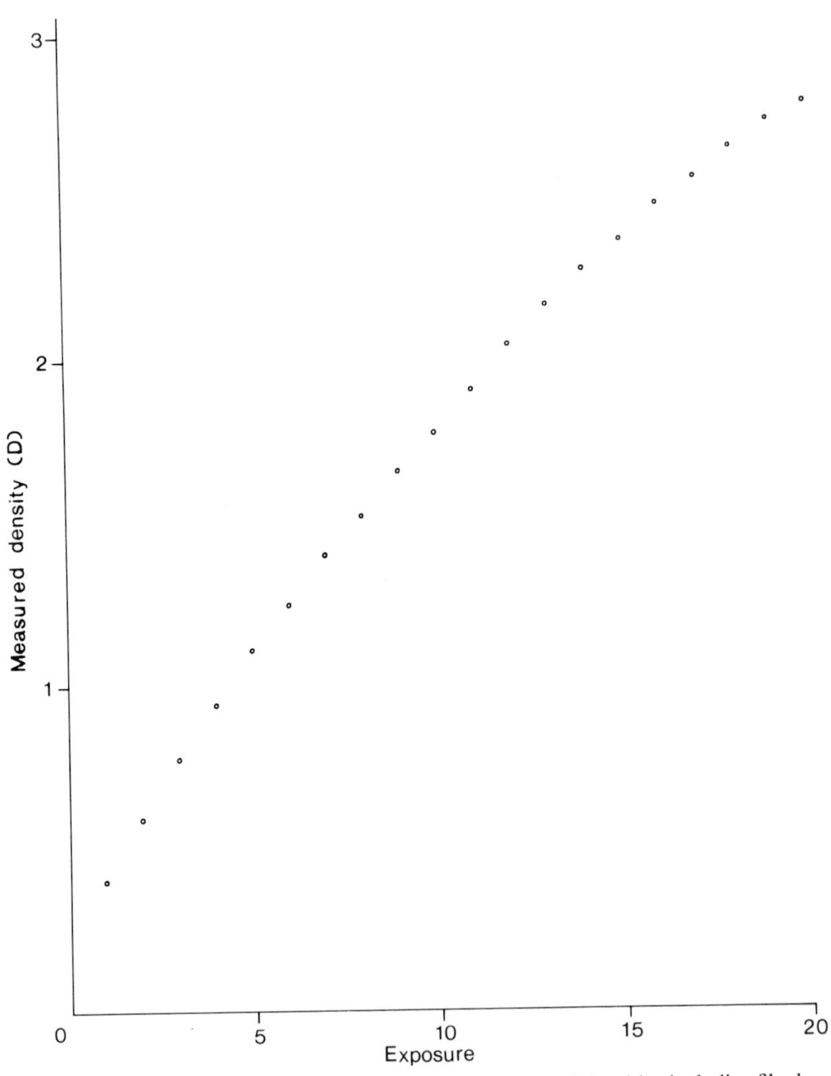

FIG. 3. A linearity plot for Reflex 25. The Scandig measured densities including film base and fog are plotted against the relative exposure times for spots on a stepped exposure strip made with Cu K_α radiation.

The higher D_{max}, the less a plot of D against E deviates from a straight line at low levels of exposure. Such a plot is shown in Fig. 3 for Reflex 25 which has a D_{max} of 4.7 given by Eq. (8). The measured densities in this plot are affected by both film and microdensitometer deviations from linearity, but comparison with Fig. 2 shows that film nonlinearity predominates.

Equation (8) is not very well adapted for use as a means of parametrizing nonlinearity. Because the data to be fitted are limited to low levels of exposure at which the deviation of a plot of D against E from a straight line is not large, the coefficient D_{max} is very sensitive to small variations in the data and has doubtful absolute significance. An alternative approach[12] involves two series expansion of Eq. (8) when some judicious approximations yield

$$D/E = S_D = S_0[1 - D/(2D_{max})] \tag{9}$$

Here S_D is the film speed at density D, and it is evident that this is expected to decrease from its maximum value, S_0 at zero density, in a linear fashion with a slope which depends upon D_{max}. In this analysis D should be the total blackening of the film and thus include the fog contribution, and E must therefore contain a contribution to represent the (hypothetical) exposure which gave rise to this fog. Figure 4 shows a plot of Eq. (9) for Reflex 25 using the data from Fig. 3 and the fact that the fog level for this film has been measured at $0.11D$. The straight line nature of the plot is confirmed, and the slope yields a value of $2D_{max} = 7.4$.

Practical Details

With a synchrotron radiation source it is practicable to expose a two-film pack for a protein crystal in less than 15 min. The carousel system of the Arndt–Wonacott oscillation camera means that this rate can be kept up for considerable periods, depending upon crystal lifetime and the oscillation range involved. Film scanning time is therefore a paramount importance, and is ultimately the limiting factor that determines the rate at which film data sets can be collected and processed. For drum scanners it is the drum rotation speed that determines the rate of film scanning. At 100-μm resolution the drum rotation speed is 8 and 12 rev/sec for the Optronics and Joyce–Loebl devices, respectively, with maximum data transmission of up to 50 kHz for both machines. At higher resolution, 50 or 25 μm, this data transfer rate is preserved by reducing the speed at which the drum rotates. Thus it is convenient to use as low a resolution as possible

[12] C. G. Vonk and A. P. Pijpers, *J. Appl. Crystallogr.* **14**, 8 (1981).

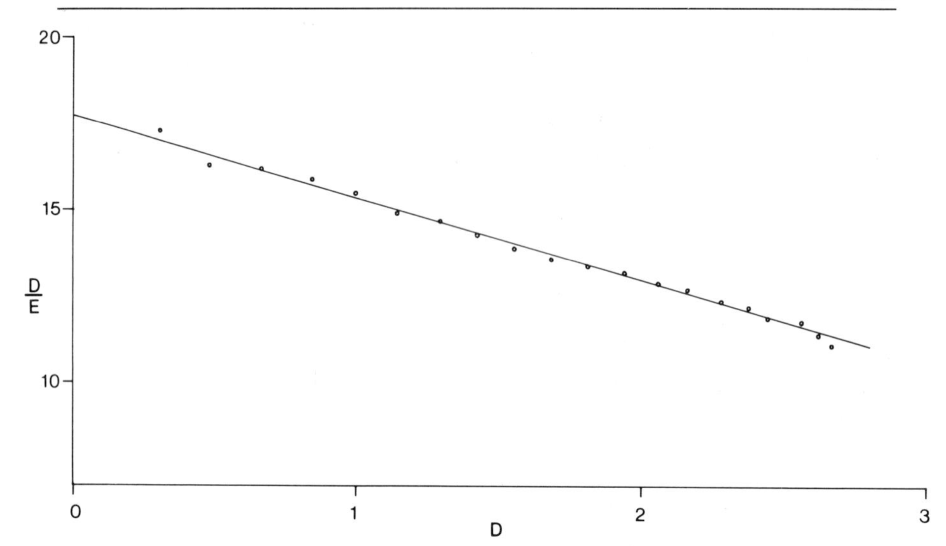

FIG. 4. Film speed (D/E) for Reflex 25 is plotted on an arbitrary scale against measured density (D) for the same data as in Fig. 3. D values include fog and are relative to film base as zero.

both in order to reduce the quantity of data that has to be handled and in order to reduce film scanning time.

In practice, although drum scanners have scanning apertures ranging from 12.5 to 200 μm, the choice is usually between 50 and 100 μm. An upper limit upon scanning aperture size is imposed by the need to avoid appreciable optical density gradients across the aperture, for the logarithmic conversion of light intensity to density means that the measured density for a steep gradient will be less than the average density.[13] This effect was investigated by Matthews[14] who concluded that the use of a 50- or 100-μm aperture led to a negligible difference in measured spot intensities for real spots, but it remains common practice to use the smaller aperture when measuring film with relatively small spots, as much to aid resolution of neighboring spots as to avoid systematic density underestimates.

It is obviously of great importance to the crystallographer to have a satisfactory method for correcting for the combined effects of film and scanner nonlinearity. If such a correction is not made then the assumption that there exists a linear relationship between film blackening measured on the microdensitometer and the intensity of the diffracted X-ray beam

[13] W. A. Wooster, *Acta Crystallogr.* **17,** 878 (1964).

[14] B. W. Matthews, C. E. Klopfenstein, and P. M. Colman, *J. Phys. E* **5,** 353 (1972).

breaks down, and the most intense reflections will be systematically underestimated. The usual method is to prepare a step wedge by exposing a piece of film of the type used for data collection to successively larger, accurately timed amounts of radiation, processing it in the standard manner, and then digitizing it on the film scanner to produce a plot of measured density versus exposure. This yields a correction curve which may be used to correct individual density values before processing, using a parametrization technique such as is described earlier.

In conclusion it can be stated that the use of a fast, low-background film together with a modern rotating drum microdensitometer provides an efficient and suitably accurate technique for the quantitative measurement of X-ray diffraction intensities from crystals of large unit cell. The problems imposed by the need to analyze the digitized data from the microdensitometer are certainly formidable, but they have been the subject of a great deal of work. An excellent survey of the computational problems and related fields is given in the book by Arndt and Wonacott.[15]

[15] U. W. Arndt and A. J. Wonacott, eds., "The Rotation Method in Crystallography." North-Holland Publ., Amsterdam, 1977.

[19] Oscillation Method with Large Unit Cells

By STEPHEN C. HARRISON, FRITZ K. WINKLER, CLARENCE E. SCHUTT, and RICHARD M. DURBIN

Oscillation photography is substantially more efficient than diffractometry for relatively large cell constants,[1] so the considerations here are in fact quite general. They are based primarily on experience with virus structures [tomato bushy stunt virus, turnip crinkle virus (TCV), poliovirus] and with influenza virus hemagglutinin.[2–5] The introduction of two-dimensional position-sensitive detectors will change a number of the specifics, but the overall approach will probably be fundamentally similar, treating the detector face as "electronic film." Some comments on area

[1] U. W. Arndt, *Acta Cryst.* **B24,** 1355 (1968).
[2] S. C. Harrison, A. J. Olson, C. E. Schutt, F. K. Winkler, and G. Bricogne, *Nature (London)* **276,** 368 (1978).
[3] J. Hogle and S. C. Harrison, *J. Mol. Biol.* (submitted for publication).
[4] J. Hogle, *J. Mol. Biol.* **160,** 663 (1982).
[5] I. A. Wilson, D. C. Wiley, and J. J. Skehel, *Nature (London)* **289,** 373 (1981).

FIG. 1. (a) Mirror benders for focused collimation. (Photo courtesy Charles Supper Co., Natick, Massachusetts, and Charles Ingersoll Corp., Waltham, Massachusetts.) (b) The design of the bender is shown in schematic cross section.

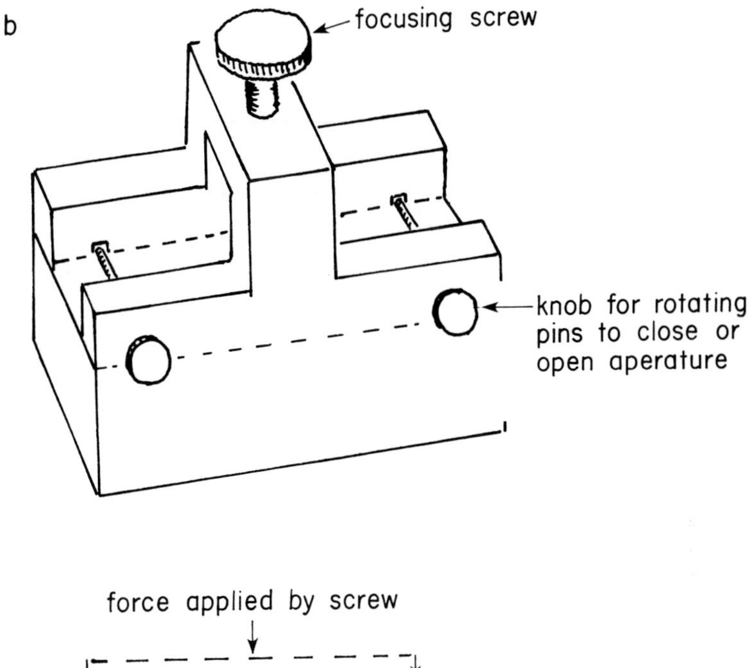

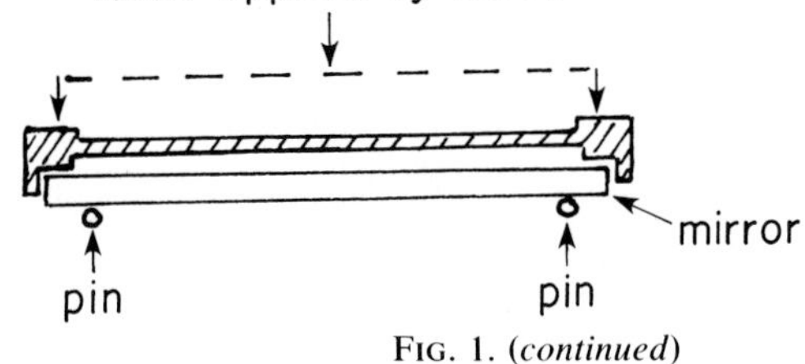

FIG. 1. (*continued*)

detectors, based largely on preliminary work with the Xentronics instrument, are included here where appropriate.

Collimation

With very large unit cells, the focusing capabilities of Franks-type optics[6] are important for satisfactory spot resolution and accurate data. The optical system consists of two perpendicular bent mirrors, providing glancing-angle reflection in horizontal and vertical planes. The mirror bender design universally employed (Fig. 1) is a modification of the original Ehrenberg bender.[7] A focusing screw depresses a spring, which in turn bears on a housing so machined that force is applied only to the ends

[6] S. C. Harrison, *J. Appl. Crystallogr.* **1,** 84 (1968).
[7] W. Ehrenberg, *J. Opt. Soc. Am.* **39,** 741 (1949).

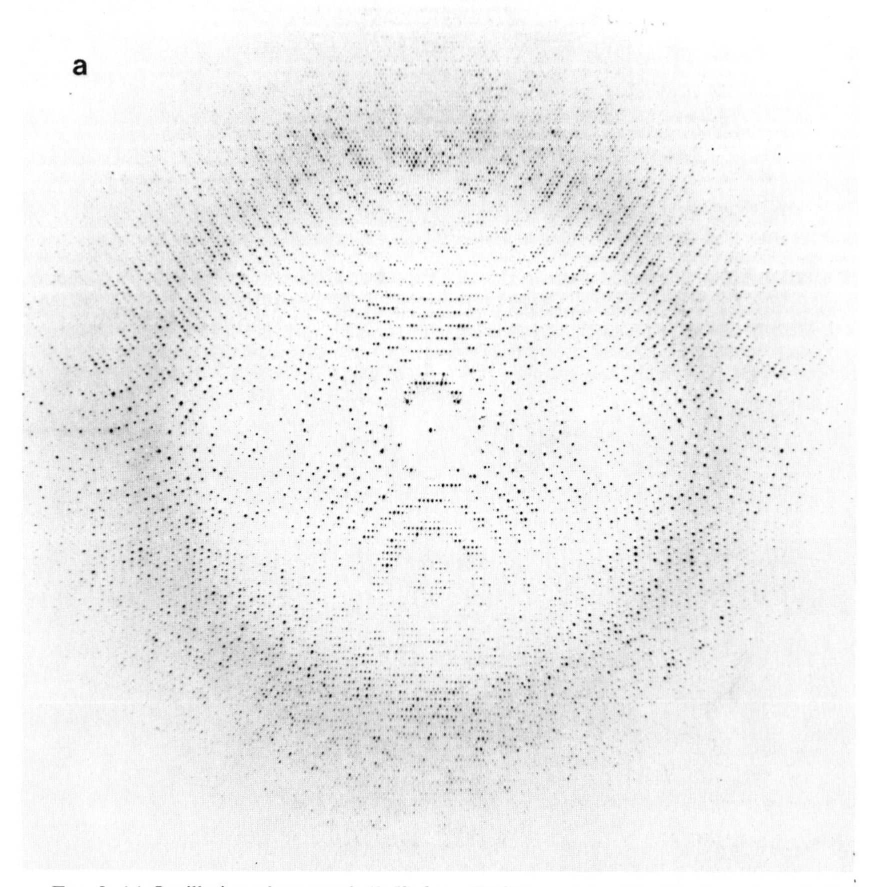

FIG. 2. (a) Oscillation photograph (0.5°) from TBSV crystal, with 101 along the spindle. (b) Example of a "mask" used in densitometry. The film is divided into regions A, B, C, D, and a different choice of pixels for spot and background is chosen for each region. The choice is illustrated for region B. It models spot elongation due to the double emulsion (and some dispersion contribution at high angle). The mask could vary with resolution, as suggested by the annuli shown, but we do not do this in practice, except to use an isometric mask for the central zone (0).

of the mirror. The latter, usually a nickel- or gold-coated optical glass flat, rests on pins that are positioned so that a couple is exerted on the glass at each end. The pins are machined flat near the axis: rotation of the pin provides a variable entrance or exit slit.

In the arrangement we use, the source is an Elliott GX-6 rotating anode generator with a 100 μm × 1 mm focus. The vertical mirror is close

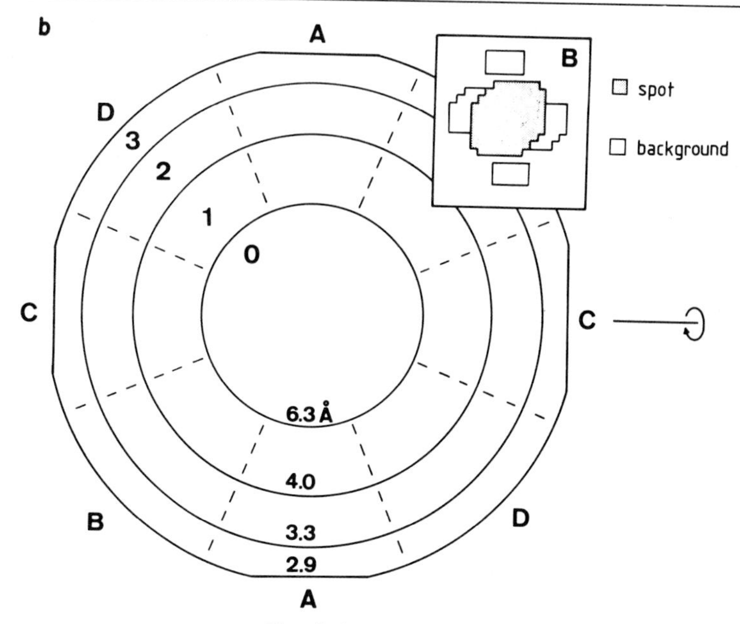

FIG. 2. (*continued*)

to the source: its position defines the effective take-off angle, which is set to about 3°. Both mirrors are adjusted to focus at the film plane. The vertical mirror "sees" an effective extended source of about 50 μm (1 mm foreshortened at 3°). Because of the asymmetric placement of the mirror, the source image at the film is magnified two to three times in the plane normal to the mirror (i.e., in the horizontal direction). The second, horizontal mirror is placed almost symmetrically. The vertical spot profile is therefore an approximate representation of the source profile itself. Very accurate adjustment of the generator filament is essential, in order to avoid substantial "wings" on the 100-μm focus. These wings would otherwise be imaged by the second mirror, degrading the spot shape at the film. It is important to place the vertical mirror next to the source, in order to be able to compensate the magnification by a suitably small take-off. The spot on the film will then be of approximately equal dimensions (15 μm). In practice, aberrations, wings, and similar problems blur the profile somewhat, and a larger area must be included in the densitometer scan of a reflection in order to measure all diffracted intensity. An oscillation film and representative spot shapes are shown in Fig. 2.

The optical system just described is designed to produce a small spot of maximum achievable intensity. A larger focus would be incompatible not only with recording diffraction from crystals with cell constants in the 300- to 500-Å range, but also with other considerations relevant to accurate data collection using film. These considerations are minimum signal-to-background ratio and maximum dose at the crystal for a given accuracy of measured intensity. The background on a film is determined by chemical fog and other film noise, by air scatter, and by scatter from interstitial solvent in the crystal, capillary, etc. The first may be minimized by correct handling of film. The second may be reduced by keeping the guard aperture close to the crystal and by setting the beam stop close to the crystal as well. The latter step eliminates some low-order reflections, but these are generally not used. Special photographs can be taken to collect this region if necessary. The recommended beam stop position minimizes the path of direct beam in air; it eliminates need for helium or other more cumbersome methods of reducing air scatter. Scatter from interstitial liquid in the crystal is, of course, unavoidable, and in a properly set up camera it accounts for well over half the recorded background. This level then determines the accuracy of intensity measurements. It is therefore important to keep the spot size as small as possible, focusing all available intensity into a minimum area. Indeed, up to a certain point it is appropriate to sacrifice total counts, which can be augmented by increasing the beam cross-fire, in order to achieve maximum counts *per unit area* (see next paragraph). These properties of intensity measurement with film have been emphasized by Arndt.[1] Conventional X-ray films have a grain size such that when one grain/μm^2 is developed, the specific optical density is about 2 (e.g., full scale on the Optronics Photoscan, OD 0–2 scale). A 100×100-μm spot with an OD of 0.1 (12 above background on a 0–255 gray scale) contains about 500 developed grains—more than enough for adequate counting statistics. With CEA film, the total background (almost entirely from liquid in the crystal) is about twice this value: if the background is estimated over an area equal to the spot size, the net measurement is $(1500 \pm 39) - (1000 \pm 32) = 500 \pm 50$, quite reasonable for a very weak reflection.

To determine the relative merits of focusing mirrors and collimators, a rough calculation is presented below. Experience confirms the rule of thumb that a focused beam is of particular advantage only for crystals with unit cell dimensions greater than ~150–200 Å. In special cases, monochromatization, with or without focusing, may be desirable: radiation-sensitive crystals or weak diffractors are the most obvious examples. A highly mosaic plane monochromator (e.g., graphite), used with a collimator, is in fact preferable in such applications, since it preserves adequate cross-fire in the beam. An important point should be made clear first. With large-unit cell crystals, efficient date collection procedures involve small-angle motions (e.g., stepped, small-angle oscillation or layer screenless small-angle precession). Absorption corrections are in many cases effectively identical for all spots on a given film (or, at worst, depend only on θ). If the scaling of films is performed properly (e.g., for stepped oscillation by reference to a perpendicular zone), these corrections are included in the scale factors. Moreover, with a monochromatic beam, absorption measurements may be made when necessary by monitoring transmission. It is therefore unnecessary to bathe the entire crystal in the X-ray beam. The following comparison of camera designs assumes that this conclusion is valid.

To obtain an estimate of the speed of a focusing camera relative to a simple collimator, consider first the collimation diagram in Fig. 3a. Huxley[8] has computed the optimum design for various constraints. In practice, one usually minimizes the total camera length, consistent with the order-to-order resolution required: $l = d = 10$ cm represent typical dimensions for large unit cells. If I_0 is the intensity that would be recorded at 1 cm from a hypothetical focus 1 cm long, the intensity at a point in the plateau, such as P' in Fig. 3, is

$$I_{P'} = I_0 \frac{f}{l + d}$$

If the width of the plateau just goes to zero, the ideal spot shape is triangular and the integrated intensity is proportional to

$$I_c = I_0 \frac{f}{l + d} \frac{fd}{l} \tag{1}$$

Since the maximum specific loading of an X-ray target increases as f decreases, I_0 is a function of f. For fixed anode tubes, $I_0 \sim 1/f$; for rotating anode tubes $I_0 \sim 1/f^{1/2}$.[9]

In the simplified ray diagram of a mirror in Fig. 3b, it is assumed that any point in the target plane T is imaged in the X-ray focal plane F, provided that the source point is within the acceptance angle of the mirror. This angle is set by the critical angle, θ_c, of the reflecting surface. The linear magnification M is approximately $(d + \frac{1}{2}m)/(l + \frac{1}{2}m)$. From the diagram, we see that the intensity at P', the brightest part in the focal plane, is equal to the intensity that would be recorded at a distance $l + d$ from a "virtual focus" in the target plane of height $f' = \alpha(l + d)$ and intensity I_P:

$$I_{P'} = I_P f/(l + d) = I_P \alpha \tag{2a}$$

[8] H. E. Huxley, *Acta Crystallogr.* **6,** 457 (1953).

[9] A. Muller, *Proc. R. Soc., Ser. A* **132,** 646 (1931).

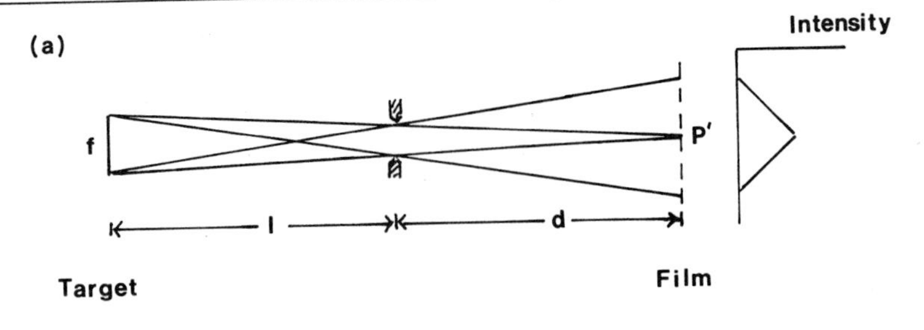

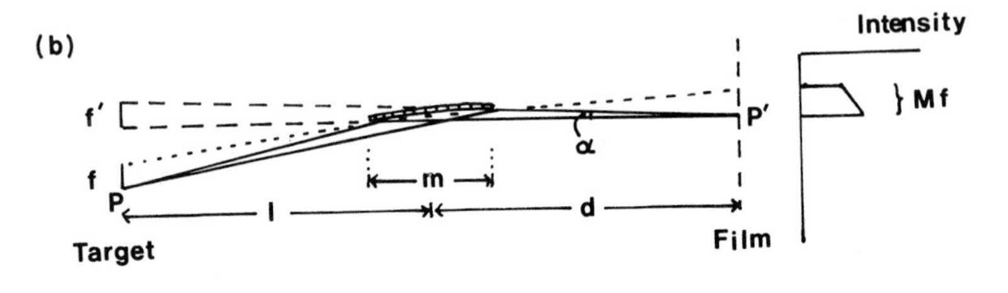

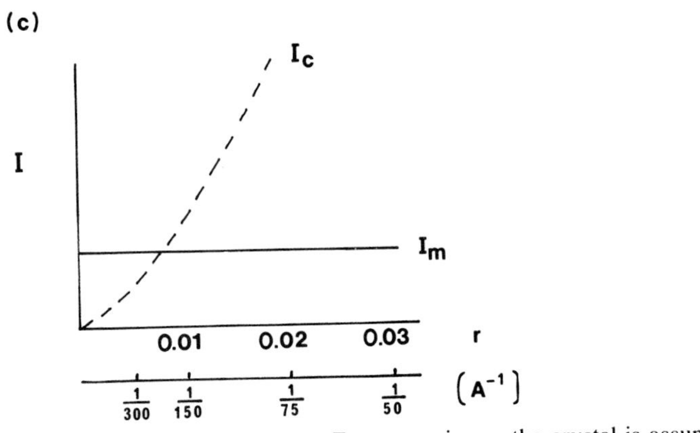

FIG. 3. (a) Collimator geometry. For convenience, the crystal is assumed to be in the same plane as the defining aperture. In practice, there are entrance and guard apertures, but the same general results will hold. The aperture dimension has been chosen to optimize resolution for a given peak intensity: the width of the intensity plateau just vanishes at P'. See Huxley.[8] Symbols: f, height of X-ray tube focus; l, focus-to-aperture distance; d, aperture-to-film distance. The intensity profile in the spot is shown at the right. (b) Mirror geometry. A single mirror is shown; in practice, there will be two mirrors with different l and d, one for each focal direction. Symbols: f, l, d as in Fig. 1; m, mirror length; P, a point in focus on X-ray target, P'; f', height of "virtual focus" that gives an intensity at P' equivalent to the peak intensity at P' in Fig. 1 (see text). The intensity profile in the spot is shown at the right. (c) I_c [Eq. (3)] as a function of resolution (for camera dimensions, see text). The computation assumes that f can be varied at will; with an Elliott GX-6 rotating anode tube, $r \sim 0.01$ for the 200-μm spot generally used for medium size proteins. I_m is shown ($l = d = 20$ cm; $m = 4$ cm; $f = 0.01$ cm, typical parameters for Franks' camera applications).

It is clearly of advantage to make the electron focus on the target of the X-ray tube as small and as intense as possible. The shape of the spot depends on the relative dimensions of the focus and the camera. If f is smaller than the aperture of the mirror, then the integrated intensity in the spot is approximately

$$I_m = I_{P'}Mf = I_P Mf\alpha \tag{2b}$$

For comparison of the two cases, consider a rotating anode source with $I_0 = Kf^{-1/2}$ (f in centimeters) and a crystal that requires an angular order-to-order resolution r.

Collimator. The angular resolution r is given by $r = 2f/d$, if we define spots as resolved if the intensity distribution just reaches background between them. A more realistic criterion is $r = 4f/d$, leaving a space between spots equal to the spot with itself; accurate background estimation requires such a region. Using this expression in Eq. (1), we have

$$I_c = k\frac{d}{l}\frac{(rd/4)^{3/2}}{l + d} \tag{3}$$

Mirror. Intense line foci of height smaller than 100 μm have not been achieved with rotating anode X-ray tubes. With a focus of this height, $I_P = 10k$. With a Ni-coated glass reflector $\theta_c \sim 23'$ or 0.007 rad. The exit aperture α is therefore given by $\alpha \simeq \theta_c m/d \simeq 0.007\ m/d$, and using Eqs. (2a) and (2b),

$$\begin{aligned} I_{P'} &= 10k\theta_c m/d = 0.07km/d \\ I_m &= 0.0007kMm/d \end{aligned} \tag{4}$$

The resolution, r, is 0.01 M/d, greater (see below) than required for almost all current applications.

In Fig. 3c, I is plotted as a function of r, taking $l = d = 10$ cm; I_m, which is independent of r, is also shown, for $l = d = 20$ cm and $m = 4$ cm. The dimensions represent typical camera parameters: the longer l and d for the mirror are due to the size of the bender. Large deviations from these values would be impractical and would in general give poorer results for reasons such as magnification by the mirror system, optical aberrations (in the real case very large even for the dimensions shown), or air scatter. One exception might be introduction of a longer *second* mirror, especially for somewhat smaller unit cells: the gain in aperture would be roughly proportional to the length of the mirror. Any gain from a longer *first* mirror would be lost by reduction in aperture of the second. Figure 3 shows that a mirror gives better results only for $r > 0.008$, corresponding to an order-to-order resolution of about 200 Å. The important general point is that if the resolution requirements are not severe, a larger integrated intensity can be achieved by letting a spot on the film "see" a somewhat more diffuse focus of greater *total* output.

Data Collection Strategy

The choice of axes and angular limits for the total oscillation range is important for efficient data collection.[10] The considerations are (1) a complete data set in a minimum total number of photographs; (2) adequate redundancy (duplicate or symmetry related reflections) for accuracy, internal scaling, etc.; (3) uniformly distributed redundancy (e.g., many reflections present in two or three symmetry equivalents rather than a small number present four or more times); and (4) convenience in crystal mounting.

As an example of finding an axis about which to oscillate that offers good distribution of redundant information throughout the asymmetric unit with the minimum total oscillation arc, consider the tomato bushy stunt virus (TBSV) space group I23. Referring to Fig. 4a, showing the Laue group m3, the volume bounded by the zone axes [100], [001], and [111] contains one of the 24 asymmetric units. By visualizing the asymmetric unit in this way, it is clear that placing $10\bar{1}$ on the spindle axis can lead to a relatively small total oscillation range along some part of the arc connecting that [010], [111], and the [101] zone axes. The idea is that the leading edge of the Ewald sphere in sweeping through this volume, photograph by photograph, will require just over 35° for a complete data set to be recorded on the upper halves of the photographs (the spindle is horizontal). Upper and lower parts of the volume passing through the Ewald sphere are not, in general, related by a symmetry operation, and therefore we expect an overall range shorter than 35° when including the data from both halves of the films. For the group m3 the "theoretical" minimum total oscillation range is 7.5° (180°/24), assuming a set of axes existed about which to collect this unique set. Note that a frequent prescription for oscillation photography, to place the axis of highest symmetry along the spindle direction, the threefold in this case, would require 60° total to guarantee a complete data set. In order to compare different film-scanning crystal orientations and total oscillation ranges, we use the film-scanning subroutine (OSCGEN) that generates a spot list for a given film to produce the entire list of reflections for a given choice. To make the computation rapid, we use a smaller unit cell for the calculation, the premise being that the fraction collected is independent of cell size. The collection of "reflections" (each representing, say, a $5 \times 5 \times 5$ or $10 \times 10 \times 10$ volume

[10] C. E. Schutt, Ph.D. Thesis, Harvard University, Cambridge, Massachusetts (1976).

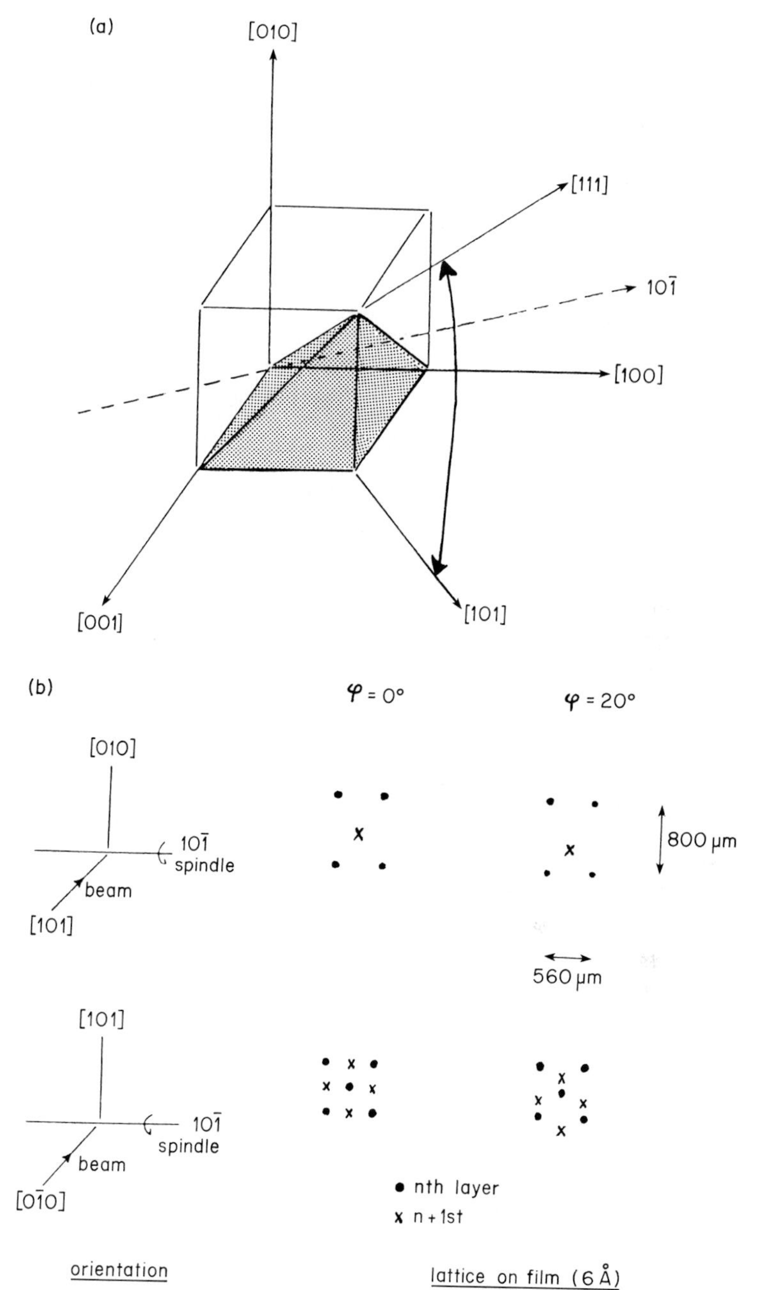

FIG. 4. Example of considerations relevant to oscillation data collection strategy. (a) Symmetry axes of the Laue group m3. The shaded region represents an asymmetric unit. (b) Geometry of projected reciprocal lattice planes as a function of crystal orientation, for the two strategies mentioned in the text. Crystal-to-film distance is 100 mm; space group is I23, a = 383 Å.

of reciprocal space) recorded for each strategy is then sorted into the asymmetric unit and distributed to show what percentage appears just once, twice, etc., or not at all. We can then compare the relative usefulness of different strategies. For TBSV, we explored several oscillation axes, and $10\bar{1}$ is indeed the best choice for orientation along the spindle. Figure 4b compares two strategies, both of which have a good distribution of symmetry-equivalent reflections over a nearly complete reciprocal-space asymmetric unit contained within 20° total oscillation. The choice between them is made by observing that reflections on adjacent lunes interleave in this space group when $10\bar{1}$ is along the spindle, due to the systematic absence of $h + k + l =$ odd. The interleaved lattices are better separated for the case in which oscillation starts with [101] along the beam direction.

Analogous strategies have proved optimal in space groups I222 (TCV), I422 (repressor DNA co-crystals), and $P4_1$ (influenza hemagglutinin). In I222, for example, each data set consisted of 88 photographs each covering a 0.5° oscillation range.[3] Of these, 72 were taken with 011 on the oscillation axis, sampling a range from 0°–45°, where 0° corresponded to [011] along the beam. In addition, 121 films and 4 films were taken with 101 and 110, respectively, along the spindle. In I422, data are best taken with 011 (or 101) along the oscillation axis.[11]

The choice of oscillation angle per film is ordinarily not included in the above calculations. There are two considerations here. An upper limit is set by the need to avoid spot overlap at the highest resolution desired. An angle near this upper limit gives the smallest number of films and largest fraction of whole spots. But for medium-size unit cells, there may be good reason to choose an angle less than the maximum. The typical reflecting range for a spot is 0.1°–0.2°. In a 1° oscillation photograph, background is recorded during the remaining 0.8–0.9°, degrading the signal-to-noise. It is our experience that oscillation angles larger than 1° are therefore unsatisfactory, despite the convenience of fewer films and fewer partially recorded reflections. The use of postrefinement to correct partial reflections makes the larger number of partials for small oscillation angles not so disadvantageous. The entire problem goes away, of course, with area detectors, and the signal-to-background improvement is precisely the advantage of such instruments. For comparable beam geometries, we can calculate that an area detector will lead to an improvement of a factor of 10 in signal-to-background for a 0.1° rocking curve width, in a case in which a 1° photograph would have been chosen, and thus to an improvement by about 3.2 in speed for comparable counting accuracy.

[11] J. A. Anderson, Ph.D. Thesis, Harvard University, Cambridge, Massachusetts (1984).

Correcting for Partially Recorded Reflections[12]

As just described, the angular range of an oscillation photograph is generally chosen by the criterion that no two reciprocal lattice points at a desired resolution limit have overlapping spots on the film. The allowable range per photograph is limited by the magnitudes of the cell dimensions, the resolution limit, and the angle over which a Bragg reflection will diffract. For example, with TBSV crystals, which have an average reflecting range (determined largely by the Franks-camera beam cross-fire) of about 10′ arc, the allowable oscillation range at 2.9 Å resolution is only 30′. In some other cases, in which larger ranges would be permitted by geometrical criteria, the oscillation is restricted in practice by accumulation of background exposure. An unavoidable consequence of a finite reflecting range is that the intensity at reciprocal lattice points near the ends of the oscillation range is recorded only partially on the film. If these partially recorded reflections cannot be used, the efficiency of the recording process is lowered. In the standard procedure using continguous photographs, intensities of complementary parts of a reflection, split between two successive photographs, are combined to give the fully recorded equivalent.[13] For large unit cell volumes, exposure times required for a good average signal-to-noise ratio on the film approach the lifetime of the crystal. In the limit, only one photograph can be obtained and partial spot addition is impossible. A method for correcting partially recorded reflections to their fully recorded equivalents, originally developed at Harvard for TBSV[10,12] and now widely implemented elsewhere,[14] relies on redundancy in the data collection scheme. A number of reflections partially recorded on a given photograph have equivalent, fully recorded reflections on the same or another photograph. The ratio of any such pair of intensity measurements is an "observed" recorded fraction, each of which is a measure of the crystal setting, unit cell, and rocking curve parameters. If these parameters are sufficiently overdetermined (that is, if the number of partials having fully recorded counterparts elsewhere in the data set is sufficiently great), a straightforward, least-squares refinement procedure can be adopted. It yields far more accurate values for the required parameters than are obtained in the usual refinement based on visually observed partial spots. The approach can be applied in principle

[12] F. K. Winkler, C. W. Schutt, and S. C. Harrison, *Acta Crystallogr., Sect. A* **A35,** 901 (1979).

[13] V. W. Arndt and A. J. Wonacott, "The Rotation Method in Crystallography." North Holland Publ., Amsterdam, 1977.

[14] M. G. Rossmann, A. G. W. Leslie, S. S. Abdel-Mequuid, and T. Tsukihara, *J. Appl. Crystallogr.* **12,** 570 (1979).

to any oscillation photograph, but since a suitably overdetermined refinement implies large numbers of partially recorded reflections on a single film, the method is restricted in practice to relatively large unit cells. This refinement is necessarily carried out after film scanning, since it depends on intensities, and it has therefore come to be called "postrefinement."

The crystal orientation notation followed here to describe crystal orientation is that used by Crawford[15] in the "Harvard System" (Arndt and Wonacott,[13] p. 140). The laboratory frame, centered at the origin of reciprocal space, is defined such that the X axis is parallel to the incident beam, the Z axis parallel to the camera spindle, and the Y axis orthogonal to these two. From one of six possible reference orientations, defined by which real axis is along Z and which reciprocal axis along X, the reciprocal lattice is brought to a required orientation by applying in turn the three rotations, ψ, ω, and ϕ around axes X, Y, and Z, respectively (Fig. 5). Note that coincidence of oscillation angle and setting angle ϕ implies zero inclination. The equations developed below would require modification for other inclination angles. The laboratory coordinates of reflection i at setting (ψ,ω,ϕ) are then obtained by applying two successive transformations represented by the matrices A and B:

$$\begin{pmatrix} x_i \\ y_i \\ z_i \end{pmatrix} = BA \begin{pmatrix} h_i \\ k_i \\ l_i \end{pmatrix} \tag{5}$$

The two matrices are

$$A = \begin{pmatrix} a^* & b^* \cos \gamma^* & c^* \cos \beta^* \\ 0 & b^* \sin \gamma^* & c^*(\cos a^* - \cos \beta^* \cos \gamma^*)/\sin \gamma^* \\ 0 & 0 & c^* \cos(c^*,c) \end{pmatrix}$$

$$B = \begin{pmatrix} \cos \omega \cos \varphi & \sin \psi \sin \omega \cos \varphi - \cos \psi \sin \varphi & \cos \psi \sin \omega \cos \varphi + \sin \psi \sin \varphi \\ \cos \omega \sin \varphi & \sin \psi \sin \omega \sin \varphi + \cos \psi \cos \varphi & \cos \psi \sin \omega \sin \varphi - \sin \psi \cos \varphi \\ -\sin \omega & \sin \psi \cos \omega & \cos \psi \cos \omega \end{pmatrix}$$

where

$$\cos(c^*,c) = (1 + 2 \cos \alpha^* \cos \beta^* \cos \gamma^* - \cos^2 \alpha^* - \cos^2 \beta^* - \cos^2 \gamma^*)^{1/2}/\sin \gamma^*$$

A orthogonalizes the reciprocal lattice, and as given above it corresponds to the reference orientation with a^* and X and c along Z.

At the time an oscillation photograph is taken we have only approximate values for the setting angles, which we call the *nominal setting angles,* ψ_0, ω_0, and ϕ_0^t. Superscript t refers to a specific value of the

[15] J. Crawford, Ph.D. Thesis, Harvard University, Cambridge, Massachusetts (1977).

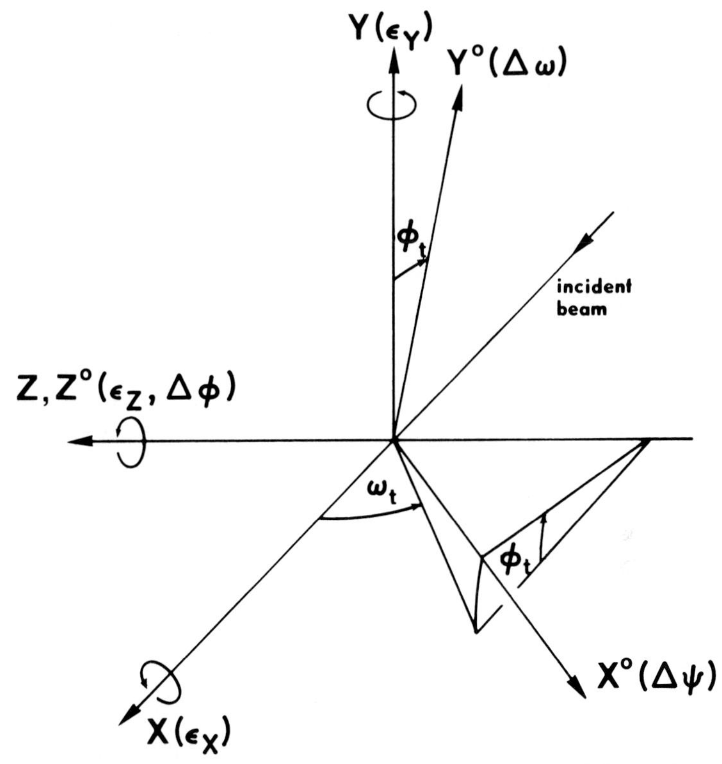

FIG. 5. Definition of the laboratory coordinate system X, Y, Z and the rotations ψ, ω, ϕ. Positive rotations are defined as anticlockwise when looking down the rotation axis. The three rotations ψ, ω, and ϕ applied in turn around these axes do not correspond to an Eulerian set of angles, since the rotation axes are defined in the fixed laboratory space and not in the rotated reciprocal space. X° is the position of an axis, initially coincident with X, after application of the rotations ω and ϕ; Y°, the position of an axis initially coincident with Y, after application of the rotation ϕ. At setting $t(\phi = \phi^{t})$, a correction $\Delta\psi$ corresponds to a rotation about X°; likewise, a correction $\Delta\omega$ to a rotation about Y°. A correction $\Delta\phi$ is always taken about the oscillation axis; Z and Z° are therefore drawn coincident. Obviously, axes for the three corrections are not orthogonal. This leads to difficulties only when ω approaches $\pi/2$ [some elements of Eq. (6) become very large], and this case must be avoided by choosing another initial reference orientation. The corrections $\Delta\psi$, $\Delta\omega$, $\Delta\phi$ are related to laboratory-frame angular adjustments by Eq. (6).

oscillation angle ϕ; for example b denotes a t value in the beginning, e a value in the end, and m a value in the middle of the oscillation range, $\phi = \phi^{e} - \phi^{b}$. More accurate values for these angles can be obtained from the orientation information present on each photograph in the form of partially recorded reflections and in the positions of fully recorded reflections (Arndt and Wonacott,[13] Chap. 8). At the time of initial evaluation of scanned intensities (either when scanning, if the full integration is done on-line, or when evaluating a digitized film off-line on a larger computer), this information is used to determine the misorientation of the crystal from its nominal setting. The Harvard SCAN12 program uses the indices of 12 partial spots and the locations of 12 whole spots to make these corrections. The partial spot indexing must be done on each film. We have simplified this indexing by arranging to display the film, rapidly digitized with a coarse raster, on a computer graphics screen, and using a cursor to identify suitable partial reflections by their shape and location. Partial reflections usually appear somewhat displaced from the row on which they lie, in a direction that depends on whether they are entering or leaving the Ewald sphere. The *misorientation angles*, ε_x^t, ε_y^t, and ε_z^t, are small rotations around laboratory-frame axes X, Y, and Z, respectively, at setting t, and are equivalent to ψ_x, ψ_y, and ψ_z in Arndt and Wonacott.[13] As shown in Fig. 5 the corrections to the setting angles, $\Delta\psi$, $\Delta\omega$, and $\Delta\varphi$ can be considered as rotations about X^0, Y^0, Z^0, and only in the special case $\omega = \phi = 0$ are the rotation axes of ψ and ε_x or ω and ε_y identical. If only small corrections have to be applied to the setting angles, they are well approximated by the transformation

$$\begin{pmatrix} \Delta\psi \\ \Delta\omega \\ \Delta\varphi \end{pmatrix} = \begin{pmatrix} \cos\varphi^t/\cos\omega & \sin\varphi^t/\cos\omega & 0 \\ -\sin\varphi^t & \cos\varphi^t & 0 \\ \cos\varphi^t \tan\omega & \sin\varphi^t \tan\omega & 1 \end{pmatrix} \begin{pmatrix} \varepsilon_x^t \\ \varepsilon_y^t \\ \varepsilon_z^t \end{pmatrix} \tag{6}$$

Rocking Curve Model. The picture of a reciprocal lattice point crossing the Ewald sphere represents an idealized experimental situation, since real crystals diffract real X-ray beams over finite angular ranges. The rocking curve is the profile of diffracted intensity, integrated over the area of a spot on the detector, as a function of the angle of crystal rotation. If we represent a reflection by a volume element around the corresponding reciprocal lattice position, each point of the rocking curve profile corresponds to the passage of an infinitesimal section of the volume element through the Ewald sphere. Over the small area traversed by the volume element of a reflection, this sphere is well represented by a plane. The one-dimensional rocking curve can therefore be generated by projecting, at each angle of rotation, the contents of the corresponding infinitesimal section onto the normal to the Ewald sphere. A partially recorded reflection will have part of its volume inside and part outside the Ewald sphere at either ϕ^b or ϕ^e, and its recorded fraction, corresponding to one of these parts, depends only on the final position and not on the actual path of the volume element during rotation. In order to obtain a one-dimensional profile it is necessary to map the contents of the volume element, described above, onto an appropriate arc that intersects the Ewald sphere. Note that the arc of actual rotation would not be a very convenient choice, since the profile width would vary with the angular position of the reflection. The obvious choice is the β arc of each reflection, defined as the arc that describes the shortest angular separation of reciprocal lattice point P_i from the Ewald sphere, and illustrated in Fig. 6. The β axes all lie in the central plane perpendicular to the X-ray beam. The advantage of this choice has been confirmed by work on TBSV, TCV, and influenza virus hemagglutinin, and by programs for the Xentronics detector.

Using the angular variable β, we define the rocking curve as follows. Let β_i be the position of the center of the profile on the β arc of reflection i, which intersects the Ewald sphere at the Bragg angle θ_i. Assuming further a symmetric profile, the recorded fraction P^i_{cal} can be expressed as

$$P^i_{\text{cal}} = 0.5\{1 \pm f[(\beta_i - \theta_i)/\gamma_i]\} = 0.5[1 \pm f(b_i)] \tag{7}$$

where $f(b_i)$ is an as yet unspecified function whose value is restricted to the range 0 to 1. The plus or minus sign indicates whether the reflection is more or less than half recorded. The function $f(b_i)$ gives the fractional intensity integrated on the rocking curve between β_i and θ_i. The half-width of the profile of reflection i, γ_i, may itself be a function of various parameters, γ_r, and of certain geometric variables. For profiles of finite width, β_i is restricted to the range -1 to 1. An important assumption in this approach is that a one-dimensional, normalized profile, specified by only a few parameters, is sufficient to describe the fractional buildup of intensity for all reflections as they pass through the Ewald sphere. A similar assumption is made in methods using learned profiles to obtain improved integrated intensities.[16,17]

We have found in practice that the cosine half-wave, yielding for the integral f

$$f(b_i) = |\sin[(\pi/2)b_i]| = \sin(\pi/2)\ |b_i| \tag{8}$$

is very satisfactory for mirror-collimated beams. Two extensions of this one-parameter function have also been considered.

1. To model anisotropy of crystal mosaicity or beam cross-fire, the half-reflecting range can be allowed to vary continuously from γ_y to γ_z, in going from meridional ($\alpha = 0$) to equatorial ($\alpha = \pi/2$) reflections:

$$\gamma_i(\alpha_i) = [(\gamma_y \cos \alpha_i)^2 + (\gamma_z \sin \alpha_i)^2]^{1/2} \tag{9}$$

The need for such a modification should become apparent if the differences between calculated and observed fractions are analyzed as a function of the angular position, α_i, of the reflections. For example, if the calculated fractions for the class of partial reflections less than half recorded near the meridian are systematically underestimated, while the

[16] R. Diamond, *Acta Crystallogr., Sect. A* **A25,** 43 (1969).

[17] D. F. Grant and E. J. Gabe, *J. Appl. Crystallogr.* **11,** 114 (1978).

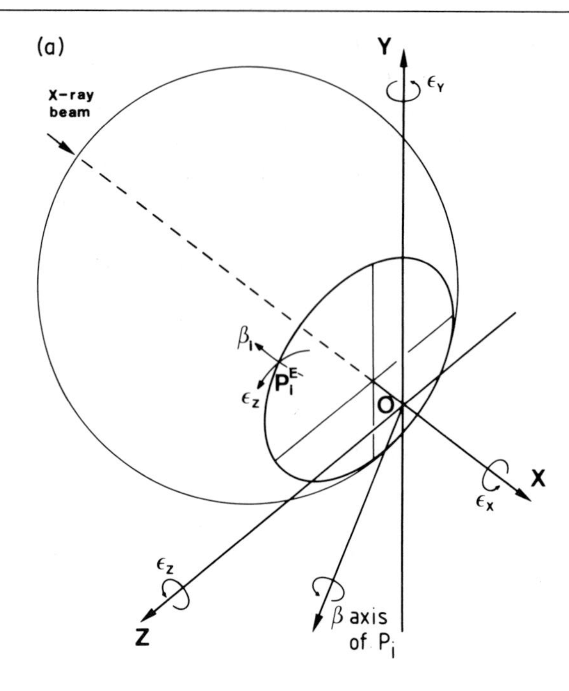

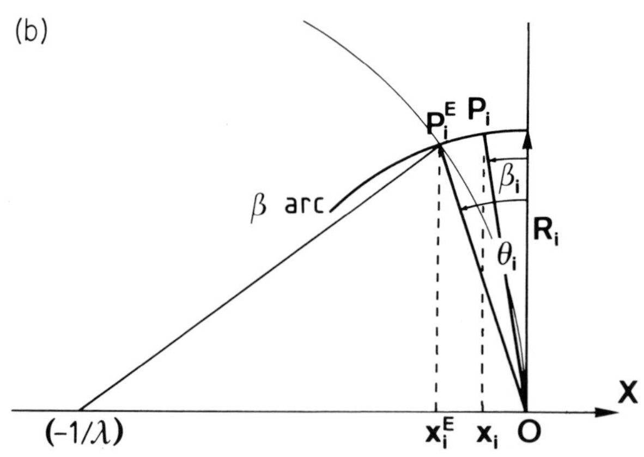

FIG. 6. Definition of β axis and β arc of reflection i. (a) The axis normal to the central plane containing the X axis and the lattice point P_i is defined as the β axis of P_i. We denote as P_i^{E} the position of P_i when it happens to lie on the Ewald sphere. Note that rotation about X causes P_i to trace out a circle on the surface of the Ewald sphere and that rotation of P_i about its β axis generates an arc (β_i) perpendicular to this circle. A positive rotation about the β axis is defined as producing a negative component on the X axis. (b) View along β axis of P_i. At P_i, β_i is equal to the Bragg angle θ_i. x_i is the x coordinate of P_i, and x_i^{E}, that of P_i^{E}.

same class is overestimated for reflections near the equator, then γ_y should be made larger than γ_z. This will be confirmed if the classes of reflections more than half recorded are found to have the opposite systematic trends.

2. To take account of the increase of the reflecting range with resolution due to spectral dispersion, γ_i can be expressed as

$$\gamma_i = \gamma_0 + \gamma_1 \tan \theta_i \tag{10}$$

where $\gamma_1 = \Delta\lambda/2\lambda$. For the case of the Cu K_α doublet, $\Delta\lambda/\lambda$ is 0.0026. These two modifications are easily combined to yield a three-parameter rocking curve. For TBSV data only the second one was needed.[12]

Refinement of Crystal Setting, Unit Cell, and Rocking Parameters. We assume a reference data set, constructed, for example, from fully recorded reflections scaled together from a number of photographs (for details, see below). For any photograph, the observations used in parameter refinement are those reflections also present in the reference data set. It is important that there is a sufficient number of observations, with a uniform distribution over the area of the film. The reflections in the reference data set (usually an incomplete data set) should therefore be uniformly distributed in space. Given n observations, the quantity minimized in the least-squares refinement is defined as

$$\phi = \sum_{i=1}^{n} w_i(\Delta I_i)^2 \tag{11}$$

where

$$\Delta I_i = I_P^i - P_{\mathrm{cal}}^i I_R^i \tag{12}$$

and

$$1/w_i = \sigma^2(\Delta I_i) = \sigma^2(I_{\mathrm{P}}^i) + (P_{\mathrm{cal}}^i)^2\sigma^2(I_{\mathrm{R}}^i) \tag{13}$$

are the intensity and estimated standard deviation (e.s.d.) of the partially recorded reflection i, and I_{R}^i and $\sigma(I_{\mathrm{R}}^i)$ are the intensity and e.s.d. of this reflection in the reference data set.

For obtaining the normal equations, which give the shifts Δp_j of the m parameters to be refined, ϕ is linearized as usual by expansion in a Taylor series with elimination of higher terms. These m equations ($n > m$) can be written

$$-\sum_{i=1}^{n} w_i \, \Delta I_i \left(\frac{\partial \Delta I_i}{\partial p_j}\right)_0 = \sum_{i=1}^{n} w_i \left(\frac{\partial \Delta I_i}{\partial p_j}\right)_0 \sum_{k=1}^{m} \left(\frac{\partial \Delta I_i}{\partial p_k}\right) \Delta p_k, \qquad j = 1, \ldots, m \tag{14}$$

or after substituting ΔI_i by Eq. (12), as

$$\sum_{i=1}^{n} w_j \, \Delta I_i I_R^i \left(\frac{\partial P_{\text{cal}}^i}{\partial p_j}\right)_0 = \sum_{i=1}^{n} w_i (I_R^i)^2 \left(\frac{\partial P_{\text{cal}}^i}{\partial p_j}\right)_0 \sum_{k=1}^{m} \left(\frac{\partial P_{\text{cal}}^i}{\partial p_k}\right)_0 \Delta p_k, \qquad j = 1, \ldots, m \tag{15}$$

The 0 subscript indicates evaluation of the partial derivatives for current values of the parameters. This evaluation requires some comment.

ROCKING CURVE PARAMETERS, γ_r. These derivatives depend on the functional form of $f(b_i)$ and of the particular parametrization of the rocking curve width. In a general way they are given as

$$\frac{\partial P_{\text{cal}}^i}{\partial \gamma_r} = \pm \frac{1}{2} \frac{\partial f(b_i)}{\partial \gamma_i} \frac{\partial \gamma_i}{\partial \gamma_r} \tag{16}$$

CRYSTAL SETTING AND UNIT CELL PARAMETERS. The discussion in the previous section has shown that the width of the rocking curve itself may vary—for example, with the angular position of a spot on the film or with resolution, Eqs. (9) and (10). However, this dependence can be neglected for calculation of partial derivatives, since the small shifts expected for orientation and unit cell parameters have almost no effect on the width of the rocking curve of a particular reflection. For the case of the symmetrical profile where

$$P_{\text{cal}}^i = 0.5 \left[1 \pm f\left(\left|\frac{\Delta\beta_i}{\gamma_i}\right|\right)\right], \qquad \text{with} \quad \Delta\beta_i = \beta_i - \theta_i \tag{17}$$

the derivatives with respect to these parameters can therefore be written as

$$\frac{\partial P_{\text{cal}}^i}{\partial p_j} = \pm \frac{1}{2} \frac{\partial f}{\partial \Delta\beta_i} \frac{\partial \Delta\beta_i}{\partial p_j} \operatorname{sign}(\Delta\beta_i) \tag{18}$$

The partial derivatives $\partial\Delta\beta_i/\partial p_j$ describe the dependence of the angular distance, $\Delta\beta_i$, of reciprocal lattice point i from the Ewald sphere on crystal orientation and unit cell parameters. Since the Bragg angle θ_i does not depend on the setting angles, the derivatives with respect to orientation parameters reduce to $\partial\beta_i/\partial\varepsilon_j$. For reflections close to the sphere, $\beta_i \simeq \theta_i$ (the Bragg angle), and these derivatives are

$$\begin{aligned} \partial\beta/\partial\varepsilon_x &= 0 \\ \partial\beta/\partial\varepsilon_y &= -z_i/R_i \cos\theta_i \\ \partial\beta/\partial\varepsilon_z &= y_i/R_i \cos\theta_i \end{aligned}$$

where R_i $(=d_i^*)$ is the length of the reciprocal lattice vector to the reflection i. As rotation around the beam direction X has no component on the β arc of reflection i, the first derivative must be zero. The other two are easily verified by considering the special cases of meridional and equatorial reflections. For the former, a rotation around the β axis is equivalent to a rotation around $Z(\varepsilon_z)$, and the derivatives have to be ± 1 and 0, respectively, corresponding to $|y_i| = R_i \cos\theta_i$ and $z_i = 0$ for such reflections. To evaluate the derivatives of $\Delta\beta_i$ with respect to unit cell parameters, p_j^c, we use the fact (see Fig. 6) that

$$\sin\beta_i = -x_i/R_i \qquad \text{and} \qquad \sin\theta_i = -x_i^{\mathrm{E}}/R_i$$

The derivatives $\partial\Delta\beta_i/\partial p_j^c$ then become

$$\frac{\partial(\beta_i - \theta_i)}{\partial p_j^c} = -\frac{1}{\cos\beta_i}\left(\frac{1}{R_i}\frac{\partial x_i}{\partial p_j^c} + x_i\frac{\partial(1/R_i)}{\partial p_j^c}\right) + \frac{1}{\cos\theta_i}\left(\frac{1}{R_i}\frac{\partial x_i^{\mathrm{E}}}{\partial p_j^c} + x_i^E\frac{\partial(1/R_i)}{\partial p_j^c}\right)$$

By setting $\beta_i = \theta_i$ and $x_i = x_i^{\mathrm{E}}$, good approximations for reflections near the Ewald sphere, we obtain

$$\frac{\partial\Delta\beta_i}{\partial p_j^c} = \frac{1}{R_i\cos\theta_i}\left(\frac{\partial x_i^{\mathrm{E}}}{\partial p_j^c} - \frac{\partial x_i}{\partial p_j^c}\right)$$

It remains to evaluate $\partial x_i/\partial p_j^c$ from Eq. (5) and $\partial x_i^{\mathrm{E}}/\partial p_j^c$, which can be done using the relation

$$x_i^{\mathrm{E}} = -R_i^2\lambda/2$$

where the length R_i of reciprocal lattice vector i has to be written as a function of unit cell parameters.

The program written to implement this procedure[12] accepts initial estimates of crystal orientation, unit cell, and rocking curve parameters. It computes improved values by the usual iterative least-squares procedure, applying shifts determined at each stage by solution of Eq. (14). After each cycle of parameter refinement, data from the photograph in question are rescaled to the reference set by equating the sum of intensities classified as fully recorded to the sum of their reference set counterparts. Any of the parameters may be kept fixed.

As discussed, the misorientation angles ε_x^t, ε_y^t, ε_z^t are defined at a particular setting t. As $\partial\beta/\partial\varepsilon_x = 0$, ε_x^t cannot be determined from partial spot information at one setting. To determine ε_x^t one needs such information at another setting, ideally differing by $\pi/2$ in the oscillation angle ϕ. In the one crystal–one photograph situation, the two settings differ by only a very small rotation of $\Delta\phi$ around z. As a consequence, ε_x cannot be determined, and it is set equal to zero. For the determination of ε_y, partial spots at ϕ_b and ϕ^e are treated identically and ϕ^t in Eq. (5) is set to the mean oscillation angle ϕ^m. However, ε_z^b and ε_z^e are refined independently. Since the oscillation range $\Delta\phi = (\phi^b - \phi^e)$ should, in principle, be known accurately from the camera setting, provision is made to adjust only ϕ^m, keeping $\Delta\phi$ fixed.

The example of ε_x illustrates in an extreme way that the recorded fractions of reflections of one single photograph are not equally sensitive to all parameters considered in the refinement. Changes in some unit cell parameters or in certain linear combinations of them may also have no significant effect on the calculated fractions or, if they do, they may be highly correlated with a correction of one of the orientation parameters. In severe cases the refinement may not converge, especially if the starting point is far from the minimum. In such cases, it is advisable to examine critically the correlation coefficients and the eigenvalue spectrum of the normal matrix. This will then indicate which parameter (or linear combination of parameters) should be held invariant. A completely revised version of this "postrefinement" program, written by P. Evans (personal communication) introduces "Diamond filtering" in order to take care of these difficulties. Use of Diamond filtering at the stage of film scanning itself has been implemented by Reeke.[18]

The parameters being refined are sensitive only to the intensities of reflections that are actually partially recorded on the film. Only these reflections should therefore be included in the calculation. At any cycle, however, classification of reflections as fully recorded ($P_{cal} = 1$), partially recorded ($1 > P_{cal} > 0$), or not recorded ($P_{cal} = 0$) is subject to errors in the current values of the parameters. Refinement is therefore in practice complicated by some variation in the population of observations included from cycle to cycle. The farther one is from the minimum, the more reflections will be misclassified, slowing convergence. It is therefore useful to include at any cycle, in addition to reflections classified as partially recorded, those that would become so classified after small shifts in the values of parameters. For a step function profile, $f(b_i) = |b_i|$, the recorded fraction, P_{cal}, is proportional to the distance of a reciprocal lattice point from the Ewald sphere. If we do not restrict the value of $f(b)$ to the range 0 to 1, P_{cal} assumes values greater than 1 for fully recorded reflections and negative values for not recorded reflections. For example, in TBSV data collection, we included reflections with $-0.2 < P < 1.2$. We can also generalize a step function profile to deal with the problem that reflections

[18] G. Reeke, *J. Appl. Crystallogr.* **17**, 238 (1984).

with $b_i < 0$ or $b_i > 1$ have undefined derivatives, $\partial f(b_i)/\partial p_j$, for a rocking curve of finite width. For the step function, $f(b_i) = b_i$, and meaningful derivatives are obtained even if b is not restricted to the range -1 to 1. Other profiles, such as the cosine in Eq. (8), cannot simply be extended, but the derivatives for fully recorded and not recorded reflections to be included in the refinement can be calculated as if a step function profile were being used. By restricting P to the range 0 to 1 in the last cycle, it can be shown that this procedure improves convergence but does not in general bias the values of the parameters.

A Practical Case: TBSV 2.9 Å Data Collection. We illustrate the methods just outlined, by describing some aspects of data collection from TBSV at 2.9 Å.[2,12,19,20] The assessment of various sources of errors as well as the evaluation of the partial-spot correction procedure depends on realistic estimates of data precision. The following account shows how we obtained these estimates, pointing out precautions taken to minimize systematic errors in intensity measurements.

All photographs were taken on a Supper oscillation camera (Charles Supper Co., Natick, Massachusetts) at a crystal-to-film distance of 100 mm, using Cu K_α radiation (Elliott GX-6, 100 μm × 1 mm focus, 40 kV, 20 mA) focused by a double-mirror system of the Franks type.[6] A complete data set could be recorded on 50 photographs, each 25–35′ oscillation, covering a total contiguous arc of 25° around the [101] axis as indicated in Fig. 4. An important advantage of this particular choice is that the part of the crystal exposed to X rays can be regarded as a plate parallel to the capillary wall, with its normal tilted by at most 25° with respect to the beam. This minimizes absorption effects. Since the beam diameter is smaller than the crystal, a significant change of the irradiated volume during oscillation must be avoided. This is guaranteed both by the very small oscillation range and by the morphology of the crystal and its relative orientation to the X-ray beam. Films were densitometered using a Optronics Photoscan (50-μm raster size) and a modified version of the program SCAN12.[15] In our current system, the program is actually run off-line, on a VAX 11/780, reading a disk file containing the digitized film.

Figure 2 shows a photograph with about 20,000 reflections, of which 25–30% are fully recorded. Details of the film-scanning procedures are described elsewhere.[12] One important feature bears mention here. With the exception of the innermost reflections, the spot shape varies considerably with position on the film. To account for variability with angular position around the beam direction four different "masks" were chosen,

[19] A. J. Olson, G. Bricogne, and S. C. Harrison, *J. Mol. Biol.* **171,** 61 (1983).

[20] J. Hogle, T. Kirchhausen, and S. C. Harrison, *J. Mol. Biol.* **171,** 95 (1983).

determining the sampling of spot intensity and background (Fig. 2, masks A to D). The close spacing of individual reflections dictated the choice of background positions; a sample rectangular box would have led to overlap of background points and adjacent spots. The size of the masks was chosen to be independent of distance from the film center, since the increase in size from 6 to 2.9 Å was not very large. A smaller rectangular box was used whenever reflections in the innermost annulus were measured. During densitometry, the intensity centroid of each spot was determined and, if necessary, the mask was shifted by up to two raster points along each scanner axis. The intensity centroid of reflections with a small recorded fraction may be displaced from the predicted center by more than this permitted limit, which is set by the close spacing of adjacent spots. In such cases the mask was set to the unshifted position.

For each reflection, the integrated intensity was output together with a standard deviation estimated from fluctuations in the spot background and from the average optical density of spot and background area (Arndt and Wonacott,[13] p. 185). Measurements from the two films in each pack were scaled by the method of Fox and Holmes.[21] In this, as in all subsequent steps in which multiple measurement were combined, a weighted mean was calculated, with weights given by the inverse variance carried along with each reflection. The variance of the weighted mean was obtained from the variances of the contributing measurements, unless there was a large discrepancy. In this case either the variance was increased or, in sets having several measurements, anomalous ones were rejected. Variances assigned during densitometry were corrected by a single factor to bring the estimated level of variation to that observed between measurements from different films in a pack. The updating of error estimates at each step of data processing in which a large number of multiple measurements are combined is useful for assessing the magnitude of various sources of errors that appear at different stages of data combination. The average absolute difference between equivalent reflections is fairly constant over the resolution range, indicating the reflections at high resolution have been measured with the same absolute precision as those at lower resolution.

The routine processing procedure was to refine ε_y, ε_z, ε_z, and γ_0 for each photograph and to correct all reflections with $P_{cal} > 0.5$ to their fully recorded equivalents. Partially recorded reflections were deweighted relative to fully recorded ones by associating a constant error of 0.05 with

[21] G. C. Fox and K. C. Holmes, *Acta Crystallogr.* **20**, 886 (1966).

each P_{cal}. The variance of the corrected intensity is then

$$\sigma^2(I_P/P_{cal}) = \frac{\sigma^2(I_P)P_{cal}^2 + (0.05)^2 I_P^2}{P_{cal}^4} \tag{19}$$

Obviously, the accuracy of corrected intensities decreases rapidly with decreasing P_{cal}, and the weight of measurements corrected by a small P_{cal} becomes negligible compared to that of the equivalent fully recorded measurement. For this reason reflections with $P_{cal} < 0.5$ were discarded. The final statistics obtained when the data from all 50 photographs were combined were very similar to those given in the table, the overall R factor being 0.13. The value given by Eq. (19) is a reasonable estimate of the accuracy of partially recorded reflections, which correlate as well as whole spots when their appropriate corrected variance is considered.

INTENSITY STATISTICS IN INTENSITY AND RESOLUTION RANGES FOR THE COMBINATION OF FULLY RECORDED, EQUIVALENT REFLECTIONS FROM 12 OSCILLATION PHOTOGRAPHS OF TBSV[a]

Intensity range	N	Q_I	R	Resolution range	N	Q_I	R	$\bar{\Delta}_I$
0–300	1839	1.0	0.60	6.00–4.73	2293	1.5	0.08	152
300–600	1106	1.1	0.20	4.73–4.12	999	1.5	0.08	196
600–900	724	1.4	0.13	4.12–3.74	1003	1.4	0.11	204
900–1200	515	1.6	0.09	3.74–3.47	776	1.2	0.13	196
1200–1500	354	1.7	0.07	3.47–3.26	761	1.1	0.21	220
1500–1800	263	1.7	0.06	3.26–3.10	657	1.0	0.30	212
1800–2100	175	1.8	0.05	3.10–2.90	309	0.8	0.32	178
2100–2400	134	2.4	0.06					
2400–2700	97	1.6	0.04					
2700–3000	64	2.6	0.05					
>3000	215	2.1	0.04					
Overall		1.3	0.11					

[a] Relative scale and exponential factors were determined from Wilson-type plots as described in the text. Intensities are on an arbitrary scale, N is the number of independent pairs of equivalent reflections in each range, Q_I is defined in the text, and $\bar{\Delta}_I$ is the average intensity difference of these pairs. The R factor is defined as

$$R = \sum_h \sum_{j=1}^{N_h} |I_{hj} - \bar{I}_h| \Big/ \sum_h N_h \bar{I}_h$$

where the reflection h has been measured N_h times and $\bar{I}_h$ is the weighted mean of each set of equivalent reflections.

For very incomplete data sets, such as can be used for difference Fourier maps when noncrystallographic symmetry is present, an extensive set of reference whole spots may not be available. We have found even with very sparse data, however, that the above procedure can converge reasonably.[20] For example, with only 4 0.5° photographs (containing 20% of the reflections to 2.9 Å) from a Gd^{3+} derivative of TBSV, it was possible to use the whole spots from those films as a reference set for postrefinement and thus for correction of the partials. Note that the high symmetry of the TBSV space group implies a relatively uniform distribution of spots in reciprocal space, even from a few films. In less favorable cases, it might be important to choose the particular photographs quite carefully.

Some Remarks on Detectors

Position-sensitive X-ray detectors, such as those developed by Burns,[22] Xuong,[23] and Arndt,[24] are likely to change preferred data collection procedures in many laboratories. The Burns device, marketed by Xentronics (Cambridge, Massachusetts), is the only current product suitable for very large unit cells. Most of the theory and practice for film described above can carry over directly, since the active detector face has a 12-cm diameter and since the point-to-point resolution is about 0.2 mm (full width, half-maximum, FWHM). We have developed programs for processing data from the Xentronics detector that combine code from film scanning and postrefinement. Individual data frames are recorded corresponding to 5′ of arc oscillation. This is a convenient choice, since it is somewhat less than the cross-fire of the direct beam (about 7–10′ of arc). Most reflections appear on two or three adjacent frames, giving us a reasonably accurate estimate of the rotation angle of maximal intensity. This ϕ centroid of the intensity distribution is information equivalent to the fractional recording of partial spots in a photograph of longer oscillation, in the sense that it gives an accurate measure of the angle at which the reflection crossed the Ewald sphere. It is used for obtaining refined values of the setting parameters, γ, etc., by an adaptation of the procedure described above. Parameters refined fall into three groups: (1) Unit cell dimensions and crystal alignment angles determine the rotation angle

[22] R. M. Durbin, R. Burns, S. C. Harrison, and D. C. Wiley, *Proc. Am. Crystallogr. Assoc. Summer Meet.* (1983).

[23] R. Hamlin, C. Cork, A. Howard, C. Nielsen, W. Vernon, D. Matthews, N. H. Xuong, and V. Perez-Mendez, *J. Appl. Crystallogr.* **14,** 85 (1981).

[24] U. W. Arndt, *Nucl. Instrum. Methods* **201,** 13 (1982).

(frame number) at which a reflection appears. (2) These parameters, together with camera parameters (e.g., specimen-to-detector distance and angular alignment of the detector face about the beam), determine where on the image plane the reflection appears. (3) The width of the diffraction profile, determined by beam divergence and crystal mosaicity, is summarized by a simple model as above. The parameters in groups 1 and 2 are refined together, weighting the error in frame number (rotation angle) much more strongly than errors in detector x,y positions. The weighting ensures that crystal parameters, which affect the rotation angle, effectively refine independently of the camera parameters, which do not affect the rotation angle. Thus, camera parameters adjust to the best fit to position on the detector face, given the adjustments in crystal parameters. This feature keeps the interactions between the two classes of parameters from giving rise to ill-determined least-squares equations.

Acknowledgments

The work that led to these methods was supported by NIH Grant CA-13202 (to S. C. Harrison).

[20] Determining the Intensity of Bragg Reflections from Oscillation Photographs

By MICHAEL G. ROSSMANN

Historical Introduction

The rivalry between film and counter methods for the measurement of diffraction effects from crystalline material goes back to the beginning of the subject. The first X-ray diffraction results were performed in 1912 by W. Friedrich and R. Knipping on film, but these were rapidly followed by the work of W. H. and W. L. Bragg with their X-ray spectrometer. The somewhat easier experimental requirements of film methods, however, helped their development, but at a possible loss of accuracy. This, in turn, encouraged the use of counters and hand-driven, three-circle diffractometers in the late 1940s. The tedium of such procedures, particularly when applied to the multitude of reflections found in proteins, led M. F. Perutz and J. C. Kendrew to the development of optical densitometers for more accurate measurements of intensities. The concurrent development by

Martin Buerger of the precession camera was opportune as all the reflections were along straight lines corresponding to their position in the reciprocal lattice. The electronics industry had by this time advanced sufficiently to permit dependable automation of diffractometers. In addition, commercially available scintillation counters provided accurate and reliable detectors. These developments took the pain out of diffractometer measurements and, hence, four-circle machines became prevalent in the mid 1960s. Indeed, that is still the case for "small-molecule crystallography" today.

Another revolution in technique was heralded by "screenless" photography. Xuong and co-workers[1,2] advocated screenless precession cameras while Arndt and co-workers[3,4] proposed redeveloping the old oscillation method. These methods were in extreme contrast to diffractometry, in which the detector must concentrate on one reflection at a time while being oblivious to all other diffraction events. The screenless concept was of great benefit to macromolecular crystallographers who had to contend with rapid radiation damage. In addition, the small crystal motion produced very favorable Lorentz factors, thus optimizing the very weak diffraction effects from large unit cells. However, this new trend was possible only due to the simultaneous development of two-dimensional film scanning devices,[5–8] thus allowing automatic data processing of reflections scattered over the surface of the film rather than along reciprocal lattice lines. In the initial stages of this development, emphasis was placed on the development of screenless precession photography,[9,10] since such cameras were widely available. However, the instrument was not ideal both because each reflection is potentially split into its entering and exiting projection and because the Lorentz factor varies rapidly for reflections at the recording edges of layer lines. Hence, the oscillation method became dominant with the development of the rotation camera by U. W.

[1] Ng. H. Xuong, J. Kraut, O. Seely, S. T. Freer, and C. S. Wright, *Acta Crystallogr., Sect. B* **B24,** 289 (1968).

[2] Ng. H. Xuong and S. T. Freer, *Acta Crystallogr., Sect. B* **B27,** 2380 (1971).

[3] U. W. Arndt, *Acta Crystallogr., Sect. B* **B24,** 1355 (1968).

[4] U. W. Arndt, J. N. Champness, R. P. Phizackerley, and A. J. Wonacott, *J. Appl. Crystallogr.* **6,** 457 (1973).

[5] C. E. Nockolds and R. H. Kretsinger, *J. Sci. Instrum.* [2] **3,** 842 (1970).

[6] B. W. Matthews, C. E. Klopfenstein, and P. M. Colman, *J. Sci. Instrum.* [2] **5,** 353 (1972).

[7] S. Abrahamsson, *Acta Crystallogr., Sect. A* **A25,** 158 (1969).

[8] A. J. Wonacott and R. M. Burnett, *in* "The Rotation Method in Crystallography" (U. W. Arndt and A. J. Wonacott, eds.), p. 119. North-Holland Publ., Amsterdam, 1977.

[9] M. Leijonmarck, O. Rønnquist, and P. E. Werner, *Acta Crystallogr., Sect. A* **A29,** 461 (1973).

[10] P. Schwager, K. Bartels, and A. Jones, *J. Appl. Crystallogr.* **8,** 275 (1975).

Arndt and A. J. Wonacott[11] and its commercial production in Europe and America.

There is now a trend to return to counters in the form of two-dimensional area detectors[12–15] associated with the oscillation method. Indeed, that had been the original concept of Arndt and co-workers.[3,16] Such instruments have the additional advantage of time resolution between reflections close together in reciprocal space. Hence, in principle, such detectors allow measurements of intensities of even larger unit cells, provided their counting elements are sufficiently close together. Unfortunately, such instruments, when commercially available, will be very expensive and might be unnecessary luxuries for intense synchrotron radiation sources.

Numerous unforeseen difficulties were encountered in the development of the oscillation method, particularly in the processing of the X-ray films. An especially distressing problem was the correct identification of partial reflections which were progressively more numerous as the angle of oscillation decreased (Fig. 1). These could be omitted, added from scaled photographs on abutting oscillation ranges, or used in "postrefinement" (see section below). However, a decade elapsed between the initial proposals of screenless methods and the routine structure determination from intensities collected on oscillation cameras. Probably the first successful structure determinations were of *Bacillus stearothermophilus* glyceraldehyde-3-phosphate dehydrogenase (GAPDH),[17] tobacco mosaic virus (TMV) coat protein,[18] and pyruvate kinase.[19] These results and the stimulus of a conference held in Gröningen in 1975[11] did much to encourage further development of the technique.

Roughly four major procedures have been developed for the processing of oscillation photographs. These are primarily based, respectively,

[11] U. W. Arndt and A. J. Wonacott, "The Rotation Method in Crystallography." North-Holland Publ., Amsterdam, 1977.

[12] U. W. Arndt and D. J. Gilmore, *J. Appl. Crystallogr.* **12,** 1 (1979).

[13] C. Cork, D. Fehr, R. Hamlin, W. Vernon, Ng. H. Xuong, and V. Perez-Mendez, *J. Appl. Crystallogr.* **7,** 319 (1973).

[14] R. Hamlin, C. Cork, A. Howard, W. Vernon, D. Matthews, Ng. H. Xuong, and V. Perez-Mendez, *J. Appl. Crystallogr.* **14,** 85 (1981).

[15] Ng. H. Xuong, S. T. Freer, R. Hamlin, C. Nielson, and W. Vernon, *Acta Crystallogr., Sect. A* **A34,** 289 (1978).

[16] U. W. Arndt, R. A. Crowther, and J. F. W. Mallett, *J. Sci. Instrum.* [2] **1,** 510 (1968).

[17] G. Biesecker, J. I. Harris, J. C. Thierry, J. E. Walker, and A. J. Wonacott, *Nature (London)* **266,** 328 (1977).

[18] A. C. Bloomer, J. N. Champness, G. Bricogne, R. Staden, and A. Klug, *Nature (London)* **276,** 362 (1978).

[19] M. Levine, H. Muirhead, D. K. Stammers, and D. I. Stuart, *Nature (London)* **271,** 626 (1978).

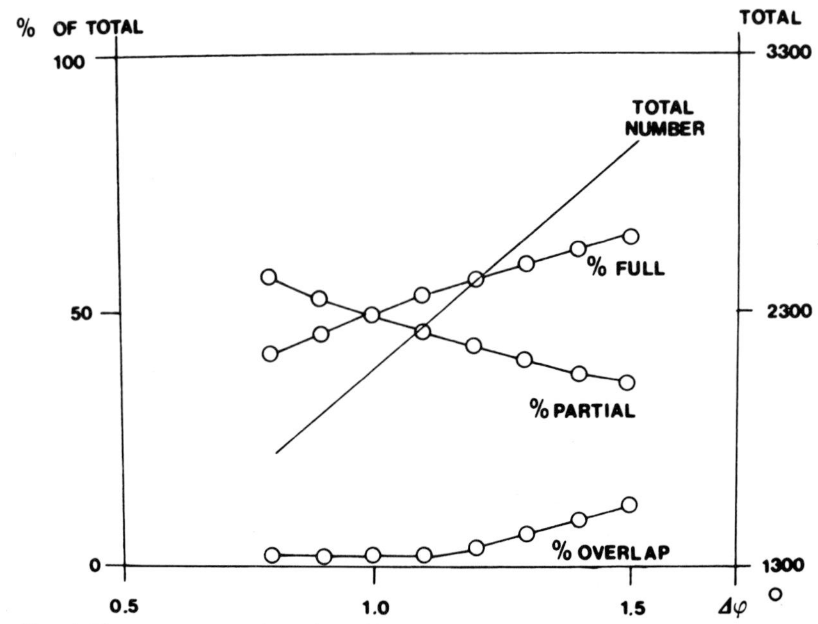

FIG. 1. The percentage and number of fully recorded, partially recorded, and overlapping reflections for phosphorylase *b* at 3 Å assuming the mosaic spread to be 0.35″. (Reprinted with permission from Wilson and Yeates.[20] © 1979 International Union of Crystallography.)

on the work of A. J. Wonacott in England, J. L. Crawford and D. C. Wiley at Harvard University, P. Schwager and K. Bartels in Munich, and M. G. Rossmann at Purdue University. K. Wilson and D. Yeates[20] have also developed a system in Oxford based in part on the Wonacott procedure. A rather different pattern recognition procedure was developed by W. Kabsch.[21] I shall describe here primarily the system in use at Purdue University,[22] but indicate where it differs from the other techniques. The Purdue techniques have been applied mostly to larger proteins (e.g., catalase of MW 240,000[23]) and viruses [e.g. southern bean mosaic (SBMV) of MW 6.6×10^6,[24] alfalfa mosaic virus (AMV)[25]]. It has been modified for

[20] K. Wilson and D. Yeates, *Acta Crystallogr., Sect. A* **A35,** 146 (1979).

[21] W. Kabsch, *J. Appl. Crystallogr.* **10,** 426 (1977).

[22] M. G. Rossmann, *J. Appl. Crystallogr.* **12,** 225 (1979).

[23] M. R. N. Murthy, T. J. Reid, III, A. Sicignano, N. Tanaka, and M. G. Rossmann, *J. Mol. Biol.* **152,** 465 (1981).

[24] C. Abad-Zapatero, S. S. Abdel-Meguid, J. E. Johnson, A. G. W. Leslie, I. Rayment, M. G. Rossmann, D. Suck, and T. Tsukihara, *Acta Crystallogr., Sect. B* **B37,** 2002 (1981).

[25] K. Fukuyama, S. S. Abdel-Meguid, and M. G. Rossmann, *J. Mol. Biol.* **150,** 33 (1981).

crystals with smaller unit cells by M. F. Schmid *et al.*[26] to be described in another chapter of this volume.[27] The reader is referred to the excellent book by Arndt and Wonacott[11] for details of the technique and a description of oscillation cameras and appropriate alignment techniques. Helliwell and Greenhough have made a very careful study of the effects of beam divergence, crystal mosaic spread, and wavelength dispersion on the partiality of reflections for conventional X-ray sources[27a] and synchrotron sources.[27b] Their procedure is similar to, but more detailed than, that of Rossmann.[22] Problems of beam polarization are severe with synchrotron radiation and this has been mentioned by Bartunik *et al.*[27c]

Film Processing

Coordinate Systems

Four different coordinate systems must be defined.

1. Film scanner coordinates in raster steps (R, S) (Fig. 2). These are coordinates of the optical densities as scanned by the film scanner. (A 50-μm raster step size was best suited for films of virus oscillation photographs obtained with double-mirror X-ray focusing devices.[28]) The corresponding vector notation, **R**, will also be used.

2. Camera coordinates in raster steps (X, Y) in the plane of the film. These use the X-ray beam as origin, with Y parallel to the spindle direction (Fig. 3). A reflection center can thus be calculated to occur at a particular value (X, Y) without consideration of which way a film was scanned or whether the film was in any way distorted during the photographic procedure. The corresponding vector notation used here is **X.**

3. Reciprocal lattice coordinates in Å^{-1} in an orthogonal system (x, y, z). y is defined as parallel to the spindle axis, z is parallel to the path of the X-ray beam, and x forms a right-handed orthogonal system with y and z (Fig. 3). Note that the camera coordinates X and Y are parallel to the reciprocal lattice coordinates x and y, respectively. A reciprocal lattice

[26] M. F. Schmid, L. H. Weaver, M. A. Holmes, M. G. Grütter, D. H. Ohlendorf, R. A. Reynolds, S. J. Remington, and B. W. Matthews, *Acta Crystallogr., Sect. A* **A37,** 701 (1981).

[27] B. W. Matthews, this volume [16].

[27a] T. J. Greenhough and J. R. Helliwell, *J. Appl. Crystallogr.* **15,** 338 (1982).

[27b] T. J. Greenhough and J. R. Helliwell, *J. Appl. Crystallogr.* **15,** 493 (1982).

[27c] H. D. Bartunik, R. Fourme, and J. C. Phillips, *in* "Uses of Synchrotron Radiation in Biology" (H. B. Stuhrmann, ed.), p. 145. Academic Press, New York, 1982.

[28] S. C. Harrison, *J. Appl. Crystallogr.* **1,** 84 (1968).

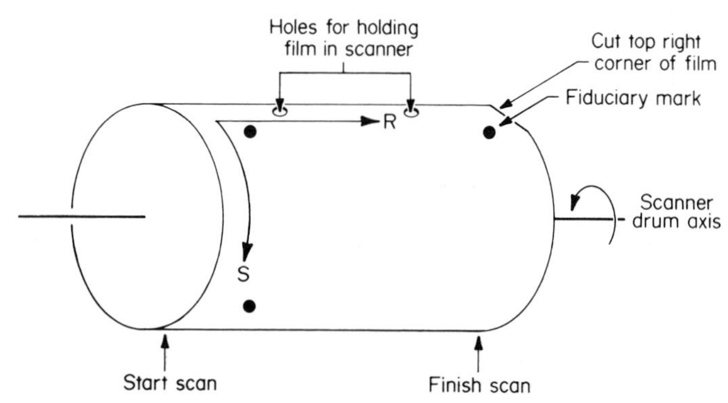

FIG. 2. Definition of the film coordinate system (R, S) for a film placed on a rotating drum scanner. R counts columns of scanned lines and S counts positions within each line where an optical density was recorded. (Reprinted with permission from Rossmann.[22] © 1979 International Union of Crystallography.)

vector will be referred to as **x**. The direction of positive rotation, ϕ, for the camera spindle axis will then be clockwise about y (Fig. 3).

4. Reflection coordinates in raster steps (p, q). The axes p and q run parallel to the film coordinates R, S but are centered at the calculated position of any given reflection rounded to the nearest raster step. The spot size can usually be confined within the limits $|p| \leq 6$ and $|q| \leq 6$ raster

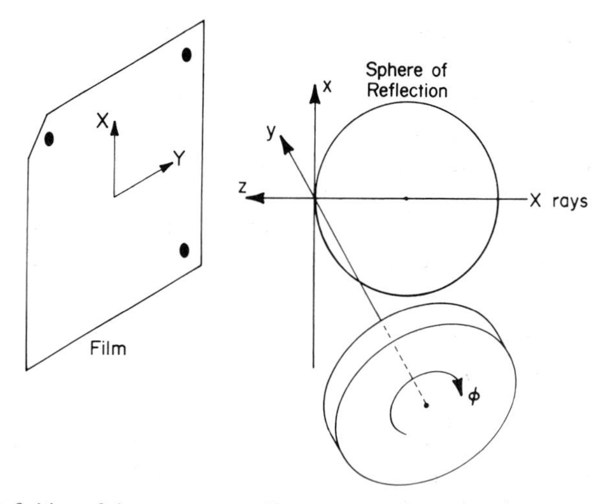

FIG. 3. Definition of the camera coordinate system (X, Y), reciprocal lattice coordinate system (x, y, z), and direction of positive rotation ϕ for the camera spindle axis. (Reprinted with permission from Rossmann.[22] © 1979 International Union of Crystallography.)

steps. If the spot appears bigger than such limits, then considerable time might be saved by scanning the film with a larger raster step.

The first three coordinate systems are connected by three sets of relationships. The first is

$$\mathbf{x} = [A]\mathbf{h} \tag{1}$$

where $\mathbf{h}(h, k, l)$ are the Miller indices of a particular reflection and $[A]$ is a 3×3 matrix which relates a given reflection to its reciprocal lattice coordinates when the crystal is in its standard position defined by setting the spindle axis angle $\phi = 0$. The second is a nonlinear transformation dependent on the camera geometry. For a flat film perpendicular to the X-ray beam (see Wonacott[29] for V-shaped and Schmid *et al.*[26] for cylindrical films)

$$X = D\sqrt{U}/V, \qquad Y = 2D\lambda y/v \tag{2}$$

where

$$U = 4\lambda^2(x^2 + z^2) - \lambda^4(x^2 + y^2 + z^2)^2$$
$$V = 2 - \lambda^2(x^2 + y^2 + z^2)$$

and λ is the wavelength of the monochromatic X-ray beam and D is the crystal-to-film distance. These formulas can be readily derived from pp. 79–84 of Arndt and Wonacott.[11] Finally, there is a linear relationship of the form

$$\mathbf{R} = [Q]\mathbf{X} \tag{3}$$

where $[Q]$ is a 3×2 matrix. $[Q]$ is roughly orthogonal, although film distortions and subsequent refinements may cause slight deviation from orthogonality. Schmid *et al.*[26] have found it useful to express the elements of $[Q]$ explicitly as rotations and translations.

The limits of the oscillation range, defined by the angles ϕ_1 and ϕ_2, can be assumed to be known with great accuracy since these angles are set by digitized shaft rotations on commercially available instruments. However, the setting of the crystal, which can be represented as rotations ϕ_x, ϕ_y, ϕ_z about the three orthogonal reciprocal axes, may well be out by a fraction of a degree. Thus, the initially given matrix, $[A]$, will have to be rotated about y (while keeping the limits ϕ_1 and ϕ_2) in order to represent the conditions of the experiment. The rotations of the crystal about x, y, and z, after it has been rotated by ϕ_1 or ϕ_2, will be defined as δx, δy, and δz.

[29] A. J. Wonacott, *in* "The Rotation Method in Crystallography" (U. W. Arndt and A. J. Wonacott, eds.), p. 75. North-Holland Publ., Amsterdam, 1977.

It is also possible that the cell dimensions are not precisely known. However, their determination from the position of reflections on a film is related in an almost linear manner to the crystal-to-film distance, D, which itself is not likely to be known with much precision. On the other hand, cell parameters also determine sensitively which reflections can occur partially or completely, thus permitting refinement of these parameters when intensities are known (see section on Postrefinement). Mosaic spread and beam divergence also influence greatly the degree of partiality of individual reflections.

General Considerations

Most procedures depend on the manual indexing and coordinate measurements of a selected dozen or so reflections. Orthogonal "stills" taken with the same crystal are particularly useful as all reflections are necessarily partial and thus determine with precision the exact crystal orientation. The Purdue procedure does not, however, require any separate still photographs but depends on a systematic convolution (comparison) of the observed diffraction picture with calculated intensity distributions for different crystal orientations. The partial reflections at the "edge of lunes" are highly sensitive to crystal orientation and, therefore, serve as a very accurate orientation indicator. This is analogous to using the intensities of edge reflections on the zero layer line of precession camera photographs to "set" a crystal.

The initial step is to search for the intense fiducial marks from which at least a crude $[Q]$ matrix can be evaluated. The additional use of the direct X-ray beam to determine the origin of the recorded reciprocal lattice will nearly always permit the correct automatic indexing of low resolution reflections. It may occasionally be necessary to guide the initial indexing by giving the R, S coordinates of a few strong reflections for crystals with cells in excess of about 600 Å.

The next step is to refine the $[Q]$ and $[A]$ matrices using all reflections considered full at relatively low resolution. Usually, data with spacing less than 7 Å are sufficient, provided there are more than about 150 such reflections. The Purdue procedure uses all reflections weighted according to a rough estimate of their intensities, while other methods use about a dozen equally weighted strong reflections, setting all other reflections to zero weight. It is usually sufficient to know the crystal orientation, given by the $[A]$ matrix, to within 1°. Hence, it is only necessary to set a crystal quite crudely using still photographs on an oscillation camera, thus avoiding excessive radiation damage.

The $[Q]$ matrix depends exclusively on the position of full reflections, whereas the rotation of the crystal about the camera X and Y axes, represented by the $[A]$ matrix, determines which reflections are fully or partially visible. The crystal ($[A]$ matrix) and film ($[Q]$ matrix) rotation about the Z axis (the X-ray beam) are, however, strongly coupled. Fortunately, provided no great error has been made in the initial crystal orientation, this is of no great consequence. Hence, the $[Q]$ matrix can be refined readily by considering the center of gravity of all the full reflections and the $[A]$ matrix will be sensitive to the occurrence of partial reflections at the edge of lunes. It is usually best to refine first the $[Q]$ matrix assuming the original $[A]$ matrix. This can be followed by an $[A]$ matrix refinement using the better $[Q]$ matrix. Two successive cycles of $[Q]$ and $[A]$ matrix refinement are nearly always sufficient.

Some consideration of the effective mosaic spread is useful at this time. Increase of mosaic spread will also increase the number of visible predicted reflections. Hence, a comparison of observed and calculated patterns gives some guidance on the value of this parameter. Simultaneously, a profile of the reflection can be calculated from a weighted mean of all full reflections, after subtraction of the background, in the inner part of the film.

All the essential parameters for intensity measurements will now have been determined and the resolution may be increased to the edge of the observable data. The profile found at low resolution can be used to compute the integrated intensity for all reflections while taking no account of profile variation over the film surface. However, during the first pass over the film, profiles can be gathered from different regions. These can be used in subsequent cycles, the profile for any given reflection being calculated by interpolation from the locally averaged profiles. However, partial reflections cannot be assumed to have the same profile and must be determined, less accurately, by integration. Overlapped reflections can be either rejected entirely or deconvoluted using a knowledge of the profile, provided they are not entirely on top of each other.

The actual program is arranged with 14 options (e.g., determine $[Q]$ from fiducial marks, refine $[Q]$, refine cell parameters, select reflections in an oscillation range within suitable resolution limits, refine δx and δy, integrate). These options can be called upon in any desired sequence, resembling the control of the well-known ORTEP program.[30] This flexibility permits the processing of a film by different routes, and determination of suitable data collection schemes and oscillation ranges or other investigations.

[30] C. K. Johnson, "ORTEP: A Fortran Thermal-ellipsoid Plot Program for Crystal Structure Illustrations," Rep. ORNL-3794 (2nd rev.). Chem. Div., Oak Ridge Nat. Lab., Oak Ridge, Tennessee, 1970.

Determination of [Q] from Fiducial Marks

The position of each fiducial mark (R_o, S_o) is determined by taking the center of gravity of the optical densities above a selected cutoff within a search box. The positioning of the search box can be estimated by measuring the positions of the fiducial marks on the film. An additional fiducial mark is also made by the direct X-ray beam on the film. The larger the search box, the easier it is to find the fiducial marks, but care must be taken not to include neighboring density unrelated to the mark. This is particularly true for the blackened area outside the beam stop around the central mark. The camera coordinates are also supplied for each mark. Thus, a linear least-squares fit between (R_c, S_c) given by

$$R_c = Q_{11}X + Q_{12}Y + Q_{13}$$
$$S_c = Q_{21}X + Q_{22}Y + Q_{23}$$

and the observed positions (R_o, S_o) provide starting elements for the $[Q]$ matrix. In general, the agreement between observed and calculated positions was found to be better than 0.3 of a raster step.

Selection of Possible Reflections in a Given Oscillation Range

When the crystal is rotated through an angle ϕ, then the reciprocal lattice point P at $\mathbf{x}$ is moved to P' at $\mathbf{x}'$ and

$$\mathbf{x}' = [\Phi][A]\mathbf{h}$$

where

$$[\Phi] = \begin{pmatrix} \cos\phi & 0 & \sin\phi \\ 0 & 1 & 0 \\ -\sin\phi & 0 & \cos\phi \end{pmatrix}$$

Thus, the position of the point P can readily be found at the extreme ends of its oscillation range by permanently evaluating $[A_1]$ and $[A_2]$ corresponding to the product $[\Phi][A]$ for both $\phi = \phi_1$ and ϕ_2, respectively. A point will have passed through the sphere of reflection, centered at S $(0, 0, -1/\lambda)$ when $P'_1S < 1/\lambda$ and $P'_2S > 1/\lambda$ on entry (if $\phi_1 > \phi_2$) or when $P'_1S > 1/\lambda$ and $P'_2S < 1/\lambda$ on exit from the sphere. These conditions can be shown to correspond to

$$d^{*2} + \frac{2z_1}{\lambda} < 0 \quad \text{and} \quad d^{*2} + \frac{2z_2}{\lambda} > 0 \quad \text{on entry into the sphere}$$

or

$$d^{*2} + \frac{2z_1}{\lambda} > 0 \quad \text{and} \quad d^{*2} + \frac{2z_2}{\lambda} < 0 \quad \text{on exit from the sphere}$$

TABLE I
CALCULATION OF DEGREE OF PENETRATION OF SPHERE, q

	Almost completely within sphere	Almost completely outside sphere	Full reflection
	Condition 1	**Condition 2**	
Entering[a]	$-1 < D_1/\eta_1 < +1$ and $D_2/\eta_2 \leq -1$ $q = \frac{1}{2}(1 + D_1/\eta_1)$	$-1 < D_2/\eta_2 < +1$ and $D_1/\eta_1 \geq +1$ $q = \frac{1}{2}(1 - D_2/\eta_2)$	$D_1/\eta_1 \geq 1$ and $D_2/\eta_2 \leq -1$
	Condition 3	**Condition 4**	
Exiting[a]	$-1 < D_2/\eta_2 < +1$ and $D_1/\eta_1 \leq -1$ $q = \frac{1}{2}(1 + D_2/\eta_2)$	$-1 < D_1/\eta_1 < +1$ and $D_2/\eta_1 \geq +1$ $q = \frac{1}{2}(1 - D_1/\eta_1)$	$D_1/\eta_1 \leq -1$ and $D_2/\eta_2 \geq 1$

[a] Where $D = d^{*2} + \delta^2 + 2z/\lambda$ and $\eta = \delta\sqrt{x^2 + y^2}$. Subscripts refer to the angle, ϕ_1 and ϕ_2, designating the end of the oscillation range. d^* and δ are independent of ϕ.

However, since the crystal contains some mosaic spread and the beam has divergence, which together give an effective mosaic spread of m, a reflection can occur (Table I) whenever

$$d^{*2} + \delta^2 + \frac{2z_1}{\lambda} - \eta_1 < 0 \quad \text{and} \quad d^{*2} + \delta^2 + \frac{2z_2}{\lambda} + \eta_2 > 0 \quad \text{on entry}$$

or

$$d^{*2} + \delta^2 + \frac{2z_1}{\lambda} + \eta_1 > 0 \quad \text{and} \quad d^{*2} + \delta^2 + \frac{2z_2}{\lambda} - \eta_2 < 0 \quad \text{on exit}$$

Symbols in the above inequalities are defined by

$$\eta_1 = 2\delta(x_1^2 + y_1^2)^{1/2}$$
$$\eta_2 = 2\delta(x_2^2 + y_2^2)^{1/2}$$

where $\delta = m/\lambda$.

Since the computing time for selecting possible reflections within any one oscillation range is critical, it is most important to rapidly reject reflections that do not occur on a given film. A useful order of tests (with R as the limit of resolution) is (Fig. 4)

1. reject if z_1' or z_2' is positive,
2. reject if z_1' or z_2' is less than $-(\lambda/2R^2)$,
3. reject if $|x_1'|$, $|y_1'|$, $|x_2'|$, or $|y_2'|$ are greater than $(4R^2 - \lambda^2)^{1/2}/2R^2$,

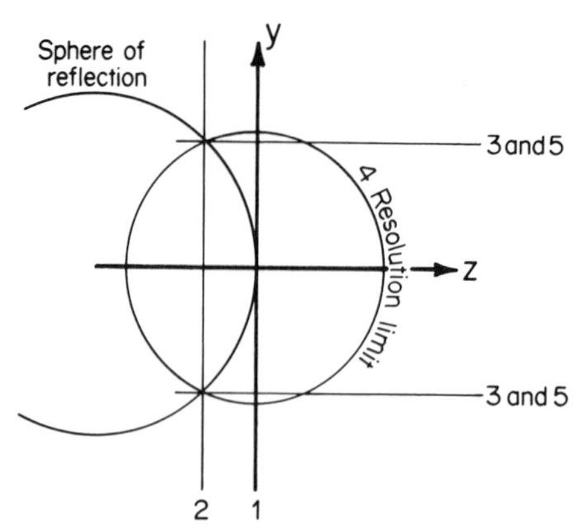

FIG. 4. The speed of selection of possible reflections within a given oscillation range can be greatly enhanced by testing whether the end points of motion, P_1 and P_2, for a given reflection lie within the limits 1–5 when applied in that sequence. (Reprinted with permission from Rossmann.[22] © 1979 International Union of Crystallography.)

4. reject if $d^{*2} > R$,
5. reject if $x_1'^2 + y_1'^2$ or $x_2'^2 + y_2'^2$ are greater than $(4R^2 - \lambda^2)/4R^4$,
6. reject if reflection cannot occur within oscillation range according to inequalities given above.

Systematic absences are omitted by generating only the nth spot along every reciprocal lattice line after having discovered the first acceptable reflection along each line. Thus, $n = 1$ for primitive lattices, 2 for body-centered lattices, or 3 for rhombohedral lattices with hexagonal indices. No attempt has been made to recognize lune boundaries.[31]

An alternative procedure for the very rapid sequential selection of reflections as they occur during the rotation of a crystal has been developed by D. Thomas.[32] This depends on selecting those reciprocal lattice points immediately in front of and approximately tangential to a line from each current reciprocal lattice point to the axis of rotation. This is particularly important when area detectors are used as it avoids a lengthy initial sorting of reflections into the order they would occur during crystal rotation.

Once the possible reflections and their camera coordinates have been listed, they can then be examined rapidly. In particular, an image of a given film can be created by use of a reflection box profile (*vide infra*) at

[31] J. Nyborg and A. J. Wonacott, *in* "The Rotation Method in Crystallography" (U. W. Arndt and A. J. Wonacott, eds.), p. 139. North-Holland Publ., Amsterdam, 1977.

[32] D. Thomas, Ph.D. Thesis, MRC Laboratory of Molecular Biology, Cambridge, England (1981).

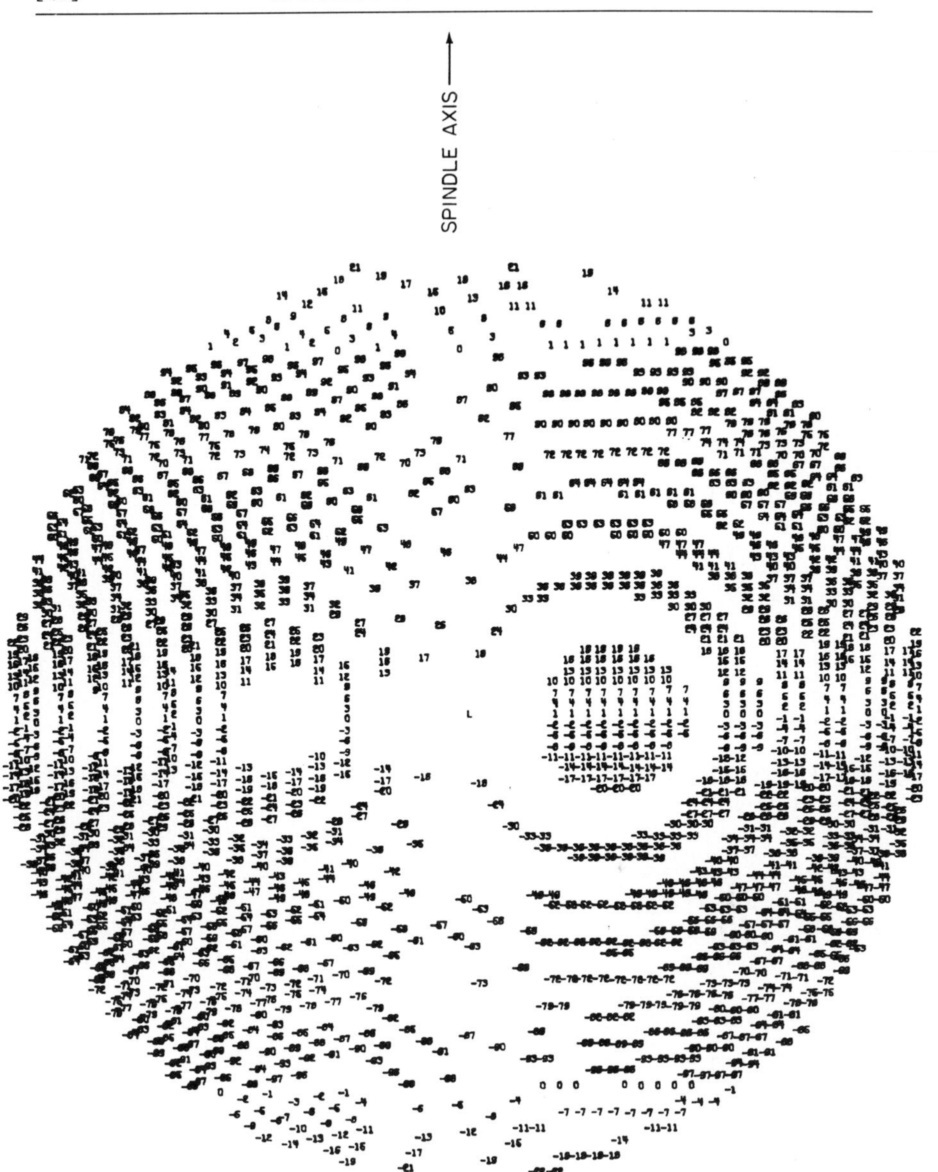

FIG. 5. Calculated 8 Å resolution image of an SBMV film oscillated through 0.6° (ϕ_1 = 25.2°, ϕ_2 = 25.8°) about the trigonal c axis (a = 337.0 Å, c = 756.0 Å). The pen-driven graphics device can plot Miller indices. Shown here is a plot of the l indices. (Reprinted with permission from Rossmann.[22] © 1979 International Union of Crystallography.)

→ R
↓ S

-1	0	3	7	10	12	10	6	4	1	-1
1	1	6	12	17	17	15	10	5	2	0
1	4	13	28	40	40	32	22	14	6	0
2	9	25	52	71	72	59	43	27	13	2
3	12	34	71	96	95	80	59	40	20	5
3	12	34	72	98	100	85	64	44	24	7 → p
3	9	28	60	85	88	76	58	39	21	6
1	6	21	44	66	69	60	44	29	16	4
0	4	13	28	42	44	37	26	17	8	2
1	1	7	14	20	21	17	12	7	3	-1
0	0	3	7	11	11	10	7	3	1	-1
					↓ q					

1	1	0	0	0	0	0	0	0	1	1
1	0	0	2	2	2	2	2	0	0	1
1	0	2	2	2	2	2	2	2	0	1
0	0	2	2	2	2	2	2	2	0	0
0	0	2	2	2	2	2	2	2	0	0
0	0	2	2	2	2	2	2	2	0	0 → p
0	0	2	2	2	2	2	2	2	0	0
0	0	2	2	2	2	2	2	2	0	0
1	0	2	2	2	2	2	2	2	0	1
1	0	0	2	2	2	2	2	0	0	1
1	1	0	0	0	0	0	0	0	1	1
					↓ q					

each reflection position. If such a film image is created by running through the list of reflections first in a forward and then again in a backward direction, all overlapped reflections can be identified. As each reflection is reached in the sequential list of reflections, a set of pixels, representing the film image, is set to one. For example, if a reflection A overlaps with a subsequently listed reflection B, then when B is reached, some pixels needed to represent B are already set to one and, hence, B is recognized to overlap some other reflection. Similarly, on reversing the list, A will be found to overlap the previously set pixels belonging to reflection B.

The actual image can be quickly produced on an electrostatic printer or more slowly on a pen-driven graphics system such as "Calcomp." The latter has the advantage of having sufficient resolution to show Miller indices (one plot per *hkl*; Fig. 5) and, if desired, representation of the intensities on a fourth plot. Comparison of these plots with the film is a useful check, after refinement of the $[A]$ matrix.

Refinement of the $[Q]$ *Matrix*

Once the reflection positions on a given film have been calculated, at least approximately, and an initial $[Q]$ matrix has been determined from the fiducial marks, then the positions of low-order reflections can be predicted to a reasonable degree of accuracy. The center of gravity of their position can be determined after subtraction of background. The difference between observed and calculated positions can then be refined in terms of the six elements of $[Q]$, thus accommodating shrinkage, rotation, and translation of the film.

In order to define the area of the reflection as well as useful positions for the measurement of background, a "reflection box" is centered on the raster coordinate nearest the calculated reflection position. Raster steps within the box are counted by the reflection coordinates p, q (Fig. 6).

The background is assumed to be a plane over the area of a reflection spot. Thus, if ρ is the optical density at the position p, q for a given reflection, then the background-subtracted optical density is given by $\rho - ap - bq - c$, where a, b, and c are constants. Hence, by setting $\Sigma_n(\rho -$

FIG. 6. Above: Reflection profile obtained from a weighted average of all reflections in a defined portion of an AMV film, scanned on a 50-μm raster. Below: The "reflection box" designating positions to be considered background and peak. The numeral 0 signifies that the corresponding optical density is neglected, 1 that it is to be used in determining background, and 2 that the position is within the designated peak area and will be used for profile fitting. The profile determination or profile fitting does not require determination of the reflection boundary nor an analytical function to describe the profile.

$ap - bq - c)^2$ to a minimum over the n background points in the reflection box (Fig. 6), it follows that

$$\begin{pmatrix} a \\ b \\ c \end{pmatrix} = \begin{pmatrix} \Sigma_n p^2 & \Sigma_n pq & \Sigma_n p \\ \Sigma_n pq & \Sigma_n q^2 & \Sigma_n q \\ \Sigma_n p & \Sigma_n q & n \end{pmatrix}^{-1} \begin{pmatrix} \Sigma_n \rho p \\ \Sigma_n \rho q \\ \Sigma_n \rho \end{pmatrix}$$

The 3×3 matrix will be the same for every reflection and need, therefore, be evaluated once only.

The observed positions of the reflection can then be rapidly evaluated as

$$R_o = \left(\sum_m \rho p - a \sum_m p^2 - b \sum_m pq - c \sum_m p\right) \Big/ I$$

$$S_o = \left(\sum_m \rho q - a \sum_m pq - b \sum_m q^2 - c \sum_m q\right) \Big/ I$$

where $I = \Sigma_m \rho - a \Sigma_m p - b \Sigma_m q - cm$. Here the sums are taken over the m peak positions in the reflection box. Again, most of the sums are common to each reflection and need be determined only once. In addition, the integrated reflection can be tentatively evaluated as

$$I = \sum_m \rho - a \sum_m p - b \sum_m q - cm$$

Refinement of the elements of $[Q]$ must then minimize $\Sigma(R_o - R_c)$ and $\Sigma(S_o - S_c)$ over all reflections, where R_c and S_c are the calculated reflection positions.

Experience has shown that the procedure generally converges from a mean displacement of over 1 raster step to about 0.3 raster steps in about five cycles. Simultaneous improvement in the profile is particularly noticeable and is an excellent guide to ascertain whether the initial $[A]$, $[Q]$, and ϕ_1, ϕ_2 values were of sufficient accuracy to home in on the reflection positions.

Determination of the Reflection Profile

Information from many neighboring reflections can be used to evaluate the integrated intensity of a given reflection by employing an empirical profile. It is reasonable to assume that, at least over a limited area of the film, the profile of each reflection is the same except for a single scale factor. Hence, only one parameter need be determined from all the optical densities for any one whole and nonoverlapped reflection within the designated peak area of a reflection box. Thus, profile fitting is intrinsically more accurate than integration, provided an even partially accepted profile has been determined (see section on Profile Fitting, below, for a more rigorous analysis).

Let $P(p, q)$ be the value of the standard profile at p, q and J be the scale factor required to fit this profile to a given reflection. Then

$$JP(p, q) = \rho - ap - bq - c$$

Hence, minimizing $\Sigma_n[JP(p, q) - \rho + ap + bq + c]^2$, it follows that

$$P(q, q) = \frac{\Sigma_N J\rho - p\,\Sigma_N Ja - q\,\Sigma_N Jb - \Sigma_N Jc}{\Sigma_N J^2}$$

for every point (p, q) within the reflection box. The summations are taken over the N significant, whole, nonoverlapped reflections within the setting resolution limit. Values of J equal to the integrated intensity will be quite acceptable for this determination.

It is useful to scale the profile values $P(p, q)$ such that their maximum is 100. Furthermore, it is equally useful to evaluate $P(p, q)$ over the whole of the reflection box since this shows clearly whether the designated background area has been well chosen. Finally, it is possible to examine the standard profile in order to omit background positions which deviate too far from the theoretical zero value. These omitted positions usually encroach on the peak area of adjacent reflections.

Variation of the profile over the whole area of the film can only be determined by examining the mean profile within different film regions. Thus, during the first pass of integration by profile fitting, the profiles can be determined in five "quadrants" over the film. "Quadrant 1" corresponds to the setting resolution limit within which the initial profile was determined. The outer four quadrants lie outside this limit. These five profiles can be used to determine a "variable profile" in a second pass over the film. If $P_1, P_2, \ldots, P_5$ are the values of the five standard profiles at (p, q), then the actual profile P at (p, q) for a given reflection can be interpolated as

$$P = \sum_{i=1,\,5} \omega_i P_i \Big/ \sum_{i=1,\,5} \omega_i$$

where $\omega_i = \exp\{-a[(R - R_i)^2 + (S - S_i)^2]\}$. Here R and S are the calculated film coordinates of the reflection, R_i and S_i are the mean film coordinates of all reflections used in determining the profile in the ith quadrant, and a is an arbitrary constant. A useful value for the arbitrary constant, a, was found to be such that $\omega_1 \simeq 0.05$ near the mean position of any of the outer profiles. Variable profile analysis was found to be most valuable and will be discussed further in the section on Profile Fitting.

Convolution of the Observed and Calculated Reflection Pattern

Small rotations δx and δy of the crystal about the reciprocal x and y axes will primarily affect only the partiality of observed reflections, not their position on the film. Thus, a comparison of the predicted pattern with the observed pattern can be used as a sensitive guide to the crystal orientation. However, the selection of possible reflections for a large variety of crystal orientations (different δx and δy values) would take an unacceptably large amount of time, and a suitable alternative strategy must be adopted.

As an initial step, reflections are selected with an excessively large mosaic spread (e.g., 0.5° instead of 0.05°) with the assumed [A] matrix and known oscillation limits, ϕ_1 and ϕ_2. The intensities of full and partial reflections can then be integrated around their corresponding predicted film positions R, S. Thus, the "intensities" of many reflections will be determined as being close to zero since they do not really exist. However, the procedure assures an intensity determination of all reflections which do exist, provided the assumed setting of the crystal is not too far off. The next step is to run down the list of reflections and tag those which would occur with the true estimated mosaic spread. It is then only necessary to run down the list and select from it those reflections which would occur with the altered crystal orientation in order to convolute the actual pattern with the calculated pattern (for a given combination of δx and δy). Particular attention is placed on those intensities which now occur but were not tagged (gained) and those which are tagged but cannot now occur (lost). A useful index of convolution can be defined as

$$C = \sum \text{intensities gained} - \sum \text{intensities lost}$$

It is easily computed since all intensities were recorded initially. Its sensitivity can be further enhanced by omitting intensities of less than one standard deviation. The particular crystal orientation adjustment for which C is a maximum should correspond to the greatest similarity between observed and calculated diffraction patterns.

If the crystal were perfectly set ([A] correct), then any rotation about x and y can only cause loss of intensity. On the other hand, if the crystal was misset, then the procedure will seek out reflections not tagged (those not predicted for the original setting) and the effect will be one of optimizing gained reflections and minimizing lost reflections.

It has been found useful to alter first the orientation about y in a search for the maximum of the criterion C. The first attempt can be done quite crudely at (for example) 0.16° intervals followed by finer (0.08°) and finer (0.04°) intervals until the maximum is determined to within 0.01°. The $[A]$ matrix can then be corrected using

$$[A_1] = [\Phi_{\delta y}][A]$$

where $[\Phi_{\delta y}]$ is the rotation matrix about y for the optimal small rotation δy. The same process is repeated about the x axis and the $[A]$ matrix is again corrected using

$$[A_2] = [\Phi]^{-1}[\Phi_{\delta x}][\Phi][A_1]$$

where $[\Phi_{\delta x}]$ is the rotation matrix about x for the optimal small rotation δx and $[\Phi]$ is the rotation $(\phi_1 + \phi_2)/2$ to bring the crystal into the given oscillation range. This procedure can be repeated again about y and x, etc., until no further refinement occurs. The process has nearly always converged after the second cycle of rotation about y and x. When large movements have occurred then the initial selection of intensities may not have been complete. Hence, if the setting matrix $[A]$ has been made to rotate more than, say, 0.4° about any one axis, it will be necessary to reselect intensities, using the excessively large mosaic spread, and to repeat the convolution search.

The convolution technique described above has been found to reset the $[A]$ matrix to within 0.01° even if it has been purposely rotated by more than 1° from its true position. A typical example of a convolution search about the y axis is shown in Table II.

The convolution process can also be used in assessing a suitable mosaic spread, m. Once the setting matrix $[A]$ has been optimized, then the reflections gained relative to a zero mosaic spread can be tabulated. As the mosaic spread is gradually increased (Table III), the convolution will find the extra partial reflections. However, this sum levels off when all reflections have been accounted for fully.

Profile Fitting

Reflections at the desired resolution limit of any given oscillation photograph can be integrated once the parameters $[Q]$, $[A]$, and m have been established. For safety, however, it is useful to increase the mosaic spread in order to make certain that no partial reflection has been included among the whole reflections. It has been found useful to take $1.5m$ as the effective mosaic spread. Starting with an accurate prediction of the film coordinates of each nonoverlapping reflection and a liberal estimate of

TABLE II
CONVOLUTION OF OBSERVED WITH CALCULATED FILM DURING ROTATION ABOUT y AXIS[a]

y (°)	Gained ΣI_g	Gained n	Lost ΣI_l	Lost n	$C = \Sigma I_g - \Sigma I_l$
−0.32	363	208	756	316	−393
−0.24	346	191	535	233	−189
−0.16	307	150	305	151	2
−0.12	265	121	182	104	83
−0.10	246	103	137	82	108
−0.08	213	88	101	60	113
−0.07	198	79	79	50	119
−0.06	178	71	59	37	120
−0.05	138	51	53	34	85
−0.04	109	38	46	30	64
0.00	0	0	0	0	0
+0.16	53	50	601	185	−548

[a] Example taken from an SBMV crystal pattern with a setting resolution limit of 8 Å. ΣI_g and ΣI_l represent the sum of the gained and lost reflections relative to the setting angle at $\delta y = 0°$. Also shown are the number (n) of gained and lost reflections. The criterion $C = \Sigma I_g - \Sigma I_l$ is maximized to find the best setting angle for the given mosaic spread and assumed rotation about x.

those which are partial, the integrated intensities can then be determined by profile fitting for the whole reflections and by integration for the partial reflections.

The quantities a, b, and c, which define the best fit of a plane through the background region, must be determined first for every reflection (see section on Refinement of the [Q] Matrix). Then a least-squares fit between the standard profile, $P(p, q)$, when scaled by the constant multiplier J, can be determined by minimizing E with respect to J where

$$E = \sum_m [JP - (\rho - ap - bq - c)]^2$$

The summation is taken over the m points (p, q) in the peak area of the reflection box. Hence,

$$J = \frac{\Sigma_m \rho p - a \Sigma_m pP - b \Sigma_m qP - c \Sigma_m P}{\Sigma_m P^2} \tag{4}$$

TABLE III
DETERMINATION OF SUITABLE MOSAIC SPREAD BY CONVOLUTION OF OBSERVED WITH CALCULATED FILM[a]

Mosaicity, m (°)	ΣI_g	n	Number of reflections found in different ranges of I									
			0–50	50–100	100–150	150–200	200–250	250–300	300–350	350–400	400–450	450–500
0.00	0	0	0	0	0	0	0	0	0	0	0	0
0.017	279	15	22	5	1	5	0	0	2	1	1	0
0.033	473	25	44	10	1	6	0	2	3	1	1	1
0.050	638	38	64	18	3	7	1	2	4	1	1	1
0.067	651	40	87	20	3	7	1	2	4	1	1	1
0.083	713	45	115	24	3	7	1	2	5	1	1	1
0.100	724	47	135	26	3	7	1	2	5	1	1	1

[a] Example taken from an SBMV crystal pattern with a setting resolution limit of 7.8 Å. ΣI_g, the sum of intensities gained relative to zero mosaic spread, omits reflections with $I < 50$. Note that increase in ΣI_g is small for $m > 0.05°$, with additional contributions primarily due to very small intensities.

The quantities ΣpP, ΣqP, ΣP, and ΣP^2 can be determined initially if a constant profile is being used, as is the case during the first pass over the complete film. However, when a weighted variable profile is computed for each reflection, time must be taken to reevaluate these quantities for each reflection.

The total integrated optical density under the peak will thus be $\Sigma_m JP$ or $J \Sigma_m P$. Hence, the integrated intensity by profile fitting is given by

$$I_1 = \frac{(\Sigma \rho P - a \Sigma pP - b \Sigma qP - c \Sigma P) \Sigma P}{\Sigma P^2} \tag{5}$$

while the actual integrated intensity will be

$$I_2 = \sum \rho - a \sum p - b \sum q - cm \tag{6}$$

It is immediately apparent that $I_1 = I_2$ if $P = 1$ over all the peak area. That is, integration is equivalent to fitting a "square profile" to the optical density and thus demonstrates the considerable superiority of the profile fitting method.[33,34]

If ε is the error on a given optical density, then it can be shown that the error in I_1 is $n\ \varepsilon P/\Sigma P^2$ and the error in I_2 is $\Sigma \varepsilon$. Thus, the total error is

[33] R. Diamond, *Acta Crystallogr., Sect. A* **A25,** 43, (1969).
[34] G. C. Ford, *J. Appl. Crystallogr.* **7,** 555 (1974).

represented by the center of mass of the weights P with ε the moment on each weight. Since the profile is peaked at its center, there will be only a few weights with large moments. In contrast, if all weights are unity (the integration process), there will probably be far more weights with large moments giving rise to a likely overall moment or larger total probable error.

The integrated error, σ_S, in fitting the profile can be determined from the differences between measured optical densities and the scaled profile. Thus,

$$\sigma_S^2 = \sigma_P + (m/n)\sigma_B^2 \tag{7}$$

where σ_P and σ_B are the standard errors in determining the peak and background regions within the reflection box, respectively. Therefore.

$$\begin{aligned}\sigma_S^2 &= \sum_m [JP - (\rho - ap - bq - c)]^2 + \left(\frac{m}{n}\right)\sigma_B^2 \\ &= J^2 \sum P^2 - a^2 \sum p^2 - b^2 \sum q^2 - cm + \sum \rho^2 \\ &\quad + 2J\left(a \sum Pp + b \sum Pq + c \sum P - \sum P\rho\right) \\ &\quad + 2a\left(b \sum pq + c \sum p - \sum \rho p\right) + 2b\left(c \sum q - \sum \rho q\right) \\ &\quad - 2c \sum \rho + \left(\frac{m}{n}\right)\sigma_B^2\end{aligned} \tag{8}$$

which can be evaluated given J from Eq. (4). The same expression can be used for partial reflections, although it will give a pessimistic estimate of σ_S since the profile should fit poorly.

Similarly, the error in fitting a plane to the n optical densities in the background region is given by

$$\begin{aligned}\sigma_B^2 &= \sum_n (\rho - \bar{\rho})^2 \\ &= \sum_n \rho^2 - \tfrac{1}{n}\left(\sum \rho\right)^2\end{aligned} \tag{9}$$

and the error between observed and calculated film coordinates will be

$$\begin{aligned}\Delta R &= \left[\left(\sum_m \rho p - a \sum_m p^2 - b \sum_m pq - c \sum_m p\right)\Big/ I_2\right] - R_c \\ \Delta S &= \left[\left(\sum_m \rho q - a \sum_m pq - b \sum_m q^2 - c \sum_m q\right)\Big/ I_2\right] - S_c\end{aligned} \tag{10}$$

With these various error estimates, reflections may be rejected (e.g., Table IV) by applying the following sequence of tests.

TABLE IV
ERROR ANALYSIS DURING FIRST PASS (CONSTANT PROFILE) AND SECOND PASS (VARIABLE PROFILE)[a]

	First pass	Second pass
Total number of reflections measured and accepted		
1. Whole reflections	4637	5085
2. Partial reflections	2673	2713
Number of whole reflections rejected because of		
1. Overloading (M = 240 OD)	683	683
2. Background too steep (C_1 = 5.0 OD/raster step)	32	32
3. Background too large (C_2 = 180 OD)	6	6
4. $\lvert(I_1 - I_2)\rvert > C_3$ ($C_3 = 150.0 + 0.1I_1$)	406	194
5. $I_1 < 0$	4272	4011
6. Too much background variation (C_5 = 2 OD)	18	46
7. Profile fitting too poor ($C_3 = 150.0 + 0.1I_2$)	68	34
8. Too large a positional error (C_4 = 4 raster steps)	976	1007
Number of partial reflections rejected because of		
1. $I_2 < 0$	2687	2666
2. Any other of the above reasons	1152	1133

[a] At 2.8 Å, 0.6° oscillation range, SBMV film. OD refers to the digitized value on a scale of 0–256 for a range up to 2 optical densities.

1. Is the reflection overloaded? That is, does any optical density within the reflection box exceed M?

2. Is the background too steep? That is, do the constants a and b exceed C_1 optical densities per raster step?

3. Is the background too large? That is, does the constant c exceed C_2 optical densities?

4. Is there too big a difference between the integrated and fitted intensity? That is, does the modulus of the difference exceed $C_3 + C_3'I_1$?

5. Is $I_1 < 0$?

6. Is there too much variation of optical density in the background region? That is, does $\sigma_B(m/n)^{1/2}$ exceed $C_5\sigma_p$ optical densities?

7. Is the profile fitting poor? That is, is $\sigma > C_3 + C_3'I_1$?

8. Is there too large a positional error? That is, do ΔR and ΔS exceed C_1?

An analysis of the mean positional rms errors of ΔR and ΔS with resolution can be used to assess the error in the determination of [Q], although the profile analysis is perhaps equally powerful. The rms error was found to increase from about 0.2 to 0.5 raster steps between the inside and the outside of an average SBMV film at 2.8 Å resolution.

Deconvoluting Overlapped Reflections

The objective of the oscillation method is to cover each film with the largest possible number of reflections to reduce the total number of films to be exposed. The greater the oscillation range, the larger will be the number of possible reflections. Sometimes this ideal must be relinquished because of increased exposure times and loss of high-resolution data. Nevertheless, there will be an increasing amount of overlap at high resolution, a situation which becomes particularly chronic for very large unit cells (Fig. 7).[34a] A method for deconvoluting overlapped reflections has, therefore, been developed.[35]

The reflections on a given film are always sorted into a sequence corresponding to the order of optical densities as they were read from the drum scanner. This is achieved by packing R_c, S_c, and the indices into a single word, with R_c as the most significant part of the word. Since the lines of optical densities are read in ascending order, the list of sorted words will follow the same order. Hence, only about 11 lines (corresponding to the width of one reflection box) need be kept in the fast-access computer storage at a given time. This list of sorted reflections is also useful to quickly determine near neighbors (those within, say, 11 raster steps) of any given reflection.

An initial estimate over all the reflections determines approximate intensities. At this stage, the area of the reflection box (Fig. 6) designating the peak area is best confined to a small central portion for profile fitting so as to avoid, as far as possible, interference from neighboring reflections. However, in the second and subsequent passes, the optical density of the N near neighbors at (R_i, S_i) can be subtracted as approximate intensity values will be available. The corrected density, ρ_c, at the point (p, q) is given by

$$\rho_c(p, q) = \rho(p, q) - \sum_N J_i P(p - R_i, q - S_i) - (ap + bq + c)$$

The corrected densities can then be used for background subtraction and profile fitting. The revised intensities provide a more accurate estimate of the integrated intensities than before. Furthermore, the area of the reflection box over which the profile is fitted can be enlarged in the second and subsequent cycle, as the effect of the neighboring reflections has been

[34a] J. E. Johnson and C. Hollingshead, *J. Ultrastruct. Res.* **74,** 223 (1981).

[35] J. E. Johnson, unpublished results.

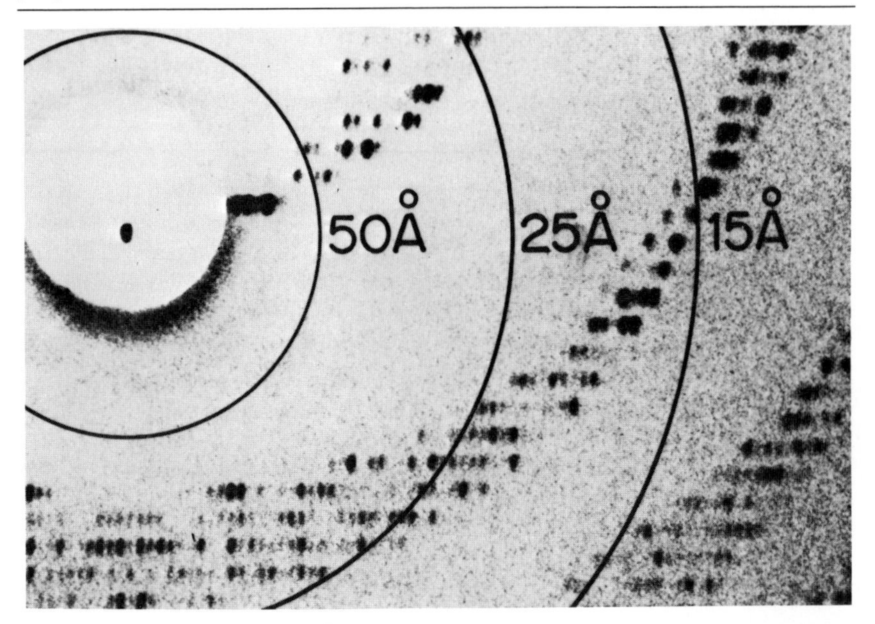

FIG. 7. Inner regions of a 2.1° oscillation picture of cowpea mosaic virus rotated about its hexagonal axis. Length of the c axis is 1032 Å. (Reprinted with permission from Johnson and Hollingshead.[34a] Copyright by Academic Press, Inc.)

mostly eliminated. Three cycles are usually sufficient to reach convergence.

Determination of the profile has to be done with some care in the initial stages to avoid the inclusion of hopelessly overlapped reflections. The profile is, therefore, best established by using "lonely" reflections which are either not overlapped or whose neighbors are weak. However, this problem is mostly eliminated in the second cycle as the neighboring overlapped reflections are at least approximately subtracted. The correction for neighboring reflections must, therefore, improve on each cycle. Three passes are normally ample to reach convergence.

The degree of permitted overlap must, however, be carefully controlled. Reflections which approach each other within some minimal distance should both be rejected as it is impossible to disentangle the distribution of one from the other. If, for instance, ρ is the optical density after background correction, then it will be necessary to determine J_0 and J_1 for the two overlapped reflections at (0, 0) and (R_1, S_1) by finding a minimum for E where

$$E = \sum_{p,q} [\rho - J_0 P(p, q) - J_1 P(p - R_1, q - S_1)]^2$$

The corresponding normal equations are

$$J_0 \sum P^2(p, q) + J_1 \sum P(p - R_1, q - S_1)P(p, q) = \sum \rho P(p, q)$$

$$J_0 \sum P(p, q)P(p - R_1, q - S_1) + J_1 \sum P^2(p - R_1, q - S_1) = \sum \rho P(p - R_1, q - S_1)$$

All summations are over (p, q) designated as the peak area in the reflection box. Thus, if

$$P(p, q) = 0 \qquad \text{whenever} \quad P(p - R_1, q - S_1) > 0$$

and

$$P(p - R_1, q - S_1) = 0 \qquad \text{whenever} \quad P(p, q) > 0$$

(i.e., the reflections are totally separated), then the off-diagonal terms are zero and the equations break down to the condition (5) in the section on Profile Fitting. However, if the reflections are totally overlapped, i.e., $R_1 = S_1 = 0$, then the determinant is zero and all that can be done is to solve for $(J_0 + J_1)$. The degree of overlap between two reflections can thus be measured with

$$U = \frac{[\Sigma\, P(p - R, q - S)\, P(p, q)]^2}{\Sigma\, P^2(p, q)\, \Sigma\, P^2(p - R, q - S)} \tag{11}$$

where $U = 0$ for no overlap and 1 for complete overlap. This function can be explored in terms of R and S and is independent of the actual intensities involved. Whenever R and S approach such values to make $U < 0.1$, for example, then the reflections are rejected. If $0.1 < U < 1$, then the reflections can be measured but at a loss of accuracy due to the ill conditioning of the above equations.

Computing Information

The oscillation processing techniques described here have been programmed entirely in Fortran and used on the CDC 6600 computer system at Purdue University. The entire program consists of five overlays which, together with the necessary core storage area, never exceeds 66K, 60-bit words. The program has also been adapted to a VAX,[26] a PDP 11 (Dr. Ivan Rayment, personal communication), and other computers. Typical computing times for an SBMV crystal oscillated by 0.6° about its trigonal axis, scanned with a 50-μm raster, containing almost 18,000 possible reflections was 0.21 CPU sec per reflection measured in terms of CDC 6500 equivalent time. In other cases, time per reflection could be halved depending on the options requested during processing.

Postrefinement

Background

The original proposal of Arndt and Wonacott had been to add partial reflections on adjoining films. This obviates the requirement for careful differentiation of full and partial reflections, provided there is a tendency to overestimate the number of partial reflections. Unfortunately, such abutting oscillation ranges do not exist when the crystal must be replaced after each exposure on account of radiation damage. This is particularly true for virus crystals which have large unit cells and, hence, weak intensities requiring long exposures. A conservative assignment of full reflections, rejecting all others, is possible provided the crystal orientation is known with sufficient accuracy. However, if the crystal has a large unit cell, then it will also require small oscillation ranges to avoid excessive overlapping and to decrease the exposure time. Thus, it might be necessary to reject more than half of all visible reflections and essentially all the reflections in the vicinity of the spindle axis.

The powerful new "postrefinement" technique was introduced by Schutt, Winkler and Harrison[36,37] to avoid these problems. Furthermore, this method is so powerful that it should replace the older addition method whether or not neighboring abutting films are taken with the same crystal. Postrefinement was first developed for the processing of data from tomato bushy stunt virus (TBSV),[38] and was then modified[39] for the analysis of SBMV data.[24] In the initial description, it depended on the matching of partially recorded, symmetry-related reflections on a given film. Later, the method was extended to comparing partial reflections on any one film with equivalent partial or whole reflections on any other film. Hence, it is also necessary to obtain an initial estimate of the relative scale factors between films. The procedure to be described here is that which has been used in the analysis of data at Purdue University.[39]

General Considerations

If F_{hi}^2 is the partial observation of a reflection h on film i and F_h^2 is the mean of all full measurements of reflection h, then the observed partiality

[36] C. Schutt and F. K. Winkler, *in* "The Rotation Method in Crystallography" (U. W. Arndt and A. J. Wonacott, eds.), p. 173. North-Holland Publ., Amsterdam, 1977.

[37] F. K. Winkler, C. E. Schutt, and S. C. Harrison, *Acta Crystallogr., Sect. A* **A35,** 901 (1979).

[38] S. C. Harrison, A. J. Olson, C. E. Schutt, F. K. Winkler, and G. Bricogne, *Nature (London)* **276,** 368 (1978).

[39] M. G. Rossmann, A. G. W. Leslie, S. S. Abdel-Meguid, and T. Tsukihara, *J. Appl. Crystallogr.* **12,** 570 (1979).

of the reflection is defined as

$$p_{\text{obs}} = \frac{F_{hi}^2}{G_i F_h^2}$$

where G_i is the inverse scale factor[40] of film i. If p_{calc} represents the calculated partiality of reflection h on film i, then it will be necessary to minimize the quantity

$$E = \sum_h \sum_i \omega_{hi}(p_{\text{obs}} - p_{\text{calc}})^2$$

where ω_{hi} is the weight to be applied to the given observation. The quantity p_{calc} will depend on the setting orientation ([A] matrix), the mosaic spread of each reflection, and the cell dimensions of the crystal. It is limited to the range $0 \leq p_{\text{calc}} \leq 1$. Once the parameters that define p_{calc} have been refined, then each partial reflection can be corrected by setting it to its equivalent whole value according to F_{hi}^2/p_{calc}. Since a similar correction must also be applied to the error on F_{hi}^2, such error becomes very large and eventually infinite as p_{calc} becomes smaller. Hence, it is usual to use only those reflections for which $p_{\text{calc}} > 0.5$ or some other specified value.

Postrefinement thus requires an initial determination of scale factors. From these, it is possible to make an estimate of the observed degree of partiality for each partial reflection and to refine these values against calculated estimates. With the revised crystal setting parameters, a better differentiation of full and partial reflections can be obtained and, hence, allows an improved estimate of scale factors. The process is iterative and found to be highly convergent. Usually, three cycles of scale factor determination followed by postrefinement are sufficient.

Preparation of the Data

The processed data from the ith film contains the Miller indices (hkl), the integrated intensity estimate (F_{hi}^2), and an error estimate, $\sigma(F_{hi}^2)$, after applying Lorentz and polarization corrections. It will be necessary to juxtapose all estimates of the same independent reflection. Thus, the given Miller indices must be transformed into $h'k'l'$ within one asymmetric unit of reciprocal space. It will then be necessary to retain for any one reflection $h'k'l'$, hkl, F_{hi}^2, σ_{hi}, an identifying number for the given film and (if anomalous dispersion Bijvoet differences are to be measured) one bit showing whether an inversion had been necessary to obtain $h'k'l'$ from hkl. The packed $h'k'l'$ indices within the asymmetric unit of reciprocal

[40] W. C. Hamilton, J. S. Rollett, and R. A. Sparks, *Acta Crystallogr.* **18,** 129 (1965).

space will then act as a "flag" and the remainder as a "tag" while sorting the reflections for each film into sequential order. The data from all the films can then be merged into a single sorted list with intensity estimates of equivalent reflections measured on any film lying adjacent to each other.

Relationship between Penetration, q, of Ewald Sphere and Partiality, p

Small differences in the orientation of domains within the crystal as well as the cross-fire of the X-ray beam will give rise to a series of possible Ewald spheres (Fig. 8). Their extreme positions will subtend an angle $2m$ at the origin of reciprocal space and their centers will lie on a circle of radius $\delta = m/\lambda$. The angle m is known as the effective mosaic spread. As the reciprocal lattice is rotated about the Oy axis, perpendicular to the mean direction of the X-ray beam Oz, a point P will gradually penetrate the effective thickness of the reflection sphere. Initially, only a few domain blocks will satisfy Bragg's law, but on further rotation there will be a

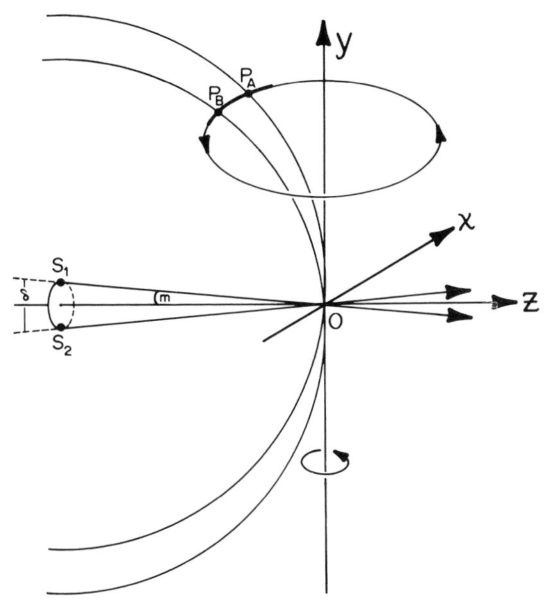

FIG. 8. Penetration of a reciprocal lattice point, P, into the sphere of reflection by rotation about Oy. The extremes of reflecting conditions at P_A and P_B are equivalent to X rays passing along S_1O and S_2O with Ewald sphere centers at S_1 and S_2 subtending an angle of $2m$ at O. Thus, in three dimensions, the extreme reflection spheres will lie with their circles on a circle of radius $\delta = m/\lambda$ at $z = -1/\lambda$. (Reprinted with permission from Rossmann *et al.*[39] © 1979 International Union of Crystallography.)

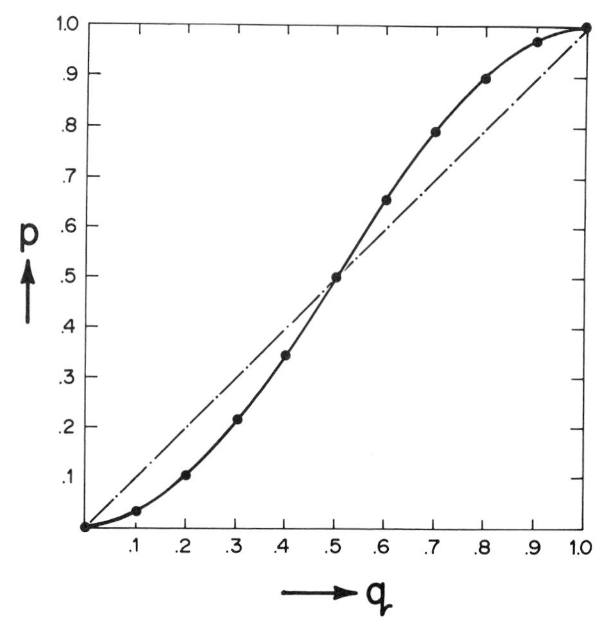

FIG. 9. Relationship between fraction of path, q, traversed by reciprocal lattice point across finite thickness of Ewald sphere and the fraction of the total scattered intensity, p. Curve is for $p = 3q^2 - 2q^3$. As an extreme case, the line $p = q$ is also shown, which would result if the reciprocal lattice point were a rectangular block. (Reprinted with permission from Rossmann *et al.*[39] © 1979 International Union of Crystallography.)

large number of blocks which are in a reflecting position. The maximum will be reached when the point P has penetrated halfway through the sphere's effective thickness, after which there will again be a decline of the fraction of the crystal volume able to diffract.

Let q be a measure of the fraction of the path traveled by P between the extreme reflecting positions P_A and P_B. A reasonable relationship between the fraction of path traveled, q, and the fraction of the energy already diffracted, p, is given by the graph in Fig. 9. The necessary conditions are that the lines $p = 0$ at $q = 0$ and $p = 1$ at $q = 1$ are tangential to the curve. Furthermore, assuming symmetry of the mosaicity in the crystal and of the flux in the X-ray beam, the curve must contain a center of symmetry at $p = 1/2$, $q = 1/2$.

A reasonable approximation to these conditions can be obtained by relating the fraction of the volume of a sphere removed by a plane a distance q from its surface (Fig. 10). It is easily shown that if p is the volume then

$$p = 3q^2 - 2q^3 \tag{12}$$

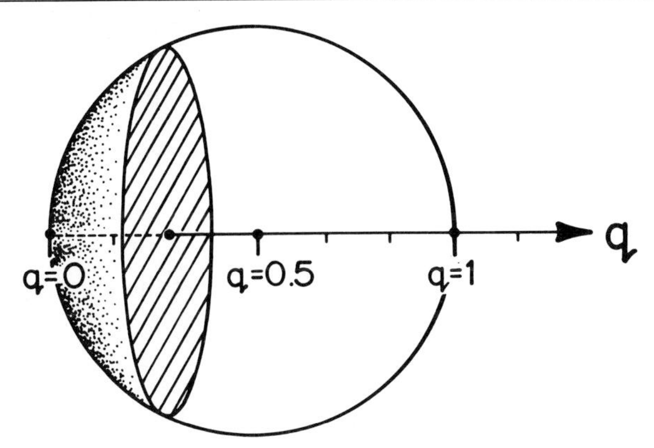

FIG. 10. The shaded volume of the sphere removed by the plane at p expressed as a fraction of the volume of the total sphere is given by $p = 3q^2 - 2q^3$. This expression is analogous to the fraction of reflected intensity as a spherical volume penetrates the Ewald sphere. (Reprinted with permission from Rossmann *et al.*[39] © 1979 International Union of Crystallography.)

This curve is shown in Fig. 9 and would correspond to assuming that the reciprocal lattice point is a sphere of finite volume cutting an infinitely thin Ewald sphere. Shown also in Fig. 9 is the line $p = q$ which would result if the reciprocal lattice point were a rectangular block whose surfaces were parallel and perpendicular to the Ewald sphere at the point of penetration. Thus, the line $p = q$ represents one extreme possibility for relating p and q. Winkler *et al.*[37] assumed a curve on the form

$$p = \frac{\pi}{2} \int_0^q \sin \pi q \, dq$$

or

$$p = \tfrac{1}{2}(1 - \cos \pi q)$$

which, in fact, differs from the curve $p = 3q^2 - 2q^3$ by less than 0.02 in p between $0 \leq q \leq 1$. It should be noted that the difference between the extreme curve $p = q$ and the more reasonable curve given by Eq. (12) differs by never more than 0.1 in p. Thus, the worst possible error which might accrue due to a poor selection of the p versus q relationship would be 10% in the estimation of partiality, but in practice should be far less.

Equation (12) has simple algebraic characteristics and will be used here. It should be noted that this is a cubic relationship with a minimum at $q = 0$, a maximum at $q = 1$, and a point of inflection at $q = 1/2$.

Degree of Penetration of the Ewald Sphere as a Function of Crystal Setting

Since q can be determined from the crystal setting parameters, p_{calc} can be derived from

$$p_{\text{calc}} = 3q^2 - 2q^3 \tag{13}$$

Assuming a right-handed coordinate system (x, y, z) in reciprocal space, as defined in Fig. 3, then it is easily shown[29] that the condition for reflection is

$$d^{*2} + 2z/\lambda = 0$$

where d^* is the distance of a reciprocal lattice point, $P(x, y, z)$, from the origin, O, of reciprocal space. Similarly, it can be shown[39] that the ends A and B of the path of the reciprocal lattice point through the finite thickness of the sphere occur when

$$d^{*2} + \delta^2 + 2z_A/\lambda - 2\delta\sqrt{x_A^2 + y_A^2} = 0$$
$$d^{*2} + \delta^2 + 2z_B/\lambda + 2\delta\sqrt{x_B^2 + y_B^2} = 0$$

where $\delta = m/\lambda$. Therefore,

$$z_A = (\lambda/2)(-d^{*2} - \delta^2 + 2\delta\sqrt{x_A^2 + y_A^2})$$
$$z_B = (\lambda/2)(-d^{*2} - \delta^2 + 2\delta\sqrt{x_B^2 + y_B^2}) \tag{14}$$

Since δ is small, it can be assumed that $2\delta\sqrt{x^2 + y^2}$ is approximately independent of the position of the reciprocal lattice point P between A and B (Fig. 11a and b).

Hence, the length of the path through the finite thickness of the sphere is proportional to

$$z_A - z_B = 2\lambda\delta\sqrt{x_P^2 + y_P^2}$$

Now, if a reflection (P) is only just penetrating the sphere, before coming to the end of the oscillation range, then (Fig. 11a) the fraction of penetration is given by

$$q = \frac{P_1A}{AB} = \frac{z_{P_1} - z_A}{z_B - z_A}$$

Substituting with Eq. (14) and simplifying, it is easy to show that

$$q = \tfrac{1}{2}(1 + D_1/\eta_1) \tag{15}$$

where

$$D = d^{*2} + \delta^2 + 2z/\lambda \qquad \text{and} \qquad \eta = 2\delta\sqrt{x^2 + y^2}$$

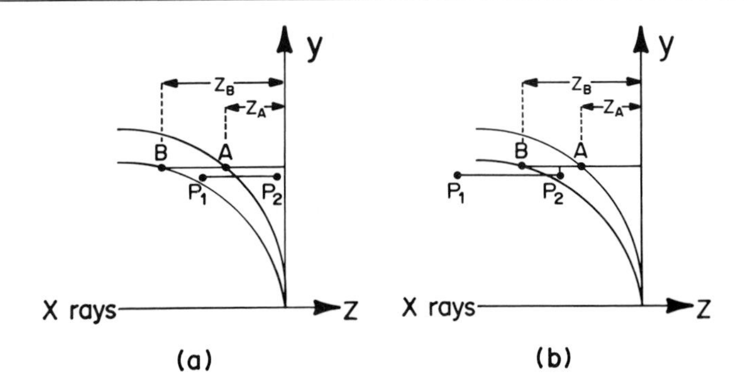

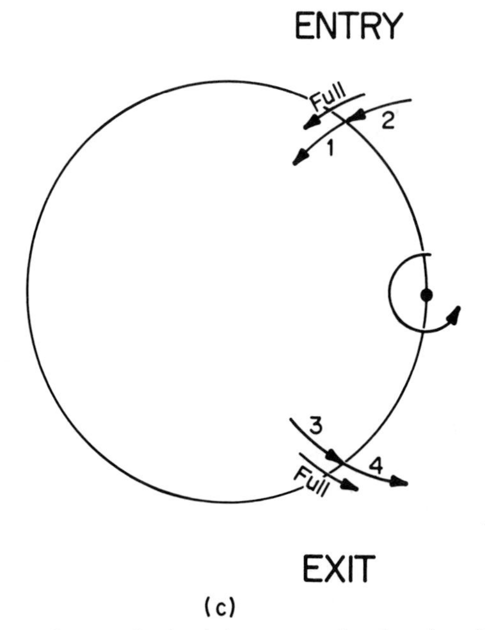

FIG. 11. If, on entering, a reflection is just penetrating the sphere (a), its motion will be from P_2 to P_1. P will not emerge into the inside of the sphere. If a reflection is almost completely within (b), its motion is from P_2 to P_1 with P never being completely outside the sphere. The four conditions 1, 2, 3, and 4 for partial reflections are shown in (c), and correspond to those in Table I. The arrow ends and heads correspond to the start and end positions of a reciprocal lattice point. (Reprinted with permission from Rossmann *et al.*[39] © 1979 International Union of Crystallography.)

The subscript 1 or 2 designates the beginning or end of the oscillation range for the partial reflection, P.

Similarly, if a reflection is almost complete within the sphere (Fig. 11)

$$q = BP_2/BA = (z_B - z_{P_2})/(z_B - z_A) = \tfrac{1}{2}(1 - D_2/\eta_2) \tag{16}$$

There are, indeed, four such conditions, two while a reflection is entering the sphere and two while exiting. It is then readily seen that $-1 < D_i/\eta_i < +1$ ($i = 1$ or 2) is the range for a partial reflection. All conditions are given in Table I as are also the limits for a full reflection.

The relationship between Miller indices and the reciprocal lattice coordinates when $\phi = \Phi$ is given by

$$\mathbf{x}' = [\Phi][A]\mathbf{h} \tag{17}$$

where $[A]$ gives crystal setting at $\phi = 0°$ (see section on Selection of Possible Reflections). Hence, by substituting ϕ_1 and ϕ_2, the coordinates at the end points of the oscillation range ($\mathbf{x}_1$ and $\mathbf{x}_2$) can be calculated and used in the evaluation of D_1/η_1 or D_2/η_2. Thus, given the Miller indices of any reflections on a film, it is easy to determine whether a reflection is full, partial, or absent on that film. Furthermore, if it is partial, then the degree of penetration, q, can be calculated using Eqs. (15), (16), and (17). The calculations, however, depend on the knowledge of the mosaicity, δ, and the setting matrix, $[A]$. The latter in turn depends on the crystal orientation relative to the camera and cell parameters.

Refinement of the Crystal Setting Parameters

A useful postrefinement criterion is to minimize the function

$$E = \sum \omega(p_{\text{obs}} = p_{\text{calc}})^2 \tag{18}$$

where the sum is taken over all partial reflections for which one or more whole reflections are also observed. An estimate of p_{calc} is given by Eq. (13). It will then be necessary to examine how p_{calc} varies with any parameter, ξ, for a given crystal. Variables are the crystal orientation angles $\delta\phi_x$, $\delta\phi_y$, $\delta\phi_z$, the independent cell constants, and the mosaic spread.

From Eq. (13),

$$\partial p_{\text{calc}}/\partial\xi = 6q(1 - q)\partial q/\partial\xi$$

But from Eqs. (15) and (16) and Table I,

$$q = \tfrac{1}{2}(1 + sD/\eta) \tag{19}$$

where $s = \pm 1$ according to the conditions of reflection. Therefore,

$$\frac{\partial q}{\partial \xi} = \frac{2q - 1}{2}\left(\frac{1}{D}\frac{\partial D}{\partial \xi} - \frac{1}{\eta}\frac{\partial \eta}{\partial \xi}\right)$$

or

$$\frac{\partial p_{\text{calc}}}{\partial \xi} = 3q(1 - q)(2q - 1)\left(\frac{1}{D}\frac{\partial D}{\partial \xi} - \frac{1}{\eta}\frac{\partial \eta}{\partial \xi}\right) \tag{20}$$

Thus, $\partial p_{\text{calc}}/\partial \xi = 0$ when $q = 0$, $q = 1/2$, or $q = 1$. This is to be expected since $p_{\text{calc}} = 0, 1/2$, or 1 at these values of q, independent of the values of any parameter.

The values of D and η and their derivatives may be evaluated from Eq. (15). From these expressions, we observe that

$$\frac{\partial D}{\partial \xi} = 2\left(x_i \frac{\partial_{x_i}}{\partial \xi} + y_i \frac{\partial_{y_i}}{\partial \xi} + z_i \frac{\partial_{z_i}}{\partial \xi} + \frac{1}{\lambda}\frac{\partial_{z_i}}{\partial \xi}\right) \qquad (i = 1 \text{ or } 2)$$

$$\partial D/\partial \delta = 2\delta$$

furthermore,

$$\frac{1}{\eta}\frac{\partial \eta}{\partial \xi} = \left(x_i \frac{\partial_{x_i}}{\partial \xi} + y_i \frac{\partial_{y_i}}{\partial \xi}\right) \Big/ (x_i^2 + y_i^2) \qquad (i = 1 \text{ or } 2)$$

$$\frac{1}{\eta}\frac{\partial \eta}{\partial \delta} = \frac{1}{\delta}$$

Values for x_i, y_i, and z_i are obtained from Eq. (17). However, since it is necessary to consider the effect of rotation of the crystal by small amounts of $\delta\phi_x$, $\delta\phi_y$, $\delta\phi_z$ (to allow for slight missetting of the crystal), the $[A]$ matrix must be modified by such rotations about the x, y, and z axes, respectively. Hence,

$$\mathbf{x}_i = [\Phi_{\delta\phi_x}][\Phi_{\delta\phi_y}][\Phi_{\delta\phi_z}][\Phi][A]\mathbf{h} \tag{21}$$

where

$$[\Phi_{\delta\phi_x}] = \begin{pmatrix} 1 & 0 & 0 \\ 0 & \cos \delta\phi_x & -\sin \delta\phi_x \\ 0 & \sin \delta\phi_x & \cos \delta\phi_x \end{pmatrix}$$

$$[\Phi_{\delta\phi_y}] = \begin{pmatrix} \cos \delta\phi_y & 0 & -\sin \delta\phi_y \\ 0 & 1 & 0 \\ \sin \delta\phi_y & 0 & \cos \delta\phi_y \end{pmatrix} \tag{22}$$

$$[\Phi_{\delta\phi_z}] = \begin{pmatrix} \cos \delta\phi_z & \sin \delta\phi_z & 0 \\ -\sin \delta\phi_z & \cos \delta\phi_z & 0 \\ 0 & 0 & 1 \end{pmatrix}$$

The derivatives $\partial x_i/\partial \xi$, $\partial y_i/\partial \xi$, $\partial z_i/\partial \xi$ can then be derived from Eqs. (21) and (22) where $\xi = \delta\phi_x$, $\delta\phi_y$, or $\delta\phi_z$.

The expression for $[A]$ depends on the manner in which the crystal is set with respect to the camera axes. For example, if the crystal is rotated about the c^* axis and if $\phi = 0$ when the X-ray beam is along the crystal b axis, then it can be shown[41] that

$$[A] = \begin{pmatrix} a^* \sin \beta^* & -b^* \sin \alpha^* \cos \gamma & 0 \\ -a^* \cos \beta^* & -b^* \cos \alpha^* & -c^* \\ 0 & b^* \sin \alpha^* \sin \gamma & 0 \end{pmatrix} \tag{23}$$

where

$$\cos \gamma = (\cos \alpha^* \cos \beta^* - \cos \gamma^*)/(\sin \alpha^* \sin \beta^*)$$

The elements of Eq. (23) can readily be rewritten to allow for the lattice properties of different crystals or different crystal mounts. For instance, for a monoclinic crystal mounted to rotate about b^*,

$$[A] = \begin{pmatrix} a^* \sin \beta & 0 & 0 \\ 0 & b^* & 0 \\ -a^* \cos \beta^* & 0 & -c^* \end{pmatrix}$$

Hence, using Eqs. (21) and (23), the derivatives $\partial x_i/\partial \xi$ etc. can be expressed in terms of the cell dimensions.

Some caution must be exercised concerning the choice of refinable parameters. In general, $\delta\phi_z$ must be kept constant since small rotations of the crystal about the X-ray beam do not significantly affect the partiality of reflections. Thus, $\delta\phi_z$ is ill conditioned during refinement. Furthermore, anisotropic alterations in cell dimensions can be in part compensated by synchronous rotations in $\delta\phi_x$ and $\delta\phi_y$. A further reasonable assumption, which improved the conditioning of the least-squares normal equations, is that the cell dimensions of each crystal are the same. Experience has also shown that mosaic spread refinement is best done only after the other parameters have converged.

Assessment of Errors

The beneficial effect of postrefinement can be assessed in a variety of ways. A plot (Fig. 12) can be made between mean p_{obs} and p_{calc} in ranges of q before and after postrefinement. The R factor (Fig. 13), representing agreement between corrected partial and the mean of corresponding reflections, can be plotted as a function of q. The postrefined parameters can be used to calculate a new representation of the film. As the occurrence of reflections close to the spindle axis is very sensitive to crystal

[41] M. G. Rossmann and D. M. Blow, *Acta Crystallogr.* **15,** 24 (1962).

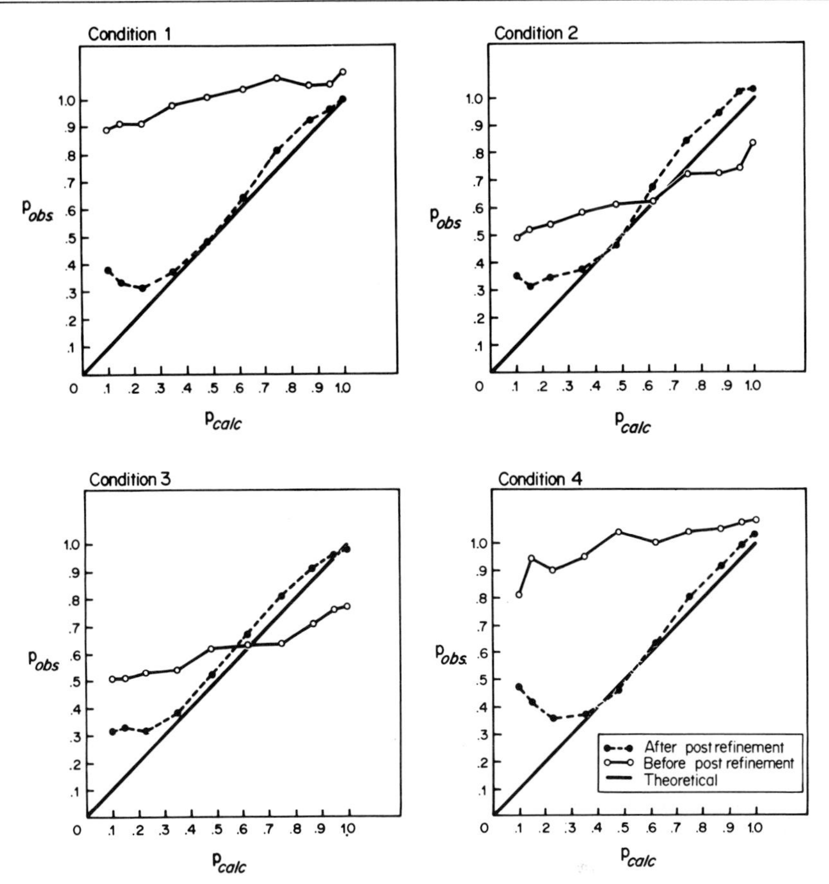

FIG. 12. Mean values of $\langle p_{obs} \rangle$ averaged in ranges of q (increasing in steps of 0.1) plotted against the corresponding value of p_{calc}. Results taken from the processing of 38 films of the PHMBS-SBMV 3.5 Å resolution (1° oscillation range) data set. Conditions refer to those given in Table I. (Reprinted with permission from Rossmann *et al.*[39] © 1979 International Union of Crystallography.)

setting and cell dimensions, there is nearly always a remarkable improvement in the visual agreement of a film with its postrefined representation. Systematic overestimation of p_{obs} (Fig. 12) when $q < 0.4$ suggests that either the small partial reflections are measured too strong or all small reflections are measured too strong. Similar and even more striking observations were made by Irwin *et al.*[42] and by Wilson and Yeates.[20] Results of scaling the reflections on the strong (A) films with those on the weak

[42] M. J. Irwin, J. Nyborg, B. R. Reid, and D. M. Blow, *J. Mol. Biol.* **105,** 577 (1976).

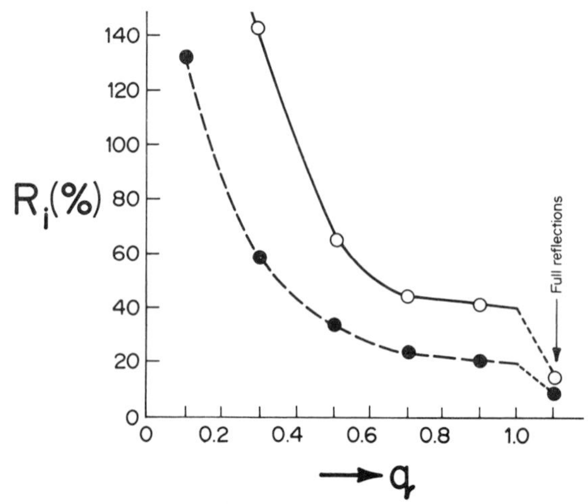

FIG. 13. R_i factors plotted in ranges of q before (○) and after (●) postrefinement. The weighted mean F_h^2 values are computed over only the full reflections. Thus, the individual F_{hi}^2 values on given films are included in the calculation of F_h^2 for full reflections but excluded for partial reflections, which accounts for the larger R values for partial reflections. As q decreases, the intensity of the partial observation decreases and, hence, the error must increase. (Reprinted with permission from Rossmann *et al.*[39] © 1979 International Union of Crystallography.)

(B) films within a film pack showed that the strong reflections on film B were systematically greater than the corresponding reflections on film A. These errors may have their origin in (1) nonlinearity of film response, (2) the Wooster effect,[43] (3) error in determination of the background around reflections (cf. Matthews *et al.*[6]), or (4) selection of only the stronger estimates of weak reflections. A parabolic correction, as suggested by Matthews *et al.*,[6] aimed at improving agreement between partial and full estimates of the same reflection, did not eliminate these problems.

Actual error determinations must be based on a weighted mean between estimates of "counting statistics", σ_S, derived from profile and background fitting [see Eq. (7)], and on the internal agreement, σ_A, between different films (see Table V).

Rejection Criteria

When scale factors have been determined and postrefinement parameters have converged, it will be necessary to calculate a final averaged F_h^2 value for each reflection. In some cases, there will be significant disagreement between observations and, hence, some system of rejection will have to be applied. The following algorithm has been found useful.

[43] W. A. Wooster, *Acta Crystallogr.* **17,** 878 (1964).

TABLE V
ESTIMATES OF ERRORS ON SCALED AND AVERAGED INTENSITIES[a]

	0 – $0.25\langle F_h^2\rangle$	$0.25\langle F_h^2\rangle$ – $0.50\langle F_h^2\rangle$	$0.50\langle F_h^2\rangle$ – $1.0\langle F_h^2\rangle$	$1.0\langle F_h^2\rangle$ – $2.0\langle F_h^2\rangle$	$2.0\langle F_h^2\rangle$ – $4.0\langle F_h^2\rangle$	$4.0\langle F_h^2\rangle$ – ∞
F^2 range	16	47	94	188	375	750
σ_S	17	18	18	18	22	31
σ_A	11	17	21	24	32	50
Number of reflections in range	42,768	51,091	48,216	39,393	24,329	5,994

[a] Comparison of estimates of error based on counting statistics (σ_S) and internal agreement (σ_A) as a function of mean F_h^2. Results are taken from scaling 81 films of native 2.8 Å resolution SBMV data.

1. All reflections for which $F_{hi}^2 < C_1\sigma_{hi}$ are rejected. A useful value for C_1 is 2.

2. All reflections for which $q < q_{\min}$ are rejected. A useful compromise between losing too many reflections and unacceptably increasing the error in F_h^2 was found to be $q_{\min} = 0.5$ for photographs overlapped by 0.1°.

3. If a reflection has been measured more than twice, the differences $D_i = |F_h^2 - F_{hi}^2/G_i|$ are calculated. If the largest difference exceeds C_R, then the corresponding reflection is rejected. The value of F_h^2 may then be recalculated and the procedure is repeated until all remaining values of D_i are less than C_R or only two observations of this reflection are retained. Here, $C_R = C_2\langle F_h^2\rangle + C_3F_h^2$ where $\langle F_h^2\rangle$ is the mean of all F_h^2 values. Useful values for C_2 and C_3 are 0.3 and 0.1, respectively.

4. If there are only two observations or if only two observations remain after applying iteratively procedure (3), then the mean of these two observations is taken as an estimate of this reflection unless $D_1 = D_2 > C_wC_R$ in which case the reflection is entirely rejected. A useful value of C_w is 3.

5. The largest estimate of any overloaded reflection is accepted when no other estimate is available. Such reflections (hopefully few) can often be phased with molecular replacement averaging and will represent important information. However, they must be tagged as being of no value to isomorphous replacement phasing.

A similar set of rejection criteria can be applied while scaling together films within a package. Here, the strong film is usually designated as A, the next film as B, and so forth. Care must be taken that reflections used for scaling are both of reasonable accuracy. Thus, one important criterion is that $I_A \gg I_B \gg I_C$ for a given reflection. Very weak reflections should only be measured on the A film as the purpose of the B and C films is primarily to supplement data which might be overloaded on the stronger films. Considerable error can be introduced due to the scaling up of inaccurate reflections on the C film to the level of the A film.

A further difficulty arises due to the variation of scale factor with changing angle of incidence, causing differing absorption as the path length increases.[29,44] The physical form of this correction is similar to that caused by variable "falloff" with resolution.

Scaling and Absorption

The postrefinement procedure depends intimately on the determination of scale factors between films. Procedures for the determination of scale factors from intersecting films were first discussed by Kraut[45] and Dickerson.[46] Problems remained, however, in determining the optimal solution for the resultant linear equations.[47] Although procedures proposed by Hamilton *et al.*[40] and by Fox and Holmes[48] are now in general use, nevertheless some care must still be used in the solution of the normal equations.[39]

Different degrees of "falloff" with increasing resolution can be corrected by using a scale factor of the form $k \exp\{-B[(\sin\theta)/\lambda]^2\}$ for each film. Here, k is the linear scale factor while B is the apparent "temperature" factor of one film relative to a standard. Crystals often exhibit anisotropic "falloff," particularly when slightly damaged by a heavy-atom substitution. Rossmann *et al.*[39] used an anisotropic variation over the surface of each film to improve the scaling.

Oscillation photography requires only a very small amount of crystal motion, and the crystal projection which would be observed from each diffracted ray at the time of reflection will be very similar, particularly for low-resolution data. Hence, the amount of absorption variation is very small for reflections recorded on one oscillation film. This is in contrast to diffractometer measurements, in which the crystal projection presented by the reflected beam will vary greatly as the crystal is rotated into a reflecting position. The small variation in absorption coefficients can be largely eliminated by using suitable scaling procedures. An anisotropic scale factor is useful but erroneously assumes a centric distribution for the absorption surface.

General procedures for removing the effect of absorption have been proposed by Katayama *et al.*[49] and by Stuart and Walker.[50] The smoothly

[44] M. G. Rossmann, *Acta Crystallogr.* **9,** 819 (1956).
[45] J. Kraut, *Acta Crystallogr.* **11,** 895 (1958).
[46] R. E. Dickerson, *Acta Crystallogr.* **12,** 610 (1959).
[47] J. S. Rollett and R. A. Sparks, *Acta Crystallogr.* **13,** 273 (1960).
[48] G. C. Fox and K. C. Holmes, *Acta Crystallogr.* **20,** 886 (1966).
[49] C. Katayama, N. Sakabe, and K. Sakabe, *Acta Crystallogr., Sect. A* **A28,** 293 (1972).
[50] D. Stuart and N. Walker, *Acta Crystallogr., Sect. A* **A35,** 925 (1979).

varying nature of absorption effects permits the representation of the scale factors by a Fourier series. Thus, if F_i is the observed structure amplitude on film i F_i' is its modified value, then

$$F_i' = F_i \left\{k_i + \sum_n \sum_m [P_{nm} \sin(n\phi + m\mu) + Q_{nm} \cos(n\phi + m\mu)]\right\}$$

where ϕ and μ are two suitably chosen angles which define the position of the diffracted beam on a film and P_{nm}, Q_{nm} are the Fourier coefficients for the integers n and m. Values for k_i, P_{nm}, and Q_{nm} can then be determined by the usual iterative procedures for finding scale factors. Such a procedure has been used very effectively by Stuart and Walker[50] in correcting data from oscillation photographs.

The alternative approach to absorption corrections for crystal mounts in capillaries has been the empirical procedure proposed by North *et al.*[51] for diffractometer data measurements. It depends on measuring the variation of transmission for a very low-order reflection as the crystal is rotated about the normal to the reflecting planes. The method was adapted for oscillation photography by Schwager *et al.*,[52] who simply measure the transmitted ray for different inclinations of the crystal with respect to the incident beam. For this reason, most commercially available rotation cameras permit the inclination of the spindle axis away from normal X-ray incidence.

Computer Information

The postrefinement and simultaneous scaling technique described here has been programmed entirely in Fortran and used on the CDC 6600 computer system at Purdue University.

Strategy for Collecting Data

The orientation of the crystal on the camera and the oscillation angle need to be carefully considered before embarking on a full three-dimensional data collection. In general, the axis of highest rotation symmetry should be placed along the spindle axis of the camera. This permits the exploration of one asymmetric unit of reciprocal space in the least number of exposures. However, if anomalous dispersion data are required, care must be taken to collect Friedel opposites on the same film. Hence, if there is no mirror plane in the Laue group perpendicular to the symmetry axis mounted along the camera's spindle axis, then an alternative crystal

[51] A. C. T. North, D. C. Phillips, and F. S. Mathews, *Acta Crystallogr., Sect. A* **A24,** 351 (1968).

[52] P. Schwager, K. Bartels, and R. Huber, *Acta Crystallogr., Sect. A* **A29,** 291 (1973).

orientation must be used. This will result in twice as much labor and time. It may be necessary to use a "bridle"[23] to achieve an optimal mount if the crystal morphology does not permit alignment of the desired axis along the capillary tube.

Large oscillation angles produce unfavorable Lorentz factors and a good deal of reflection overlap. Nevertheless, this may be a good and fast method for collecting low-resolution data, which are also far more intense. A typical example when this type of strategy paid off was in the case of the uranyl derivative of SBMV.[53] It was possible to collect the 30 "A/B" film pairs in 12 days. (The "A" and "B" films refer to the top and bottom films of a two-film pack.) A further 7 days were required to process and scale the A films and these data were used to calculate a 5 Å resolution difference Fourier map, using phases based on mercury and platinum derivatives. The difference map was immediately interpretable, and demonstrated that the compound gave well-substituted sites at positions which were distinct from those of the mercury and platinum derivatives. This result justified collecting and processing high-resolution 0.5° oscillation films, which required a further 3 months. While the high-resolution data collection was in progress, the B films of the 4 Å data set were processed and the complete 4 Å data set was used to refine the occupancy and positional parameters of the major uranyl sites. This in turn led to the location and characterization of further minor sites.

Although this strategy of data collection requires a larger number of photographs than for the high-resolution data collection alone, it is well worth the additional effort in order to be confident that the derivative is a useful one. In addition, there is the advantage that the B films of the high-resolution data set do not have to be processed, since the intensity values for all the strong reflections (which would be overloaded on the A films) are already available for the low-resolution data set. In the case of SBMV, this meant that only 140 films in total required processing, whereas 160 films would have had to be processed if only the high-resolution films had been collected. Furthermore, the accuracy of the low-resolution data is improved since all but the strongest reflections are measured on both sets of films. The statistics of the high- and low-resolution data sets are presented in Table VI.

A large number of data sets have been collected at Purdue University using the procedures described in this chapter. The procedure has also been used in a modified form elsewhere.[26] The best results were obtained for native crystals of the MoFe protein of *Clostridium pasteurianum* nitrogenase.[54] The data were collected to 2.4 Å resolution. The R factor (see

[53] A. G. W. Leslie and T. Tsukihara, *J. Appl. Crystallogr.* **13,** 304 (1980).

[54] T. Yamane, M. S. Weininger, L. E. Mortenson, and M. G. Rossmann, *J. Biol. Chem.* **257,** 1221 (1982).

TABLE VI

THE STATISTICS OF THE HIGH- AND LOW-RESOLUTION DATA SETS

	Total number of films	Total number of reflections with $I > 1\sigma$	Number of independent reflections with $I > 1\sigma$	Number of overloaded reflections	R factor[a] (%)	Resolution limit (Å)	Oscillation angle (°)	Film overlap (°)
Low resolution data								
A films	26[b]	141,230	84,266	943	12.6	4.0	1.25	0.25
A/B pairs	30	167,913	90,098	85	13.0	4.0	1.25	0.25
High-resolution data								
A films only	77	530,612	281,143	8699	13.6	2.8	0.5	0.1
Combined data	—	358,606	280,857	85	14.4	2.8	—	—

[a] $R = 100\,[(\Sigma_h \Sigma_i |F_h^2 - F_{hi}^2/G_i|)/(\Sigma_h \Sigma_i F_h^2)]$, where F_h^2 is the best estimate for reflection h derived from the observation values F_{hi}^2 and G_i is the inverse scale factor for film i.

[b] Four films were rejected due to poor scaling results. These films were retaken at a later stage and are included in the 30 A/B pairs.

Table VI for definition) was 8.0% after rejecting all reflections less than 2σ but including all reflections with $p_{calc} > 0.6$. The R factor decreased to 7.7% when all partial reflections were omitted. A total of 92,505 full and partial measurements were reduced to 59,651 independent reflections.

A full account of the strategy used in the SBMV data collection has been given by Abad-Zapatero *et al.*,[24] for beef liver catalase by Murthy *et al.*,[23] and for alfalfa mosaic virus by Abdel-Meguid *et al.*[55,56a]

Acknowledgments

The methods described in this chapter were made possible by the vision of Dr. Uli Arndt in reintroducing the oscillation camera to crystallographers and to Dr. Alan Wonacott in the beautiful execution of this task. I am also most grateful to Dr. Uli Arndt for sending me copies of chapters from the book *The Rotation Method in Crystallography*[11] at an early stage and later for sending me a copy of the printed book. I also appreciated receiving two excellent manuscripts from Drs. John Helliwell and Trevor Greenhough prior to publication.

A program written by Dr. Geoffrey C. Ford at Purdue University gave a lot of experience in the processing of oscillation photographs. These provided the basis of the new procedure described in section on Film Processing. I am most grateful for the numerous helpful discussions and participation by Drs. Andrew G. W. Leslie and Ivan Rayment, and to encouragement and further developments by Drs. John E. Johnson, Sherin S. Abdel-Meguid, M. R. N. Murthy, and R. Michael Garavito.

The concept of postrefinement was first described by Schutt and Winkler.[36] I was most grateful to Dr. Clarence Schutt for sending me his Ph.D. thesis[56] in which the idea was further developed. Furthermore, I had the opportunity to discuss the method with Dr. Stephen Harrison on a number of occasions. I am also most grateful for the outstanding work and ideas of Dr. Andrew Leslie and for the excellent support given to me by Drs. Sherin S. Abdel-Meguid, Tomitake Tsukihara, and Rikkert K. Wierenga.

I wish to thank Sharon Wilder for the meticulous manner in which she prepared my manuscript for publication. The work was supported by the National Science Foundation (Grant No. PCM78-16584) and the National Institutes of Health (Grants No. GM 10704 and AI 11219).

[55] S. S. Abdel-Meguid, K. Fukuyama, and M. G. Rossmann, *Acta Crystallogr., Sect. B* **B38,** 2004 (1982).

[56] C. E. Schutt, Ph.D. Thesis, Harvard University, Cambridge, Massachusetts (1976).

[56a] A variety of advances have occurred in the period since this manuscript was submitted. These can be summarized as follows. (1) The extensive use of synchrotron radiation where there may be significant beam divergence and wavelength dispersion [reference 27b; T. J. Greenhough *et al., J. Appl. Crystallogr.* **16,** 242, 1983; M. G. Rossmann, *in* "Biological Systems: Structure and Analysis" (G. P. Diakun and C. D. Garner, eds., pp. 28–40. SERC, Daresbury, 1984; K. S. Wilson *et al., J. Appl. Crystallogr.* **16,** 28, 1983; R. Usha *et al., J. Appl. Crystallogr.* **17,** 147, 1984]. (2) The use of mis-set crystals in synchroton beams to beat the effects of radiation damage (M. G. Rossmann and J. W. Erickson, *J. Appl. Crystallogr.* **16,** 629, 1983). (3) The treatment of overlapped reflections (M. G. Rossmann, unpublished). (4) The availability of super-computers for film processing permitting greater precision of the required calculations in a reasonable time (M. G. Rossmann, unpublished). (5) The use of Laue photography for rapid synchrotron data collection (K. Moffat *et al., Science* **223,** 1423, 1984).

[21] A Rotation Camera Used with a Synchrotron Radiation Source

By R. FOURME and R. KAHN

Introduction

Synchrotron radiation (SR) emitted by high-energy electron or positron storage rings is now a tool which is commonly used for X-ray macromolecular crystallography.[1] Clearly, the practical advantages of these new sources more than counterbalance the inconvenience of distant and shared facilities with a strict time schedule.

In this field, the property of SR sources which is currently of highest relevance is the *high spectral brilliance,* defined as the number of photons emitted at a wavelength λ in a given relative bandwidth $\delta\lambda/\lambda$, per second, per unit area of source, per unit solid angle. The spectral brilliance of an SR source is two to three orders of magnitude higher than that of any existing rotating anode tube, even at the wavelength of the K_α line, with the further advantage of *tunability* since the SR spectrum is a smooth continuum. This comparison applies to time-integrated spectral brilliances; in fact, the synchrotron emission is pulsed, with typically 0.1- to 1-nsec pulses spaced by 10–1000 nsec, so that the sample under study is submitted to short and very intense bursts of X rays. Finally, the *polarization* of SR must be considered in the design of diffraction apparatus and in the correction of observed intensity data.

SR is used at present for (at least) three types of experiments in macromolecular crystallography. A discussion of each of these follows.

Collection of High-Resolution Diffraction Data

This is the most usual reason to use SR.[1] Such experiments are routinely performed at the facilities of LURE-DCI (Orsay, France),[2,3] EMBL Outstation (Hamburg, West Germany),[4] SRS (Daresbury, United Kingdom),[5] CHESS (Cornell University, New York, United States), and

[1] H. D. Bartunik, R. Fourme, and J. C. Phillips, *in* "Uses of Synchrotron Radiation in Biology" (H. Stuhrmann, ed.), p. 145. Academic Press, New York, 1982.

[2] R. Fourme, *in* "Synchrotron Radiation Applied to Biophysical and Biochemical Research" (A. Castellani and I. F. Quercia, eds.), p. 349. Plenum, New York, 1979.

[3] R. Kahn, R. Fourme, A. Gadet, J. Janin, C. Dumas, and D. André, *J. Appl. Crystallogr.* **15,** 330 (1982).

[4] H. D. Bartunik, P. N. Clout, and B. Robrahn, *J. Appl. Crystallogr.* **14,** 134 (1981).

[5] J. R. Helliwell, T. J. Greenhough, P. D. Carr, S. A. Rule, P. R. Moore, A. W. Thomson, and J. S. Worgan, *J. Phys. E* **15,** 1363 (1982).

TABLE I
DIFFICULTIES ENCOUNTERED IN MACROMOLECULAR CRYSTALLOGRAPHY[a]

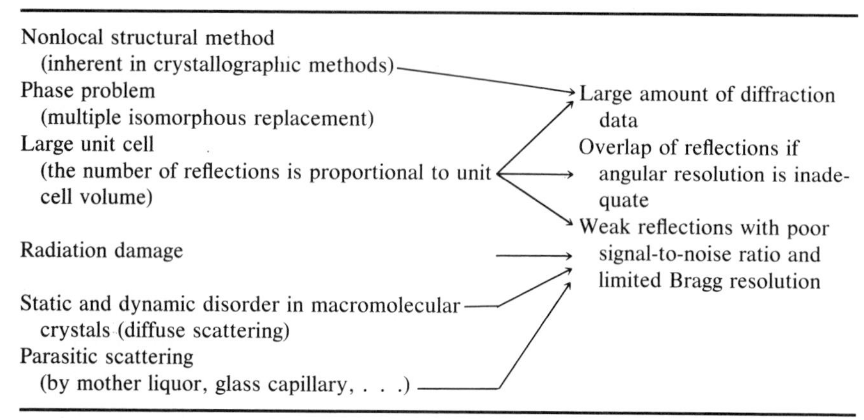

[a] After Fourme and Kahn.[5a]

SSRL (Stanford University, United States), and they are planned in other laboratories as well. In effect, various difficulties are encountered in collecting diffraction data from macromolecular crystals. The nature and the origin of these difficulties are summarized in Table I.[5a] The most obvious way to avoid the overlap of individual Bragg reflections and enhance the signal-to-noise ratio of each of these reflections is an intense, sharply collimated and monochromatic incoming X-ray beam, which requires a source of a high spectral brilliance such as a storage ring. A crucial point is the fact that radiation damage is less for short, intense exposures than long, weak (or interrupted) exposures corresponding to the same total dose of radiation. This effect has been observed with conventional sources[6] and for most crystals submitted to SR.[1,3,7,8] The reasons of this improvement are not quite clear: this is presumably because radiation damage involves slow, diffusion-controlled processes such as migration of chemically active radiation products.[9]

[5a] R. Fourme and R. Kahn, *Biochimie* **63,** 887 (1981).

[6] M. F. Schmid, L. H. Weaver, M. A. Holmes, M. G. Grütter, D. H. Ohlendorf, R. A. Reynolds, S. J. Remington, and B. W. Matthews, *Acta Crystallogr., Sect. A* **A37,** 701 (1981).

[7] J. C. Phillips, A. Wlodawer, M. M. Yevitz, and K. O. Hodgson, *Proc. Natl. Acad. Sci. U.S.A.* **73,** 128 (1976).

[8] K. S. Wilson, E. A. Stura, D. L. Wild, R. J. Todd, Y. S. Babu, J. A. Jenkins, D. A. Mercola, L. N. Johnson, R. Fourme, R. Kahn, A. Gadet, K. S. Bartels, and H. D. Bartunik, *J. Appl. Crystallogr.* **16,** 28 (1983).

[9] T. L. Blundell and L. N. Johnson, "Protein Crystallography." Academic Press, New York, 1976.

The tunability allows selection of the wavelength which is optimal for each experiment. Except with very small samples for which the optimal wavelength may be longer than the standard 1.54 Å, the use of shorter wavelengths is beneficial: absorption is reduced and the lifetime may be increased; the signal-to-noise ratio is improved, for a given resolution, by increasing the crystal-to-detector distance since the intensity of the monochromatic focused beam falls off only slowly with distance while the background obeys the inverse square law; data of a higher resolution may be collected with a detector of a fixed size (within the limits set by the overlap of reflections). At the EMBL Outstation and at LURE respectively, X rays of 1 Å and 1.2–1.4 Å are commonly used.

Optimized Anomalous Scattering Experiments

Both the phase and amplitude of the X-ray scattering factor of an element are wavelength dependent. This variation is largest in the region of an absorption edge. Since the wavelengths of absorption edges are different for each element, it is possible to alter selectively the contribution of one element to the diffraction pattern while that of the rest of the macromolecule remains essentially constant. The effect is analogous to that obtained in the multiple isomorphous replacement method (MIR). In principle, data collection at several wavelengths from one crystal will permit the location of the anomalous scatterers in the unit cell and, eventually the solution of the phase problem. It is no longer necessary to prepare several heavy-atom derivatives and isomorphism is perfect.

With an SR source, it is possible to get an intense X-ray beam at any selected wavelength; anomalous scattering experiments may then be performed under optimal conditions. The resonance peaks (the so-called "white lines") in absorption spectra are a matter of particular interest because the variation of anomalous dispersion terms in the white line region is very large. These new possibilities have contributed to stimulate both theory and experiments (for a review, see Bartunik *et al.*[1]).

Since the success of methods using anomalous scattering and their field of application to complex structures are largely dependent on the accuracy of intensity measurements at several, accurately determined wavelengths, improved methods of data collection are necessary. The standard photographic rotation method may be modified to get an energy distribution within each Bragg spot; this gives rise to the possibility of a polychromatic diffraction experiment allowing a sampling of an absorption edge for optimized anomalous dispersion studies.[10,11] As pointed out

[10] U. W. Arndt, T. J. Greenhough, J. R. Helliwell, J. A. K. Howard, S. A. Rule, and A. W. Thompson, *Nature (London)* **298,** 835 (1982).

[11] T. J. Greenhough, J. R. Helliwell, and S. A. Rule, *J. Appl. Crystallogr.* **16,** 242 (1983).

by Hoppe and Jakubowski,[12] this is also a strong incentive to the development of electronic area detectors using the highly monochromatic and tunable radiation from an SR source.[13–15]

Kinetics and Dynamics

Studies of intramolecular dynamics at various temperatures based on group temperature factor analysis[16,17] are made more easily and more systematically using SR sources, due to the dramatic reduction in exposure times.[18] Information about molecular dynamics might also be derived from the study of diffuse scattering; the intensity and the good collimation of monochromatized SR are ideally suited to such experiments. In the case of crystals of phosphorylase *b*, it has been reported that the intensity of thermal diffuse scattering is dependent on the time structure of the SR source.[8]

Stepwise investigation of enzyme kinetics, using low temperature to stabilize transient states,[19–21] may take advantage of fast data acquisition rates.

A first application of time-resolved protein data collection on a submillisecond time scale to a study of structural changes in carbonmonoxy myoglobin induced by debinding of the ligand through laser photolysis has been reported by H. D. Bartunik.[1] Using the time structure of SR, a higher time resolution, in the subnanosecond range, might be obtained.[1]

In our laboratory, a rotation camera devoted to macromolecular crystallography has been used by various groups since mid-1976. This instrument has introduced routine high-resolution data collection by the rotation method using a high-intensity SR source.

[12] W. Hoppe and U. Jakubowski, *in* "Anomalous Scattering," p. 437. Munksgaard, Copenhagen, 1975.

[13] R. Kahn, R. Fourme, B. Caudron, R. Bosshard, R. Bouclier, G. Charpak, J. C. Santiard, and F. Sauli, *Nucl. Instrum. Methods* **172,** 337 (1981).

[14] R. P. Phizackerley, C. W. Cork, R. C. Hamlin, C. P. Nielsen, W. Vernon, Ng. H. Xuong, and V. Perez-Mendez, *Nucl. Instrum. Methods* **172,** 393 (1980).

[15] T. D. Mokulskaya, S. V. Kuzev, M. Yu. Lubnin, G. E. Myshko, M. A. Mokulskii, E. P. Smetanina, S. E. Baru, G. N. Kulipanov, V. A. Sidorov, and A. G. Khapakhpashev, *Eur. Crystallogr. Meet. 6th, 1980* Abstract 2B, p. 14 (1980).

[16] P. J. Artymiuk, C. C. F. Blake, D. E. P. Grace, S. J. Oatley, D. C. Phillips, and M. J. E. Sternberg, *Nature (London)* **280,** 563 (1979).

[17] H. Frauenfelder, G. A. Petsko, and D. Tsernoglou, *Nature (London)* **280,** 558 (1979).

[18] J. Walter, W. Steigemann, T. P. Singh, H. Bartunik, W. Bode, and R. Huber, *Acta Crystallogr., Sect. B* **B38,** 1962 (1982).

[19] P. Douzou, G. Hui Bon Hoa, and G. A. Petsko, *J. Mol. Biol.* **96,** 367 (1974).

[20] A. I. Fink and A. I. Ahmed, *Nature (London)* **263,** 294 (1976).

[21] T. Alber, G. A. Petsko, and D. Tsernoglou, *Nature (London)* **263,** 297 (1976).

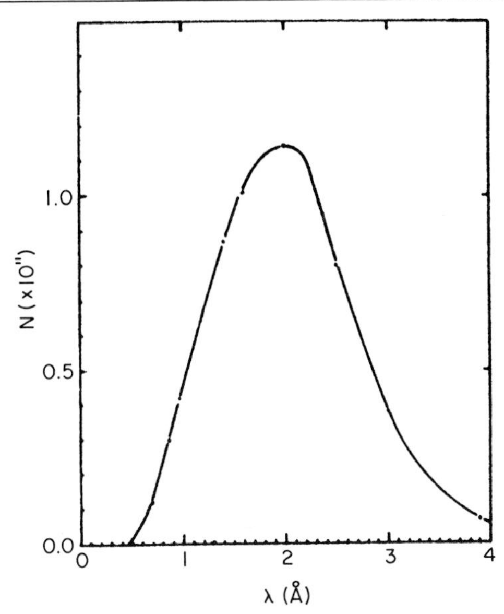

FIG. 1. Plot of $N(\lambda)$ versus λ, in the X-ray range, after the graphite filter and the beryllium windows of the beam line D1 at LURE-DCI. $N(\lambda)$ has the units number of photons $\sec^{-1}$ (horizontal mrad)$^{-1}$, for a bandpass $\delta\lambda = 10^{-4}$ Å. The storage ring is operated at 1.72 GeV and 300 mÅ. (Courtesy of Kahn *et al.*[3])

We describe in this chapter the source and the D12 instrument. The level of performance is discussed with a few examples. Some details about the operating procedure, seen from the user's point of view, are also given.

The Synchrotron Radiation Source

The experiments are performed at LURE, a laboratory devoted to the applications of synchrotron radiation. The electron–positron storage ring DCI is used as the X-ray source. This accelerator is partly dedicated to high-energy physics and partly (25%) to the production of SR. During the so-called "shifts" which are allotted to LURE, the energy of orbiting particles (positrons) is usually 1.72 GeV. With this relatively low energy combined with a short magnetic radius (3.82 m), the critical wavelength[22] of the SR spectrum is 4.2 Å. This value is quite suitable for X-ray crystallography because the flux is maximal in the 1–3 Å range (Fig. 1), whereas the flux of radiation with wavelengths below 0.6 Å is low, so that shielding

[22] G. K. Green, *Brookhaven Nat. Lab.* [*Rep.*] *BNL* **50522** (1976).

against hard radiation and the rejection of higher order harmonics in the monochromatized beam are more easily achieved.

The single bunch of positrons completes the orbit within 280 nsec and takes about 1 nsec (full width half-maximum, FWHM) to pass any given point in the orbit. The apparent transverse dimensions of the source are 5.8 mm wide and 2.4 mm high (FWHM). Each positron radiates into a sharp cone whose axis lies in the forward direction; for photons of 1.4 Å, the opening of this cone is 0.3 mrad. The average positron trajectories in the bending magnet are nonparallel and a further angular spread in these trajectories is caused by betatron oscillations. As a result, the total divergence of the SR in the vertical plane, including the angular spread of positrons and photons, is about 0.6 mrad.

The geometry of the beam line D1 limits available radiation to a fan of 20 mrad in the horizontal plane, which is shared between six experimental stations. The D12 station, located 15.5 m away from the source, is devoted to macromolecular crystallography. Data are collected during 24-hr shifts at an average rate of 40 shifts per year. In a shift, 30 min are required to inject 300–330 mA of positrons at 1 GeV and ramp the energy to 1.72 GeV. Since about half of the particles initially stored are lost after 30–40 hr, at most one further injection is normally required during a shift. After each injection, the beam, as observed at the D12 station, is stable in the vertical direction to better than ±0.5 mm. The machine parameters are usually reproduced from one injection to another and so is the beam position. In practice, the user must fit his experiments in 12- to 24-hr cycles during which the intensity of the X-ray source steadily decays. The average efficiency of the shifts (usable beam time/planned beam time) turns out to be about 80%. As such, these conditions are reasonably well tailored to the requirements of a carefully prepared macromolecular crystallography experiment.

Description and Operation of the D12 Instrument

Design Features

The D12 camera was designed for the main purpose of fast, high-resolution data collection by photographic techniques. The screenless rotation method was chosen in view of its efficiency.[23] The rotation camera, a modified commercial instrument, is installed on a rotating frame for easy alignment after each change of wavelength. A curved crystal mono-

[23] U. W. Arndt, J. N. Champness, R. P. Phizackerley, and A. J. Wonacott, *J. Appl. Crystallogr.* **6,** 457 (1973).

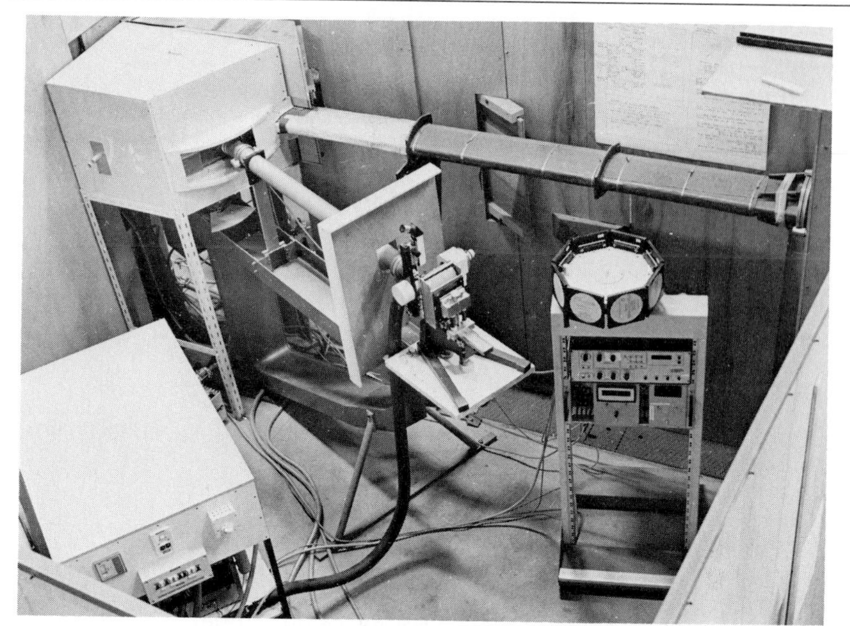

FIG. 2. View of the D12 instrument in the shielded enclosure. (Courtesy of Kahn *et al.*[3])

chromator provides an intense, line-focused, tunable beam in the wavelength range of 1–2.6 Å. A beam monitor and a cooling system are installed on the camera. In the design of this instrument, shown in Fig. 2, simplicity and minimal constraints were considered as essential; in fact, a trained crystallographer is able to use the camera and its ancillary equipment after a short introduction by a member of the local staff.

Monochromator

The principle of the monochromator is shown in Fig. 3. Focusing in the horizontal plane is obtained by reflection of SR on a plane-parallel germanium plate asymmetrically cut at an angle α to (111) planes and curved to a portion of a cylinder. The focus is a narrow vertical line; in the horizontal plane, the diffracted beam is convergent; in the vertical plane the divergence of SR is preserved. As any wavelength change (θ rotation, Fig. 3) sweeps the monochromatized beam in the horizontal plane, the camera has to be moved and aligned each time the wavelength is modified; these adjustments are simplified by the setting of the camera on a rigid frame which can rotate about the same vertical axis as the germanium plate. The wavelength, in angstroms, of the reflected beam is displayed on the control panel; the circuitry includes an angle-to-sine con-

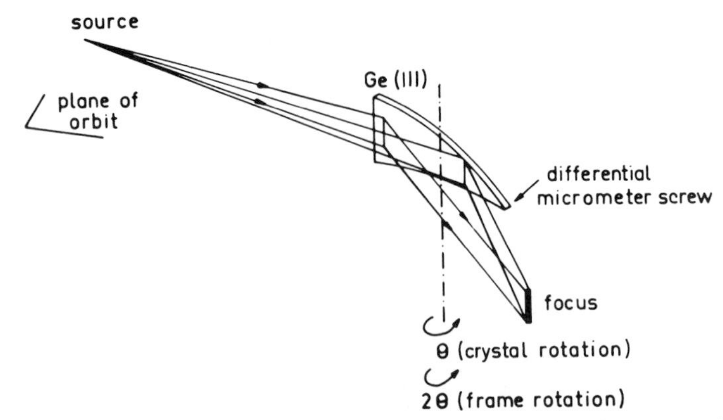

FIG. 3. Principle of the curved crystal monochromator of the D12 instrument. (Courtesy of Lemonnier *et al.*[24])

verter mounted on the θ rotation axis, for an analog calculation of Bragg's law.

This monochromator introduced a crystal-bending principle which is now of common use.[24] The germanium plate has a triangular shape and is mounted as a cantilever. The basis of the triangle is secured to a steel plate and the free end can be displaced from the rest position by a differential micrometer screw. The effect of the linear increase of the bending moment along the plate is compensated by the linear increase of the section so that the stress and hence the curvature are constant at any point. An adjustable cylindrical curvature is then obtained in a simple and effective way. The mechanical design is shown in Fig. 4. The curvature is remotely controlled and the position of the micrometer screw is permanently displayed.

The horizontal and vertical divergences of the primary and monochromatized beams are adjusted by two pairs of tantalum slits. The entrance slits are connected to the beryllium window of the beam port. The exit slits are fixed on a rotating shaft; they are connected on one side to the body of the monochromator by steel bellows and on the other side to a pipe closed by a kapton window. The whole optical path is under vacuum so as to minimize parasitic scattering and absorption.

This optical system does not provide vertical focusing and the rejection of higher order harmonics is effective only at wavelengths shorter than 1.6 Å. A single bent glass mirror has been recently tested at LURE

[24] M. Lemonnier, R. Fourme, F. Rousseaux, and R. Kahn, *Nucl. Instrum. Methods* **152,** 173 (1978).

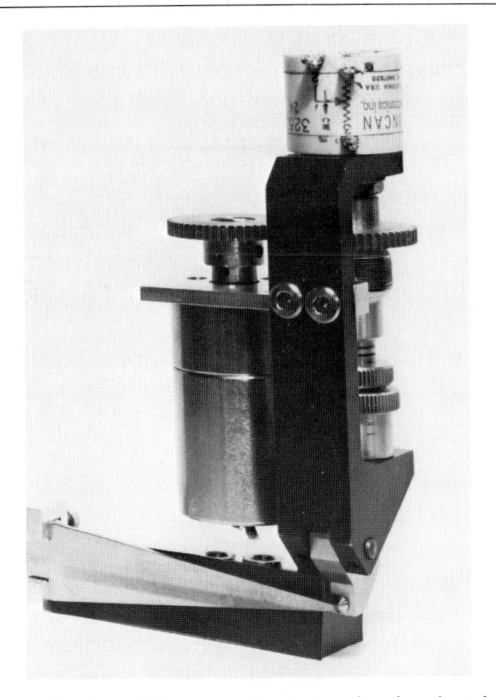

FIG. 4. Bending mechanism of the monochromator showing the triangular Ge crystal. The differential micrometer screw is driven by a direct current motor and connected to an helical potentiometer. (Courtesy of Kahn *et al.*[3])

by M. Lemonnier and J. Goulon to perform these functions; the prototype reflector is made from a triangular gold-coated piece of inexpensive float glass, 35 cm long, using the same bending principle as the monochromator. With a similar device installed on the D12 instrument between the monochromator and the rotation camera, we anticipate a reduction in exposure times by a further factor of about 3–4. At EMBL-Hamburg and at SRS-Daresbury, respectively, segmented quartz mirrors[25] and a bent rectangular platinum-coated quartz mirror[5] are used for vertical focusing on similar instruments.

Rotation Camera

An Enraf-Nonius camera is used with the multifilm technique. This camera stands on a platform which is movable along an optical bench fixed on the rotating frame. A few modifications have been made to the commercial instrument.

[25] M. Hendrix, M. H. J. Koch, and J. Bordas, *J. Appl. Crystallogr.* **12,** 467 (1979).

1. A unique feature of SR is the steady decay of intensity. In order to get diffracted intensities at the same relative scale for all Bragg spots on a given pack of films, the duration of one oscillation of the sample crystal has to be a small fraction of the characteristic decay time of SR (for example, 0.5% or less, i.e., 9–12 min in our case). The connections of the speed selector have been modified to provide repeated oscillations at high scanning speeds (typical conditions are 10 oscillations per photograph with each oscillation performed at an inverse angular velocity of 30 sec per degree). Multiple oscillations are also useful to smooth out the effect of low-frequency fluctuations of the SR beam.

2. A special collimator with adjustable horizontal and vertical slits has been supplied by P. B. Sigler (University of Chicago). This collimator (designed after Schmid *et al.*[6]; adapted from Love *et al.*[26]) is used for the final shaping of the monochromatic beam in each specific case, especially for samples with unit cell parameters in excess of 300 Å.

3. In addition to standard flat cassettes, V-shaped cassettes or curved cassettes with 96 mm radius have been used. Nonplanar cassettes using 175 × 125-mm films, and especially cylindrical cassettes, have obvious advantages over a flat cassette using 125 × 125-mm films for high-resolution data collection. These cassettes are compatible with the commercial carousel, and fiducial marks are made by allowing X rays from an auxiliary generator to enter narrow holes. Another cylindrical cassette with 110 mm radius, supplied by P. B. Sigler, is designed to collect data for crystals with long cell edges (300–1000 Å). This cassette is fixed on a special mount which fits on the camera's optical track (Fig. 5). It incorporates a pair of metal pins in the same relative positions that they occupy on the film scanner. Then, when the film is mounted on the film scanner, it is aligned in precisely the same way that it was in the cassette during exposure. A beam stop which is easily swung out of the way is used to produce an attenuated photograph of the undiffracted beam; this spot is used as a reference mark on the film to define the origin in the horizontal direction.

4. With SR, the thermal load produced by the absorption of X rays by the sample is much larger than usual and cooling is a necessity. The crystal cooling system is based on the design of Marsh and Petsko.[27] Temperatures down to 263 K stabilized to 0.1 K are obtained. Owing to the large section (30 mm) and the turbulence of the cold gas stream, temperature gradients along the capillary containing the crystal are negli-

[26] W. E. Love, W. A. Hendrickson, J. R. Elliott, E. E. Lattman, and G. L. McCorkle, *Rev. Sci. Instrum.* **36,** 1655 (1965).

[27] D. J. Marsh and J. A. Petsko, *J. Appl. Crystallogr.* **6,** 76 (1973).

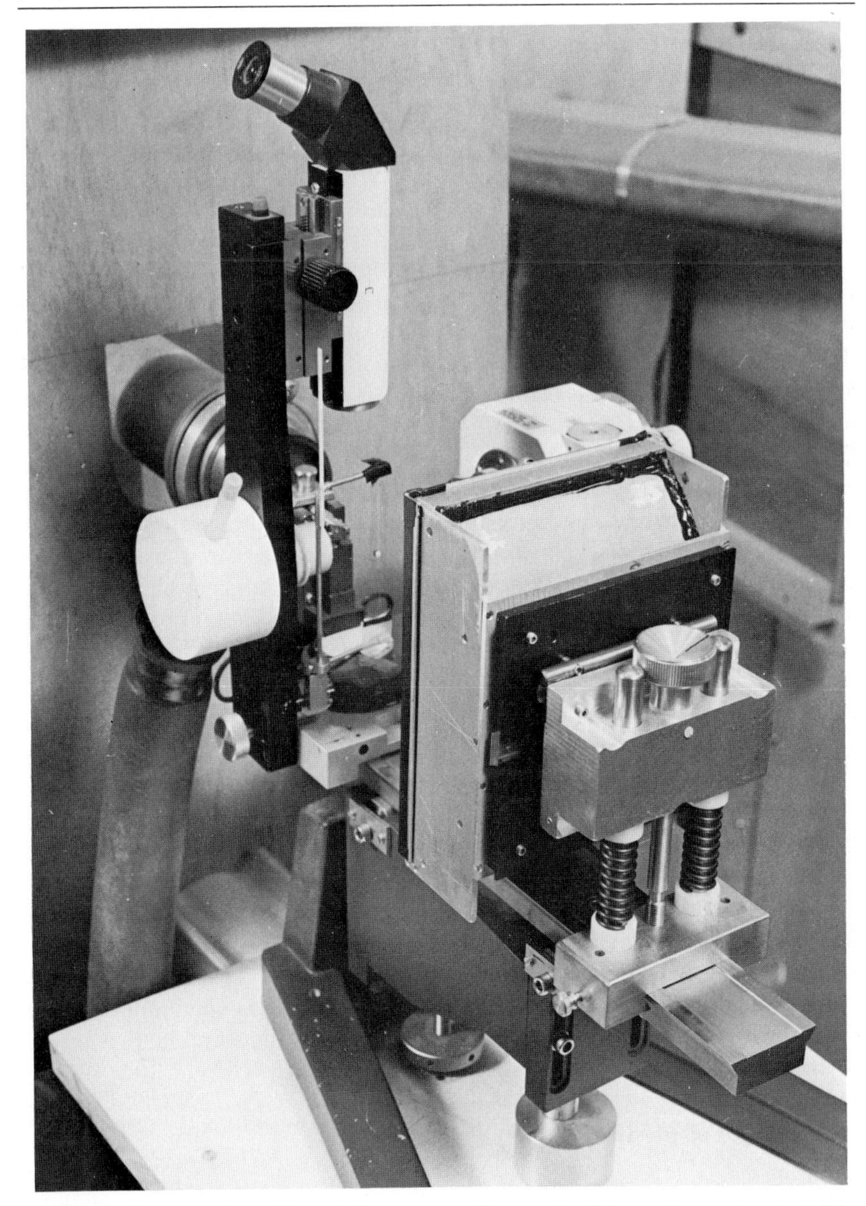

FIG. 5. Close-up view of the rotation camera. The nozzle of the cooling system is visible. A cylindrical cassette (P. B. Sigler, University of Chicago) is mounted on the camera's optical track.

gible, thus avoiding distillation of mother liquor. In most cases, the cooling improves quite significantly the crystal lifetime.

5. A tiny ion chamber (a gift of H. D. Bartunik), placed in the collimated beam, is connected to a charge integrator.[27a] It is used as an intensity monitor during the alignment of the camera.

Experimental Area

The instrument is located in a 2.5 × 3-m enclosure with walls made of a sheet of lead sandwiched between two sheets of wood (Fig. 2). Two keys are required to open the door of the enclosure and to get access to the monochromator; these keys are normally inserted in a control panel located outside and they cannot be removed from it without closing the beam port shutter. Camera controls are actuated from the panel.

As the monochromator is separately shielded and the beam line is wrapped with lead, there is no radiation in the enclosure even when the beam port shutter is open, as long as the camera shutter remains closed. Local staff members are allowed to stay in the enclosure in order to align the camera. The alignment procedure, which is fast and simple due to the high collimation of SR, uses the beam monitor and visual observation of the trace of the collimated beam on a fluorescent screen located at 2 m from the collimator; the operator is protected against scattered radiation by means of plastic tubing and lead glass.

Handling of Films

With this camera it is possible to expose up to 50–140 packs of three films during a 24-hour shift. It has been found convenient to have a stock of fresh films kept in a lead-lined box at the disposal of teams which use the D12 instrument. The brand of film is Reflex 25 manufactured by CEAVERKEN (Strängnäs, Sweden) and has been chosen for a low background level.[28] Two formats are available, 125 × 125 mm and 125 × 175 mm. Films are processed using thermostated baths; the developer tank is provided with periodic nitrogen bubbling. Twelve films can be processed at the same time.

Performance of the Instrument

Characteristics of the Monochromator

All relevant parameters of the monochromator are shown in Fig. 6. (after Lemonnier *et al.*[24] and Kahn *et al.*[3])

[27a] R. Bosshard, R. Kahn, and R. Fourme, *J. Appl. Crystallogr.* (submitted for publication).
[28] S. Abrahamsson, O. Lindqvist, L. Sjölin, and A. Wlodawer, *J. Appl. Crystallogr.* **14,** 256 (1981).

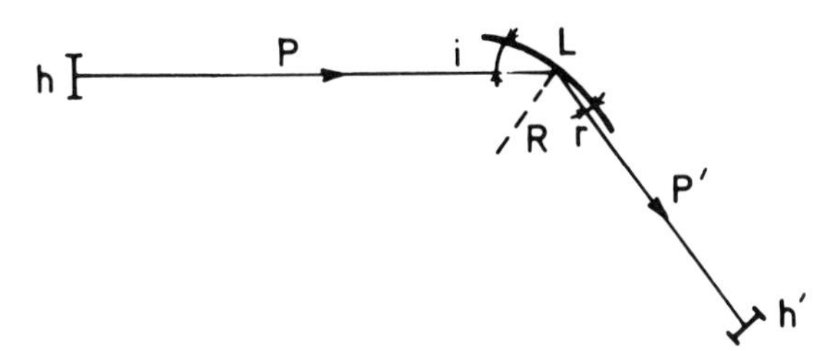

FIG. 6. Relevant parameters in the description of the characteristics of the monochromator: h and h', horizontal extensions of the source and focus; p and p', source-to-crystal and crystal-to-focus distances; R, bending radius of the crystal; L, length of the crystal; $i = \theta + \alpha$; $r = \theta - \alpha$. The angle of asymmetry α is positive when the incident beam makes an angle larger than θ to the crystal face (which is the actual case). ω_s is the angular width of dynamic diffraction in the symmetric Bragg-case geometry; ω_a is the angular width of acceptance in the asymmetric case, $\omega_a = \omega_s[(\sin r)/(\sin i)]^{1/2}$. (After Kahn *et al.*[3])

The *total spread of wavelengths* at the focus of the monochromatic beam $\delta\lambda/\lambda$ is a convolution of the crystal rocking curve with the geometrical factors of the source size and the variation of the angle of incidence along the curved crystal. The result is

$$\frac{\delta\lambda}{\lambda} = \left[\left(\frac{h}{p} + \frac{L}{2}\left|\frac{\sin r}{p'} - \frac{\sin i}{p}\right|\right)^2 + \omega_a^2\right]^{1/2} \cot\theta$$

The best resolution is obtained for $(\sin i)/p = (\sin r)/p'$ (Guinier setting) and depends essentially on the horizontal source size.

The *width of the line focus* is a convolution of the aberrations of the optical system[29] (negligible), the angular dispersion in the diffracted beam, and the source demagnification (hp'/p). Again, the latter dominates the convolution.

The *horizontal angular apertures* of the incident and diffracted beams are $L(\sin i)/p$ and $L(\sin r)/p'$; they are equal at the Guinier setting.

In high-resolution studies, the optimal adjustment of the monochromator requires a crystal with an angle of asymmetry corresponding to the Guinier setting at the selected wavelength: for instance, $\alpha = 10°$ for $\lambda = 1.4$ Å (Table II).

The properties of a curved crystal monochromator may be used in an unconventional way for special applications with SR sources, as discussed by Greenhough *et al.*[11] $\delta\lambda/\lambda$ is not what the sample receives if the sample size is less than the focus width. In particular, there is a photon energy gradient along the focus due to the finite source size; hence a narrow slit will provide a fine spectral resolution beam incident on the

[29] F. W. Martin and R. K. Cacak, *J Phys. E* **9,** 662 (1976).

TABLE II
D12 CAMERA MAIN CHARACTERISTICS[a]

Source-to-monochromator distance	15500 mm
Monochromator-to-focus distance	1675 mm
Crystal Ge (111), $\gamma = 10°$, cylindrically bent with an adjustable radius of curvature	
Bragg angle	12.36°
Demagnification ratio	1/9.25
Horizontal divergence of beam both entering and leaving the monochromator (adjustable)	1.7 mrad
Vertical divergence	0.6 mrad
Size of line focus (length is adjustable)	0.7 × 7 mm
Intensity at focus (DCI operated at 1.72 GeV, 300 mA)	1.5×10^{11} photons sec^{-1}
Wavelength spread, $\delta\lambda/\lambda$	0.0017

[a] Main characteristics of the monochromator in the most used configuration (angle of asymmetry $\alpha = 10°$, $\lambda = 1.40$ Å).

sample. Away from the Guinier setting, when the curvature contribution to the spread of wavelengths is significant, spots on the films are drawn out into streaks; the energy gradient along the streak gives rise to the possibility of a polychromatic diffraction experiment allowing a sampling of an absorption edge for optimized anomalous dispersion studies.[10]

Considerations of Exposure Times and Angular Resolution

The intensity at the focus of the monochromator was measured at 1.54 Å with a solid-state detector. The beam was attenuated with calibrated metal foils; the slits were adjusted to limit the vertical extension of the focus at 4.5 mm. The storage ring was operated at 1.72 GeV and various currents. The average result, extrapolated at 300 mA, was 1.5×10^{11} photons sec^{-1}. After a collimator with a 300-μm pinhole, the result was 3.6×10^9 photons sec^{-1}, corresponding to about 0.5×10^{11} photons sec^{-1} mm^{-2}. The addition of a mirror, as mentioned previously, will hopefully improve the latter result by a factor of 3–4.

Using an Elliott GX-6 rotating anode tube operated at 1.6 kW (Ni-filtered Cu radiation), Harmsen *et al.* have reported 9×10^8 sec^{-1} mm^{-2} after a similar collimator.[30] The *practical* gain obtained with the SR source is in fact significantly higher than the ratio of the intensities, 55, owing to the improved signal-to-noise ratio produced by the highly monochromatic, focussed beam.

In the study of crystals with a large unit cell, the D12 instrument should be compared to a conventional double-focusing camera producing

[30] A. Harmsen, R. Leberman, and G. E. Schulz, *J. Mol. Biol.* **104,** 311 (1976).

a beam of comparable divergence. A typical example may be found in the study of crystals of a complex of actin and profilin (space group $P2_12_12_1$, a = 38.6 Å, b = 71.6 Å, c = 172.2 Å), performed by C. Schutt (MRC Laboratory of Molecular Biology, Cambridge, UK). At Cambridge, the camera incorporated a pair of nickel-coated mirrors (a 60-mm vertical mirror and a special 200-mm horizontal mirror) and the X-ray source was an Elliott GX-13 rotating anode tube (Cu anode, 100-μm focusing cup, run at 40 kV and 60 mA); this equipment provided a beam with a vertical divergence of 0.8 mrad and an horizontal divergence of 1.7 mrad, close to the values in Table II. Data to 2.5 Å resolution were collected at a rate of 30,000–40,000 sec per degree. At Orsay, with the storage ring run at 1.72 GeV and 200 mA (60% of the nominal flux), times of exposure of 300–400 sec per degree were sufficient to get diffraction spots of a comparable intensity in the low to medium resolution range. On visual inspection of films spurious streaks were visible in the low-angle region of films obtained at Cambridge, due to the imperfect monochromatization of the beam by mirrors. As a net result, it was possible to collect at Orsay data of a higher resolution; in that case, the improvement came essentially from the dose rate effect.

When the full cross-fire of the beam is used together with a standard collimator, the angular resolution is adequate for spacings up to about 300 Å. For longer parameters, the entrance and exist slits of the monochromator are adjusted to reduce the horizontal divergence, and the collimator with adjustable slits is used for the final, accurate shaping of the beam. In the case of Δ^5-3-ketosteroid isomerase (space group $P6_122$ or $P6_522$, a = 65.5 Å, c = 504.0 Å, for which data were collected by E. M. Westbrook and P. B. Sigler (University of Chicago), the horizontal divergence was reduced from 1.7 to 0.6 mrad and the collimator slits were adjusted for an optimal resolution of spots along the c^* axis, as shown in Fig. 7. On the basis of these results, J. E. Johnson (Purdue University) has tackled, with essentially the same equipment, high-resolution data collection for crystals of cowpea mosaic virus (space group $P6_122$ or $P6_522$, a = 454 Å, c = 1044 Å. A typical photograph is shown in Fig. 8; in this case, the exposure time is fairly long, due to the size of the unit cell and to the extreme collimation of the direct beam.[31]

Data Processing and Data Reduction

The D12 camera has been used mostly with the monochromator adjusted for a minimum bandwidth (Guinier setting). In that case, films are

[31] R. Usha, J. E. Johnson, D. Moras, J. C. Thierry, R. Fourme, and R. Kahn (1984). *J. Appl. Crystallogr.* **17,** 147–153.

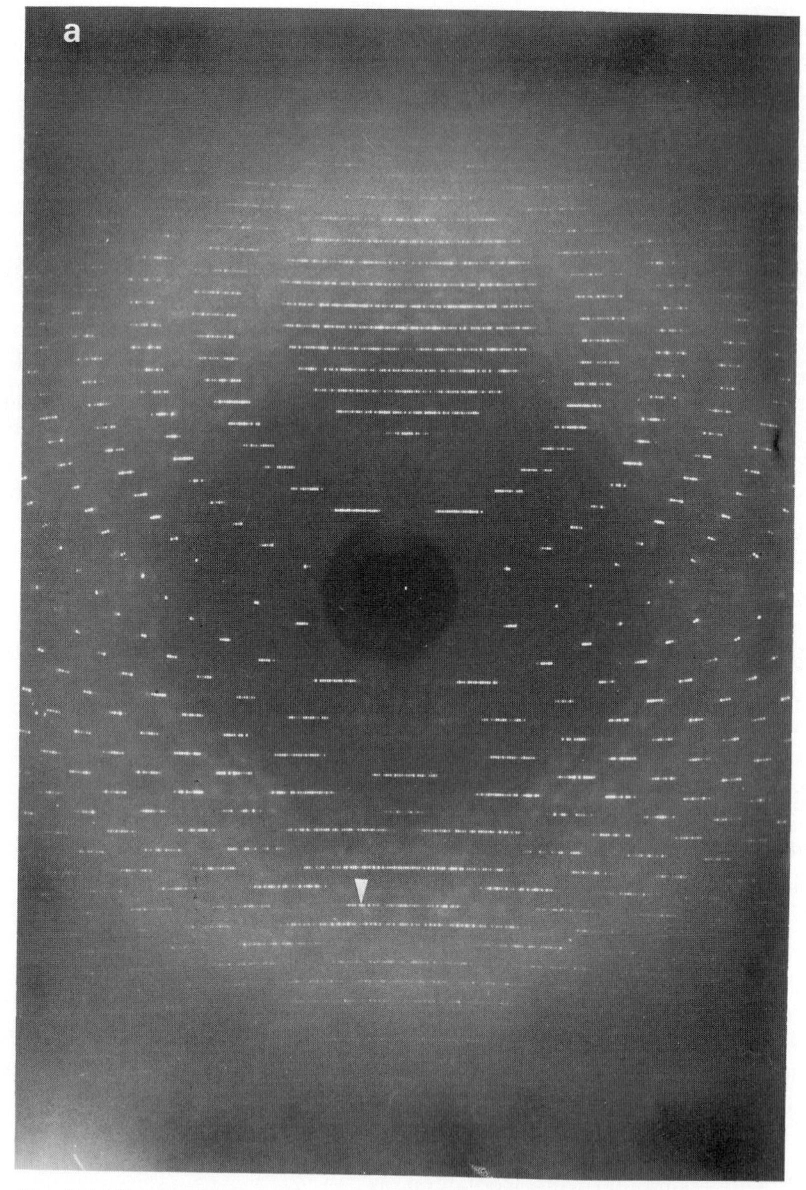

FIG. 7. (a) Rotation photograph of a crystal of Δ^5-3-ketosteroid isomerase mounted along the c^* axis (c = 504 Å); 2.2° sector, exposure time 83 min with the storage ring operated at low current (100 mA). Cylindrical cassette, λ = 1.4 Å. Data extend to a resolution of 2.6 Å where the Lorentz factor is favorable. (b) Enlarged and inverted portion of the previous photograph. (Both courtesy of P. B. Sigler and E. M. Westbrook.)

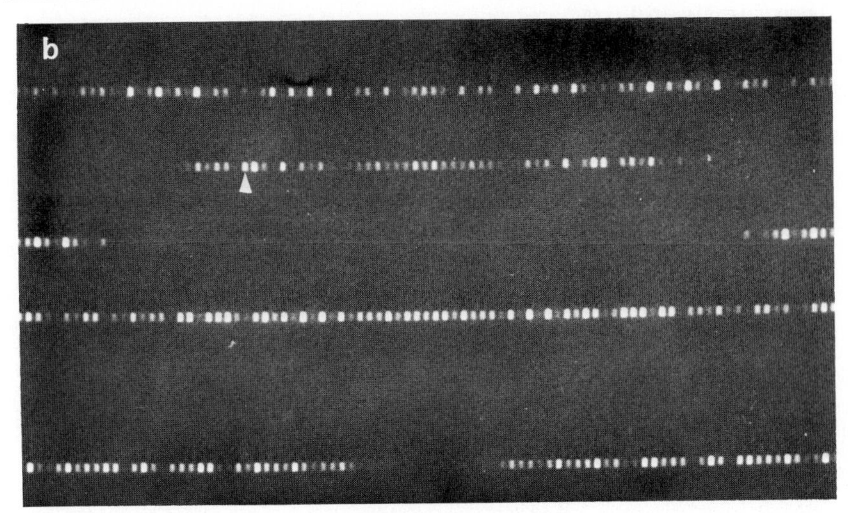

FIG. 7. (*continued*)

evaluated in essentially the same way as films obtained from a conventional focused camera. In particular, the cross-fire and the wavelength spread lead to small-size reflections on the films, which is approximately given by either the size of the crystal or the collimator aperture: a small raster scan (25–50 μm) must be used for densitometry. Greenhough and Helliwell[32] have given, for conventional sources, reflecting range formulas which lead to expressions from which partiality of reflections on films can be calculated; these expressions are applicable to the D12 camera at the Guinier setting after introduction of the asymmetric beam cross-fire parameters.

Greenhough and Helliwell, in another paper,[33] have derived equations for the general case, i.e., when the monochromator is away from the Guinier setting; in addition to the asymmetric cross-fire, the variable spectral bandwidth and the correlation between the direction of the incident ray and its wavelength are taken into account to get the appropriate flagging of partial reflections.

The calibration of films at unconventional wavelengths and for a given chemical processing of films is important for accurate measurements. As suggested by P. B. Sigler, the Cu K_α line (1.39 Å) should conveniently simulate the 1.40 Å radiation commonly used at LURE.

[32] T. J. Greenhough and J. R. Helliwell, *J. Appl. Crystallogr.* **15,** 338 (1982).
[33] T. J. Greenhough and J. R. Helliwell, *J. Appl. Crystallogr.* **15,** 493 (1982).

FIG. 8. Rotation photograph of a crystal of cowpea mosaic virus mounted along the c^* axis (c = 1044 Å); 0.4° sector, exposure time 200 min with the storage ring operated at 170 mA. Flat cassette, specimen-to-film distance 175 mm, λ = 1.4 Å, resolution of photograph 4.3 Å. Direct beam size with a square cross section, 0.2 × 0.2 mm². (Courtesy of J. E. Johnson.)

The intensity I of a reflection is related to the experimentally determined intensity I_{obs} by the relation

$$I = I_{obs}/LP$$

where P and L are the polarization and the kinematic factors, respectively. The SR beam is almost completely linearly polarized, in contrast to the beam from a conventional source. The polarization correction P for the rotation camera placed on a horizontally dispersing curved monochromator has been derived by Kahn *et al.*[3]

Conclusions

At the moment, three-dimensional data have been collected with the D12 instrument for about 35 different macromolecular crystals, including in most cases one or several heavy-atom derivatives in addition to the native crystal. The situation of each of these projects is very variable. The final stages of structure analysis have actually been reached in a few cases: a typical example is the high-resolution structure of human deoxyhemoglobin.[34] This is not surprising, since problems tackled with SR are more complex or more ambitious than the average, with unusually large data files. The high spectral brilliance permits exposure times which are dramatically shorter, thus saving crystallographer's time and allowing the use of crystals from the same batch within a short time interval. The radiation damage is comparatively reduced, for a fixed total dose, when the dose rate is increased: typically two to five times more data than with a rotating anode source can be collected from one and the same crystal within a given resolution range. The increased useful lifetime of the crystal may permit collection of higher angle reflections which are most sensitive to radiation damage. In cases in which only one rotation photograph is obtainable from one crystal with a conventional source, several may be obtained using SR. This increases the efficiency of the screenless rotation method, allowing the partially recorded terms to be used by summation of the parts from adjacent films, as well as the fully recorded terms; at the same time, the number of crystals required in these studies is reduced and/or smaller crystals may become usable. In cases where a crystal larger than the exit pinhole of the collimator is available, the dose rate effect, short exposures, and tight collimation of SR all favor acquisition of multiple exposures from different parts of this sample.

As a net result of improved signal-to-noise ratios and dose rate effects, the instrument has been found useful to get data to a resolution which would have been otherwise very tedious to obtain if not impossible, even for crystals with modest cell parameters such as small polynucleotides or polypeptides. It is a real breakthrough for most demanding studies, i.e., small crystals with a large unit cell, poor crystalline order or samples which are highly radiation sensitive. Those difficulties are often met with crystals of high biological relevance, such as protein–nucleic acid complexes, membrane proteins, ribosomal fragments, and viruses.

[34] G. Fermi, M. F. Perutz, B. Shaanan, and R. Fourme (1984). *J. Mol. Biol.* **175,** 159–174.

[22] X-Ray Sources

By WALTER C. PHILLIPS

Most X-ray diffraction studies in molecular biology use characteristic X rays produced by rotating anode or sealed-tube X-ray sources. Although the monochromatic X-ray beams from these "conventional" sources are generally 10–100 times less intense than those from synchrotron sources, they are adequate for the majority of investigations. In addition, conventional sources are relatively simple, reliable, and inexpensive laboratory instruments that can be operated more or less continuously. Yoshimatsu and Kozaki[1] have co-authored an excellent review article that provides a historical perspective and a detailed treatment of many technical aspects of rotating anode generators. Their article has been an important source of the information and ideas used in preparing this chapter. The first part of the chapter describes characteristics of rotating anode and sealed-tube sources, and presents some general considerations that are encountered when operating, purchasing, or designing experiments that use conventional sources. The second part deals with the operation and maintenance of rotating anode generators.

Conventional Sources

Source Characteristics

Some specifications for currently available sealed-tube and rotating anode generators that are appropriate for macromolecular diffraction studies are listed in the table. The maximum target loading and power per unit area, or brightness, are given for Cu anodes.[2] The advantages of rotating anodes are immediately obvious: (1) small source widths (0.1–0.2 mm) with very high brilliance are possible only with rotating anodes; and (2) for large source widths (greater than about 0.4 mm), rotating targets can sustain 7–45 times more power loading than sealed tubes. Thus, many experiments are feasible only with a rotating anode source. The advan-

[1] M. Yoshimatsu and S. Kozaki, "Topics in Applied Physics, Vol. 22. X-Ray Optics" (H.-J. Queisser, ed.), p. 9. Springer-Verlag, New York, 1977.

[2] Sealed tubes are available from several manufacturers, including Philips Electronic Instruments, Mahwah, New Jersey; Rigaku USA, Danvers, Massachusetts; AEG Telefunken (Molecular Data, Cleveland, Ohio); Enraf-Nonius, Bohemia, New York; Macroni Avionics/Elliott, Peabody, Massachusetts; and Nicolet XRD, Cranford, New Jersey.

SPECIFICATIONS FOR FIXED AND ROTATING ANODE SOURCES

Source	Focal spot size[a] (mm)	Maximum load (kW)	Maximum brightness (kW/mm^2)	Maximum target velocity (m/sec)
Standard focus sealed tube	1.0 × 10	2	0.2	—
Long, fine-focus sealed tube	0.4 × 12	1.8	0.4	—
Fine-focus sealed tube	0.4 × 8	1.4[b]	0.4	—
Very fine-focus sealed tube[c]	0.15 × 8	0.8	0.7	—
Elliott GX-20	0.1 × 1.0	1.2	12.0	28
rotating anode	0.2 × 2.0	3.0	7.5	28
(4 kW)	0.3 × 3.0	4.0	4.4	28
Rigaku (RU-200)	0.1 × 1.0	1.2	12.0	31
414BH	0.2 × 2.0	3.0	7.5	31
(12 kW)	0.3 × 3.0	5.4	6.0	31
	0.5 × 10.0	12.0	2.4	31
Elliott GX-13	0.1 × 1.0	2.7	27.0	108
(4 kW)	0.2 × 2.0	4.0	10.0	108
Rigaku FR	0.1 × 1.0	3.5	35.0	130
Elliott GX-21	0.1 × 1.0	1.2	12.0	28
(15 kW)	0.3 × 3.0	4.5	5.0	28
	0.5 × 5.0	10.0	4.0	28
	0.5 × 10.0	15.0	3.0	28
Rigaku RU-1000 (60 kW)	1.0 × 10.0	60.0	6.0	52
Rigaku RU-1500	0.1 × 1.0	3.5	35.0	130
(90 kW)	1.0 × 10.0	90.0	9.0	130

[a] Other sizes are also available for most rotating anode generators.
[b] Typical value.
[c] Manufactured by AEG Telefunken and distributed by Molecular Data Co., Cleveland, Ohio, item AMLCo.

tages of sealed-tube sources are not obvious in the table: (1) the initial and maintenance costs of a sealed-tube generator are much less than for a rotating anode generator; and (2) the downtime for a sealed-tube machine may be significantly less than for a rotating anode source, although this is not necessarily the case. Many experiments can be successfully carried out with sealed-tube sources.

Heating of the target at the focal spot by the electron beam limits the maximum anode loading. Heating produces three effects: surface roughening, target melting (at sufficiently high loading), and thermal stress, which is caused by differential expansion of target material at the edge of the focal spot. By continuously moving the target with respect to the electron beam, electron heating is distributed over an area which is large compared to the focal spot, allowing operation at a much higher electron beam power (for a given focal spot size) than for a stationary target. Various schemes of accomplishing this motion have been employed, but the scheme universally used today for focal spot sizes of interest here is the cylindrical rotating anode. Assuming the focal spot is a narrow rectangle with short dimension (width) d in the direction of anode rotation, and long dimension L parallel to the anode axis, the maximum loading for small focal spots is roughly proportional to

$$\text{Load}_{\text{max}} \approx L(dDn)^{1/2}(kC\rho)^{1/2} \quad (1)$$

where D is the diameter, n the rotational frequency, k the thermal conductivity, C the specific heat, and ρ the density of the anode. For small focal spots the ultimate loading limit occurs when the anode surface at the focal spot melts. The practical loading maximum is reached well before this point, usually at 60–70% of the melting load, at a point when the anode surface roughness produced by electron heating begins to increase noticeably over a time short compared to the filament lifetime (for a sealed tube) or normal maintenance period (for a rotating anode). For larger focal spots on rotating anodes, the loading limit for most target materials usually occurs at considerably lower power, when the thermal stresses become greater than the shear strength of the target.

The practical consequences of Eq. (1) are that (1) small focal spots, $d \approx 0.1$ mm, can be much more brilliant than larger spots, and (2) for a given spot size, anode loading approaches an upper limit because D and n are constrained by both the tensile strength of the anode itself and the characteristics of the seals and bearings. Practical limits for n and the anode surface velocity are ~ 170 sec^{-1} and ~ 150 msec^{-1}, respectively. The relationships between maximum load, maximum brightness, and target velocity for today's rotating anode generators are given in the table.

For Elliott generators, purchase of a GX-13 rather than a GX-20 can probably be justified only if a 0.1-mm focal spot is required, and purchase of a GX-21 only if a focal spot >0.3 mm is required. For Rigaku generators, the large 1000-mA RU-1000 and 1500-mA RU-1500 can be justified only if very large focal spots are required. The maximum brightnesses for anodes constructed of elements other than Cu are less than or equal to those for Cu, because of the high thermal conductivity of Cu. It is possible, although not usual, to increase the specific loading, or decrease electron-induced surface roughness, by alloying Cu with an element such as Cr, e.g., $Cu_{0.99}Cr_{0.01}$, which has a higher shear strength than pure $Cu_{1.00}$.

General Considerations

Optimum Operating Voltage. Assuming the anode surface is clean and the tube vacuum is good, the cathode–anode potential V which maximizes the characteristic K_α output at a given power depends on (1) the anode material and (2) the take-off angle.[1,3,4] For a particular anode material, the important parameters are the electron range and the X-ray path length in the target. The range of 10–50 KeV electrons, which is approximately the distance an electron travels before coming to thermal equilibrium, is roughly proportional to V^2 and inversely proportional to the anode density.[5] The path length of X rays in the anode is a function of the depth below the surface at which the X ray originates (and is therefore a function of the range) and the take-off angle. Smaller take-off angles and greater ranges give rise to longer path lengths and greater X-ray absorption.

The K_α yield $N(V)$, i.e., the number of K_α photons produced per incident electron, is primarily a function of the (e, γ) cross section for K_α production and the electron recoil ratio of the target metal.[3] For a target with no absorption, the expression for $N(V)$ derived by Webster and Clark[6] in 1917,

$$N(V) = K(Z)(V - V_k)^n \tag{2}$$

is still generally found in the literature, where $K(Z)$ is a constant which depends on the atomic number Z of the target and V_k is the K excitation voltage. Values for n from 1.4 to 1.7 have been given for $10 \leq V \leq 50$ kV.[1,3,7,8] In the most recent study, Green[3] found $n = 1.63$. When absorption within the target is considered, the number of K_α photons emitted per unit solid angle at a take-off angle ϕ is

$$N(V, \phi) \approx f(\phi, V, Z)K(Z)(V - V_k)^n \tag{3}$$

The absorption term f approaches unity as ϕ approaches 90°.[1,3] The K_α output at a given power is proportional to $N(V, \phi)/V$. Parrish and Spielberg[4] have measured the voltage dependence of the intensity of char-

[3] M. Green, *in* "X-Ray Optics and X-Ray Microanalysis" (H. H. Pattee, Jr., V. E. Cosslett, and A. Engström, eds.), p. 185. Academic Press, New York, 1963.

[4] W. Parrish and N. Spielberg, *Meet. Am. Cryst. Assoc. & Miner. Soc. Am.* Abstract B-5, p. 25 (1964).

[5] R. D. Evans, "The Atomic Nucleus," p. 622. McGraw-Hill, New York, 1955.

[6] D. L. Webster and H. Clark, *Proc. Natl. Acad. Sci. U.S.A.* **3,** 181 (1917).

[7] A. H. Compton and S. K. Allison, "X-rays in Theory and Experiment," p. 81. Van Nostrand-Reinhold, Princeton, New Jersey, 1935.

[8] A. Guinier, "Théorie et technique de la radiocrystallographie," p. 12. Dunod, Paris, 1956.

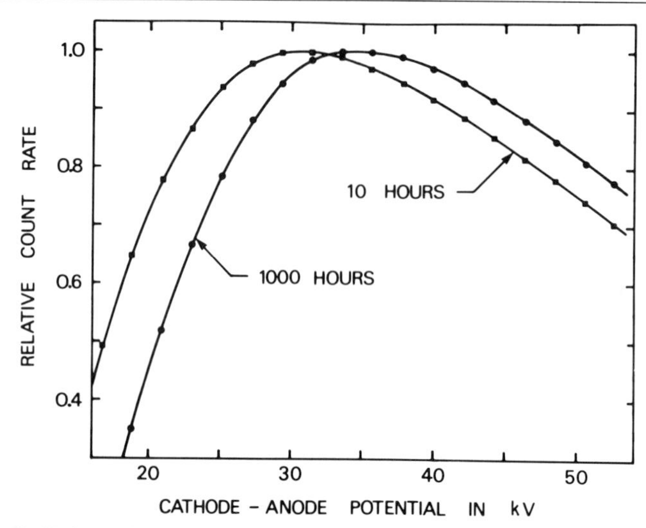

FIG. 1. Cu K_α intensity vs cathode–anode potential at constant power for a 2.5° take-off angle at 10 and 1000 hr (7 and 750 kW-hr) running time after the anode surface was polished.

acteristic radiation for a number of elements from 10 to 54 kV for ϕ between 1 and 10°. The results of recent measurements in our laboratory[9] of $N(V, \phi)/V$ are shown in Fig. 1. These were obtained using an Elliott GX-6 operated at 750 W, a 0.1 × 2.0-mm focal spot, a vacuum $\sim 10^{-6}$ Torr maintained by an oil diffusion pump, a Cu target, and $\phi \approx 2.5°$.

In time a layer of contaminants builds up on the surface of the anode. The thickness and composition of the absorbing layer depend on details of the tube operation, e.g., filament temperature, oil contamination, vacuum, and filament–anode distance. Electrons lose energy in the contaminant layer, and so enter the target metal with less than the applied cathode–anode potential. Thus, the operating voltage must be increased with time in order to maintain the maximum K_α intensity at constant power. The time dependence of the increase is a function of the particular operating conditions and geometry of the tube and filament. Figure 1 shows the K_α intensity shift after 1000 hr for the conditions described above. Over this 1000-hr time frame, an operating voltage of 33 kV would be about optimal. Operating at this relatively low voltage would reduce voltage instabilities caused by flashover due to less than ideal vacuum conditions. As ϕ is increased, the best voltage will also increase.

Optimum Focal Spot Size. Three considerations determine the optimum focal spot size for a particular experiment, or more generally, for the

[9] W. C. Phillips, unpublished results (1982).

kinds of experiments to be performed with an X-ray generator. These are the photon detector, the order-to-order resolution required, and the sample size.

PHOTON DETECTION. In principle, both photosensitive emulsions and photon counters, either scintillator–photomultiplier counters or gas-filled counters or area detectors (with some exceptions for television-based systems), convert each incident X-ray photon into a detectable signal. However, there is no practical method of extracting this quantized signal from X-ray film. To do so would essentially require that the location of every developed grain to recorded. Instead, the average number of photons incident in a given area of film, usually 600 μm^2 or larger, is estimated by measuring the film optical density (OD) with a scanning microdensitometer (1 X-ray photon/μm^2 produces an optical density of ~1.0). The optical density is roughly proportional to the incident intensity.

Both counter and film measurements are affected by intrinsic background, i.e., counts or optical density which occur in the absence of an X-ray beam. Generally, the intrinsic background of counters and area detectors is very low compared to the signal, while X-ray film has a very significant intrinsic background, ~0.1–0.3 OD. This chemical fog, i.e., developed grains in an unexposed emulsion, can be minimized by using fresh film and developing solution, processing for relatively short times, and using low-background film such as CEA Reflex 25, but probably cannot be made <0.1 OD.

The measured parameter for film is the OD above background OD of a given area, whereas for a counter or area detector, the total number of photons is the significant measured quantity. This difference in the method of intensity measurement gives rise to different optimum X-ray source characteristics. Consider, for example, recording a Bragg reflection from a single crystal for a given number of photons incident on the sample, with a given spot size at the film or detector. If the area of the spot increases, the OD recorded on the film decreases, and the accuracy of the measurement decreases correspondingly, while the detector measurement, i.e., the total number of photons diffracted, is unchanged. Thus, for film recording, the source brightness is very significant; the brighter the source, the greater the recorded optical densities. For diffractometers or area detectors, the source power is a more important parameter (the other considerations discussed below being equal). As more one- and two-dimensional position-sensitive detectors become available, it will be possible to take advantage of higher power rotating anode sources with larger focal spots. Data collection with TV area detectors at high counting rates represents a situation intermediate between film and other electronic detectors, and corresponds to a film with a low intrinsic background and

~100-μm resolution. A TV detector has been developed[10] which can count individual photons at low counting rates (see this volume, Arndt [29], and Kalata [30]).

RESOLUTION. In principle, instrumental resolution is not limited by the focal spot size, or by the corresponding size of the beam at the film (or detector). The instrumental resolution is determined by the ratio of the size of the beam at the film to the sample-to-film distance. But in practice, the effective spot size does limit the resolution, because neither the sample-to-film distance nor the film or detector area can be increased beyond certain practical limits. The limits are set by the cost of increasing the size of focusing monochromators, mirrors, collimators, diffractometer or camera, detector or film scanner, etc., as well as by finding enough space for the apparatus. The resolution requirements of a given experiment on a particular instrument will limit the optimum size of the focal spot. For most crystallographic studies with sample-to-film distances of <1 m, source widths greater than about 0.5 mm will not provide adequate resolution.

SAMPLE SIZE. It is never advantageous to have the beam size at the sample larger than the sample, as this can only decrease the signal-to-background ratio. If possible, the best way to reduce the beam size at the sample is to reduce the focal spot width appropriately, because then the source brightness can be increased (see section on source characteristics). Limiting the beam from a focal spot with a large width using slits or apertures will give fewer photons at the sample. When the intensities are being recorded on film, it may be advantageous to reduce the source size still further to get a brighter beam.

Take-Off Angle and Surface Roughness. In order to get high power from a small source, focal spots are rectangles (line sources) with $L/d \geq 10$. To obtain an approximately square effective source, the focal spot is viewed along the direction of the line at a small angle to the anode surface, the take-off angle ϕ. For most protein diffraction studies, an angle ϕ between 2° and 6° is required to get the the most useful effective source size. The smaller ϕ, the greater the absorption of the X-ray beam by protrusions on the anode surface, or surface roughness, and by self-absorption in the target of X rays emitted from below the anode surface. The increase of surface roughness with time is a strong function of anode loading. Surface roughness is particularly troublesome in sealed tubes because the anodes cannot be polished. For the usual ϕ angles, the total number of photons in the beam over the life of a sealed tube, or between anode polishing in a rotating anode, will be greater if the generator is run

[10] K. Kalata, *Proc. Soc. Photo.-Opt. Instrum. Eng.* **331,** 69 (1982).

at less than the maximum target loading given by the manufacturer. For any particular configuration, the optimum loading can only be determined in a controlled experiment. Generally, operation at about 80% of the maximum loading is determined to be best.

Focal Spot Profiles. Under optimum electron-beam focusing conditions, an intensity profile of the X-ray beam across the focal spot width (perpendicular to the line) of a rotating anode looks like a flat-topped Gaussian with long tails, for small (0.1–0.3 mm) spot widths. About 90% of the intensity is usually within the nominal width. The tails can be eliminated with slits placed as close as practical to the tube window. An intensity profile across the beam from a 0.4-mm-wide sealed tube typically has maxima near the edges and a local minimum near the center, with the minimum-to-maximum intensity ratio ~0.5. Many tubes have even more nonuniform and complex profiles. For some experiments and focusing cameras, these focal spots are not useful, and a significant portion of the source must be masked off in order to obtain a more uniform spot with which to work.

Rotating Anodes in Practice

Vacuum

The most common source of operating problems in rotating anode machines is poor vacuum within the tube. Typical symptoms of this condition are high-voltage discharges, short filament lifetimes, blown high-voltage transformer primary fuses, X-ray current saturation, and often high readings on vacuum gauges. Common causes of poor vacuum include vacuum leaks at O rings, rotating seals, Be windows, forepump lines and the anode, contaminated diffusion pump fluid, and outgassing of deposits within the tube.

The most efficient way to minimize vacuum-related problems is to have a regular, comprehensive preventative maintenance program. This includes a complete overhaul of the vacuum and related systems at about 18- to 24-month intervals. Additional service performed between overhauls is outlined below. The time spent in doing this maintenance will be significantly less than the time used finding and repairing vacuum problems as they arise. The overhaul should include a complete disassembly of the tube and thorough cleaning of all interior surfaces, replacement of every O ring in the generator, disassembly and cleaning of the diffusion pump(s), replacement of deteriorated forepump lines, cleaning of the mechanical pump, replacement of any worn drive belts or noisy motor bearings, and repair or replacement of unreliable thermostatic valves, relays, interlocks, switches, or circuitry. Viton O rings should be used exclusively, never rubber O rings. With few exceptions, O rings can be purchased locally. O rings should be lightly coated with silicon high-vacuum grease.

Locating small vacuum leaks at O rings, Be windows, the cathode, or the anode is facilitated by having a He leak detector or the considerably less expensive Varian "Smart Gauge" leak detector.[11] The cost of a "Smart Gauge" can easily be justified in a laboratory with several high-vacuum instruments. In well-maintained rotating-anode generators, high-vacuum leaks generally occur at the cathode O ring(s) which operates at a relatively high temperature, and at O ring joints which have recently been reassembled. When poor vacuum is indicated and no obvious leaks are found, the diffusion pump oil may be contaminated. There are two sources of hydrocarbon oil which lead to contamination of the silicone diffusion pump oil: mechanical pump oil and seal oil. Backstreaming of mechanical pump oil into the diffusion pump occurs when the generator is switched off without venting the vacuum system, or when the mechanical pump drive fails. Oil leaking past the rotating anode seal often migrates into the diffusion pump. The use of ferromagnetic seals eliminates this source of contamination.

A small fraction of the vaporized diffusion pump fluid escapes from the pump and enters the tube vacuum chamber, where it reduces the vacuum, is broken down by the electron beam, and finds its way to the anode and filament. Diffusion pump backstreaming can be reduced significantly by installing a cold baffle above the pump. Filament lifetime is increased when an efficient baffle is installed. Replacement of the diffusion pump with a turbomolecular pump eliminates the problem.

Turbomolecular pumps have several advantages over oil diffusion pumps: they produce higher tube operating vacuum, shorter pump-down times, and "clean" pumping, unaffected by backstreaming of oil into or out of the pump. The drawbacks of turbomolecular pumps are their high initial and maintenance costs, and more frequent maintenance requirements. Turbomolecular pumps are definitely superior high-vacuum pumps for rotating anodes, although oil diffusion pumps do provide adequate service at reduced cost.

The designs of rotating anode tubes do not include the most efficient configuration of the high-vacuum pump. Wyckoff[12] has improved the design of a Rigaku RU-200 by adding a second pumping port and diffusion pump, making a substantial improvement in the tube vacuum and filament lifetimes.

[11] Varian Inc., Santa Clara, California.

[12] H. Wyckoff, private communication (1982).

If a good vacuum is to be maintained, the surfaces of the anode, focusing cup, recoil shield, and tube housing should be cleaned regularly, usually whenever the tube is opened.

It is important to have a reliable high-vacuum gauge. Penning gauges are, on the whole, unreliable. Ionization gauges are superior in that they provide more reliability, accuracy, and longer service life between tube changes. It is not productive to have a questionable vacuum gauge. Rigaku and Elliott place the gauge-head port close to the pump, so that the vacuum in the region of the filament, the vacuum of most interest, is not sensed by the gauge, but must be deduced from the gauge reading provided. For a rotating anode operating properly, the gauge vacuum should be $\sim 1-2 \times 10^{-6}$ Torr with the generator running at full power, and $<10^{-6}$ with the X-ray power off. Even when the gauge vacuum reading is satisfactory, tube vacuum may not be. Vacuum gauge heads respond differently to different molecules. Penning gauges, for example, are not very sensitive to water. Small water leaks by O rings or through a porous anode may not be detected. The most obvious sign of a vacuum problem is high-voltage flashover. If flashover occurs below ~40 kV, tube vacuum is very likely poor and the condition should be remedied as soon as possible.

Vacuum Seals

Considerations. Two types of vacuum seals are now commonly used on the rotating anode shaft: oil seals[13] and magnetic fluid seals.[14] In oil seals, oil with a low vapor pressure forms a vacuum seal and provides lubrication between a hardened polished sleeve on the anode shaft and a flexible rubber or plastic ring fitted tightly around the sleeve. In the magnetic seals, a suspension of small magnetic particles in low vapor pressure oil is trapped by a magnetic field in the small gap between the sleeve and annular pole pieces surrounding the sleeve. The steel sleeve and pole pieces are magnetized by a permanent magnet (or magnets). Oil seals can be operated at higher circumferential velocities than magnetic seals of the present design, and this is the major advantage of the oil seal. Oil seals have been successfully run at velocities of 18 $msec^{-1}$, magnetic seals at ~ 7 $msec^{-1}$.[1] At higher velocities frictional heating may cause a chemical breakdown and/or significant lowering of the viscosity of the magnetic particle suspension oil.

Oil seals offer the additional advantages of lower initial and rebuilding costs, and a simpler rebuilding procedure. Ferromagnetic seals have sub-

[13] A. Taylor, *J. Sci. Instrum.* **26,** 225 (1949).

[14] First designed by Ferrofluidics Corp., Nashua, New Hamphsire.

stantially longer lifetimes, eliminate seal-oil contamination of the vacuum chamber and anode surface, and allow a higher tube vacuum to be maintained, thus prolonging filament lifetimes. For most workers using Elliott GX-6 and GX-20 or Rigaku 12-kW generators, the very significant reduction in machine downtime possible on average with magnetic seals will be the most important consideration. Both types of seals can be removed and replaced quickly; any difference in the time required for refurbishing is not a significant fraction of total machine downtime. Users who choose to rebuild their own magnetic seal assemblies, rather than returning them to the manufacturer for recharging, will need an electromagnet large enough to accommodate the assembled seal between the pole pieces in a field greater than about 1.0 T, because most seal units must be remagnetized after assembly.

Expected lifetimes for seals under correct operating conditions are still finite. For oil seals running at 5 $msec^{-1}$ or less, lifetimes are between 1 and 2 khr (kilohours). For oil seals at 7 $msec^{-1}$, lifetimes are 0.7–1.4 khr. These values may be somewhat conservative, as some users have reported longer average lifetimes. For properly assembled magnetic seals at 4 $msec^{-1}$, expected lifetimes are 3–6 khr. Lifetimes $>$ 10 khr have been reached in the author's laboratory. At 7 $msec^{-1}$, lifetimes are 2–4 khr.

Optimizing Seal Performance

OIL SEALS. Seal lifetime is dependent on having the correct fit between the seal lip and the sleeve. The fit is determined by the inner diameter of the lip, the elasticity and composition of the rubber, and to a lesser degree the lip compression provided by a removable metal spring within the seal. Experience has shown that all of these factors are variable, even for seals from the same batch from the same manufacturer. It is, therefore, very useful to measure the inner diameter of seals, and to reject those incorrectly sized. Seals with the correct nominal dimensions are generally available from several manufacturers through local distributors. Many users find these seals, which have somewhat different composition and are intended for use in automobiles, etc., superior to those sold by Elliott. (The steel spring provided in some should be replaced by a bronze spring, or removed.)

Lifetime will be extended if seals are restarted very slowly. Turning the anode by hand for several revolutions before starting the drive motor will accomplish this. To avoid any chance of severe wear which can occur at start-up, the anode can be driven continuously at low speed when the the X-ray power is off. This will most likely prolong seal lifetime and reduce oil leakage into the vacuum chamber.

Seal performance and lifetime may be significantly reduced if the cooling-water temperature falls below ~20° or rises above ~35°, as the physical characteristics of the rubber change appreciably over this water temperature range. Some users have found that seal performance has improved when the viscosity of the seal oil (usually Apiezon C) was increased. Well-outgassed Apiezon J can be mixed with C oil to increase viscosity.

Two successful methods for improving the tube vacuum and reducing oil contamination of the tube chamber have been worked out. The first method, suggested by Wilson[15] and developed by Hackert,[16] utilizes differential pumping of the seal oil reservoir. The space between the two seals is evacuated by a mechanical pump. In this way the pressure differential across the inner oil seal is effectively eliminated, and oil is no longer forced past the seal and into the tube chamber. The second method is even more straightforward: the oil is simply removed from the reservoir. The rubber and sleeve are coated with oil when the seal is assembled, but no additional oil is added to the reservoir. No decrease in seal life is reported, and far less oil enters the tube.

MAGNETIC SEALS. Unlike oil seals, magnetic seals are not affected by rapid starting. Therefore, the most obvious way to increase seal lifetime is to turn off the anode drive whenever the X-ray power is off. Bearing and magnetic fluid lifetimes decrease as the anode speed increases. This should be kept in mind, because in many instances operating at reduced speed (power) does not restrict data collection.

Magnetic seals operate best when the anode cooling water temperature is between 15° and 35°. If the water temperature exceeds ~40°, seals may fail rapidly as the seal fluid deteriorates. An interlock which turns off the anode drive when the recirculating water temperature exceeds ~35° can prevent seal failures. Under proper operating conditions, the bearing grease must be replenished before the magnetic fluid. If the bearings are not regreased, they will generally fail before the vacuum seal. Bearings should be regreased at 1- to 2-khr intervals with ~20 mg of grease per bearing. A good high-temperature vacuum grease is available from Elliott, BP Energrease HTB2. In the seals designed for Elliott machines, the bearings can be regreased without removing the seal from the anode plate. Therefore, when the tube is disassembled for a filament change or other maintenance, the bearings should be regreased.

Two considerations about the bearings are important for users rebuilding their own seal assemblies. First, precision bearings, ABEC 5 or preferably ABEC 7, will maximize bearing lifetime. The additional cost of high-precision bearings can thus easily be justified. Second, bearings must

[15] R. R. Wilson, *Rev. Sci. Instrum.* **12,** 91 (1941).
[16] M. Hackert, private communication (1976).

be greased with the correct amount of lubricant; too little will cause premature bearing failure, and too much will cause seal failure when the excess grease is spun out of the bearings and contaminates the magnetic fluid. Grease mixed with the particle-containing oil produces an increase in viscosity and causes rapid seal failure.

The axis of the bearings and the pole pieces must be concentric, because the dimensions of the small gap (~60 μm in seals designed for the Elliott) must be closely maintained. To ensure concentricity, bearings and pole pieces should be a snug fit (light press fit) in the rotor plate. This will also ensure that the bearing outer races are held stationary. In order to maintain the correct tolerances, care should be exercised when reassembling the seals. Having the proper tools will prove to be worthwhile.

With magnetic seals there is a pronounced reduction of the contaminant layer buildup on the anode surface. The layer on anodes in use for several thousand hours can be removed in a few minutes with a very fine (4/0) polishing paper while the anode is slowly rotated in a lathe. When new, a magnetic seal will maintain a pressure differential of 2–3 atm. The greatest drawback with the magnetic seals is catastrophic failure, which occurs at the moment when the seal is no longer capable of supporting a pressure differential greater than 1 atm. If this happens rapidly enough, the inrushing air badly oxidizes the filament and often the focusing cup, and may deteriorate the diffusion pump fluid. Such failures can usually be predicted, and therefore prevented, if the tube vacuum is regularly monitored. Many hours before total vacuum failure, seals will generally start to leak sporadically. This will produce fluctuations of $\sim 1\text{–}5 \times 10^{-6}$ Torr in the tube vacuum. Vacuum will also deteriorate rapidly, $\sim 1\text{–}5 \times 10^{-5}$ Torr, when the anode is started after being stationary for a few minutes or longer. In order to prevent catastrophic failure, it is advisable to closely monitor, or rebuild straightaway, a seal with those vacuum leaks.

Filaments

In order to achieve electron currents greater than about 20 mA from tungsten filaments, a helical coiled filament must be used, rather than a straight-wire filament. Emitted electrons are focused by an electrode (focus cup) surrounding the filament, to which a bias voltage (usually negative) can be applied. The bias voltage and the shape of the electrode determine the spatial distribution of electrons (space charge) within the cup. The space charge modifies the trajectory of electrons emitted from the filament. The filament is positioned in an aperture in the cup in order to limit electron emission to the exposed surface of the filament. These electrons have initial velocities which can give rise to focused trajectories.

Although filaments occasionally break due to local stresses or imperfections, most failures are caused by vaporization or erosion of tungsten, or by chemical modification of the tungsten produced by interaction with contaminants in the vacuum chamber. Stresses and dislocations can be reduced with the proper heat treatment, but flaws beyond the users' control will produce some failures. However, by far the most common causes of filament failure are poor vacuum and positive-ion erosion. To realize long lifetimes the vacuum should be $\sim 1\text{–}2 \times 10^{-6}$ Torr or less above the diffusion (or turbomolecular) pump at full X-ray power, and oil contamination of the chamber reduced as much as possible (see preceding sections). For the Elliott GX-20 generator operated at currents of 20–30 mA, expected filament lifetimes are ~2 khr, and at 40–50 mA, ~1.5 khr. For the Rigaku 12 kW generator, expected lifetimes at 200 mA can be >3 khr.[12]

When a high vacuum is maintained, fine-focus filaments usually fail when positive ions erode the tungsten in a thin line running lengthwise along the filament. The ions are produced at the surface of the anode by electrons, and accelerated toward the filament by the electric field. Filament lifetimes can be increased substantially if positive-ion bombardment of the filament is eliminated. This can be accomplished by deflecting the electron beam with a magnetic field produced by a magnet on the outside of the tube, as shown in Fig. 2.[17] When the electron beam (focal spot) is deflected along the anode circumference by slightly more than half the filament width, the ion beam goes beyond the edge of the filament and through the aperture in the focus cup. The trajectory of the ions is practically unaffected by the magnetic field, because the ion mass is very much greater than the electron mass. A magnetic field of ~4 mT measured between the focus cup and anode surface produces a focal spot displacement ~1 mm at 40 kV. With a magnet optimally positioned, there is no evidence of ion bombardment of filaments run longer than 5 khr at 40 mA on Elliott GX-6 machines. In our laboratory, the use of deflecting magnets has increased filament lifetimes, on average, by more than a factor of 2.

Focal Spot Shape

At a given ϕ, the beam flux from a rotating anode can be increased by increasing the length L of the focal spot line to L', and increasing the loading by L'/L. To regain the effective focal spot size, ϕ must be reduced by L/L'. When ϕ is reduced, beam attenuation from both self-absorption in the target and surface roughness is increased. The actual ϕ and the condition of the anode surface determine whether there will be a net gain

[17] W. C. Phillips, *J. Appl. Crystallogr.* **13,** 338 (1980).

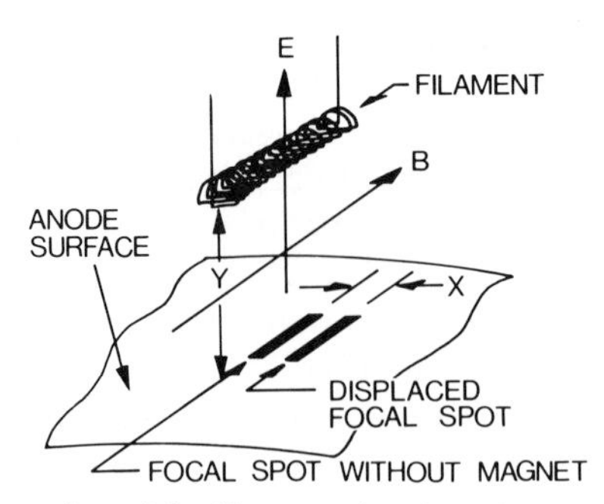

FIG. 2. Relative orientation of the filament, electric and magnetic fields, and focal spot with a beam-deflecting magnet in place.

in beam flux. Using a Cu anode, focusing double-mirror camera, and Elliott GX-6, the original focused-beam size at the film can be recovered with an increase in flux of ~35% in going from a 0.2 × 2.0-mm spot (X-ray power at 30 kV × 30 mA) to a 0.2 × 3.0-mm spot (X-ray power at 30 kV × 45 mA), for ϕ equal to 2.3° and 1.5°, respectively. Elliott 0.1 × 1.0 and 0.2 × 2.0-mm focusing cups can easily be modified to give ~0.1 × 2.0 and 0.2 × 3.0-mm focal spots.[17] The width of the 0.2-mm spot can be reduced by ~25% to 0.15 mm by adjusting the bias voltage, focus cup-to-anode distance, and filament height in the cup. Thus, for example, the 0.2 × 2.0-mm focal spot can be changed to a 0.15 × 2.6-mm spot which would be more useful for many experiments.

Anodes

The anode should be cleaned when a noticeable accumulation of surface contaminants is visible in the focal spot ring, usually whenever the filament or seals are replaced. Cleaning should be with a mild abrasive; coarse polishing papers can leave deep scratches or remove material unevenly, causing possible surface eccentricity. A satisfactory procedure for cleaning ~10-cm diameter anodes is to hold the anode shaft in a collet and turn the anode slowly in a lathe while holding an emery polishing paper against the anode surface with a flat metal bar. Emery polishing papers grade 2/0, 3/0, and 4/0, used in succession, starting with 2/0, with or without propanol solvent, work well. Care should be taken to prevent large pieces of dirt from falling onto the anode while it is being polished. Anodes should not be cleaned in ultrasonic cleaning machines, because anode metal will be preferentially removed from flaws and regions of high dislocation density by the process, leading to possible microchannels through the anode which would allow water vapor to enter the tube.

If the surface at the electron beam track becomes badly pitted or deeply scratched, or the anode becomes eccentric, a Cu or other solid metal anode can be resurfaced by removing ~50–100 μm from the surface on an accurate lathe with a diamond cutting tool. The anode must be held in a good collet with the driving shaft axis accurately on the axis of the lathe.

The run-out at the anode surface should be measured regularly using a precision dial indicator with the anode assembled in its bearings in the anode rotor plate. Run-out should be ~10 μm or less when small focal spots are used. Anode run-out increases the apparent focal spot size. A 1-mm long spot viewed at $\phi = 3°$ has a projected size of 50 μm. A run-out of 25 μm (0.001 in.) will thus increase the effective spot size by 50%. As pointed out above, operating at only ~80% of the manufacturer's maximum loading will significantly reduce roughening of the anode surface, leading to greater X-ray output at small ϕ over extended operating periods.

Other Operating Procedures

The time to pump down will be considerably shortened if only a minimum volume of air is allowed into the vacuum chamber. A simple way to accomplish this is to bleed nitrogen or argon into the chamber when the vacuum pumps are off, and maintain a slight positive bleed-gas pressure when the cathode, anode plate, etc. are removed for service. Blanking plates can be fitted in place of the cathode and anode plate, and the vacuum pumps turned on while the anode and cathode are being serviced. A relatively long time is required to pump out the water vapor which is adsorbed from the air onto surfaces in the vacuum system. After the anode or tube chamber are cleaned, a final washing with methanol will remove most adsorbed water. Deposits on the Be windows should be removed occasionally by chemical polishing. If an appreciable tungsten layer builds up on the window, X-ray transmission will be significantly reduced.

The breakdown of the insulating potting compound used in cathode guns can be prevented if the cathode temperature is kept low. For air-cooled cathodes, a small fan circulating air over the cathode is effective in cooling the cathode.

After a new filament has been brought up to operating power, pinhole photographs of the focal spot should be taken as a function of cathode rotation, focus cup-to-anode distance (where allowed by the cathode design), and bias voltage, in order to optimize the focal spot shape. Unless these adjustments are made and the results observed photographically, the spot shape may deviate very significantly from the nominal shape. Occasionally, a new filament will be sufficiently distorted by internal stresses, or by the recrystallization procedure, that a satisfactory focal spot cannot be produced. A very convenient method for making the photographs is the Polaroid XR-7 cassette with Polaroid 52 film. Care should be taken not to overexpose the film.

[23] A Systematic Method for Aligning Double-Focusing Mirrors

By WALTER C. PHILLIPS and IVAN RAYMENT

Introduction

Double-focusing mirrors are now commonly used for X-ray diffraction studies requiring high order-to-order resolution. These systems, which utilize total reflection from two bent mirrors to focus the X-ray beam, have several advantages over systems using either Bragg diffraction from mosaic crystals or simple collimators, particularly for investigations of crystals with large unit cells.[1] Double-mirror systems provide small, intense X-ray beams with a narrow angular divergence, uniform beam profile, and low parasitic background intensity. A double-focusing mirror device for X rays was first described by Kirkpatrick and Baez.[2] Franks[3] constructed a system for X-ray diffraction, and Harrison[4] first used double-focusing mirrors for X-ray studies of crystals with large unit cells.

We present here a systematic method for optimizing the alignment of a double-mirror system. This procedure has been developed over the past 7 years in our laboratory where presently nine such systems are in use. The method is illustrated with reference to mirror assemblies designed and constructed at Brandeis University[5] and a rotating anode X-ray generator with the effective anode surface horizontal (i.e., the cathode in the vertical position). However, it is straightforward to extend the method to other mirror assemblies and anode configurations.

[1] U. W. Arndt and R. M. Sweet, *in* "The Rotation Method in Crystallography" (U. W. Arndt and A. J. Wonacott, eds.), p. 59. North-Holland Publ., Amsterdam, 1977.

[2] P. Kirkpatrick and A. V. Baez, *J. Opt. Soc. Am.* **38,** 766 (1948).

[3] A. Franks *Proc. Phys. Soc, London Sect. B* **68,** 1054 (1955).

[4] S. C. Harrison, *J. Appl. Crystallogr.* **1,** 84 (1968).

[5] Double-mirror assemblies have been constructed in the Brandeis University Machine Shop for more than 25 other laboratories. Improved assemblies based on this design are now available from Charles Supper Co., Natick, MA 01760.

Mirror Assemblies

A typical double-mirror assembly is shown in Fig. 1. The nickel-coated glass optical flats are held in two perpendicular housings which allow each mirror to be translated vertically and horizontally, bent for focusing, and rotated with respect to the incident beam. A differential screw provides the necessary fine angular adjustment of the mirror position. The mirror curvature is adjusted by a fine-threaded screw. The mirrors rest on rods which in our design are slotted and can be used as beam-defining apertures. Slits between the source and the first mirror serve to define the horizontal extent of the source image. Slits beyond the second mirror define and guard the doubly reflected beam. The entire apparatus is fastened to a block or plate which is screwed to the generator bench. A labyrinth of nearly concentric lead cylinders forms a radiation

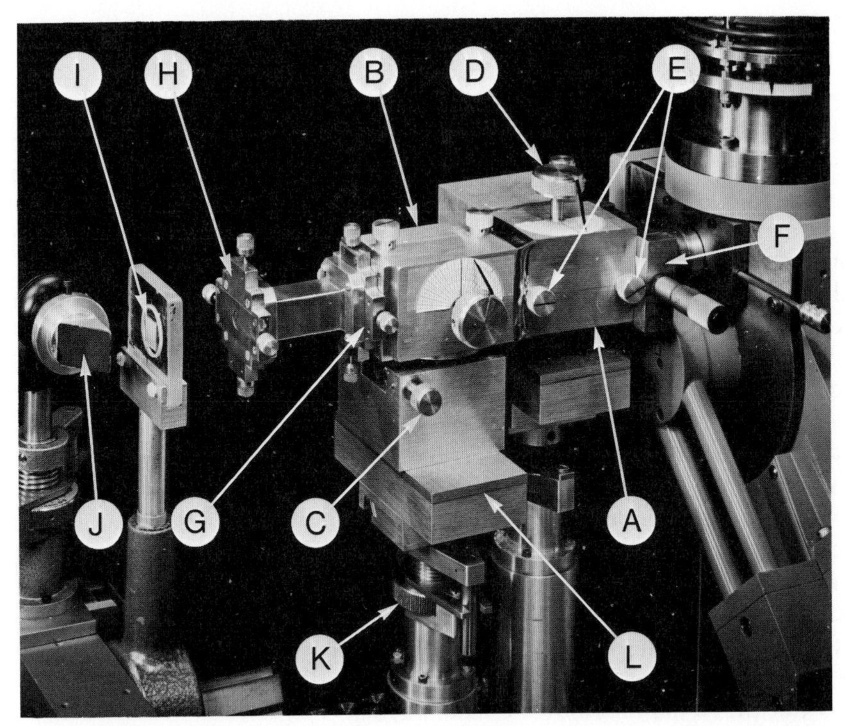

FIG. 1. Double-mirror assembly mounted on an Elliott X-ray generator: (A) first mirror housing; (B) second mirror housing; (C) mirror rotation-angle adjusting screw; (D) mirror curvature (focus) adjusting screw; (E) knobs on slotted rods; (F) source defining slit; (G) second defining slits; (H) guard slits; (I) transparent fluorescent screen mounted on leaded glass; (J) microscope with 90° prism; (K) mirror height adjustment; (L) mirror-translating slide.

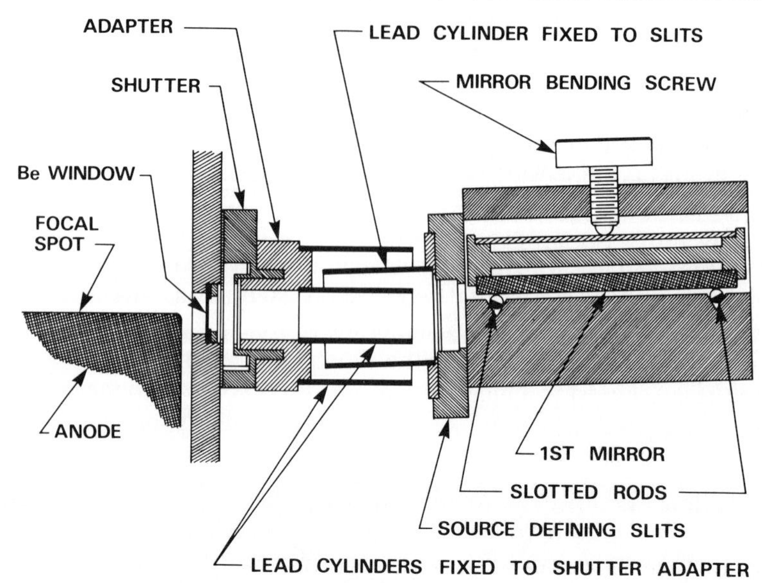

FIG. 2. Cross section of the first mirror housing and the coupling between the source-defining slit and shutter.

shield between the port and first set of defining slits (Fig. 2). The cylinders allow for some side-to-side motion and rotation of the first mirror with respect to the port.

Each mirror acts effectively in one dimension like a simple lens. Thus the magnification of the source is given by the ratio of the distance between the mirror and the film to the distance between the mirror and the source. The image of the source at the film will be most magnified by the mirror nearest the source. Therefore, it is generally advantageous to place closest to the source (first mirror) the mirror which defines the take-off angle (the angle between the anode surface and the beam incident on the mirror). Then, by decreasing the take-off angle, the effective size of the source seen by this mirror is decreased, and thus the image of the source at the film. With the second mirror, the only way to decrease the effective source size is to move the mirror further from the source along the beam axis, or move the film closer to the mirror.

Monitoring the Process

It is important to have a reproducible method of seeing what one is doing. We use two methods: (1) photographs and (2) a semitransparent fluorescent screen[6] viewed with a traveling microscope.[7] Both techniques lead to the same endpoint. Overexposure and short development times (15–30 sec), and the use of an X-ray film such as Kodak SB, which clears rapidly in the fixer, considerably shorten the film processing time. In order to accurately evaluate the beam profile image, it is necessary to keep the maximum optical density below 3 by controlling the development time.

We use a low-energy X-ray G-M counter with calibrated absorbers to measure the beam intensity. The counter is particularly useful when adjusting the guard slits and checking the final beam flux.

Method

The method is summarized in the flow chart in Fig. 3. A detailed explanation of the individual steps is given in the following sections. It is expected that the filament is well aligned in the focusing cup, the bias optimized, and the cathode rotation reasonably well adjusted using pinhole photographs. No amount of effort will compensate for a poorly defined focal spot.

Preliminary Alignment

Position the mirror assemblies by eye so that, when all the translation adjustments are at the center of their ranges, a line extending out from the focal spot line position goes down the center line of both mirror surfaces.

When aligning the mirrors for the first time, start with only the first mirror assembly (with radiation shield) in place. Rotate the slotted rods so that there is a maximum clearance between the mirror and the slots, and open all slits. Place a fluorescent screen beyond the mirror assembly. Run the generator at low power and rotate the mirror until two beam lines can be seen on the screen. (Our mirror can be rotated coarsely by hand after the locking screw has been released, or finely using the rotation screw.) If no lines are found, raise the mirror a few millimeters (i.e., increase the take-off angle by ~1°) and search again for the beams. X Rays passing straight through both slots form the upper (direct) beam; X rays reflected from the mirror surface form the other (reflected) beam. Once the two

[6] Our fluorescent screens are made of $La_2S_2O(Tb)$ deposited on glass substrates to a thickness of ~7 mg/cm^2. They are semitransparent and are viewed from the back (i.e., the side away from the source). These screens are about four times brighter than 0.3–0.5 mm thick aluminized CsI(Th) scintillating crystals.

[7] Our microscopes have ~2× power objectives with about a 7-cm working distance, an 8× or 10× eyepiece, and a right-angle prism in front of the objective (see Fig. 1).

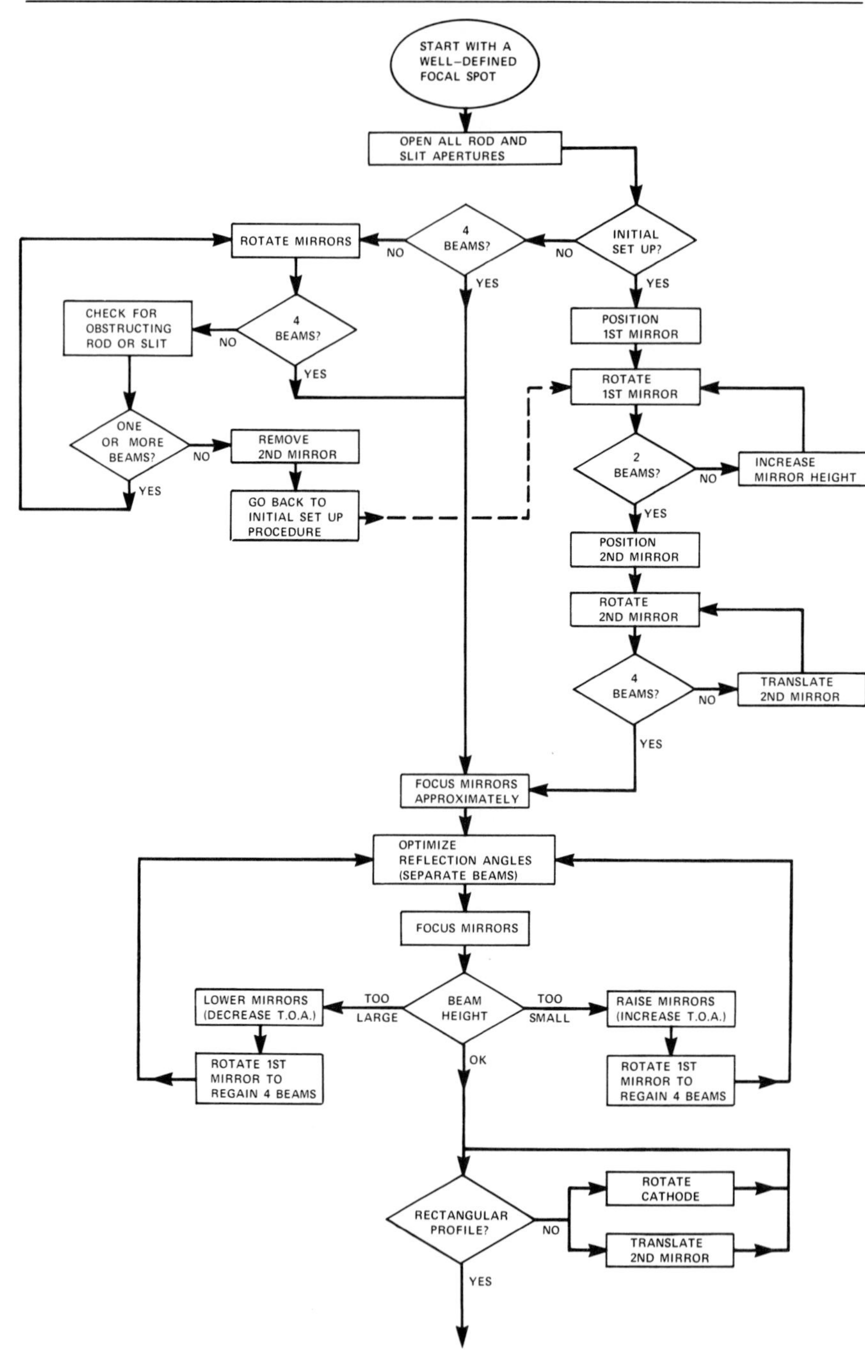

FIG. 3. Flow chart of the method.

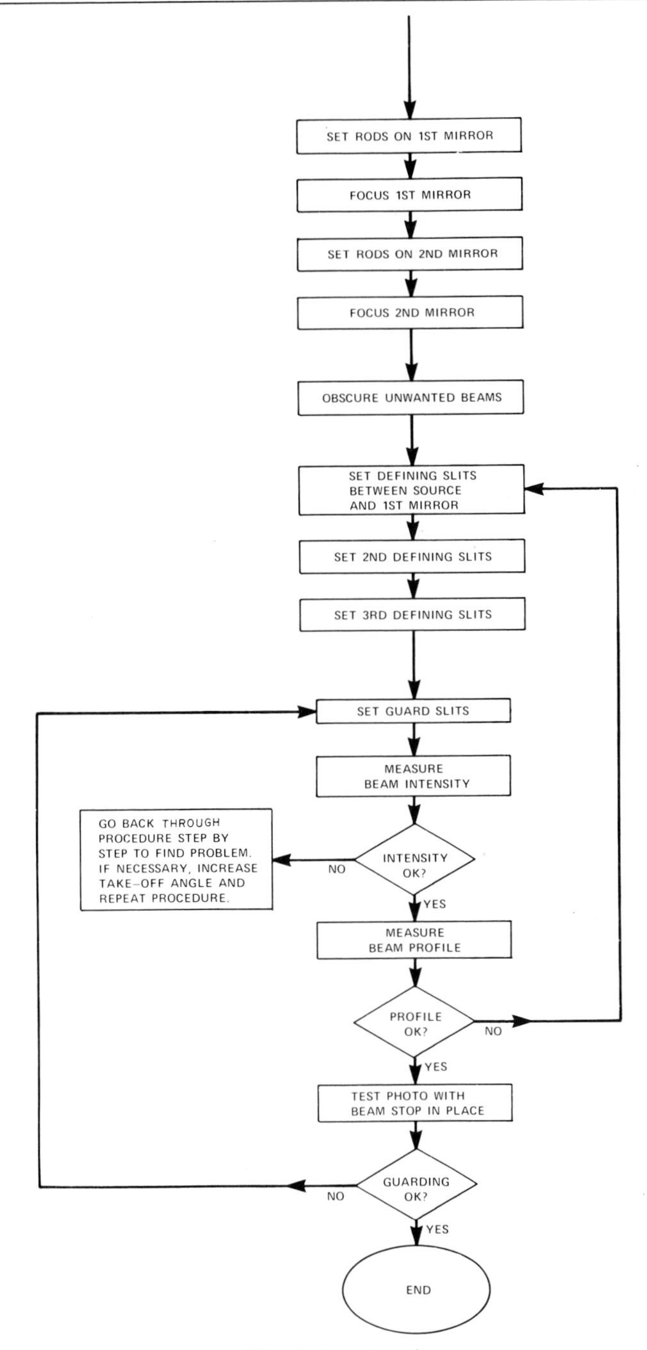

FIG. 3. (*continued*)

beams are found, put the second mirror in place, with a gap of ~5 mm between the two mirror assemblies. Rotate the slotted rods for maximum clearance and open all slits. By adjusting the height of the second mirror, align the center line of the mirror with the beam reflected from the first mirror. Place a fluorescent screen after the second mirror and increase the generator output to the normal operating power. Rotate the second mirror; at the appropriate angle four beam spots should be visible on the screen. These arise because the transmitted and reflected beams emerging from the first mirror are each transmitted and reflected by the second. (For safety, a thin lead sheet should be wrapped around the gap between the two mirror housings to eliminate scattered radiation.)

Focusing the Beam

The X-ray beam is focused by bending the mirrors, using the screws perpendicular to the mirrors. Ideally, each mirror surface is formed into an ellipse with the source and the film at the two foci. In order to optimize the curvature, the beam shape must be monitored by viewing the double-reflected beam at the film position (focal plane). Decide on the camera length, place either a fluorescent screen (or scintillating crystal) or test film at the film position, and adjust the tension on each mirror until the smallest spot is found. Using a screen or crystal, the image is viewed through a microscope in a darkened room. Using film, a series of beam photographs is recorded as the mirror tension is changed by small increments. (A scale fixed to the mirror-bending screw aids in reproducing the tension settings.)

When the mirror is asymmetrically located between the source and image plane, the mirror should in principle be asymmetrically positioned on the rods so that each end of the mirror is bent to a different curvature. In practice, for cameras with focal lengths less than 30 cm, the improvement in the size of the focused beam when the mirror is positioned asymmetrically will be noticeable only for the first mirror. The optimum displacement of the mirror from the symmetric position will be only a fraction of a millimeter, and can be determined by focusing the beam as described above for a series of mirror positions.

Adjusting the Take-Off and Reflection Angles

As the take-off angle is increased by increasing the height (distance above the bench) of the first mirror, the apparent length of the focal spot seen by the mirror increases, resulting in a corresponding increase in the vertical dimension of the source imaged by the mirror at the focal plane. The intensity of the beam incident on the mirror also increases as the take-off angle increases. This is because, at small take-off angles, X rays produced at the source are partially absorbed by the anode, an effect due primarily to surface roughness. The relationship between intensity and take-off angle depends on the quality of the surface and the amount of tungsten deposited on the surface. The intensity of K_α radiation from a Cu anode versus take-off angle after ~100 kW-hr running time is shown in Fig. 4a. Setting the take-off angle amounts to making a compromise between the intensity and height of the focused beam. The take-off angle

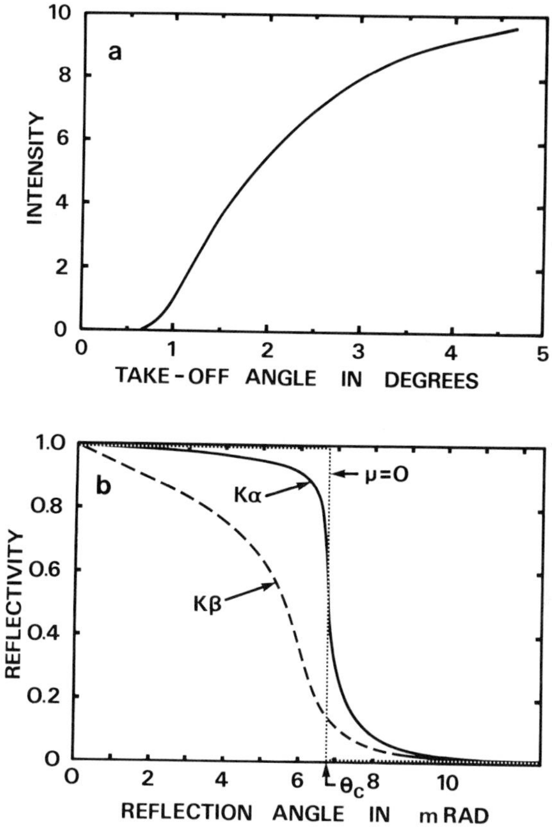

FIG. 4. (a) Intensity of the focused doubly reflected beam vs take-off angle, measured by counting X rays diffracted into a small solid angle from a polyethylene sample, using a proportional counter set to record the Cu K_α wavelength. The generator was operated at 30 kV with a source size of 0.1 × 2.0 mm, and the Cu anode had run ~100 kW-hr since it was polished. (b) Reflectivity calculated for Cu K_α and K_β wavelengths from a flat Ni-coated mirror vs reflection angle. We assume a perfect reflecting surface and perfectly collimated, monochromatic incident beam. The reflectivity at the K_α wavelength for a hypothetical Ni mirror without absorption is also shown. Measurements of the reflectivity at the K_α wavelength for Pyrex optical flats with Ni coatings of ~2000 Å thickness are in excellent agreement with the calculation.

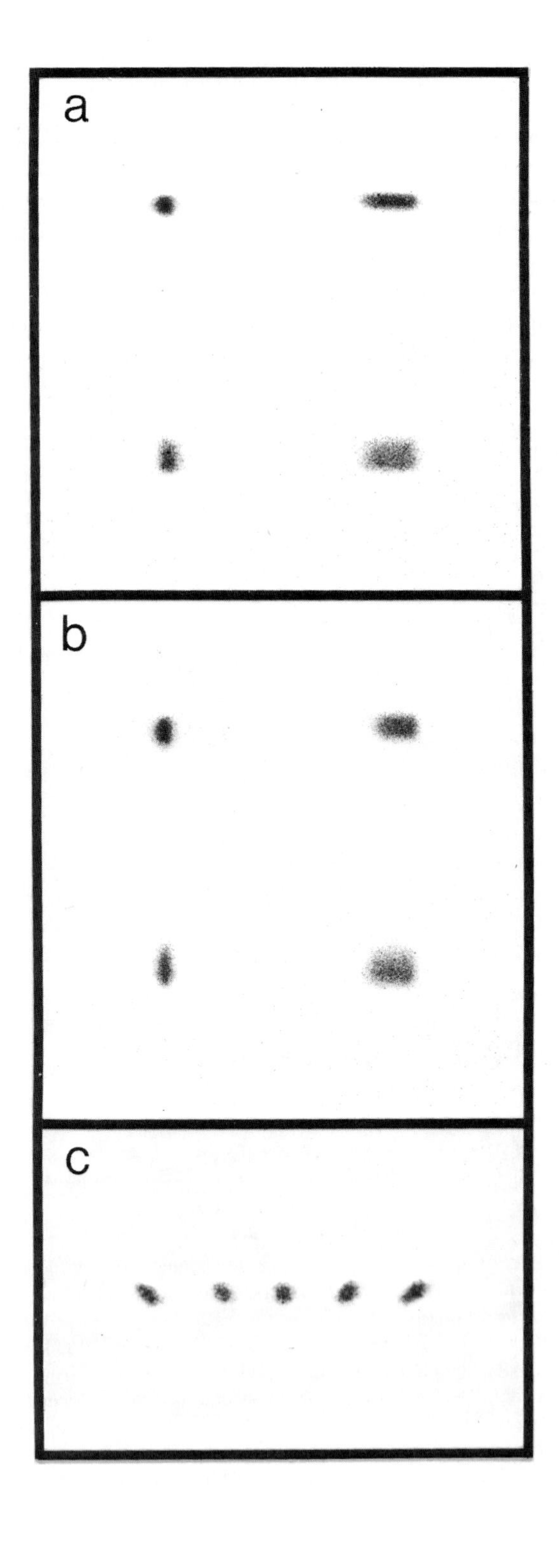
a
b
c

adjustment is usually only necessary the first time the mirrors are set up, or thereafter if a smaller beam or greater intensity is required.

The reflection angle is the grazing incidence angle of the beam onto the mirror, i.e., the angle between the mirror surface and the source to mirror direction. The reflectivity of the mirror depends on the reflection angle, the incident wavelength, the flatness of the glass substrate surface, and the quality, thickness, and atomic number of the coating. The reflectivity of Ni-coated mirrors for Cu K_α and K_β radiation is shown in Fig. 4b, together with the reflectivity calculated for a mirror without absorption. With finite absorption, the change in reflectivity occurs over a range of angles on either side of the critical angle θ_c. As the reflection angle increases from 0°, the solid angle from the source subtended by the mirror increases, and the intensity of radiation reflected increases until the critical angle is approached. Thus, the optimum setting for the mirror rotation is just below the critical angle, at which point the K_α intensity and the K_α/K_β intensity ratio will be maximized. Although the Ni-coated mirrors do act as Cu K_β filters, we find that for most experiments an additional Ni foil filter placed between the source and first mirror is necessary to further reduce the K_β radiation.

We assume the preliminary adjustment of the mirrors as described in the section on Preliminary Alignment has been completed and all slits and slotted rods are wide open. Four beams should be visible on a screen or film placed at the image plane. Bring the reflected beams into approximate focus as described in the section on Focusing the Beam. Adjust the mirror rotation angles, i.e., separate the direct and reflected beams as far as possible without a significant loss of intensity in the reflected beams, as described above (see Fig. 5a and b). Refocus the reflected beams. If the height of the doubly reflected beam is smaller than necessary for the order-to-order resolution desired, a more intense beam can be obtained by increasing the take-off angle and readjusting the reflection angle of the first mirror. Conversely, if the beam height is too large for the resolution required, the height can be reduced by decreasing the take-off angle and resetting the first-mirror reflection angle. It is best to change the take-off angle by small increments. Raise (or lower) the mirror assemblies by approximately 0.5 mm and reset the reflection angle. Repeat the procedure until a satisfactory compromise between beam size and intensity is

FIG. 5. (a) Photograph of the four beams separated and focused, at a 1.5° take-off angle, with all rod and slit apertures fully opened. The apparent relative intensities of the beams have been photographically altered for illustrative clarity. (b) The four beams at a 2.5° take-off angle. (c) Doubly reflected, focused beam photographed as a function of cathode rotation angle at 1° increments, with the three unwanted beams masked off.

reached. We generally use take-off angles between 1.5° and 3° for cameras with ~20 cm sample-to-film distances.

Beam Shape Optimization

The shape of the focused beam is a parallelogram whose included angle is determined by the position of the second mirror relative to the axis of the source line. A rectangular beam image is formed when the second mirror views the source directly down the axis of the focal spot line, i.e., the surface of the mirror is in the plane defined by the perpendicular to the anode surface and the source line. A rectangular profile is usually desirable, especially when crystallographic film data are to be processed.

The tilt of the beam shape parallelogram can be changed by either (1) rotating the focal spot line by rotating the cathode (and thus the filament), or (2) translating the second mirror. Although rotating the cathode is the quickest method, it will clearly be satisfactory only for the first camera port to be aligned. Before any cameras are put in place, it is generally advantageous to adjust the cathode rotation using pinhole photographs of the source so that the focal spot line axis is within a few degrees of the line between the centers of the two opposite ports.

In order to find the cathode position which produces a rectangular spot, make a series of photographs of the focused beam versus the cathode rotation angle. An example of this process is shown in Fig. 5c. It is convenient to fix a scale and pointer to the cathode in order to provide an angular calibration. Alternatively, the beam can be viewed directly as the cathode is turned using a microscope and fluorescent screen or scintillating crystal.

In order to find the horizontal position of the second mirror which produces a rectangular spot, translate the mirror and make a series of photographs of the focused beam versus the mirror position (or observe the beam using a screen and microscope). Note that, because the reflection angle is also a function of the mirror position, the second mirror rotation angle must be reset each time the mirror is translated ~1 mm. (If the second mirror is translated several millimeters, it may also be necessary to translate the first mirror to ensure that the two first-mirror beams strike the second mirror.)

Defining and Guarding the Beam

The objectives of these adjustments of the rods and slits are to define the edges of the beam and to remove unwanted diffuse scatter. The result is a more useful focused-beam profile and a lower camera background level, particularly at small angles. The procedure is outlined in the flow chart of Fig. 3.

For assemblies with slotted rods, the first step is to set the two rods on the first-mirror assembly. The rod nearest the source should be rotated in order to remove as much of the direct beam as possible, but without reducing the intensity of the reflected beam. The second rod can then be rotated slightly to further remove scattered radiation. The most convenient way to visually monitor the rod adjustment is with a fluorescent screen and microscope. Care should be taken to avoid cutting into the reflected beam. Once the rods are adjusted, the mirror should be accurately refocused (see section on Focusing), because rotation of the rods may significantly defocus the reflected beam. On assemblies with a slit between the mirror rods, adjust the slit to remove as much unwanted radiation as possible.

Adjust the rods or slit on the second mirror in an analogous way. It is unlikely that the beam which is not reflected by the second mirror can be totally attenuated by the rods without reducing the intensity of the doubly reflected beam. In order to carry out the subsequent adjustments, it is useful at this point to eliminate the remaining part of the unreflected beam by using one of the horizontal slits beyond the second mirror. Advance the slit just far enough to cut out the unwanted beam, but not enough to reduce the intensity of the doubly reflected beam.

The next step is to limit the horizontal extension of the source which is seen by the mirrors using the defining slits between the source and first mirror. The source itself has ill-defined edges because not all the electrons emitted by the filament are sharply focused. If this radiation from either side of the prominent source line is not removed by the slits, it will be imaged by the second mirror and give rise to horizontal "tails" on the focused beam. The slits should be adjusted so that they remove the tails but do not appreciably cut into the main beam. Generally about 4–8% of the X rays emitted from the target originate in the tails. There are three ways to assess how the slits are defining the beam edges: (1) if the observer is thoroughly dark adapted, the tails can be seen directly on an efficient fluorescent screen viewed through the microscope; (2) a series of photographs of the focused beam can be made as the slits are incrementally advanced into the beam; (3) the beam intensity can be monitored with a radiation counter as the slits are adjusted.

The final step is to adust the defining and guard slits beyond the second mirror. In order to remove parasitic small-angle scatter originating primarily at the surfaces of the mirrors, rods, and slits, and in the air, at least one set of slits (a horizontal and vertical pair) is required, two sets are certainly desirable, and a third set, spaced 5–30 cm from the second, is necessary for critical small-angle measurements. The more slits, the lower the background and the better defined the beam. When correctly set, none of the slits will appreciably diminish the beam intensity. For oscillation or precession photography, the crystal camera collimator serves as a final slit set. The slit nearest the specimen is the guard slit; the other slits are defining slits. These are adjusted sequentially, starting with those closest to the second mirror.

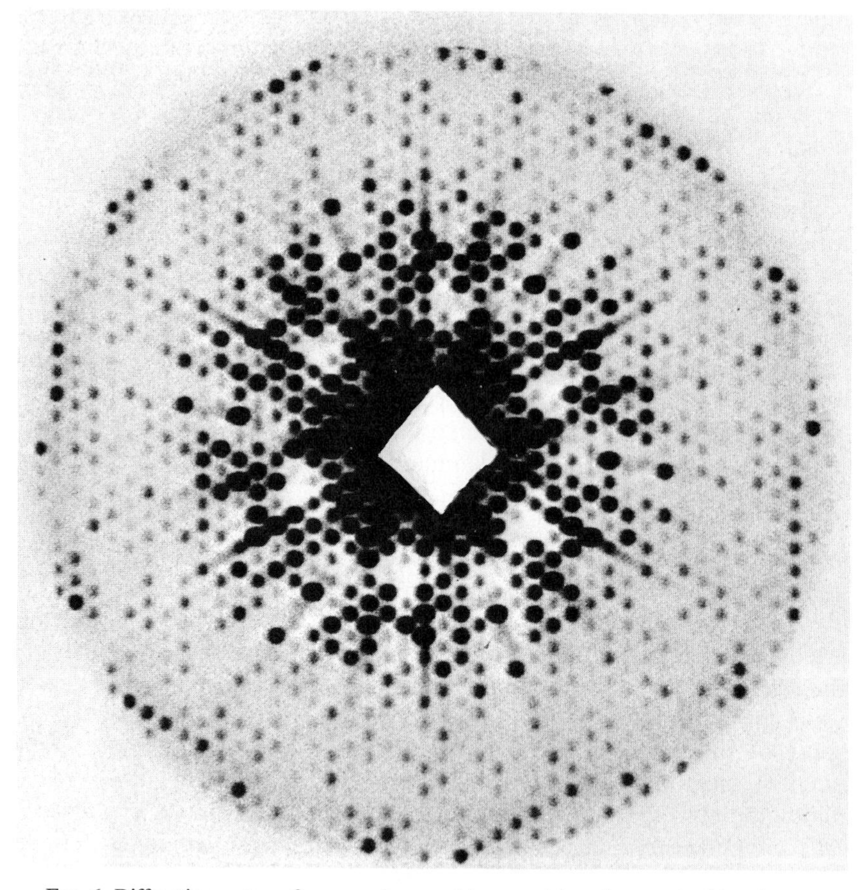

FIG. 6. Diffraction pattern from a polyoma virion crystal made on a double-mirror camera with a well-defined, well-guarded beam. The source focal spot size is 0.15 × 3.0 mm, crystal-to-film distance 150 mm, and exposure time 46 hr at 30 kV and 40 mA. The crystal is oriented along the [111] direction, and the precession angle is 2°. The minimum spacing between reflections is 404 Å.

The defining slits should be set so that they just intercept the edges of the beam. If they cut into the beam appreciably, intensity will be lost and there will be appreciable small-angle scattering from the slits. If they do not come up to the beam edges, the focused beam may be ill defined. The guard slits should be set so that they come up just to the edge of the beam defined by the defining slits, but do not quite cut into the beam. The guards should remove scatter around the beam, but not generate any additional scatter.

The defining and guard slits can be adjusted using a radiation counter, or, if the observer is dark adapted, using a bright fluorescent screen placed as close as possible to the slit being set and viewed through a microscope.

After all slits have been set, the focused-beam intensity, shape, and guarding should be checked. The intensity is easily measured using a counter and calibrated absorbers. Alternatively, a timed photographic exposure made with a standard specimen and film-to-specimen distance can be compared to similar exposures made previously on this or other cameras. The beam profile and guarding should be checked photographically. Usually a 30-min photograph with the beam stop in place, and without a sample, will allow a satisfactory assessment of the guarding.

To go through the entire alignment procedure we have outlined usually takes several hours, although the first time through it will probably take 2 days. The method yields reproducible beam profiles and beam intensities. After the filament has been replaced and the mirrors and slits completely realigned, we find that the beam intensity returns to within 10% of the value measured when the system was previously aligned (provided the take-off angle and operating power remain the same). The diffraction pattern from a polyoma virus capsid crystal (Fig. 6) made with an Elliott GX-6 rotating anode generator and 200-μm focusing cup at a 15-cm crystal-to-film distance shows that the process can be worthwhile.

Acknowledgments

Eaton E. Lattman was instrumental in systematizing the alignment method. Much of the procedure described here comes directly from his ideas and effort. Charles Ingersoll, Jr. did most of the design and development work on the mirror assemblies. His ingenuity and sense of practical design were invaluable. This work was supported by a grant from the National Science Foundation Grant No. PCM77-16271 to Dr. Donald L. D. Caspar.

[24] Diffractometry

By H. W. WYCKOFF

Introduction

Comparison of Diffractometry with Other Methods

There are several obvious, and several more subtle, similarities and differences between film methods and single-detector diffractometry. The most obvious difference is that the film records many reflections or partial reflections at one time. But in both methods a given reflection is recorded completely only if the crystal is moved during the measurement or if the source is large enough to provide a stationary crystal integration. In both methods a single X-ray photon gives rise to a detectable event. In the case of film, a silver halide crystal is activated such that later chemical amplification can produce a detectable grain on the film. In the diffractometer the primary event is either gas ionization or molecular excitation in a scintillator. The subsequent gas amplification or burst of light with subsequent electronic amplification leads to a detectable event in this case also. The not-so-subtle difference here is that the chemical fog on the film, that is, the grains that would be observed after development in the absence of X-ray exposure, is equivalent to a large background count, depending on the size of the spot. A fog optical density of 0.04, with a commonly used film, would correspond to a background count of 10,000 if the spot size for a reflection were 0.5 mm in diameter. If the spot is only 50 μm then the equivalent background count of 100 would compare reasonably with a diffractometer.

The diffractometer has another appreciable advantage with respect to background: the detector does not need to be exposed to all of the stray radiation scattered from air and glass near the crystal as is the case with a film. The entrance pupil to the diffracted beam tunnel can shield the detector. Also, the detector aperture can be adjusted critically and moved during the recording of a given reflection on the diffractometer. And finally, the detector can be further from the crystal on a diffractometer and this is useful in reducing background without losing any diffraction data.

The diffractometer allows collection of a limited resolution three-dimensional set of data more easily than film. A preselected set of reflections that are strong on native crystals can be measured on a derivative if desired.

Radiation damage is manifested differently in the two techniques. On a film all reflections are recorded with accumulative damage. On a diffractometer some reflections are measured after more damage than others and then decay corrections are applied. In either case the data are not from pristine crystals. The damage is monitored during a diffractometer run and can be limited by mounting a new crystal when needed and if available.

Surveying for heavy-atom derivatives is classically done with film methods by looking at one plane of data. With a diffractometer a three-dimensional set of 6 Å data can be used instead.

Temporal experiments based on repeated observation of a limited set of reflections can be done with a diffractometer.[1] Binding isotherms can be more easily examined. Diffusion into a crystal can be monitored. Reversible or nonreversible heavy-atom reactions can be followed. Drift of a unit cell can be detected and followed.

Anomalous dispersion data can be better collected on a diffractometer.

Film has the great advantage that simultaneous diffractions are recorded, and much greater efficiency with regard to data vs radiation damage is obtained at high resolution. As implied above, this is partially offset by the chemical fog on a film and the poorer ability to screen against scattered radiation.

The postion-sensitive detector diffractometer systems discussed in other chapters have many of the advantages of both methods. The inability to shield against much of the background pertains to these systems as in the case of film, since the position-sensitive detector is in essence a low-resolution, low-"fog" electronic film.

Single-Crystal Diffractometry: An Overview

Since protein, nucleic acid, and virus crystals have large unit cells, the reflections to be measured are numerous, weak, and closely spaced. Many slight variants on the same crystal form are examined for scientific and technical reasons, such as studying ligand or affector binding on the one hand or screening heavy atoms for phasing on the other. Severe radiation damage limits the data available from each crystal. Background scatter contributes significantly to error and inefficiency.

Resolution of reflections from one another, efficiency and speed of data collection, radiation damage, and background scatter are topics of special concern in single-crystal diffractometry of biological macromole-

[1] H. W. Wyckoff, M. Doscher, D. Tsernoglou, T. Inagami, L. N. Johnson, K. D. Hardman, N. M. Allewell, D. M. Kelly, and F. M. Richards, *J. Mol. Biol.* **27,** 563 (1967).

cules. These lead to special considerations of beam geometries, data collection algorithms, and empirical correction procedures. In addition, there are special topics such as low (or high) temperature techniques, flow cell arrangements, and high-speed arrangements with rotating anode generators, multiple detectors, linear position-sensitive detectors, and area detectors to consider.

The beam geometries are central to considerations of resolution, radiation damage, background scatter, and scan mode design. These will be treated in some detail.

The task of collecting data to higher and higher resolution is influenced by four factors. (1) The total number of reflections and the number needed to increase the resolution of the electron density map by a given percentage increase in proportion to the inverse of d^3, where d is the Bragg spacing. (2) The intensity observed decreases by a fundamental and unavoidable factor which is independent of the structure, called the Lorentz factor, as the resolution increases. The observed intensity is proportional to the inverse of the sin 2θ, where 2θ is the Bragg angle. (3) The intensity is proportional to F^2, the structure-dependent factor, and one of the terms in the expression for F is $\exp\{-B[(\sin\theta)/\lambda]^2\}$ which decreases with increasing resolution. B is related to thermal motion or disorder and is appreciable in protein crystals. (4) The atomic structure factor decreases with increasing resolution due to the diffuse nature of the electron density distribution within the atom. This factor can be approximated well by the sum of a constant and two Gaussian functions of $(\sin\theta)/\lambda$. In terms of the Bragg spacing d, therefore, the raw intensity equation contains one direct factor of d and two involving d, $\exp[-2B/(2d)^2]$, and the atomic structure factor. Figure 1 plots these factors and the resultant effects on measured intensities.

The time required to measure a complete data set or increase resolution by a given percentage with a given statistical accuracy thus varies inversely with the fourth power of the resolution, and additionally by the factors $\exp[B/2d^2]$ and the square of the unitary atomic structure factor at d. Assuming $B = 14$, a task that takes 1 hr at 10 Å will take 1 day at 5 Å, 23 days at 3 Å, and 2 years at 2 Å. With a 10-fold increase in speed these times become 6 min, 2.5 hr, 2 days, and 10 weeks.

Two factors which affect low-order reflections markedly, and reflections up to 4 Å resolution noticeably, were omitted from the considerations above. First, nonuniform distribution of matter in the unit cell accentuates low-order intensities compared to the random distribution assumed. Second, the presence of interstitial liquid reduces this factor, depending on the degree of electron density match between the molecule and the liquid.

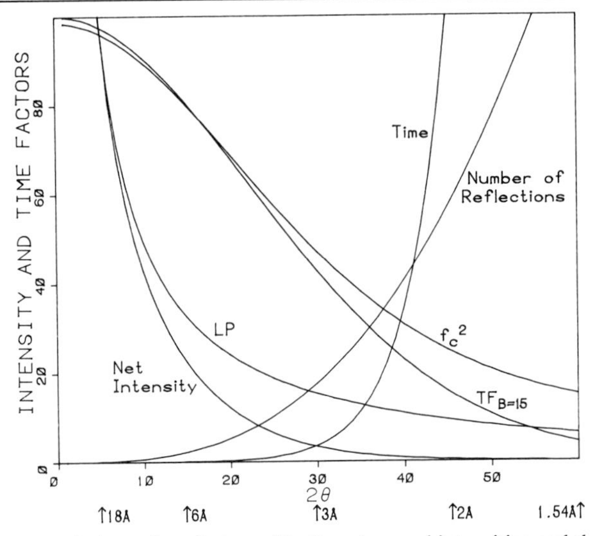

FIG. 1. Structure-independent factors affecting observed intensities and the implications regarding time of data collection at various resolutions. f_c is the atomic structure factor for carbon. TF is the temperature factor term with a B value of 15.

Background scatter from the glass capillary used to enclose the crystal increases to 3.5 Å and then decreases somewhat. Thus the real problems increase more from 9 to 3 Å than implied above.

Note that the familiar curves of $\langle F \rangle$ vs resolution are deceptive in terms of the primary experimental measurements since the intensities are proportional to F^2 and further reduced by the Lorentz factor.

Essential Features of the Four-Axis, Single-Detector Diffractometer

The essential features of the basic four-axis, single-detector diffractometer are illustrated in Fig. 2.

The basic mechanism holds the source, the crystal, and the detector in a horizontal plane, and provides a circular translational motion for the detector and orientational motions for the crystal. The detector arm and χ circle rotate on a common vertical axis, thereby providing the so-called 2θ and ω motions. The ϕ head which holds the crystal support moves along the χ circle and provides a ϕ rotation about an axis passing through the instrument center. The ϕ and χ mechanism together constitute an Eulerian cradle which can further be rotated on the ω axis. The whole system is computer controlled and data collection is largely automatic. Initial setup for a run involves considerable human interaction.

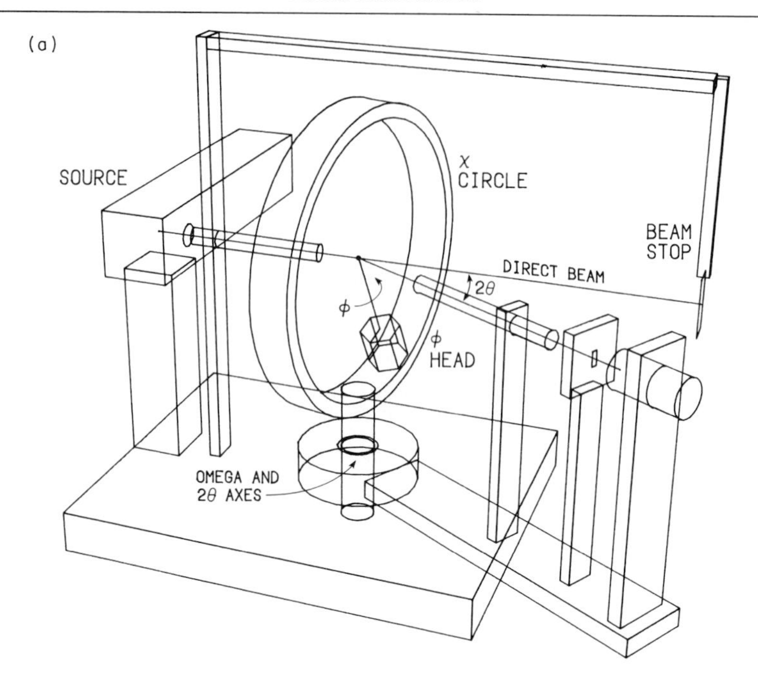
(a)
SOURCE
χ
CIRCLE
BEAM
STOP
DIRECT BEAM
2θ
ϕ
ϕ
HEAD
OMEGA AND
2θ AXES

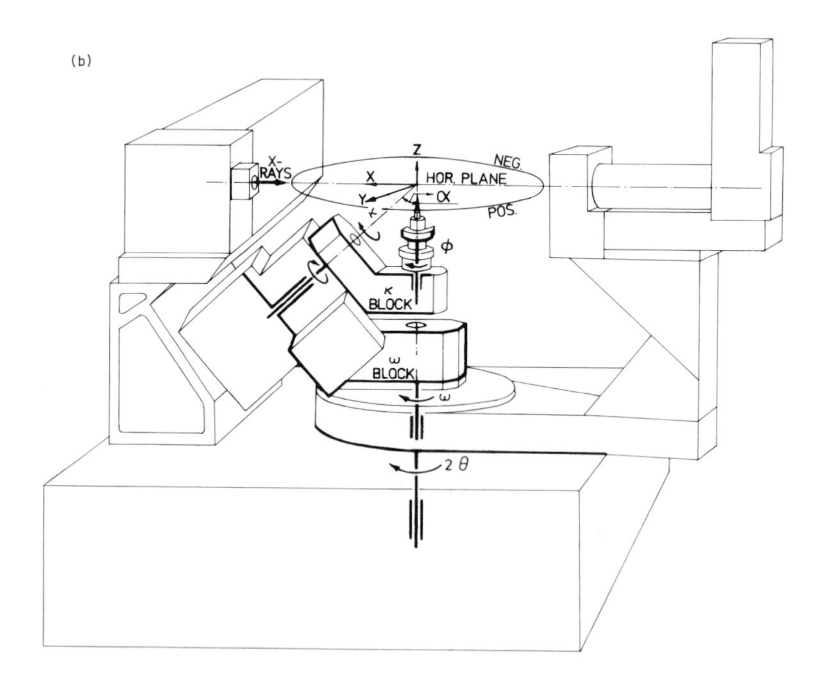
(b)
X-
RAYS
Z
NEG.
X
HOR. PLANE
Y
α
κ
POS.
ϕ
κ
BLOCK
ω
BLOCK
ω
2 θ

Two beam tunnels are added to block direct rays from the source to the detector and to minimize scattered rays reaching the detector. They, and the other components introduced next, are discussed in more detail below. Beam tunnel details are diagrammed in Fig. 12. The incident beam tunnel minimizes the hazard to the operator and others by allowing only a very small beam to emerge. In some arrangements both beam tunnels, and more commonly the diffracted beam tunnel only, are evacuated or filled with helium to reduce absorption by air which is 1% per centimeter and thus a significant loss with a typical crystal-to-detector distance of 50 cm. A nickel filter is commonly used with Cu radiation to substantially absorb the undesired K_β radiation while transmitting as much of the desired K_α radiation as possible. Proteins and nucleic acid crystals are usually mounted in a glass (or fused-quartz) capillary to maintain a controlled atmosphere, thus contributing to scatter and absorption. A shutter is provided for safety and to protect the crystal from unnecessary radiation damage.

The source take-off angle determines the apparent size of the source and may be variable. A central beam stop is incorporated for safety and to protect the detector. The detector aperture is variable both horizontally and vertically. The detector may be a scintillator–photomultiplier combination or a gas discharge tube operating in the proportional counter mode. The crystal capillary is held in a goniometer head (not shown) which allows accurate translational centering in three directions and optional orientation on two arcs with the aid of a telescope (not shown) mounted on the χ circle.

κ Geometry

One commercial four-axis diffractometer uses a different arrangement to position the ϕ head at various orientations (Fig. 2b). Instead of the χ circle motion the ϕ head is held on an arm which has a single rotary articulation 50° from the vertical. The base of this arm rotates on the θ axis and the motion is identical to the ω facility of the more conventional system. The upper portion of the arm carries the ϕ head with the ϕ axis

FIG. 2. (a) Diagram of the basic four-circle diffractometer. The detector arm and the χ circle rotate on the same axis with motions called 2θ and ω. The ϕ head carries the crystal holder and a goniometer head (not shown), and provides the ϕ rotation about an axis through the center of the instrument. This head is moved along the χ circle. The two tubes surrounding the incident and diffracted beams are beam tunnels discussed in the text and illustrated in more detail in Fig 11. (b) Diagram of a diffractometer using an alternative carriage for the ϕ head, called a κ mechanism. (Patented by Courtesy of Enraf-Nonius Co., Holland.)

inclined 50° to the swivel. A combined motion of ω and the swivel, κ, can elevate the ϕ head to the equivalent of 100° on the χ circle and to all intermediate values. The advantage of this system is the ability to swing under the direct beam and access more of the χ–ω space than the conventional system, or at least access it more easily without removing and replacing parts. This advantage is of little importance in macromolecular crystallography. The system is precise and flexible and has no particular disadvantages. The design of crystal cooling systems is different in the two systems if coaxial air flow is desired.

Crystal Geometry and the Orientation Task[2]

Crystal and Reference Frames

A crystal can be represented as a rigid three-dimensional array of atoms placed repetitiously, by translation, at real lattice points specified by the expression

$$n_1\mathbf{a} + n_2\mathbf{b} + n_3\mathbf{c}$$

where n_1, n_2, and n_3 are integers and **a, b,** and **c** are principal lattice vectors.

In the diffractometer the reference frame for this description of the crystal can be any one of several, as indicated in Fig. 3. There is the base box or laboratory frame, the χ circle frame rotated by $\omega + \theta$ in the laboratory frame, the ϕ head frame tilted by the angle χ, and finally the innermost frame rotated within the ϕ head frame by ϕ. A Bragg plane can be drawn through any three lattice points and a set of planes parallel to this will then contain all of the lattice points, as illustrated in Fig. 4. Diffraction will occur when X rays of wavelength λ are incident on these planes at an angle θ given by Bragg's law, $\lambda = 2d \sin \theta$ where d is the distance between planes within the set. The diffracted ray will make the same angle with the planes such that the deflection is 2θ and the plane of the incident and diffracted rays will be perpendicular to the Bragg plane.

Orienting Bragg Planes

The orientation task for observing, for example, the (112) plane of an orthorhombic crystal mounted with the c axis along the ϕ axis and the a

[2] W. C. Hamilton, *in* "International Tables for Crystallography," Vol. IV, p. 276. Kynoch Press, Birmingham, 1974; W. R. Busing and H. A. Levy, "Angle Calculations for 3- and 4-Circle X-ray and Neutron Diffractometers," ORNL 4054. Oak Ridge Natl. Lab., Oak Ridge, Tennessee, 1976.

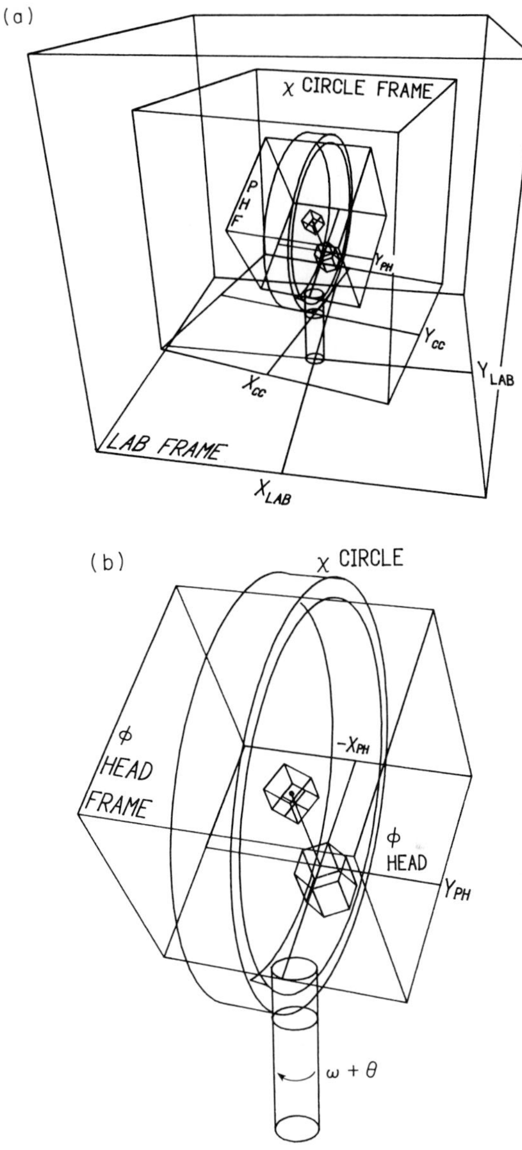

FIG. 3. Crystal planes and reciprocal lattice vectors can be represented in any of several reference frames. These are represented here with origins displaced along the respective Z axes to separate the boxes and to relate them to the mechanical components of the diffractometer. The lower drawing (b) is extracted from the drawing on the top (a). Successive computational rotations starting with the innermost axis system and proceeding to the outer frames will produce the orientations obtained by a combination of ϕ, χ, and ω motions. CC, χ circle frame; PH, ϕ head frame.

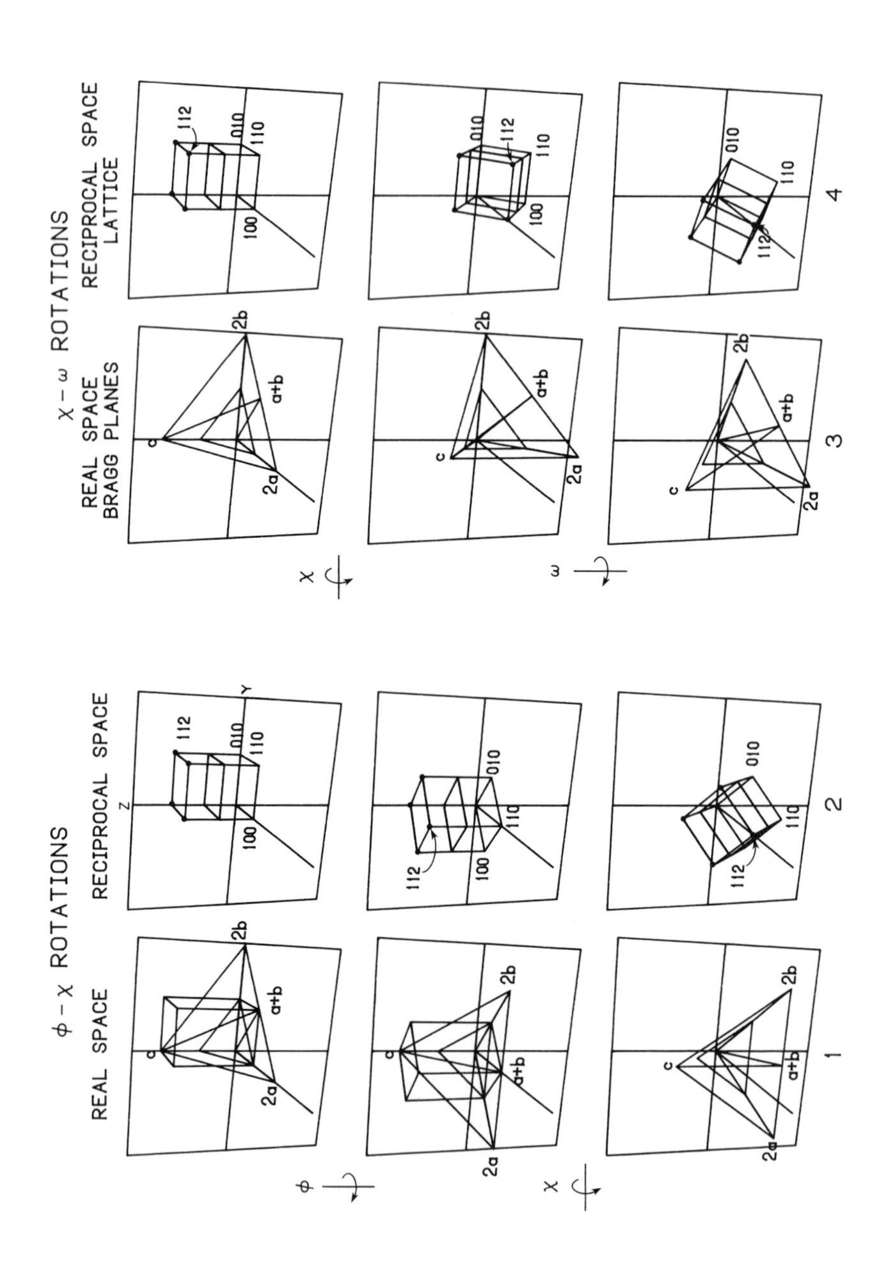
φ − χ ROTATIONS
REAL SPACE
RECIPROCAL SPACE
χ − ω ROTATIONS
REAL SPACE
BRAGG PLANES
RECIPROCAL SPACE
LATTICE
c
2a
2b
a+b
Z
Y
100
010
110
112
φ
χ
ω
1
2
3
4

axis initially in the plane of the χ circle and the ϕ head at χ zero is illustrated in Fig. 4.

ϕ–χ *Rotations*. Starting with the detector arm at the proper 2θ value and with the plane of the χ circle bisecting unit vectors from the crystal toward the source and the crystal toward the detector, several schemes can be used to orient the crystal for diffraction. One procedure, illustrated in the left panels of Fig. 4, would be to use the ϕ motion to bring the vector from 2**a** to 2**b** parallel to the χ axis and then use the χ motion to set the vector from **c** to (**a** + **b**) vertical. This will set the plane defined by the point 2**a**, 2**b**, and **c** parallel to the *YZ* plane as required. These operations are illustrated in the first panel. *Y* is the original χ axis and *Z* is vertical.

χ–ω *Rotations*. An alternative set of motions, illustrated in the third panel, would first rotate on χ to bring the line from 2**a** to **c** vertical and then on ω to make the line from 2**b** to **c** also parallel to the original *YZ* plane. The ϕ axis (coincident with **c**) is not coincident with the ω axis after the first motion and thus cannot provide the specified motion. The χ axis is not coincident with the original *Y* after the second rotation. The Bragg plane ends up in the desired orientation in both cases (panels 1 and 3) but the direction of the line from 2**a** to **c** is different, corresponding to a rotation around the Bragg plane normal. This rotational parameter is called the azimuthal angle.

Bragg reflection will occur at all azimuths but secondary effects such as absorption within the crystal, scatter from the glass capillary, anisotropic mosaic spread effects, and potential double reflection will be variable. Some programs allow combined use of ω and ϕ to optimize absorption, for example, but most assume the χ circle to be in the bisecting position, except for small variations in ω to scan a reflection and/or allow for slight missetting. Mechanical collisions restrict the available ω–χ space.

Azimuth Scan. One particular use of an azimuthal scan is to evaluate the absorption profile for use in later corrections. One specific technique is to orient the crystal with a strong reflection in the polar position, using the manual adjustments of the goniometer head. The Bragg plane is thus

---

FIG. 4. Panels 1 and 2 show the ϕ and χ motions of crystal planes and the corresponding reciprocal lattice point that would bring the (112) plane into reflecting position. In the starting orientation the crystallographic *c* axis is vertical along the ϕ axis of the diffractometer, with χ at zero. The *YZ* plane is the reflecting plane determined by the incident and diffracted beams defined by the source, the crystal center, and the detector, preset to the proper Bragg angle. Panels 3 and 4 show an alternative pair of motions on the χ and ω axes. The crystal is positioned for reflection in the bottom row of each panel. The azimuthal orientation is different in 1 and 2 compared to 3 and 4.

perpendicular to the ϕ axis, and then the ϕ rotation can be used for an azimuthal scan.

Reciprocal Lattice, the Ewald Sphere, and the Sampling Vector Related to the Diffractometer

Panels 2 and 4 of Fig. 3 show the use of the reciprocal lattice concept to achieve the same positioning of the Bragg plane as in the panels 1 and 3, respectively. The reciprocal lattice construct

$$\mathbf{R}(hkl) = h\mathbf{a}' + k\mathbf{b}' + l\mathbf{c}'$$

is simple and easy to use once the basis vectors $\mathbf{a}'$, $\mathbf{b}'$, and $\mathbf{c}'$ have been determined. The resultant vector given by this construct is perpendicular to the Bragg plane with Miller indices h, k, and l, and thus setting this vector colinear with the bisectrix of the crystal-to-source and crystal-to-detector unit vectors is synonymous with the proper orientation of the Bragg plane. This is one of the beauties of the reciprocal lattice concept. Additionally, the length of the $\mathbf{R}$ vector is equal to $1/d$ which in turn is $2(\sin\theta)/\lambda$.

Figure 5 illustrates the sampling vector $\mathbf{R} = (\mathbf{s} - \mathbf{s}_0)/\lambda$ and its relationship to the diffractometer. The sum of unit vectors toward the source, $-\mathbf{s}_0$, and toward the detector, $\mathbf{s}$, is in the direction required for the Bragg plane normal and of magnitude $2\sin\theta$. Scaling the $\mathbf{s}$ vector by $1/\lambda$ the reciprocal lattice construct thus defines the 2θ setting as well as the crystal orientation required for reflection. The observation vector, $(\mathbf{s} - \mathbf{s}_0)/\lambda$, can be visualized by thinking of a mechanical linkage completing the parallelogram specified by adding $\mathbf{s}/\lambda$ to $-\mathbf{s}_0/\lambda$ as indicated in Fig. 5b. An actual mechanical linkage can be used to keep the χ circle in the bisecting position. The pivot point on the source to crystal line is then the origin of the Ewald sphere and the reciprocal lattice is centered in the crystal and tied to its orientation. As in the left part of the figure the diagram can be scaled to place the origin of the sphere at the source. Alternatively, the diagram can be scaled and shifted so that the Ewald sphere is centered in the crystal and the reciprocal lattice origin is at the point of intersection of the direct beam and the detector arc, indicated by the beam stop position in the figure. The observation vector then goes from this origin to the current detector position and the diffracted beam passes through the observation point thus defined. In this latter abstraction it must be remembered that the reciprocal lattice orientation is still tied to the crystal orientation. The latter construct is the common one used when thinking of film patterns. Both are useful in diffractometry.

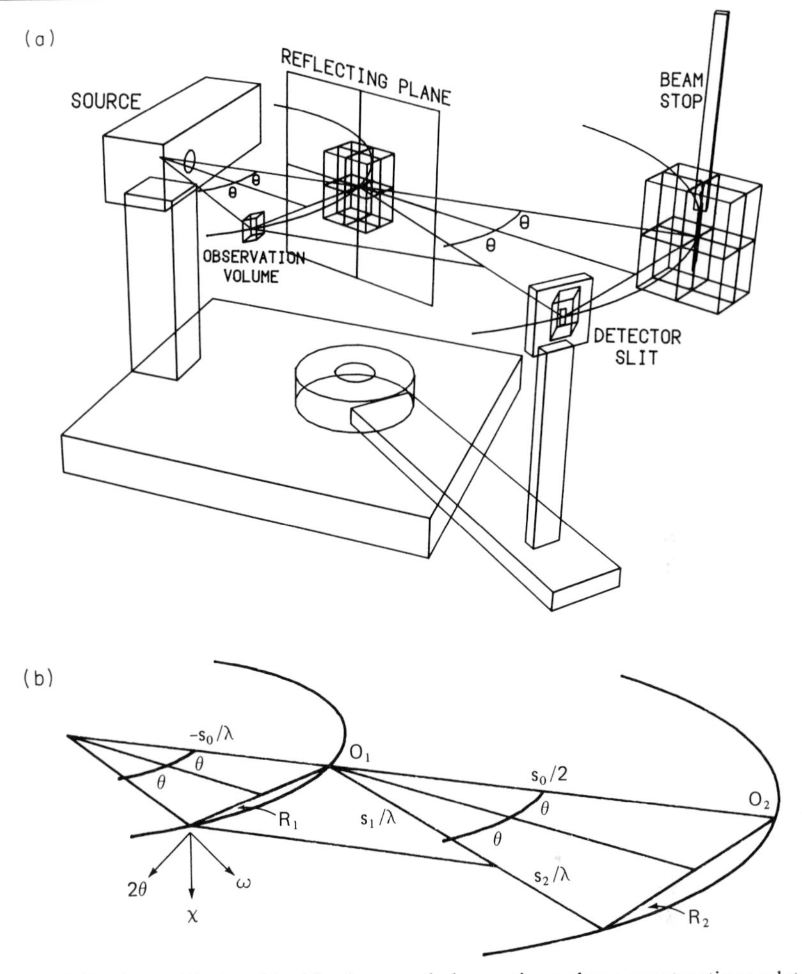

FIG. 5. Reciprocal lattice, Ewald sphere, and observation volume constructions related to the diffractometer (a). Two alternative origins (b) of reciprocal space are indicated, O_1 and O_2, and the constructs are scaled to the diffractometer: 2θ is the total deflection angle; $\mathbf{s}_0$ and $\mathbf{s}$ are unit vectors in the directions of the incident and diffracted beams; λ is the wavelength; $\mathbf{R}_1$ and $\mathbf{R}_2$ are reciprocal space vectors from the alternative origins. The lattices drawn represent a portion of the reciprocal lattice positioned at either origin. The arcs are intersections of the Ewald spheres with the horizontal plane. The observation volume is a volume rather than a point since the source and the detector are both areas rather than points.

It should also be noted that there is not a single sampling vector in the real case with finite source, crystal, and detector. Rather there is a sampling volume with different weights at various points within it. Each point within this volume is defined by one or more triples of points, one on the source, one within the crystal, and the third within the detector aperture. This concept is elaborated below in the text and in Fig. 10.

With the χ circle in the bisecting position the observation point, that is, the center of the observation volume, falls on the plane of the χ circle. At a given 2θ value the Ewald sphere intersects this plane in a vertical circle (not shown). The entire plane can be observed by a combination of 2θ and χ motions. If a reciprocal lattice plane is coincident with the χ plane then all the reflections in the plane can be observed by changing these two parameters. Chi motion provides a circular scan which devolves to a vertical displacement of the lattice point when it is near reflecting position. The 2θ motion is radial and the ω rotation provides motion of the lattice point orthogonal to both χ and 2θ motions. At $\chi = 0°$ the ϕ motion is equivalent to ω but at $\chi = 90°$ the ϕ motion is azimuthal and will not move a lattice point if it happens to lie on the pole. To bring a lattice point which is very near the pole into the χ plane may require a large ϕ rotation but only a small ω motion. It is often helpful in thinking about diffraction geometry to plot the ϕ and χ values of a number of reflections on the surface of a foam plastic sphere with colored pins, with longitude and latitude circles drawn on the sphere as on a globe of the earth.

Orientation and Unit Cell Matrix[3]

In an orthorhombic system $\mathbf{a}'$, $\mathbf{b}'$, and $\mathbf{c}'$ are parallel to $\mathbf{a}$, $\mathbf{b}$, and $\mathbf{c}$ but in general $\mathbf{a}'$ is perpendicular to the *bc* plane, $\mathbf{b}'$ is perpendicular to the *ac* plane, and $\mathbf{c}'$ is perpendicular to the *ab* plane. The reciprocal lattice vector $\mathbf{R}(hkl)$ is perpendicular to the corresponding Bragg plane. The length of the $\mathbf{R}$ vector is inversely related to the spacing. In vector notation $\mathbf{a}' = \mathbf{b} \times \mathbf{c}/V$ where V is the unit cell volume. The magnitude of $\mathbf{b} \times \mathbf{c}$ is the area of the *bc* face and the area divided by the volume is $1/d(100)$, where $d(100)$ is the spacing between the *bc* faces rather than the length of a.

The diffractometer geometry is naturally treated in polar coordinates while the reciprocal lattice is in linear skew coordinates related to the unit cell as given above and to the orientation of the crystal. It is convenient to introduce a Cartesian coordinate system X, Y, Z related in a natural way to each of these to be used in an intermediate step in converting between them. This system is orthogonal with its origin at the center of the diffractometer circles and with Z coaxial with the ϕ rotation axis. This is the innermost frame in Fig. 3. X and Y are perpendicular the ϕ axis and X is at

[3] W. C. Hamilton, *in* "International Tables for Crystallography," Vol. IV, p. 280. Kynoch Press, Birmingham, 1974.

$\phi = 0$, which is an arbitrary point of the ϕ rotation. Any reciprocal lattice point can be located in this intermediate system by the equations

$$\mathbf{R}(hkl) = h\mathbf{a}' + k\mathbf{b}' + l\mathbf{c}', \qquad \mathbf{S}(hkl) = \lambda\mathbf{R}(hkl)$$

where $\mathbf{a}'$ is designated by the triple $X(a)$, $Y(a)$, $Z(a)$, and likewise for $\mathbf{b}'$ and $\mathbf{c}'$. Then

$$\begin{aligned} X(hkl) &= h\,X(a) + k\,X(b) + l\,X(c) \\ Y(hkl) &= h\,Y(a) + k\,Y(b) + l\,Y(c) \\ Z(hkl) &= h\,Z(a) + k\,Z(b) + l\,Z(c) \end{aligned}$$

or in matrix notation

$$\begin{pmatrix} X \\ Y \\ Z \end{pmatrix} = \begin{pmatrix} \text{cell and} \\ \text{orientation} \\ \text{matrix}\ldots \end{pmatrix} \begin{pmatrix} h \\ k \\ l \end{pmatrix}$$

The conversion of these coordinates to polar form specifying the 2θ, ϕ, and χ settings needed to observe the reflection on the diffractometer is

$$\begin{aligned} \sin\theta &= S/2, \qquad RR = XX + YY + ZZ, \qquad S = R\lambda \\ \tan\phi &= Y/X \\ \cos\chi &= Z/R \end{aligned}$$

One fact in the above is that reflections can be located in terms of a "reciprocal" lattice, R or S, tied to the crystal orientation. Knowledge of the unit cell and the crystal orientation and even the X-ray wavelength is not needed to determine the S lattice experimentally. Given the reciprocal lattice and λ, the unit cell and orientation are derived parameters. If the 2θ, ω, χ, and ϕ settings for three identified reflections are determined, these points can be converted into X, Y, and Z for each of the three hkl triples, and then the X, Y, and Z values for each of the basis vectors $\mathbf{a}'$, $\mathbf{b}'$, and $\mathbf{c}'$ can be extracted by solving the nine simultaneous equations for the nine unknown coefficients from the nine input coordinates.

More on the Reciprocal Lattice: Symmetries and Centering

At this point restraints can be introduced to force known relationships characteristic of the known Bravais lattice such as axes orthogonality or cell edge equalities if desired. The unit cell parameters can also be calculated without restraint and the quality of agreement with the expected values can be used to indicate the quality of the matrix or to flag blunders such as misindexing. More than three points can be measured and least-squares procedures used to obtain the best matrix. It should also be noted that some of the angles are much more accurately determined than others.

In particular, χ values are poorly determined at low 2θ values and are very sensitive to translational miscentering of the diffraction center of the crystal. We will come back to this topic later.

It should be noted that a lattice can be defined in terms of an infinite number of sets of basis vectors, and therefore conventions are needed to assure that various investigators will specify a given lattice in one way. The basic conventions are that the cell edges will be as short as possible and the cell angles will be equal to or greater than 90° in real space. In reciprocal space the angles will be 90° or less. Ancillary conventions also exist.[4] The unique axis of a monoclinic cell is usually called *b* but sometimes *c*. The polar axis of an hexagonal, trigonal, or tetragonal crystal is called *c*. The longer of the two nonunique axes in a monoclinic cell is preferably called *a*.

There can be, and more often than not are, symmetry relationships among the atoms within a unit cell. In order to make concomitant lattice vector relationships more obvious and the symmetry axes and operations easier to specify, further conventions are usual, namely, to employ the Bravais lattices as applicable. These lattices include "centered" cells, either face centered or body centered. The centered lattice points of the primitive set are omitted from the new conventional lattice and the primitive motif is expanded to include the atoms omitted either explicitly or implicitly. The conventional cell is larger by a factor of 2 (or 4) than the primitive cell. In reciprocal space this gives rise to twice (or four times) as many *hkl* triples but clearly no new reflections are produced. Half (or three-fourths) of the nominal reflections are systematically absent. The primitive cell determines the observable lattice points. Two trivial consequences of this are that a reject routine is needed to avoid wasting time measuring absent *hkl* values and that the observed primitive reciprocal lattice will not reveal the true symmetry without a little extra thought. A more substantial consequence is that reflections are systematically redundant. Thus only a limited asymmetric "quadrant" of reflections need be measured but redundancies are available if desired.

Screw axes will also give rise to systematic absences. These will be limited to the corresponding reciprocal lattice axes. With a 2-fold screw only even orders will be allowed; with a 3-fold screw every third reflection is allowed, and similarly for 4- and 6-fold screw axes. It is these absences that signal the existence of a screw axis as distinct from a rotation axis which would produce the same symmetries of intensities. These screw axis absences prevent observation of the lowest order reflections which

[4] J. D. H. Donnay and G. Donnay, *in* "International Tables for Crystallography," Vol. II, p. 99. Kynoch Press, Birmingham, 1959.

otherwise make principal axes comparatively easy to locate and identify. These absences also reduce the number of higher order axial reflections which are convenient for precise unit cell and orientation determination, but this is not serious since the intensities of the remaining reflections tend to be stronger than otherwise and furthermore axial reflections are not needed to determine the matrix.

Locating Principal Axes and Matrix Input Reflections

How one finds and identifies a principal axis or a known *hkl* and determines the precise parameters has not been discussed. In practice, especially in protein crystallography, one usually knows the 2θ values within narrow limits before the crystal is mounted since the unit cell of a derivative or complex normally (or at least hopefully and usefully) varies 1% or less from the parent crystal. It is also usually known which reflections will be rather strong or weak and the approximate angular relationships between the low-order reflections. If the morphology is clear the position of the principal axes are known within a few degrees. In a blind case the axes can in fact be located by a systematic search without too much difficulty. The low-order reflections have easily distinguishable 2θ values, are far apart in ϕ and χ, tend to be quite strong, and for reasons discussed below can be detected at χ values quite far from optimum (up to 10° away). The angle between the incident beam and the Bragg plane (instrument θ plus ω) must be quite precise since the full width at half-maximum (FWHM) in this direction is only 0.1° or 0.2°.

After locating a low-order reflection 2θ can be increased and a higher order reflection found along the same radial row. Careful automatic centering of this yields precise orientational and *d* parameters. Since 2θ, ω, and χ are orthogonal near a reflection, they are the parameters scanned in this centering operation. Phi is left fixed since as noted above at $\chi = 0°$ it is equivalent to ω, and at χ values near 90° large rotation would be needed to produce small motions of the lattice point.

Unit cells of a given crystalline form of a protein often vary by as much as 1% with varying ligand conditions and sometimes by as much as 5%. The 2θ values for a given *hkl* would change by 0.02° at 2° 2θ and 0.3° at 30° for the 1% change in cell. Since the 2θ width of reflection is on the order of 0.5° there is little chance of missing a reflection during a search for this reason. Since the center of a reflection, once it is found, can be readily determined to 0.02° precision, comparatively small changes in unit cell can be detected at low 2θ and precisely determined at moderate 2θ.

The possibility of misindexing based on 2θ alone is another matter, except at low 2θ. Take for example a trigonal cell with $a = b = 44.0$ Å and

$c = 97.0$ Å. A portion of the reciprocal lattice for such a cell is drawn in Fig. 6a. The 2θ values for the 100, 010, and 001 reflections are 2.32°, 2.32°, and 0.91°. The 002 and 101 reflections are at 1.82° and 2.49° and the 110 is at 4.01°. These are easily distinguished and if the c axis is a 3-fold screw the first nonzero 001 reflection is the 003 at 2.73°. The 2θ values for the 10,0,0 and 10,0,1 reflections on the other hand differ by only 0.018° and an error could be made trying to base an identification on these values.

The angular separation of the 100 and 110 reflections is 30° in the ϕ–χ space and the angular separation of the 100 and 101 is 21.4°. The corresponding angular separation between the 10,0,0 and 10,0,1 reflections is only 2.25° which is ultimately easily resolvable but could lead to misindexing during the initial setup.

The intensities of the 100 and 010 reflections will be the same in this case (except for possible anomalous dispersion differences) and thus six equivalent reflections can be found. The c axis will be perpendicular to the plane of these and on this basis can readily be located even if the low-order intensities happen to be weak. Often a screw axis is most easily pinpointed by locating equivalent reflections related by the rotation.

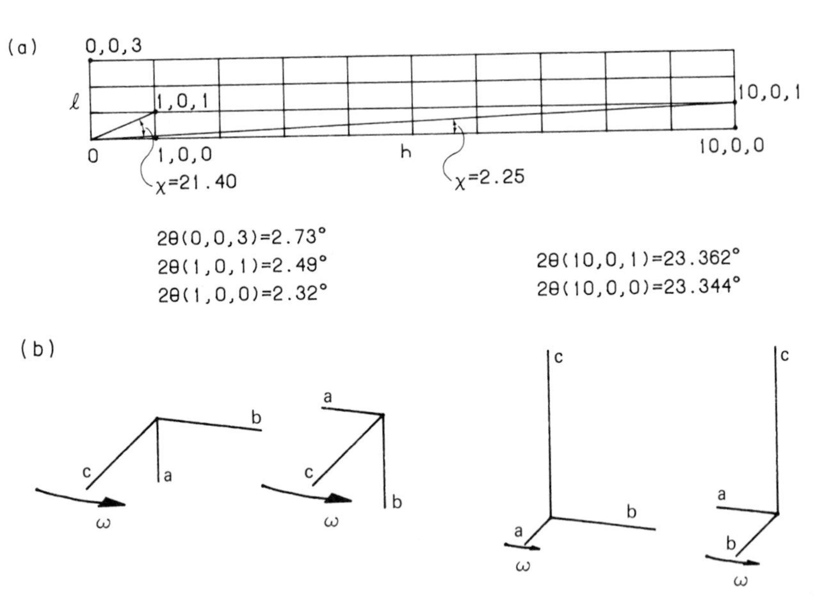

FIG. 6. (a) A portion of the reciprocal lattice for a crystal with $a = 44$ Å and $c = 97$ Å, illustrating the χ and 2θ values for several reflections. (b) Illustration of the ω scan direction for a crystal positioned in several directions useful in determining the cell and orientation matrix. Axis c is along the ϕ axis. In the first two panels $\chi = 90°$ and in the latter two, $\chi = 0°$.

If the crystal morphology is clear and the crystal is aligned optically and the operator is experienced and knowledgeable about a given crystal he can proceed directly to higher order reflections. Precise orientation of a polar reflection ($\chi = 90°$) can be achieved with a strong low-order reflection since the ω width of these is just as sharp as at higher order reflections. By changing the ϕ value by 90° precise orientation in two orthogonal directions, along the major and minor arcs of the goniometer head, can be made quickly, typically within 0.05° if desired (see Fig. 6b).

If the cell angles are well known as in the cases determined by symmetry to be exactly 90° or 60°, then the cell orientation can be accurately determined from low-order reflections even though the χ values cannot be accurately determined directly. Determining the ω values of the polar reflection, c', as illustrated in Fig. 6b, with a' or b' horizontal will indirectly specify the χ values for a' and b', respectively. The ϕ values for these axes can be directly determined precisely. If the α or β were other than 90° they could not be determined accurately from the low-order data with the crystal in this setting since the ω range is limited by mechanical hindrances.

Real Diffraction Geometry: Finite Beams, Crystals, and Apertures

The ray optics discussed above neatly define the narrow geometric conditions for reflection but fail to deal with real intensities. Infinitesimal sources and crystals yield infinitesimal intensities. In addition, the precise Bragg relationship defines a single diffraction angle for a monochromatic beam whereas diffraction actually occurs over a finite range of angles for each wavelength in a real beam which is in fact polychromatic. The fundamental diffraction breadth is determined by crystal perfection, mosaic block size, strain, and mosaic block orientation distribution. To deduce these properties the distribution of intensity within the narrow confines of a real reflection is needed, but these parameters are not of concern except to the extent that they impact on the proper measurement of the integral intensity which is the quantity of interest. The square of the structure factor is related to this integral, which is to be taken over the entire fundamental breadth and ideally over a narrow, specified wavelength band. In practice, in the case of total immersion of the crystal in the beam, one also integrates over all of the crystal. In standard small-molecule crystallography and in many macromolecular experiments care is taken to rotate the crystal about the ω axis sufficiently to integrate over the entire source for each portion of the crystal, each wavelength, and the total range of mosaic spread. If the source is uniform and large enough, the integration over the source is not essential and certain efficiencies in terms of total time, signal-to-noise ratio, and signal-to-radiation-damage ratio can be achieved. The advantages and dangers of this method will be discussed below after further discussion of the geometry of real beams in a typical diffractometer.

The Acceptance Cone

One aspect of real diffraction which often is not appreciated is the fundamental difference in geometric considerations in the horizontal and vertical directions. With the Bragg plane vertical and the source–crystal–detector plane horizontal only a limited angular range of rays in the horizontal direction will be diffracted from a stationary crystal even if the source is large. On the other hand a ray from any point on a vertical line within the source will be incident at essentially the same Bragg angle and thus diffract the same as any other point on the line.

Above it was pointed out that a Bragg plane could be set at different azimuths and diffract equally well. Here, with the crystal in a fixed position, various azimuths are illuminated by different rays from the source. The situation is illustrated in Fig. 7. Rays from any points between the two curves drawn on the source plane, passing through the rim shown, could diffract. Only those from within the actual source will produce diffracted rays, through the corresponding points on the opposite edge of the rim to the corresponding positions at the detector. In the case of an ideal point crystal with monochromatic radiation all possible rays meeting the Bragg conditions would lie on a conical surface and define an acceptance cone. In the real case this conical surface becomes a conical shell with rays fanning out in both horizontal and vertical directions from each point in the crystal. The variety of rays through the cylindrical rim shown are meant to illustrate this. The real beam is thus a fan, the shape of which is determined by the source and the crystal.

As long as a ray from the top of the source through the bottom of the crystal can enter the detector aperture, and likewise from bottom to top, vertical integration is automatically achieved. Vertical nonuniformity is thus no problem. Doubling the height of the source will double the integrated intensity, provided this can be done without reducing the brightness. When the crystal is rotated on a vertical axis the acceptance fan will move across the source. If the source is nonuniform in this direction, or uniform but too small, insufficient rotation will lead to errors in the integral intensity. If the source is uniform and large enough, rotation is not needed and all of the observation time can be spent at full diffracted intensity, thus exploiting these conditions.

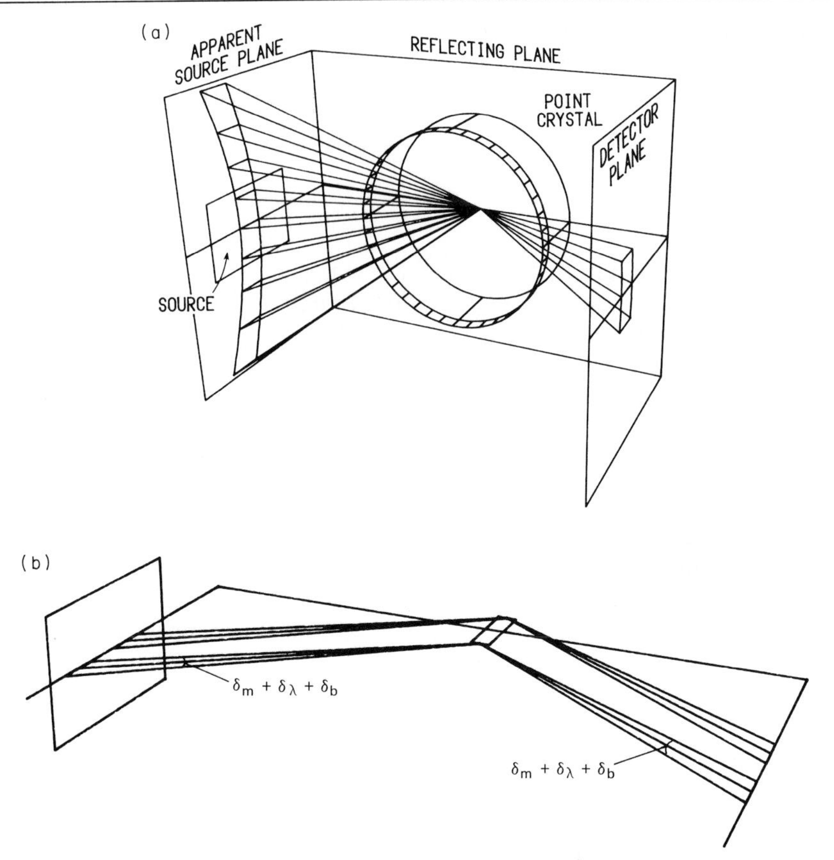

FIG. 7. (a) Illustration of the acceptance cone and the difference in horizontal and vertical beam geometry considerations. Rays from various positions on the apparent source plane approach the point crystal at substantially the same Bragg angle and there reflect to form a fan-shaped diffracted beam. The actual source would be to the left on a surface inclined at nearly 90° to the apparent source plane, at a take-off angle usually between 2° and 8°. (b) The equatorial section of (a) with an additional feature incorporated, namely a sizable crystal instead of a point crystal. The divergence angles from each point within the crystal are determined by the mosaic spread, δm, the wavelength dispersion, $\delta\lambda$, and fundamental broadening, δb, resulting from small crystallite size or/and disorder within the crystallites.

As a consequence of the asymmetry of the acceptance cone and diffracted beam, neither the source nor the detector aperture need be or should be round for optimum signal and minimum noise. For the complete ω scan the source should be short in the horizontal direction while it should be long for the stationary integration method. The detector aperture should be rectangular and adjustable in both directions to minimize acceptance of unwanted diffusely scattered radiation while not blocking the diffracted beam.

Another consequence of the asymmetric shape of the beam is that resolution of one reflection from another is not the same in all directions.

Vertical Beam Profiles

Considerations of the beam height and profile of intensity are somewhat simpler than those in the horizontal direction. Figure 8 illustrates these considerations. The upper diagram shows a situation in which the source is larger than the crystal and the detector is 50% farther from the crystal than the source is. The beam is in fact bent at the crystal and the diagram is thus bent, with the left portion being in the vertical plane containing the incident beam and the right portion being the diffracted beam. Alternatively the right portion can be considered the mirror image of the diffracted beam reflected by the Bragg plane.

The intensity profile is the convolution of the appropriate projections of the source and crystal onto the detector surface. If the intensity profile

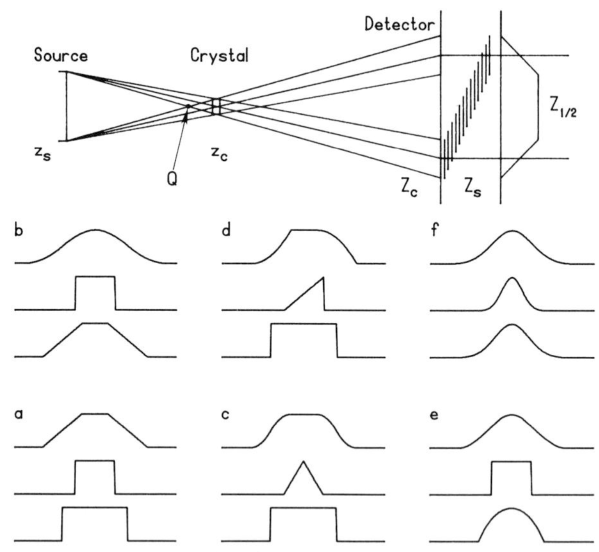

FIG. 8. Vertical beam convolutions for various profiles of source (lower curve in each triple) and crystal (middle curve in each triple) resulting in the combined profile (upper curve in each triple). (b) convolutes a schematic mosaic spread profile with the resultant shown in (a). All of the profiles in (f) are Gaussian. The lower profile in (e) is a segment of an ellipse.

of each is a "square wave," the resultant profile is trapezoidal as illustrated in the upper portion of the diagram and in panel a below. The total height is the sum of the individual components. The full height at half-maximum intensity is the height of the taller profile. The plateau height is the difference between the respective heights.

If the crystal is anything other than a transparent truncated prism the crystal profile will not be "square." The profile of a vertical bipyramid will be a quadratic cusp. Absorption will modulate this further. The profile will change with crystal orientation and in general will be more complex than in either of these cases. A simple triangular crystal profile would lead to a beam profile the edge of which would be quadratic up to the half-intensity point and have a quadratic approach to the plateau as illustrated in Fig. 8c. The full height, half-intensity height, and plateau would be the same as with a "square-wave" crystal. If the crystal is wedge shaped a crystal projection and beam profile as shown in Fig. 8d would result. The width and plateau are the same as in Fig. 8c but the edge profiles are different and asymmetric.

The source profile is not square in practice either. One possible shape is illustrated in Fig. 8e with the crystal still a square wave. The convolution looks like a Gaussian. A Gaussian profile for both the source and crystal projections is used in Fig. 8f and the Gaussian convolute is, in fact, almost identical to the beam profile shown in Fig. 8e. If desired for evaluation purposes the source profile could be determined with a small test crystal and a small detector height. The crystal profile could then be obtained by deconvolution. Assuming a source profile would obviously be risky.

Three points are of interest. (1) The total height is the sum of the individual components in every case. (2) The intensity profile is complex and is a convolution of all components. (3) Extracting any one component is a complex operation. The total height of the diffracted beam will increase as though it were diverging from a point Q, as illustrated, with an angle called the "cross-fire" which is the sum of the angular height of the source as seen from the crystal and the crystal as seen from the source.

If mosaic spread is significant it will have a greater effect at higher 2θ as discussed later. An appropriate mosaic profile, projected to the detector, would then be convoluted with the combined source and crystal size profile. The total heights would add and the half-intensity heights combine in a more complex way depending on the characteristics of the individual profiles. An illustration is given in Fig. 8b to show how quickly the effect of multiple convolutions, even of square waves, approaches a "bell-shaped" curve.

Horizontal Beam Geometry

Several aspects of horizontal beam geometry are illustrated separately and in various combinations in Fig. 9. In each the source is considered to be a point. Panel 1 shows diffraction of a monochromatic ray from a transparent crystal with no mosaic spread and no fundamental line broadening. Panel 1a is with crystal thin in the *Y* direction and 1b with a thicker crystal. The diffracted beam width is determined by the thickness in this direction.

Wavelength Dispersion. Panel 2a of Fig. 9 illustrates the situation with two wavelengths such as the Cu $K_{\alpha 1}$ and Cu $K_{\alpha 2}$ with a crystal thin in the *Y* direction and with the wavelength spread greatly exaggerated. Different portions of the crystal diffract at the corresponding Bragg angles and the diffracted rays converge as shown and then diverge again. If the crystal is thickened in *Y* as shown in 2b the diffracted beam size at the convergence distance is determined by the *Y* dimension as it was in panel 1b. The area between the rays shown would diffract if intermediate wavelengths were present as is the case with a real source. The ratio of wavelengths shown is 1.5. The actual ratio of the $K_{\alpha 2}$ to $K_{\alpha 1}$ wavelengths is 1.00245. The corresponding spread of the rays reflecting in the case illustrated in 2a

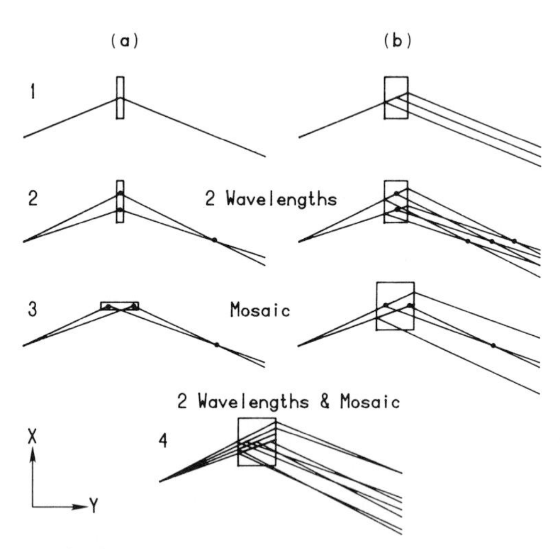

FIG. 9. Ray traces in the horizontal plane. Row 1 is for monochromatic X rays and perfect crystals. Row 2 is for two wavelengths differing by a factor 1.5. With a wavelength continuum of this bandwidth the area of the crystal between these rays would diffract accordingly. Row 3 depicts the effect of mosaic spread with an angular range of 5°, with a monochromatic source. Row 4 shows a combination of these two conditions. *X* is vertical and *Y* is horizontal.

would depend on the actual 2θ value as well. At 2° (45 Å), 15° (6 Å), 30° (3 Å), and 45° (2 Å) and with a source-to-crystal distance of 300 mm, the corresponding acceptance widths at the crystal would be 0.012, 0.097, 0.197, and 0.304 mm, respectively.

Mosaic Spread. Returning to a monochromatic beam but introducing the concept of mosaic spread in crystallite orientation, the situation in panel 3a of Fig. 9 is observed. The crystal is modeled as a collection of semi-independent blocks, or crystallites, with internal coherence only and with slight misorientation possible. So-called Bragg focusing produces a convergence such that the Y thickness does not produce beam broadening at this distance as illustrated with a crystal thick in Y but thin in the X direction. Thickening in the X direction produces beam broadening as shown in panel 3b. Again a limited area of the crystal is diffracting under these conditions. The mosaic range illustrated is 5°. For a real crystal with, for example, 0.01° spread and 300 mm source-to-crystal distance the width of the accepted beam would be 0.05 mm.

Panel 4 shows the combined effect of wavelength dispersion and mosaic spread with a crystal thick (but transparent) in both directions. The wavelength dispersion does not broaden the diffracted beam at the convergence point.

Fundamental Broadening. Small crystallite size and disorder within the crystallite both can give rise to significant relaxation of the Bragg's law restrictions on θ and 2θ. A crystallite thin in the X direction will permit diffraction over a wider range of 2θ values than a thick crystallite, since destructive interference between rays to the front and back sides will not occur as rapidly as 2θ deviates from the strict Bragg condition. Likewise, a crystallite that is thin in the Y direction can be rotated more than a broad crystallite before an edge atom will move one-half of the Bragg spacing and thus diffract out of phase with a central atom.

Disorder reduces the coherence distance within a crystal or crystallite and has an effect similar to (but distinguishable from) particle size broadening. Fundamental broadening for either reason, particle size or imperfection, thus has manifestations similar to mosaic spread and wavelength dispersion combined. In the simplest case with isotropic particle size or disorder, both components are present and independent of crystal orientation. Anisotropy of particle size or mosaic spread is physically reasonable and does occur. In these cases the effect on beam width correspondingly depends on crystal orientation.

The Effect of a Finite Source on the Diffracted Beam. Moving the source upward in Fig. 9 (backward in Figs. 2 and 7) would move each of the rays upward without changing their directions except in the case illustrated in panel 3a of Fig. 9. In that case with a flat mosaic crystal, the incident rays would reflect from positions further to the left and thus move downward at any given point in the diffracted beam. A ray to the center of the crystal would make a shallower angle of approach and reflect from a mosic block tilted clockwise and thus reflect further downward in this figure. The apparent 2θ value as recorded at the detector would be larger.

The effect of using a real source of appreciable width is similar to that of moving a point source up and down. The beams shown would be broadened accordingly until the top and bottom edges of the crystal became partially limiting. If the source is made still larger the situation depicted in Fig. 7 is obtained and the horizontal diffracted beam size does not depend on the source size but rather on the crystal size and properties and the source wavelength dispersion. Adjacent mosaic blocks in the crystal will diffract rays at a fixed Bragg angle from different point on the source depending on their tilt. At the same time a given crystallite will diffract different wavelengths from different positions on the source and at different Bragg angles. These effects will convolute with each other and with the crystal size to define the acceptance fan, and the diffracted beam will have a base size and divergence accordingly.

The diffracted beam profile will be affected by further factors. The Cu $K_{\alpha 1}$ and Cu $K_{\alpha 2}$ spectral lines are not of equal intensity but rather in a 2:1 ratio and thus the beam shape is not symmetric once this separation becomes apparent. The wavelengths are 1.54056 and 1.54439 Å with a weighted average of 1.5418 Å. The line shapes are Cauchy rather than Gaussian. The full widths at half-maximum intensity are 0.00058 and 0.00077 Å. The angular separation of $\alpha 1$ and $\alpha 2$ on the 2θ scale is $2(\delta\lambda/\lambda) \tan \theta$. The physical separation of points on the source diffracting simultaneously from a single point on the crystal is the same as the separations at the crystal of rays from a point source diffracting as stipulated above.

The source is not completely uniform since the center of the filament will be hotter than the ends, unless a tapered winding is used, and thus the center will emit more electrons than the ends. As the filament ages, tungsten evaporates, the center thins more rapidly than the ends, and the nonuniformity is accentuated. There may also be surface damage and contamination as the source ages and the latter can reduce the local intensity considerably by retarding the incident electrons before they reach the copper. This is particularly noticeable with rotating anode generators. There is one additional cause of apparent nonuniformity of the source. The actual electron target area is elongated and viewed from a shallow angle to make it brighter but apparently smaller. Typically the source is 10 × 0.5 mm viewed at 5.73° or 2.9° to make it look 1 or 0.5 mm wide, respectively. One end of the source is actually 10 mm farther from the crystal than the other and this reduces the vertical angular height and concomitantly the vertically integrated intensity by as much as 5%. This effect can be seen with a pinhole projection or when scanning with a small crystal.

Profile Scans

The shape and intensity profiles of the diffracted beam have been discussed above. When a crystal is scanned on the ω or χ axes the observed profile of recorded intensity depends on the detector aperture also. One more function needs to be convoluted with the beam profile. If the detector aperture is wide open and a small, good-quality crystal diffracting at moderate 2θ, such as a basic beryllium acetate test crystal, is used, an ω scan will probe the source profile. If the source is large and the detector aperture small the ω scan will sweep a small beam across the detector aperture and detect the edges quite sensitively. If the source and detector widths are comparable, both will affect the half-intensity points observed. Iterative centering will locate both and also determine the crystal orientation.

A χ scan will sweep the diffracted beam up or down and the amount that it will move at the detector aperture will be proportional to the distance from the detector to the central beam and thus linearly related to $2 \sin \theta$. The angular width of the beam in the vertical direction as seen from the origin of reciprocal space, at the beam stop in the discussion above, is inversely related to $\sin \theta$. Considering either of these causes of the breadth of the χ scan, it is apparent that the full width at half-intensity will be very broad at low 2θ and much narrower at high 2θ. For this reason low-order reflections are easy to find but hard to position accurately in the χ direction. At 2° 2θ, the χ width is typically several degrees, as many as 10°. At 18° 2θ, the center can be located within a few hundredths of a degree.

Profile scans should (could) be done routinely on each sample crystal at several orientations, and scanned on ω, 2θ, and χ to test for cracks, slightly misoriented parasitic crystal growths, severe mosaic spread, and potential bending. The χ range should be automatically adjusted inversely to 2θ for these scans.

To locate a reflection precisely for input to the cell and orientation, matrix calculation scans on ω, 2θ, and χ are usually used. Alternatively one machine uses two 2θ scans with a narrow detector slit oriented at ±45° from vertical to locate both the horizontal and vertical position of the diffracted beam.

In practice the crystal is moved during the measurement of a given reflection; the various scan modes will be discussed in a following section.

The Observation Volume

Before discussing scan modes and resolution it is useful to expand on the concept of the observation volume, which was introduced in Fig. 5 and is expanded in Fig. 10. As stated above, one is not observing a single point in reciprocal space in a real situation, but rather a continuum of points within a limited volume is being observed. A horizontal line on the source, coupled with a given point on the detector, will trace out a line in reciprocal space for any given crystallite within the crystal. Likewise a horizontal line on the detector will trace out a line in reciprocal space for the same crystallite. Both lines will be skew to the central sampling vector normally considered. They will convolute with each other to map out a diamond-shaped horizontal area as shown. The shape of this figure in the horizontal plane is a function of 2θ. If the angular widths of the source and detector slit are different, the sampling area is modified as shown in Fig. 10c and d.

A given vertical line on the source will map to a vertical line in reciprocal space, positioned at a point determined by the point within the crystal and a point on the detector. A colinear line will be generated by a vertical line on the detector for a given point on the source line. In reciprocal space these vertical lines will convolute in the same way as illustrated in Fig. 10 for the vertical components of the diffracted beam. Each point on the diamond-shaped horizontal area will spawn a vertical line of weighted observation points defining and filling the observation volume. The vertical component is independent of the diffraction angle.

Each point within the crystal will be observed a little differently since it is positioned differently with respect to the source and detector. The composite observation volume in reciprocal space must take this into account, and this can be done by further convolution of the crystal volume with the observation figure derived for a point crystal. This is indicated by the dotted lines in Fig. 10b, where an outer limit and inner plateau are delineated.

These positional geometric factors can be treated separately from the crystallite transform and mosaic spread effects and then recombined as a group in considering the combined effect on resolution, background, and scan mode. The wavelength dispersion can be incorporated in the observation volume concept or in a modified reciprocal space concept. The latter will be used here since it relates more closely to what is seen on a film or in a diffractometer scan. The wavelength spread is included by expanding the crystal transform radially instead of contracting the sampling volume and its position radially. The sampling volume and the modified reciprocal space, $R\lambda$, are both in dimensionless units with $S = 2 \sin \theta$.

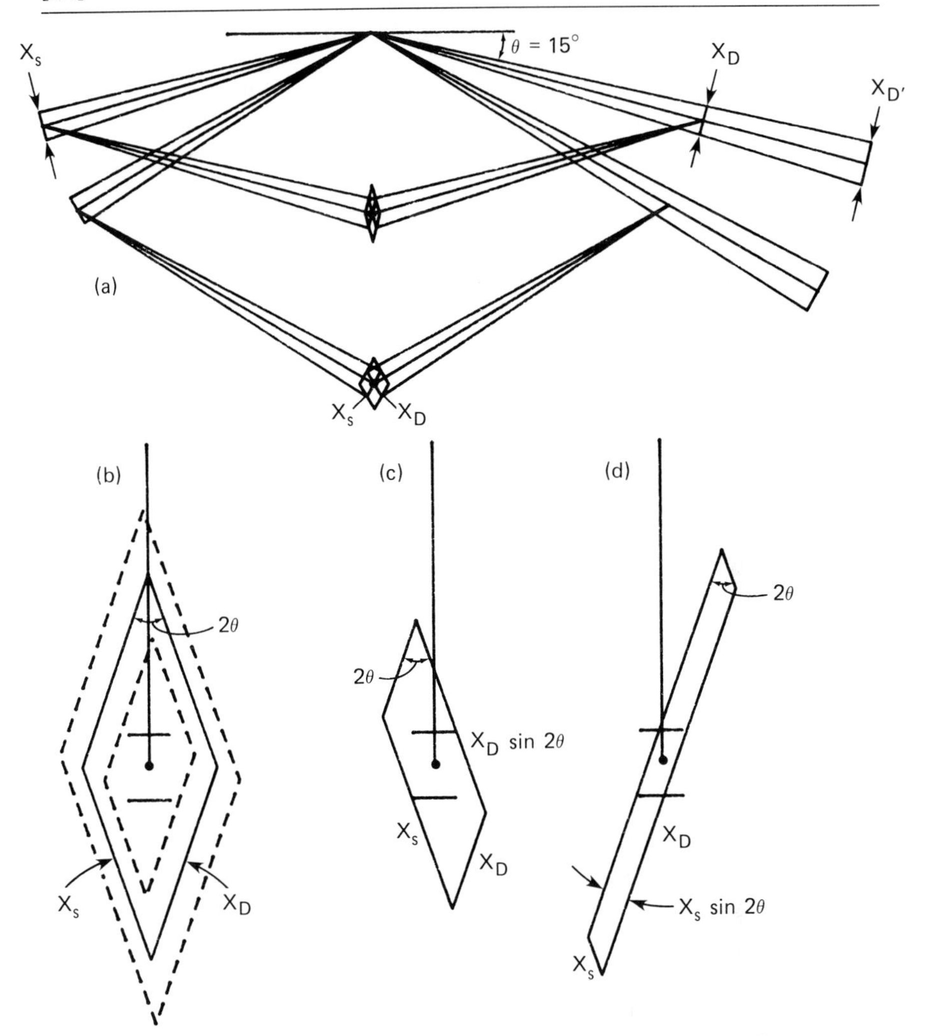

FIG. 10. Horizontal sections of the observation volume in $R\lambda$ space. (a) Two cases are illustrated, one with $2\theta = 30°$ and one with 90°. X_s is the horizontal dimension of the apparent source. X_d is the detector aperture width projected to be equidistant from the crystal. The ω range is constant. The S range is proportional to $\cos \theta$. The height perpendicular to the page is constant and therefore the χ range is proportional to $1/\sin 2\theta$. (b) Expanded view of part of (a) with the effect of the crystal size added. A diagram of the lattice point arced by mosaic spread and expanded radially to account for the two wavelengths is superimposed. The outer dotted contour shows the extent of the observation volume and the inner dotted contour the plateau region within which the weighting function is constant (assuming a uniform source). (c) The shape of the cross section with unequal X_s and X_d. (d) Similar to (c) with a small source and large detector; ω scans move the crystal transform sideways, and 2θ scans move the observation volume radially.

The modified transform of the crystallite takes into account the remaining factors; particle size, disorder, mosaic spread, and the wavelength band. Fundamental broadening spreads the transform in all directions. Mosaic spread expands a point onto a spherical surface, thereby producing a cap centered at the origin of reciprocal space. The $\alpha 1$ and $\alpha 2$ peaks expand the transform into two overlapping shells.

An ω scan moves the crystal transform sideways. A 2θ scan moves the sampling volume radially. A χ scan moves the lattice figure up and down from the page. Since the vertical profile of the sampling volume is independent of 2θ, the χ range does vary with 2θ. The tangent of the maximum χ is simply the maximum height of the sampling volume divided by 2 sin θ.

The area of the horizontal cross section is proportional to the source width, the detector width, and sin 2θ since it is a skew system. As a consequence the integral intensity in reciprocal space, which is the integral related by theory to the square of the structure factor, is proportional to the integral observed on the diffractometer multiplied by sin 2θ. This is one way of deriving the somewhat enigmatic Lorentz factor.[5]

In terms of these concepts the choice of source dimensions, detector slit dimensions, and scan mode center around observing all of the transform of the crystal with equal weighting; that is, proper integration is required. At the same time inclusion of background scatter, diffracted white radiation, and fringes of adjacent reflections within the observation volume during any portion of a scan must be minimized.

Data Collection

Integral Intensity: Scanning during Data Collection

The reasons for scanning, that is, rotating the crystal and optionally moving the detector during the measurement of a single reflection, are to integrate over the source, the crystal volume, the mosaic spread, the fundamental broadening, and the wavelength band and to compensate for setting errors, and to provide a measurement of the background scatter.

Since scanning is done for many reasons, one or another of the above factors may be most significant. Therefore there has been considerable controversy and uncertainty about the best mode of scan in a particular situation. Young[6] presented a particularly useful discussion of this matter

[5] B. E. Warren, "X-Ray Diffraction," p. 43. Addison-Wesley, Reading, Massachusetts, 1969. See also ref. 26, p. 405.

[6] R. A. Young, *Trans. Am. Crystallogr. Assoc.* **1,** 42 (1965).

and concluded that an ω scan is best at low 2θ and a 2θ scan at higher 2θ. Burbank[7] pointed out the great similarity between these two scans, and emphasized that the motion of the crystal relative to the source is the same in both cases and that the occurrence of diffraction depends on this relationship and not the detector position. Furnas[8] presented useful photographic records of what is "seen" during a scan as well as traces of the count rate during the various scans.

Vertical integration of the source and crystal occur automatically if the detector slit is high enough as explained above. One can therefore consider only ω and 2θ motions and combinations of these.

At times thinking in terms of real beams and slits is most useful and at times thinking in terms of reciprocal space and the observation volume is more instructive. Rotating the crystal sweeps the acceptance cone across the source or sweeps the crystal transform across the observation volume.

In an ω scan the detector is stationary with respect to the laboratory frame. Both the source and detector move with respect to the crystal and in the same sense. In a 2θ scan the detector is moved at twice the angular velocity of the crystal. Again both detector and source move with respect to the crystal, but in this case they both move either toward or away from the Bragg plane, and the result is a change in Bragg spacing while maintaining the same reflecting plane orientation. The sampling volume moves radially. In a coupled ω–2θ scan the crystal and detector are moved at the same rate. (The χ circle rotates at half the rate of 2θ when ω is constant. The second half occurs as ω, as defined here, is changed.) The result is equivalent to moving the source alone relative to a stationary crystal and detector.

A stationary integration can be obtained if the source is large enough to span the total acceptance cone (in the horizontal direction) and is uniform when integrated vertically. All parts of the crystal must be illuminated at all angles that comply with the mosaic spread, the wavelength spread, and the fundamental broadening resulting from crystallite size or internal disorder. The detector aperture needs to be large enough to encompass the diffracted beam at the central setting of the crystal, but no wider. No motion needs to be taken into account. The situation with real beams is illustrated in Fig. 7b and in terms of the sampling volume in Fig. 10b. In the latter case the crystallite transform, spread out by the wavelength spread and mosaic spread, must fall within the inner dotted contour which couples the source and detector profiles with the overall crystal

[7] R. D. Burbank, *Acta Crystallogr.* **7**, 434 (1964).
[8] T. C. Furnas, *Trans. Am. Crystallogr. Assoc.* **1**, 67 (1965).

size effect. Moving a small source over the range encompassed by this large source would give an identical overall result. As noted above this is the effective motion in a coupled ω–2θ scan.

Either the source or the detector may be larger in principle, and two shapes of the observation volume that might result are depicted in Fig. 10c and d. In either case an ω or a 2θ scan would allow proper integration. In both cases some of the white radiation streak and some of the background on either side of the reflection would be included in the measurement. If the observation volume is optimally shaped, the radial scan will overlap the white radiation streak but handle the α_1–α_2 separation best and thus may be favored with a good crystal at high 2θ. The pure ω scan would be best with high mosaic spread and at low 2θ. In the figures as drawn, the shortest scan would be perpendicular to the longer edge and these directions of scan would be combinations of ω and 2θ. The scan direction could also be selected to minimize possible overlap with adjacent reflections for a given crystal.

The white radiation streak in the vicinity of one reflection comes from that reflection and from radially adjacent reflections. If there are no nearby reflections the white radiation component need not be excluded nor subtracted out since it is proportional to the desired integral. If the radial row is densely populated then white radiation overlap from adjacent strong reflection can be serious, especially for weak reflections, and it is desirable to be able to correct for this fringe overlap. But if the row is too densely populated a radial scan will produce overlap of the extremes of the sampling volume with the next reflection, that is, lack of resolution. Just where the radial scan may be favored for one reason it may be disfavored for the other. The most serious source of background in protein crystallography is from the Cu K_α radiation scattered from the crystal and from the capillary, not from diffracted white radiation.

Perhaps the most compelling reason for use of the ω scan is that errors in ω are more likely to occur than in 2θ and that errors in ω are also more serious. If the crystal slips, ω can change appreciably but 2θ will not change at all. Also, changes in the unit cell which might occur during a series of measurements due to radiation damage, pH shift, or continuing heavy-atom reactions tend to change orientations more seriously than actual 2θ values. If orientation changes do occur they can be monitored at each reflection when an ω scan is used and corrections can be made to the calculated orientation continuously. With this capability the slits can be narrowed more critically and the scan range can be more securely limited than with a 2θ scan.

The question still remains whether a stationary integration or a broad ω scan with a small source is preferred. There is no question that the scan

can give more reliable results in the sense of reducing potential systematic errors. The stationary integration will give more counts, and thus more precision, for the same total time and the same radiation damage. The gain is as much as a factor of 2. The scan can provide a local background value and, coupled with profile fitting,[9,10] could regain some of the precision lost by lost net count and increased background incorporation in the total count. Systematic errors that may occur in the stationary integration method can be corrected, if they can be detected, by calibration or by scaling procedures. The systematic error due to an inadequate source size compared to that needed at high 2θ values to span the wavelength spread is an example of this situation. The source cannot easily be enlarged but a lack-of-integration correction based on a limited set of integrating scans on selected strong reflections could be made as a function of ϕ, χ, and 2θ (or h, k, and l).

A Stationary Integration–Limited Step Scan Procedure.[1,11] We have used stationary integration–limited step scan procedures for many years. Our first system was a sealed-tube diffractometer operating with angles calculated on a mainframe computer, transmitted with IBM cards, and stored in the diffractometer controller in banks of relays reflection by reflection. The program was wired into a telephone stepper relay. Our second system employed two sealed tubes, facing each other and reflecting from opposite sides of the same Bragg plane into two detectors. The current workhorse system obtains X rays from a 10 kW rotating anode generator and is controlled by a PDP8L computer with 12K of memory, a decktape drive, and no disk. Symmetry R factors of 2–4% are obtained and crystal-to-crystal R factors of 4–7% are obtained measuring 500–1800 reflections/hr on crystals with cells up to 200 Å.

The basic design and design parameters are illustrated in Fig. 11, in which the beam tunnel configuration is given.

The data collection algorithm employs a limited step scan. The source is 1.2 mm wide to provide integration with crystals up to 0.7 mm. The detector aperture is typically 1.2 mm wide × 3.5 mm high for reflections up to 3 Å resolution and 1.6 mm wide at higher 2θ. The steps are normally 0.01° or 0.015°, depending on the observed profile. The full width at half-maximum is about 0.10° on an ω scan.

The limited step scan is used to smooth out any fine structure in the source, to eliminate error due to the slope of top of the complete profile,

[9] R. Diamond, *Acta Crystallogr., Sect. A* **A25,** 43 (1969).

[10] H. C. Watson, D. M. Shotten, J. M. Cox, and H. Muirhead, *Nature (London)* **225,** 806 (1970).

[11] J. M. Sowadski, B. A. Foster and H. W. Wyckoff, *J. Mol. Biol.* **150,** 245–272 (1981).

and to update the setting calculations for each reflection so that maximum efficiency can be maintained.

A minimum of five counts are taken for each reflection. Ultimately, the highest four contiguous counts are summed and logged as the intensity of that reflection. First, two measurements are made one step to either side of the calculated peak, corrected by an offset determined from the previous reflections. The next three fill in the raster of five. If the fractional slope of the first two indicates a significant error in the original center point, one of the end points is omitted and one added at the other end. The end points of these five are tested and if the fractional slope indicates that more points are needed more are taken but otherwise the first five are sufficient. The offset calculated from the final end points is printed out with the reflection data so that the operator can track the performance, and it is carried over to the next reflection as stated above.

If the reflection is weak the apparent error in ω may be due to random fluctuations in count and therefore a weighting function is applied to the calculated shift required. This avoids wasting time drifting on weak reflections and initial missetting of the next reflection. Occasionally this will cause several reflection to be missed if weak reflections follow a sudden change in the region of *hkl* space being measured. Care is taken in the data collection pattern to minimize the frequency of this occurrence and in the data processing to detect resultant missed reflections. The weighting function used is

$$\mathrm{Wt} = \frac{1}{1 + [2(I_2 + I_1) + 100]/(I_2 - I_1)^2}$$

and was chosen to be reasonable and to be easy to calculate with the primitive computer hardware and software employed. The 100 was added to minimize the damage done by an occasional spurious batch of counts introduced by the electronics.

If the crystal moves more than 0.1° or 0.2°, a new set of measurements of the reflections used for the cell and orientation matrix may be in order. Some diffractometer systems will do this automatically but there is a danger that translational displacement has occurred as well as the orientational slippage that is detected. In our system, indicated offsets greater than an operator-specified limit will result is several retrys on the current reflection to check for an erroneous indication and then data collection is stopped with the shutter closed to avoid radiation damage. We have not implemented a system to telephone the user when this happens nor can he call into the computer to find out how the run is proceeding.

Crystal shape and anisotropic mosaic spread can produce a variable beam profile. As long as this does not exceed the size of the source no error will result.

The calculation used to calculate ω displacements has somewhat arbitrary constants and has a form which may not be best for a given crystal. These factors will not give rise to errors in general since they are only used to assure that low points on the sides of a reflection are not included in the sum of the highest four, which could occur if an insufficient range were collected.

The equation used to estimate the center point of a reflection from any two measurements is based on the properties of a Gaussian curve. The slope of a Gaussian at any point is directly proportional to the magnitude at that point, to the distance from the center, and to the reciprocal of σ squared. The distance to the center is therefore proportional to the fractional slope times a constant. This is strictly true for the exact slope at any point but not for the apparent slope measured by two points some distance apart. The quality of the calculation is quite good for separations equal to the half-width at half-maximum even when the center point of the two measurements is halfway down the slope. With a separation of the full width at half-maximum, the estimated center point would still be one-third of the way to the peak for an observation made at points at 10% of the peak height. The algorithm finds strong peaks and climbs up to the top from levels even below this. If the diffraction profile has edges that are sharper than Gaussian or if a shift constant which is too large is used, the algorithm will occasionally oscillate back and forth over the top in a never-ending loop. If the constant is too small some reflections may be underestimated but the method seems quite robust.

Accurate reflection centers could be found more precisely by making more measurements on the sides of the reflection but these are wasteful and unnecessary. The rationale for omitting some measurements is that the value at the peak should represent the proper integral, values on the slopes are subject to excessive error due to positional jitter, values on the slopes are not proper integrals, and the lower count values on the slopes have a higher background-to-signal ratio than at the center and therefore are not the place to spend time and cause radiation damage.

Resolution

The term *resolution* or *resolving power* of a system is used in several contexts. It may refer to the ability to measure the first-order reflection of a large spacing or a high-order reflection corresponding to a small Bragg spacing. Here we will consider a third meaning, the ability to measure the intensity of one reflection without any overlap from an adjacent reflection. This depends on the size and shape of the source, the crystal, and the detector, and the distances from the source to the crystal and the crystal to the detector. It also depends on the wavelength spread, the mosaic spread, and the scan mode. The resolution is not the same in all directions. Both the resolution and the anisotropy are functions of 2θ.

In terms of the observation volume in $R\lambda$ space and the crystallite transform spread by mosaic spread and wavelength dispersion, the task is to prevent the observation volume required to measure one reflection from overlapping the transform of an adjacent reflection any time during the scan. The observation volume changes shape with 2θ as discussed above. Mosaic spread moves the crystallite transform in ω and χ directions, forming a cap of constant angular spread and thus of increasing spread in R space as R is increased. The wavelength dispersion expands and contracts the $R\lambda$ diagram and thus produces greater spread at high 2θ. In the absence of mosaic spread the vertical extent of the observation volume is independent of 2θ, and thus vertical resolution is also independent of 2θ in this case.

In $R\lambda$ space the width of the observation volume in the ω direction, B_{OM}, is $G \sin \theta$; the 2θ extent, B_{TH}, is $G \cos \theta$; and the range in the χ direction, B_{CH}, is G_V where G and G_V are the horizontal and vertical geometric factors. They equal half the sum of the angular size of the source as seen from the crystal plus the angular size of the crystal as seen from the source plus the size of the detector as seen from the crystal plus the size of the crystal seen from the detector. The size of the crystal enters twice. In addition, the size of the source and the crystal determine the required size of the detector and the required scan range. In the stationary integration method the size of the source required is variable, depending on the crystal, but usually it is not varied in practice. Introduction of a variable slit near the detector might be worthwhile in this regard. In the ω direction, in the absence of mosaic spread, the adjacent reflection will not be sampled if the distance between lattice points is just greater than B_{OM}. The resolution will be $\lambda/(G \sin \theta)$. In the 2θ direction the requirement is that the sampling volume must not touch the adjacent reflection at any significant portion of the wavelength band. The resolution in this direction is thus $\lambda/[B_{TH} + \delta(2\theta)/2]$, where $\delta(2\theta) = 2[(\delta\lambda)/\lambda] \tan \theta$. The wavelength band taken into account has an arbitrary bound since the spectral line profile never goes to zero. It asymptotes to zero and blends smoothly with the white radiation. In practice the width can safely be taken as the sum of the α_1–α_2 separation and the full width at half-maximum of each line. In terms of fractions of λ these values are 0.00246, 0.00038, and 0.00050, which yield a resolution of $\lambda/(B_{TH} + 0.00334 \tan \theta)$. In the vertical direction the resolution is λ/B_{CH}. A selected set of calculated resolutions are given in Table I for a system with dimensions as in Fig. 11.

Systems with an incident beam smaller than the crystal are not considered here but are useful in some cases.

Monitors, a Part of Data Collection Strategy

During data collection, monitor reflections are normally inserted periodically. These provide a running check on the crystal alignment, radiation damage, and any spurious peculiar behavior. Decay corrections have been discussed separately by Fletterick.[12] Here it can be mentioned that these reflections should be strong so that background and count statistics are not serious. They should be over a range of 2θ so that the correction parameters can reflect the substantial difference in radiation damage at high and low resolution. Preferably they should not be sensitive to spurious effects such as produced by changes in salt concentration within the crystal or changes in pH or continuing heavy-atom reactions. Since new matrices are occasionally needed during a run, due to crystal slippage, it is prudent not to include a polar reflection in the list since the ϕ value may change markedly at this point and require different absorption corrections. One should also not waste time driving long distances back and forth. Monitors every 200–400 reflections or 5–10 times during a run are ample.

In order to monitor crystal misorientation it is desirable to have two reflections at high χ differing 90° in ϕ, and one reflection at low χ so that the ω scan of these will probe all three coordinate directions (see Fig. 6b).

Data collection is often done in shells. Care should be taken not to waste time driving long distances across the open center.

It is well to measure each reflection twice so that an internal evaluation of the data can be done routinely. Anomalous dispersion is often useful and Freidel pairs are therefore desired. These can be taken at $\pm 2\theta$ or by rotating 180° on ϕ. The latter is accomplished by measuring h,k,l and $-h,-k,-l$, both with θ positive, and serves to distribute the radiation damage more uniformly within the crystal. Absorption may be as high as 60% and this can produce significantly nonuniform damage if the crystal is rotated only a limited range of ϕ.

Absorption

Absorption corrections cannot be made easily by calculation of the absorption from crystal geometry, as is commonly done with small molecules, because of the presence of glass and water in the beam and because

[12] R. J. Fletterick and J. Sygusch, this volume [25].

TABLE I
CALCULATED RESOLUTIONS[a]

Crystal size (mm)	Detector height (mm)	2θ (°)	Source width (mm)	Detector width (mm)	Resolution (Å) ω direction	2θ direction	χ direction	Comments
0.5	2.6	15	1.2	1.2	2606	313	323	Fixed source and detector
		30			1314	293	323	
		60			680	264	323	
		15	0.76	0.94	3336	391	323	Optimum source and detector
		30	1.03	1.39	1340	297	323	
		60	1.65	2.42	474	204	323	
1.0	3.3	15	1.26	1.44	1903	234	260	Optimum source and detector
		30	1.53	1.89	838	199	260	
		60	2.15	2.92	336	156	260	

[a] Sample resolutions calculated for the stationary integration method on a diffractometer with a source-to-crystal distance of 35 cm and crystal-to-detector distance of 50 cm. The source height is 0.5 mm. Wavelength dispersion is taken into account but not mosaic spread.

of the odd shapes of many crystals used in practice. An approximate method is therefore used, as delineated by North *et al.*[13] An azimuthal scan of a polar or near-polar reflection is measured, and the variation in intensity is considered to be due to absorption. The correction is applied as a function of ϕ. The main variation in absorption is well taken care of by this method. Even the absorption by the capillary, which would seem to be very different for equatorial reflections with the crystal on the near wall or the far wall, is handled properly. There are no χ or 2θ variations per se taken into account, but at low χ there are obviously two rays of concern, the incident and diffracted rays that are at different ϕ values. This χ-dependent 2θ adjustment to the ϕ values used must be included.

The transmission curve should be symmetric with a 180° periodicity. It is well to measure the full circle since asymmetries do sometimes occur and reveal problem situations. Either the crystal is bent or it is off center and the beam is being vignetted by an off-center guard aperture.

[13] A. C. T. North, D. C. Phillips, and F. S. Matthews, *Acta Crystallogr., Sect. A* **A24,** 351 (1968).

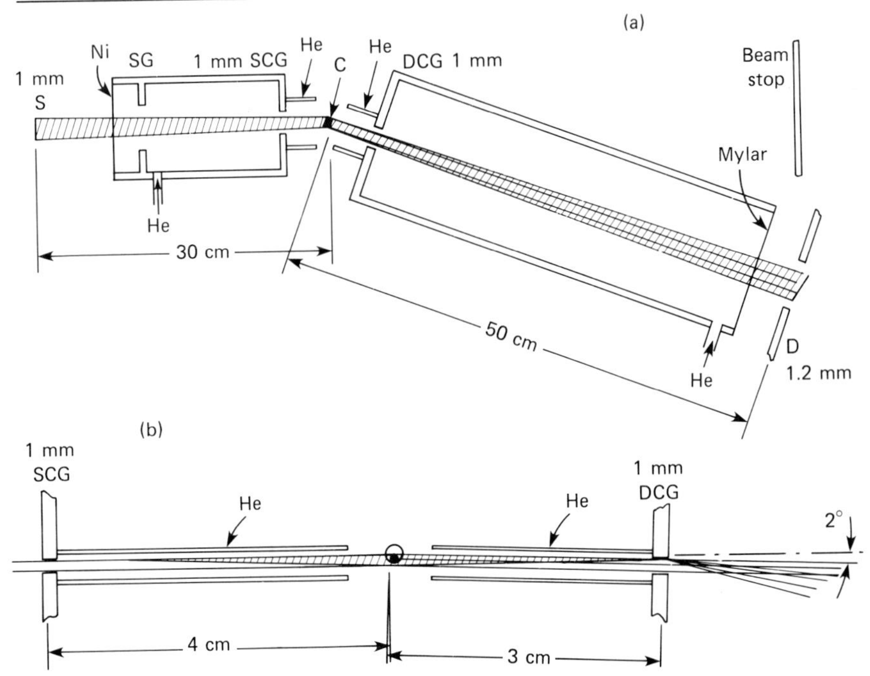

FIG. 11. Beam tunnel design. (a) The overall design, greatly exaggerated in the vertical direction. S, source; C, crystal; D, detector; SG, guard aperture near the source; SCG, guard aperture near the crystal on the source side; DCG, guard aperture near the crystal on the detector side. The shaded area represents the useful rays from the source to the crystal on the left, and the diffracted beam on the right. A nickel filter closes the incident beam tunnel on the left. Helium flows through the tube and out of the exit pupil, SCG, and on through an added extension tube. The exit aperture of the diffracted beam tunnel, on the right, is closed with a Mylar window and the helium flow is indicated. (b) The central portion of the beam tunnel system drawn to scale. 2θ is at 2° where parasitic scatter from the detector side guard aperture, DCG, is just occurring. The shaded area represents the volume of air and helium which is illuminated by the source and "seen" by the detector. The mass of air involved can be comparable to the mass of the crystal in the absence of helium.

The transmission curve is not necessarily sinusoidal, and in fact it can have real Fourier components as high as the fourteenth order, as shown in Fig. 12a and b. This condition exists with a thin crystal because of its absorption and because it is mounted on a glass wall such that the path within the glass varies with ϕ. A correction by more than a factor of 2 is common in this situation. The reflection used to obtain this curve can be at low 2θ and often is quite strong so that statistics are good. A curve-fitting procedure such as a Fourier analysis or a spline fit is still useful. Ultimately a correction to the absorption correction can be obtained in

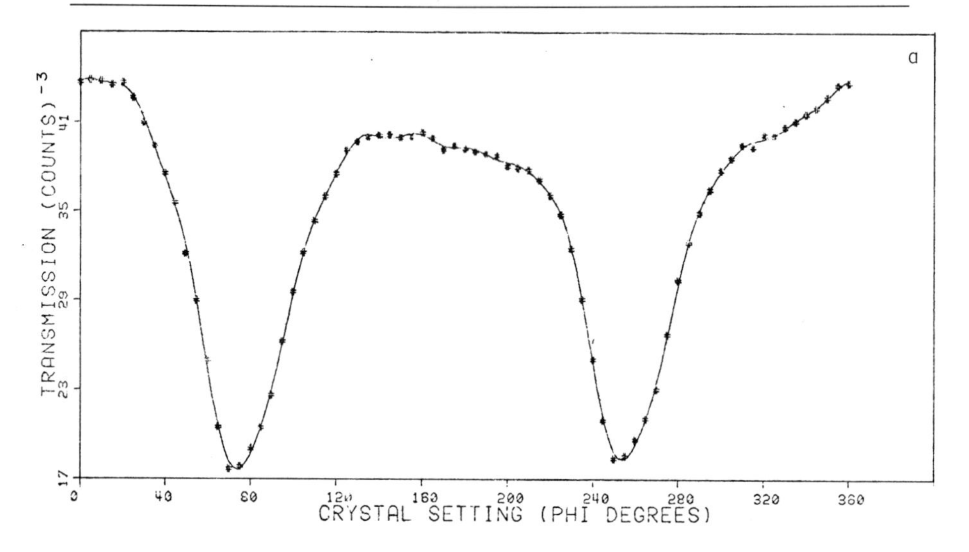

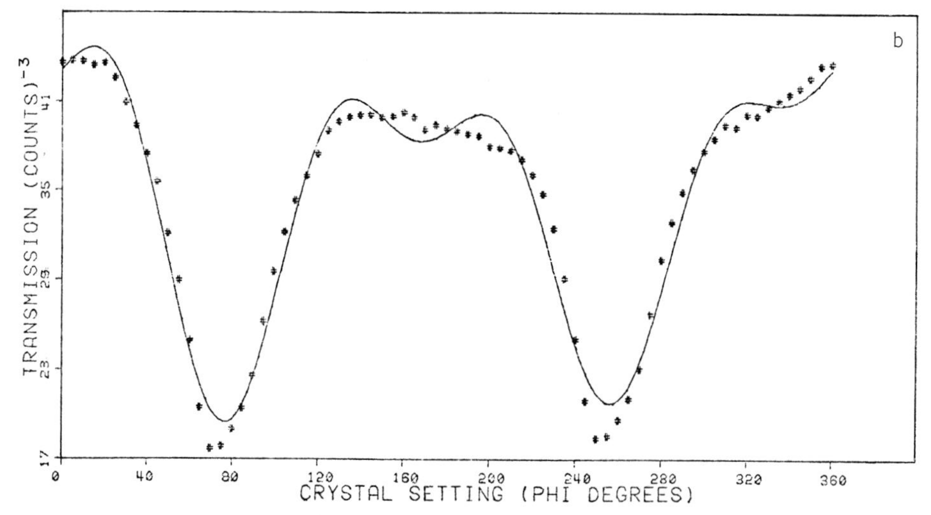

FIG. 12. Samples of transmission curve and background curve. (a) and (b) show the same transmission curve fit by a 14-term Fourier analysis or a 4-term fit, respectively. The angular range is 360°. The crystal was flat against the wall of a 1-mm capillary. (c) The background vs ϕ curve. A curve for glass alone was almost identical. (d) The radial distribution of scatter from air alone and air and glass combined. (e) A background curve exhibiting a spike from the fringe of a strong reflection. (f) A background vs 2θ plot with the beam tunnel design shown in Fig. 11. The sharp rise at 1.5° is due to parasitic scatter from the edge of a guard aperture. In (d) the lower curve is with helium replacing most of the air.

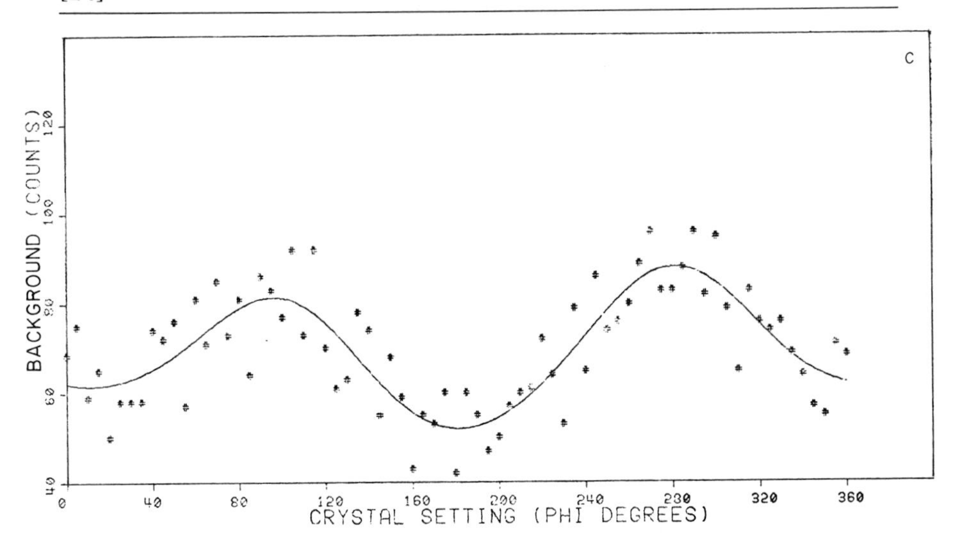

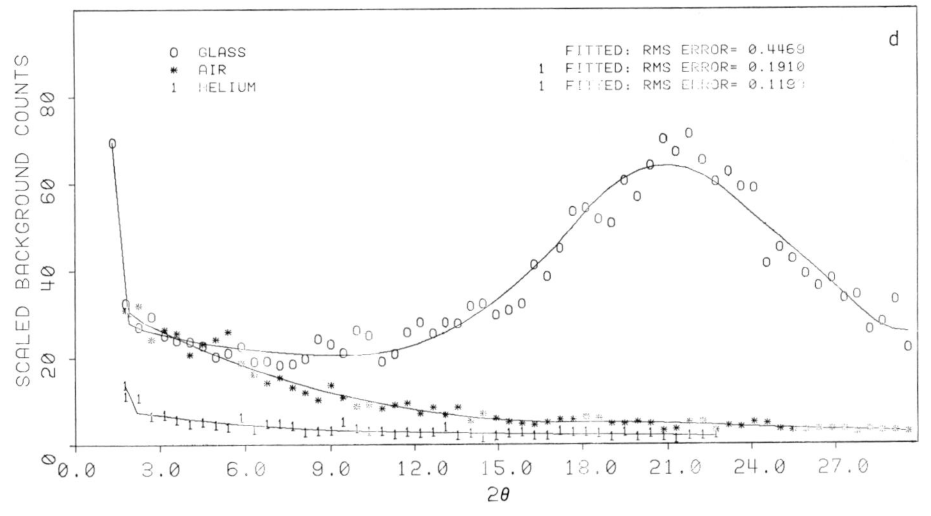

FIG. 12. (*continued*)

scaling procedures when comparing various sets of data. When trying to get the most out of anomalous dispersion data this is necessary and in general is good practice.

Background

Background corrections are commonly approximations also. Background is usually measured as a function of 2θ if an appreciable shell

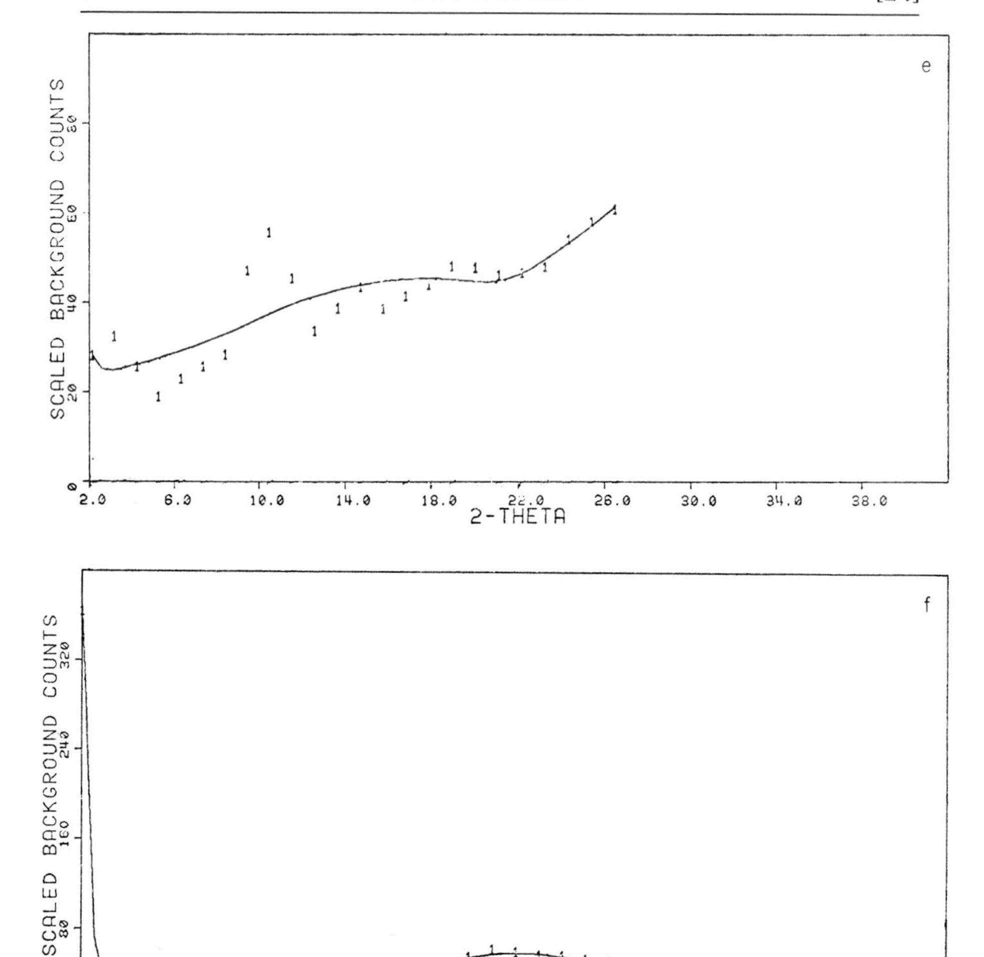

FIG. 12. (*continued*)

thickness is involved. It should also be measured as a function of ϕ since, as Stroud's group showed,[14] the amount of glass irradiated and seen by

[14] M. Krieger, J. L. Chambers, G. G. Cristoph, R. M. Stroud, and B. L. Trus, *Acta Crystallogr., Sect. A* **A30,** 740 (1974).

the detector is a function of ϕ, independent of the fact that the absorption of the diffuse scatter from the crystal is ϕ dependent. Furthermore, at low 2θ the air scatter or parasitic scatter from the guard apertures is significant and the absorption of this by the glass is ϕ dependent. The amount of this scatter reaching the detector is least when the most glass is in the beam and thus when the amount of scatter from the glass is maximal. The transmission of scatter from within the crystal is maximal at this point also, since the crystal usually lies flat on the surface. These variations are mainly due to the fact that the crystal is mounted at the edge of the capillary and is minimized by the use of small capillaries.

The background vs 2θ curve can be measured by fixing ω at some angle such as 0.9° and doing a dummy data collection run along an equatorial row of reflections. Since the background is hopefully low and some reflections in this row may be quite strong, a weak fringe of some of these may distort the background curve. This can be minimized by being adequately far from the reflection and can also be handled by editing the observed curve if large positive excursions occur. The number of data points in this curve do not have to be large, and therefore more time can be spent on each datum than is employed for the diffraction data. This will reduce the error, and curve fitting can further reduce the statistical fluctuations.

The background vs ϕ curve can be measured near a weak polar reflection by an azimuthal scan with a fixed ω offset. This should be done at a 2θ value in the middle of the shell of the main data or at the peak of the scatter curve which is in the 20°–30° range. For data below 5°, for most diffractometers, a separate curve is needed. These curves can be smoothed the same way as the transmission curve.

Commercially available fused-quartz capillaries are much stronger than the glass capillaries but they scatter significantly more. Glass capillaries are more likely to produce a change in pH than fused quartz. The commercial glass capillaries are not made of Lindeman glass as was at one time commonly thought (such glass would have absorbed less and probably scatter less). Perhaps the technical problems are more difficult in frabrication, which is possibly true also with thinner fused quartz. One of the keys to working with smaller crystals at a synchrotron might be development of a better crystal mount.

Residual Systematic Error and Scaling

When the objective of an experiment involves the comparison of two data sets it is theoretically possible to make all of the corrections mentioned above without making the measurement discussed for one of the data sets. This can be done by scaling procedures provided only that there are enough data and the sets are similar enough, as is often the case. The only caution is that the local scaling must not mask the ultimate data of interest (not usually the case).

Due to the large slope in some regions of the absorption curve shown in Fig. 12a and b, residual errors are likely to occur. The background can also be systematically in error. When comparing two data sets block scaling vs ϕ and χ and 2θ and time of data collection may thus be in order. Multiplicative scaling will not take care of the background error but an additive term can be included when comparing intensities. We have used block scaling, of progressively more complexity, for many years and have used the additive term for more than a year. Usually the latter is not demonstrably necessary but it has been useful in some cases.

Count Statistics, Error Propagation, and Time Parsing

In the absence of background, error statistics are simple and propagate to the electron density map with a somewhat surprising result. The variance (σ^2) in a stochastic process, such as a diffraction intensity measurement, is the number of events, where σ is the square root of the count and the fractional σ is the reciprocal of the square root of the count. Strong reflections are more accurately measured than weak reflections if the count time is constant. But what about the error in F, which is itself proportional to the square root of the intensity I? If the errors in the intensities are not large compared to the intensities, then the fractional errors in F are equal to half the fractional errors in I. (The range of F is less than the range of I.) The error in F is thus $F\,\frac{1}{2}(1/\sqrt{I})$ but F is $\sqrt{I}$ and the errors becomes 0.5 on this scale whether the the count is high or low. For example, if the count is 100, the σ count is 10, the fractional error is 10%, F is 10, the fractional error in F is 5%, and the error in F is 0.5. If the count is 10,000, the σ count is 100, the fractional error in I is 1%, F is 100, the fractional error in F is 0.5%, and the error in F is, again, 0.5.

Time parsing with equal time on each reflection is reasonable. The scale factor between the count and F^2 includes the Lorentz factor adjustment. To obtain the same error in terms of electron density for low-order and high-order reflections more time should be spent on the latter. There are many fewer low-order reflections, and in fact not much time would be saved by reducing the time per reflection for these reflections.

The background count necessarily included in the measurement of a reflection adds to the variance of the net count. If equal time is spent on the background measurement and the reflection, the variance in the corrected intensity will be the variance in the raw count for the reflection plus the variance in the background estimate. The variance is thus the net

TABLE II
FRACTIONAL ERROR (%) FROM COUNT STATISTICS FOR SELECTED VALUES OF NET COUNT AND BACKGROUND COUNT[a]

	Background count						
Net	0	50	100	200	400	1000	10,000
50	14.1	24.5	31.6	42.4	53.3	90.6	283.2
100	10.0	14.1	17.3	22.4	30.0	45.8	141.8
400	5.0	5.6	6.1	7.1	8.7	12.2	35.7
1,000	3.2	3.3	3.5	3.7	4.2	5.5	14.5
10,000	1.0	1.0	1.0	1.0	1.1	1.1	1.7

[a] Assuming the same measurement time for background and reflection data collection, the fractional σ (%) is given by $\sigma = 100\ (\text{Net count} + 2\ \text{Background count})^{1/2}/\text{Net count}$.

count plus twice the background count. The resulting fractional δ values for various combinations of net count and background are given in Table II. The very high background counts listed are not experienced in diffractometry but are relevant to film measurements with spot diameters of 0.5 mm.

If the background for a given reflection is estimated from a collection of adjacent reflections, one of the variances can be reduced appreciably but the background component of the reflection measurement, and the variance due to this, cannot be removed. If the background is estimated from generalized functions of 2θ and ϕ as outlined above the random errors are reduced by the smoothing functions and by spending more time on the sample background measurements than on individual reflections. Systematic error is substituted for local statistical error.

Bent Crystals

Several times bent crystals have been referred to. Bending can occur during growth or subsequently. The result is that a 180° rotation on ϕ and corresponding adjustment to χ may not give the expected result of identical intensities. This is particularly apparent in an absorption curve taken with a polar reflection rotating on ϕ at $\chi = 90°$ (and is the only explanation that I can think of for this phenomenon). The explanation is that the acceptance cone is no longer symmetrical because the back of the crystal does not have the same orientation as the front. In one case the cone may spread so much that it is not fully illuminated; that is, it extends beyond the end of the source. Rotation by 180° will then present a converging cone toward the source and thus pick out the brightest portion. If the angular range of the source viewed from the crystal is different from the angular range of the detector, then this asymmetry in the acceptance cone will not be compensated by partial exclusion of the diffracted beam, and the observed intensities will be different even if the source is uniform. Thus if the absorption curve is not symmetric, the crystal probably is bent and data taken at different orientations will have substantial systematic errors that are potentially correctable but not simply related to any one variable, ϕ, χ, or 2θ.

More Details of the Instrument

Zero Point Adjustments and Reflection Centering

The source position relative to the goniometer and the detector position can be accurately determined by iterative ω and 2θ centering using the half-intensity method or an alternative method. The ω and 2θ zero points are thus well defined. To adjust these settings, the source can be moved in a sealed-tube system or the diffractometer moved in a rotating anode system. Alternatively the zero points could be redefined in the software.

In a rotating anode system the whole diffractometer can be mounted on a milling machine bed and translated and rotated so that it is pointing properly at the source. A three-screw support can also be used, as in many camera mounts, and then a slight rocking motion, raising one corner and lowering another, can be used to aim the diffractometer. A sideways displacement of the support farthest from the source can be used as a fine adjustment, or a displacement of the front end can be used for coarser settings.

If the source is high or the crystal is below the instrument center, the reflection will be low at the detector when the Bragg plane is vertical. The center of a χ scan will accordingly be displaced from the true χ value for that reflection. This fact can be used to evaluate and adjust the source height relative to the diffractometer axis defined by the center of the various circles and the ω–2θ axis. The χ zero point is defined by these mechanical factors. Any given Bragg plane can be observed at four distinct χ values with ω restrained to zero. Two are obtained at opposite ends of the plane normal by a 180° rotation on χ. The other two are obtained by rotation by 180° on ϕ and adjusting χ as required. These positions are symmetrically placed around the true $\chi = 0°$ and $\chi = 90°$ marks. The average of two observed χ values thus indicates where the true χ zero point is, provided the crystal is centered. Once the source height is adjusted with a well-centered crystal, the centering of another crystal can be checked by this measurement. At low 2θ the χ value is very sensitive to crystal miscentering in the vertical direction.

If a test crystal of known lattice is used, the 2θ zero point can be checked with one measurement. For the beryllium acetate 111 reflection the proper 2θ value is 22.585°. This reflection can be observed directly at full power if an appropriate attenuator is inserted to prevent saturation of the detector. With a protein crystal on the machine the zero point can be checked by measuring a given low-order reflection at $\pm 2\theta$. The average should be zero if the crystal is translationally centered in the horizontal direction. Conversely if the machine is well aligned this measurement can be used to test and even adjust the crystal position.

In general, observation of the orientation parameters and 2θ at eight positions, the combinations of $\pm 2\theta$, $\pm\chi$, and ϕ and $\phi + 180°$, will determine the instrument errors independently of crystal mispositions and unit cell knowledge. Hamilton[15] has given the details, and a modern program system should include routines to make these measurements and calculate the adjustments to be made. The shorter version outlined here is relatively quick and useful.

The meaning of "center" is not unambiguous and the best definition depends on the purpose of the measurement. For several reasons the highest point on the curve should not be taken to be the center. Statistical fluctuations in the count determine the highest point observed. A slight slope in an otherwise fairly flat-topped curve would shift the point to one end. Fine structure in the source could lead to an erroneous value. The best evaluation for precise lattice constants might be a centroid[16] or a curve-fitting algorithm designed to account for partial α_1–α_2 separation. If the centroid were used for later positioning of the crystal, a skewed profile could lead to some truncation by the detector aperture. For setting purposes as well as ease of determination, the center of the two half-intensity points is often used. If partial α_1–α_2 separation occurs one of the half-intensity points may be ill defined. In this case a three-quarters intensity point might locate the α_1 peak and a one-quarter intensity point would better identify the combined α_1–α_2 beam position. At 3 Å resolution or less this is not usually a problem.

One very simple way to locate two points at a somewhat arbitrary fractional value is to get an approximate peak value from several mea-

[15] W. C. Hamilton, *in* "International Tables for Crystallography," Vol. IV, p. 282. Kynoch Press, Birmingham, 1974.

[16] I. Tickle, *Acta Crystallogr., Sect. B* **B31,** 329 (1975).

surements near the peak and then to step off in moderately large steps in ω until the intensity falls below half of this value. The scan direction is then reversed and the increment halved, and steps are taken until the intensity is above target. The direction is reversed again and so on until adequate convergence is achieved. Setting ω to its center, the 2θ half-intensity points are similarly located and finally the χ center is determined. With peak counts as low as 100, approximately 43 total steps will determine all three centers quite well. An improved algorithm could probably halve this number. Since the 2θ and ω scans both depend on the source and detector aperture, in practice they do interact. An erroneous ω setting will give rise to a 2θ error. At least one repeat of the complete determination is required for final matrix input values.

Size Considerations, Scale of the Diffractometer, and Wavelengths

The scale of a diffractometer is important with respect to the resolution of one reflection from another, with respect to the signal-to-background ratio, and with respect to radiation damage. Ideally the total system should be scaled to the crystal size.

The advantages of moving the detector further from the crystal are most easily understood. The diffracted beam is not diverging as rapidly as the diffuse background scatter of Cu K_α radiation from materials surrounding the crystal. The detector aperture does not have to be enlarged in direct proportion to the distance, and the signal-to-noise ratio improves, provided that the diffracted beam is not reduced by absorption. This is the case for quite large distances when the beam tunnel is evacuated or filled with helium but not air. The divergence of the beam is different in the horizontal and vertical directions. In practice, once the horizontal size of the diffracted beam is twice the size of the crystal, there is not much to be gained by further enlarging the system. The optimum size depends on the crystal size and the divergence. The latter depends on the properties of the crystal, the wavelength band of the source, the 2θ setting, and other factors discussed in other sections. If the resolution of the detector is limited as in some of the area detector systems, a larger distance may be optimum. Focused-beam systems are not included in this discussion but clearly the size considerations are different when a focused beam is used.

The considerations in choosing a source-to-crystal distance are more complex. As a fixed source is moved back, the irradiation of the crystal which leads to radiation damage and to background scatter is reduced proportionally to the square of the distance. The diffracted intensity is reduced because the vertical cross-fire is less in linear proportion to the distance. If the source is larger than the acceptance cone in the horizontal direction, there is no loss of diffracted intensity due to the reduced horizontal cross-fire. The net result is an increase in efficiency with respect to background and radiation damage at the sacrifice of data collection speed. An increase in resolution is obtained. The improvement in signal-to-background ratio is further enhanced by the reduction in detector slit height needed to admit the full diffracted beam. Commercial units targeted for small molecule use typically have distances of 15–25 cm. We use 35 cm in our systems.

As the source is moved back it may be required or desirable or acceptable for it to be larger in either or both directions. The apparent horizontal size can be increased by changing the "take-off angle" or by choosing a source with a longer target area. The relationship of total intensity of a source, the brightness, the apparent size, and the optimum tube voltage is discussed in another section. If the actual source is made longer, more power can be applied without exceeding the specific loading limits imposed by heat dissipation problems. If the source is made higher the total loading can be increased but not in direct proportion to the height, again because of heat dissipation considerations. The advantage of a rotating anode X-ray generator is that the specific loading can be increased, giving rise to a higher brilliance.[17] The inherent troubles are not worth it if the total power is only increased by a factor of 2 or 3, but a factor of 6 or 10 can be useful. By doubling the investment in money and effort a factor of 10 in capacity can be obtained in practice.

The optimum crystal size is dictated in part by absorption of X rays within the crystal and thus is wavelength dependent. Any portion which is thicker than the reciprocal of the linear absorption coefficient will not increase the diffracted intensity but actually decrease it compared to the optimum thickness. For proteins, this thickness is approximately 0.7 mm with copper radiation. For the crystal as a whole there is no optimal thickness, as pointed out by Bond,[18] since the total intensity increases with size as long as the crystal remains completely bathed in the beam. He recommended sizes up to twice as large as the value given above for spherical crystals. If the crystals are always small, a small, more brilliant, source could be used. If large crystals are available the total rate of data collection benefits and the amount of radiation damage occurring per useful photon is distributed over more molecules.

[17] W. C. Phillips, this volume [22].

[18] W. L. Bond, *in* "International Tables for Crystallography," Vol. II, p. 300. Kynoch Press, Birmingham, 1959.

Copper radiation is almost universally chosen for biological macromolecule diffraction but some arguments can be made in favor of either longer or shorter wavelengths. A number of factors are involved and deserve reconsideration in the context of expensive modern systems and special situations such as at synchrotron sources. The equations normally cited suggest that the diffracted intensity from a given structure is proportional to λ^3. As Arndt pointed out,[19] the Lorentz term is often left out of these considerations and the observed intensity is actually proportional to λ^2. Since absorption is also reduced at low wavelength some intensity is regained, depending on the original transmission factor. However, it is not the power of the diffracted beam that determines statistical fluctuations, but the number of photons. This factor favors longer wavelengths as long as the detector noise level is not approached. Film appropriate for copper K radiation is not suitable for molybdenum K radiation since it does not absorb enough of the beam. The same is true for gas proportional counters. Radiation damage depends on absorption and thus is less at short wavelengths. Residual errors after absorption correction are less at low wavelength. The balance of pros and cons seems to leave copper as a good choice on these grounds; in addition, copper tubes function well. Molybdenum (0.71 Å), iron (1.94 Å), and chromium (2.29 Å) sources are available. At a synchrotron all wavelengths are avilable.

Beam Tunnel Design and Operational Considerations

Beam tunnels are provided for the incident and diffracted beams to shield the detector and operator from scattered radiation. The position of these is indicated in Fig. 2 and a more detailed representation is given in Fig. 11. The entrance pupil to the incident beam tunnel is usually comfortably large such that it does not block any useful rays from the source to the crystal but does block spurious rays from stray electrons within the source or secondary fluorescent rays. This aperture also blocks rays that might hit the walls of the beam tunnel and then scatter and produce undesired rays. It could be used to tailor the source size in the horizontal direction to the size of each crystal to minimize radiation damage and background scatter. The exit aperture of the incident beam tunnel performs a more important function in reducing the illumination of air and glass near the crystal that would then give rise to unnecessary background by scattering from this matter. The smaller this aperture the better, providing it does not block any useful rays under any circumstances. Typically this aperture may be 1.2 mm in diameter for a 0.7-mm crystal with a

[19] U. W. Arndt, *J. Appl. Crystallogr.* **17,** 118 (1984).

source size of 1 mm. It could be smaller with a smaller source such as that used with a complete ω scan data collection or when partial illumination of the crystal is acceptable. Some systems provide a variable or easily replaceable aperture at this point. It is best to be round so that if vignetting does occur with a large crystal the amount of loss will be a smooth function of χ. If this aperture is close to the crystal it will be most effective at high 2θ but will give rise to observable scatter from its edge at low 2θ values. Placing it nine-tenths of the way from the source to the crystal is a reasonable compromise between the objectives of reducing background at very low 2θ, so that low-order reflection can be found easily, and reducing background at higher 2θ where most of the data is collected. Reducing the background at low 2θ also simplifies the background correction procedure. The entrance pupil to the diffracted beam tunnel is equally important in preventing background scatter from air and glass from reaching the detector and the same design considerations apply. The mass of air and glass in the beam and "seen" by the detectors can be comparable to the mass of the crystal. With a source-to-crystal distance of 350 mm, which is a little longer than normal, and 1.2-mm guards 40 mm from the crystal, no parsitic scatter from these apertures is seen above 1.5° 2θ.

A convenient way to check on the centering of the incident beam exit aperture and to adjust it is to mount a stiff metal rod on a goniometer and observe the shadow on a fluorescent screen a few millimeters away. If the rod diameter is three-quarters the size of the beam the relative amount of light showing on each side of the shadow will be a very sensitive indicator of proper centering, if the rod itself is centered. The rod is easily precentered with the crystal centering telescope. Vertical and horizontal centering of the guard can be monitored with the rod at $\chi = 90°$ and 0°. The same rod can be used to center the entrance pupil to the diffracted beam tunnel by removing the detector and sighting down the beam tunnel to the rod with a well-illuminated white card in front of the incident beam tunnel. In this technique the operator does not have to be close enough to the rod to focus on the shadow very clearly, and only needs to judge the amount of light on each side.

As noted above the beam tunnels can be evacuated or filled with helium to reduce absorption. If a slow flow of helium is used no windows are needed near the crystal and this source of scatter can be eliminated. To further reduce air scatter, which is a nuisance at low 2θ, helium can be introduced between the beam tunnels. Tulinsky[20] has done this by enclosing the space within the χ circle with thin Mylar film and flooding this with helium. Alternatively, the beam tunnels can be extended beyond the

[20] A. Tulinsky, personal communication.

guard apertures and helium allowed to flow out through these extensions. If they are a few millimeters in diameter they will not give rise to any parasitic scatter which can reach the detector.

The three apertures discussed here do not affect the meaningful illumination of the crystal or affect the diffracted beam and thus are not collimators in this sense. The source, the crystal, and the detector aperture define the volume of reciprocal space being observed at any moment.

Since the entrance aperture of the incident beam tunnel is small compared to the source window and the shutter plane opening normally provided, a fast, reliable auxiliary shutter capable of millions of operations with 40-msec response time can be conveniently installed at this point. This shutter can be used to protect the crystal from radiation damage at any moment when data are not being collected, such as during the drive time taken to move from one reflection to the next.

Systems with the incident beam smaller than the crystal are not considered in detail in this chapter. In such systems the background can be reduced compared to the diffracted beam and the resolution can be increased. Systematic variation in the volume of the crystal being irradiated and variation in the absorption correction with crystal orientation must be taken into account. Residual systematic error might occur in such a method.

Detectors and Pulse Height Discrimination[21,22]

There are two kinds of detectors commonly used on diffractometers. One is a scintillation detector and the other, a gas proportional counter.

Scintillation Detector. In the scintillation detector each X-ray photon is converted into a burst of light photons which are then detected by a photomultiplier tube. The scintillator is usually a thalium-activated sodium iodide crystal ~2 mm thick and 15 mm or more in diameter. The material is deliquescent and therefore must be hermetically sealed and deterioration with age may occur. The X rays enter through a beryllium window and are completely absorbed in the first few micrometers of the iodide with copper radiation, and at a somewhat greater depth for molybdenum radiation. The burst of photons occurs within 0.5 μsec. Some pass directly through the glass backing and into the photomultiplier, others reflect from a reflector on the beryllium, and some scatter from white material placed around the edges. The number of photons generated is

[21] U. W. Arndt and B. T. M. Willis, "Single Crystal Diffractometry." Cambridge Univ. Press, London and New York, 1966.

[22] W. Parrish, *in* "International Tables for Crystallography," Vol. III, p. 144. Kynoch Press, Birmingham, 1962.

stochastic and the number ultimately causing emission of primary electrons within the photomultiplier is stochastically reduced from this. The primary electrons are amplified by as much as a million, depending on the tube design and applied voltage. The ultimate burst of electrons is on the average proportional to the energy in the original X-ray photon but individually they have quite a range of amplitudes in accordance with the Poisson distribution for the random number of primary electrons generated by a given burst of photons. The full width at half-maximum on the corresponding pulse height distribution is typically 40 or 50% of the average pulse height. This is not far from the theoretical limit so further improvements are not to be expected. The consequence of this is that pulse height discrimination can be used to suppress the noise from thermal electrons ejected from the photocathode and to suppress bursts from radioactivity in the materials of the detector or from cosmic rays or from hard X rays from the source. But the pulses from the Cu K_β radiation or white radiation of wavelengths near the Cu K_α cannot be well distinguished from the desired pulses.

The discriminator levels of the single-channel analyzer used are therefore set quite wide to accept most of the K_α pulses. If they are set symmetrically a slight drift in any component of the amplification chain will not produce systematic error. Since the photomultiplier and other components may age the discriminator settings should be checked yearly. The scintillator sensitivity should also be checked but this is more difficult. The quantum efficiency is normally close to 100% but moisture penetration of the beryllium window can destroy the front of the scintillator and result in the loss of as much as 90% of the count, the loss occurring gradually over a long period so that it is not immediately obvious. The pulse heights observed may seem normal as though a given X-ray photon is either lost in a damaged layer or passed through intact.

Gas Proportional Counter. The properties of a proportional counter are very similar to those of a scintillation detector. In this detector the X-ray photon loses its energy in a series of ionization events in an absorbing gas such as argon. These primary electrons are accelerated toward a fine wire, and as they approach the wire the field increases, ultimately to a point where sufficient energy is obtained between collisions with the gas to produce secondary ionization. A cascade then occurs and produces substantial amplification. The voltage is set so that a spark does not develop and a quench gas such as methane is included in the mix to prevent this from happening. The burst of current reaching the wire is proportional, on average, to the energy of the original X-ray photon but as in the scintillation detector it varies stochastically. The resolution is better than in the proportional counter (18% full width at half-maximum) but not enough better to resolve the Cu K_β radiation. The gas counter is not as useful with molydenum radiation, which is often used in small-molecule crystallography, because of the reduced absorption and therefore is not chosen for most commercial units. Gas counters age because of degradation of the quench gas. Some detectors are rechargeable and others use constant flow. Most are sealed and expendable.

Audio Speaker

It is useful to have an audio amplifier and loudspeaker attached to the output of the pulse height analyzer. By listening to this a reflection can be detected during a fairly rapid sweep when searching for initial reflections. The speaker also gives a warning if the X-ray shutter is open when it is not intended to be.

X-Ray Filters

Since the detectors cannot reject the K_β radiation, filters or monochromators are used for this purpose.[23] The filter is most common among macromolecular crystallographers. It is the nature of all elements that the absorption coefficient increases with wavelength up to the K absorption edge and then drops precipitously when the energy is too low to eject a K electron. The absorption then increases until the L edges are reached. The increases are proportional to the cube of the wavelength. The drop at the K edge varies smoothly with atomic number, by a factor of 8.3 for nickel and 5.6 for gold.[24,25] Curves of the absorption coefficient are familiar but it is the transmission that counts in a filter. These curves depend on the thickness of the filter and are rarely shown in the textbooks.[26] Figure 13 presents transmission curves based on these general properties, in terms of percentage transmission and in terms of values normalized to the K_α transmission. The K_β radiation from any source is at 0.9 times the K_α wavelength and it can be absorbed 100-fold while allowing the K_α through by inserting a material with an absorption edge between the K_α and K_β wavelengths. Nickel is the material of choice for Cu radiation. The removal of the β radiation is at the expense of transmission of the α radia-

[23] B. W. Roberts and W. Parrish, *in* "International Tables for Crystallography," Vol. III, p. 73. Kynoch Press, Birmingham, 1962.

[24] B. Koch and C. H. MacGillavry, *in* "International Tables for Crystallography," Vol. III, p. 157. Kynoch Press, Birmingham, 1962.

[25] A. H. Compton and S. A. Allison, "X-rays in Theory and Experiment," p. 538. Van Nostrand-Reinhold, Princeton, New Jersey, 1935.

[26] V. E. Cosslett and C. W. Nixon, "X-Ray Microscopy," p. 239. Cambridge Univ. Press, London and New York, 1960.

tion. Approximately 60% of the desired radiation is commonly sacrificed. The filter cannot absorb the half-wavelength white radiation any more than the K_α and below this point the transmission is actually greater than at the K_α line. Fortunately the pulse height analyzer following the detector can eliminate these wavelengths.

The filter is placed near the source rather than near the detector so that radiation damaging to the crystal will be reduced and so that fluorescent X rays from the filter will not enter the detector directly.

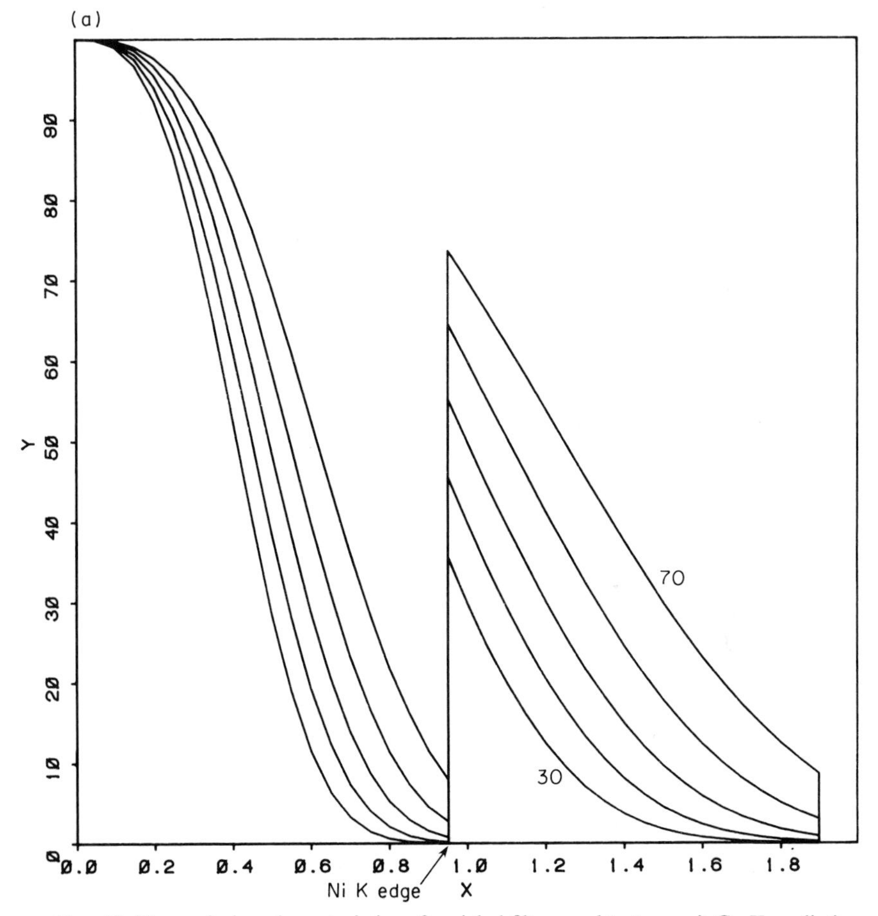

FIG. 13. Transmission characteristics of a nickel filter used to transmit Cu K_α radiation while absorbing the undesired K_β emission. Thicknesses transmitting 30–70% of the K_α line are assumed. (a) Fractional transmissions vs fraction of the K_α wavelength. (b) Transmission normalized to the K_α transmission. Below the half-wavelength the transmission of the white radiation is enhanced compared to the K_α peak. The curves are essentially universal for the X-ray filters if scaled to the K absorption edge wavelength.

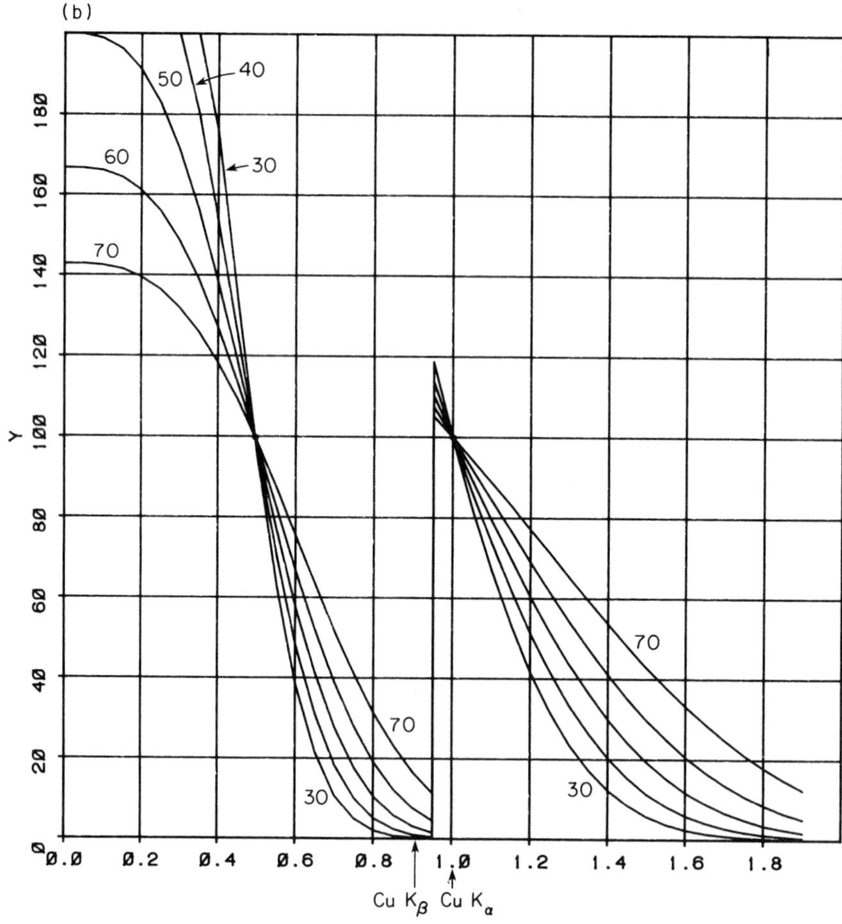

FIG. 13. (*continued*)

Monochromator

If a monochromator is used, there is considerable loss in intensity but some of this is gained back by removal of the β filter. The net intensity with a graphite monochromator may be as much as 70% of that obtained with a filter.[27] The filter cannot remove the white radiation just on the long wavelength side of the α peak, whereas the monochromator can. Along densely packed principal axes this is a significant gain but for the vast

[27] A. W. Moore, *in* "Chemistry and Physics of Carbon," Vol. II, pp. 69–187. Marcel Dekker, Inc., New York, 1973.

majority of the data it is not critical. One potential problem with a monochromator is nonuniformity of the beam at the crystal. Other factors that must be considered are the effect on the apparent source size, the cross-fire, and dispersion within the desired wavelength band. Furnas[28] pointed out that if the monochromator disperses the beam horizontally then effects on reflections at $\pm 2\theta$ will be different. He recommended vertical dispersion by the monochromator to remove this effect. The diffracted beam outline becomes tilted slightly in this case. In the former case when the monochromator and the sample crystals deflect the beam in opposite directions the reflections will be sharpened at intermediate 2θ values.

Monochromators have been reported to improve the lifetime of some crystals. Clearly the irradiation is reduced by the monochromator but collimation, that is, reduction of cross-fire, may be equally important. Some crystals have a considerable temporal component in the radiation damage independent of irradiation rate and this would diminish the benefit of the monochromator.

Operational X-Ray Source Considerations

Sources are considered in a separate chapter[17] but one observation needs to be added. If a larger source is desired for the diffractometer compared to film systems then a larger take-off angle may be desired. In this case the optimum voltage is higher than reported in that chapter. The reason is connected with the depth of penetration of the electrons into the target and self-absorption as the rays exit at a shallow angle.[29] The X-ray source is shallow enough so that photons emitted from the deepest portion still have a high probability of being transmitted if they are traveling perpendicular to the surface. The main consequence of this is that the brightness of the source will increase when looked at obliquely, since a greater apparent depth will be viewed. The source will get smaller, and the total intensity will decrease. If the source were uniform as a function of depth the brightness would never decrease. The high-energy electrons entering the surface are scattered significantly and thus curl around and lose energy more as a function of depth the deeper they penetrate. The source is inherently brighter at some depth, depending on the incident voltage, than at the surface. The less bright surface layer still absorbs X rays and the net effect is that at some shallow angle the brightness is reduced by this absorption more than it is increased by the greater apparent depth. This angle is voltage dependent. The inherent efficiency of X-

[28] T. C. Furnas, U.S. Patent No. 3394255 (1968).

[29] M. Green, *in* "X-Ray Optics and X-Ray Microanalysis" (H. H. Pattee, V. E. Cosslett, and A. Angstrom, eds.,), p. 185. Academic Press, New York, 1963.

ray production increases with voltage at constant power due to the threshold excitation value and the 1.63 power dependence on voltage above this level.[29,30] High voltage is favored until the depth factor outweighs these considerations. At a take-off angle of 7.5° the brightness is still increasing at 50 kV while at 2° the optimum is near 35 kV. Target absorbtion will also affect the white radiation spectrum but that is not of prime importance.

[30] A. H. Compton and S. A. Allison, "X-rays in Theory and Experiment," p. 81. Van Nostrand Reinhold, Princeton, New Jersey, 1935.

[25] Measuring X-Ray Diffraction Data from Large Proteins with X-Ray Diffractometry

By Robert J. Fletterick and Jurgen Sygusch

Introduction

Film or area electronic counters are the obvious detection systems for the measurement of X-ray diffraction data from large macromolecules. The single counter of an automatic diffractometer can be used successfully and perhaps to advantage in special cases in which crystal supply is abundant or the irradiated lifetime is long. The decision to use single-counter diffractometry for a particular case depends on the available alternatives and the amount of data and accuracy required to determine the structure. This chapter addresses some of the points that should be considered for diffractometer setup and some procedures for successful data collection. Other details will be found elsewhere in this volume and in cited literature. Specific procedures from our data measurements on the enzyme glycogen phosphorylase *a* are given. The crystals are $P4_32_12$, a = 128.3, c = 116.5. The asymmetric unit has a single polypeptide chain of 97,400 daltons. There are 55,000 unique reflections in the 2.1 Å data set.

Radiation Decay Corrections

The diffractometer is an instrument which is ideally suited for monitoring and providing the means for correction for radiation damage to protein crystals. The exposure of protein crystals to X rays necessarily damages the structure of the crystal, and the damage is manifest as a loss in diffracted intensity. Generally, the measurable intensity for reflections is decreased with increasing exposure time, though this diminution of scattered intensity is pronounced with increasing scattering angles. Blake and Phillips[1] have estimated that 160 molecules of myoglobin are destroyed by each absorbed X-ray photon. Even though the mechanism of the destruction of scattering material is largely uncharacterized, it is possible to monitor the damage and empirically correct the scattering amplitudes.

A number of schemes have been proposed to correct for radiation damage by rescaling the observed intensities. Blake and Phillips suggested three components of scattering material be considered. These are A_1, an undamaged fraction; A_2, a highly disordered fraction; and A_3, a fraction which scatters incoherently.

Fraction A_3 is considered to contribute significantly only at low scattering angle. Transitions from states 1 to 2 or 3 and from 2 to 3 are allowed. The disordered fraction is described by an increased smearing of the ordered fraction characterized by a mean square positional displacement of $D/8\pi$. The relative intensity, therefore, is given by equation:

$$I(t)/I(0) = A_1(t) + A_2(t) \exp[-D(\sin \theta)/\lambda]^2 \tag{1}$$

The intensity correction for the glycogen phosphorylase diffraction data used by Fletterick *et al.*[2] modified this model slightly. With the interpretation of the transitions between the states as successive conformational transitions, the reaction kinetics are first order and the rate model then becomes

$$A_1 \xrightarrow{k_1} A_2 \xrightarrow{k_2} A_3$$

The total time-dependent diffraction intensity is expressed as

$$I(t) = \sum_{i=1}^{3} A_i(t) I_i \tag{2}$$

where the I_i values are related by Fourier transform to the electron density for the ith state. This model differs from that of Blake and Phillips by disallowing direct transitions between A_1 and A_3.

[1] C. F. Blake and D. C. Philips, *in* "Biological Effects of Ionizing Radiations at the Molecular Level," p. 183. IAEA, Vienna, 1962.

[2] R. J. Fletterick, J. Sygusch, N. Murray, N. B. Madsen, and L. N. Johnson, *J. Mol. Biol.* **103,** 1 (1976).

The distribution of electron density among the three states can be quantitated by the distribution function $h(r)$ defined by

$$\rho_2(r) = \int_V \rho_1(r')h(r - r')\, dr' \tag{3}$$

Here ρ_1 represents the electron density distribution of the native state while ρ_2 is that for the intermediate disordered state. For the smearing of function h, a Gaussian model is assumed, i.e.,

$$h(r) = (1/B) \exp[-\pi r^2/D] \tag{4}$$

where D is the average mean square smearing of density ρ_1 from its starting value.

The variation of the diffracted intensity as a function of time, on Fourier transformation of the electron densities, becomes

$$I(t) = I(0)\{\exp(-k_1 t) + \exp(-8\pi B) \frac{\sin^2\theta}{\lambda^2} \frac{K_1}{k_1 - k_2} [\exp(-k_2 t) - \exp(-k_1 t)]\} \tag{5}$$

where $I(0)$ is the intensity at time $t = 0$. If k_1 is approximately equal to k_2, this expression reduces to Eq. (6).

$$I(t) = I(0)[\exp(-kt) + \exp(-8\pi B) \frac{\sin^2\theta}{\lambda^2} kt \exp(-kt)] \tag{6}$$

Figure 1 shows the rate of observed intensity drop as a function of time for four reflections at different scattering angles. The experimental observation points are analyzed by nonlinear least-squares determination of the radiation decay parameters and are not fit as straight lines.

Hendrickson[3] has generalized the effects of radiation damage to protein crystals by the path shown below.

$$\begin{array}{ccc} \text{native } (A_1) & \xrightarrow{k_3} & \text{amorphous } (A_3) \\ & k_1 \searrow \quad \nearrow k_2 & \\ & \text{disordered } (A_2) & \end{array}$$

By setting up the differential equations subject to appropriate boundary conditions, the system of differential equations can be solved to give the expression

$$\frac{k_1}{k_1 + k_3 - k_2} = \exp(-k_2 t)\{1 - \exp[-(k_1 + k_3 - k_2)t]\} \exp(-Ds^2) \tag{7}$$

[3] W. A. Hendrickson, *J. Mol. Biol.* **106,** 889 (1976).

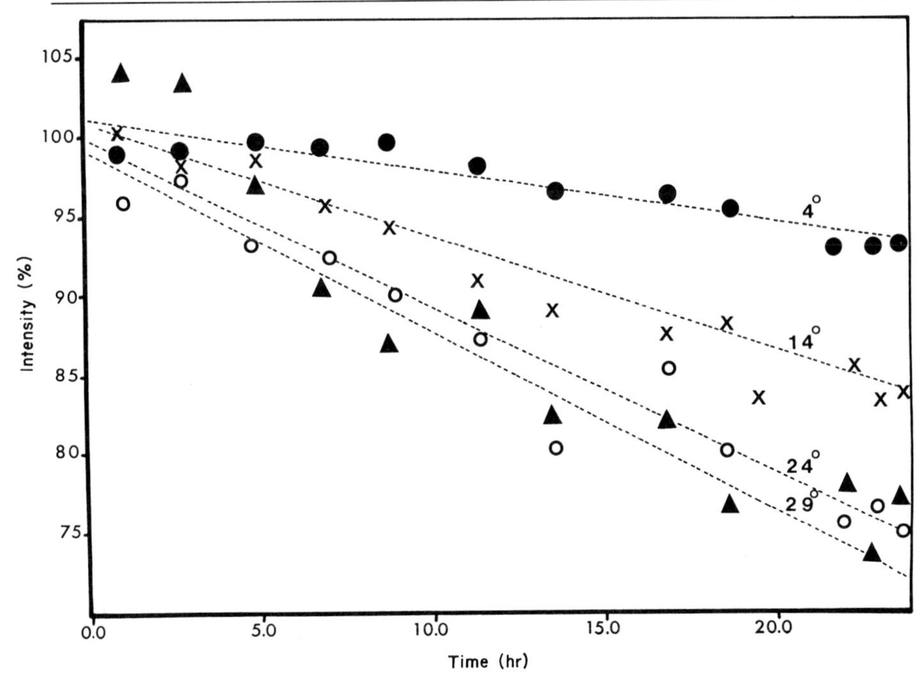

FIG. 1. The percentage of intensity measured, relative to time $t = 0$, as the ordinate is plotted against the time of the measurement in hours. Four monitored reflections are shown at 2θ of 4° (●), 14° (×), 24° (○), and 29° (▲). The lines fitting the observation points result from a least-squares determination of the radiation decay parameters for the model as described in the text [Eq. (5)]. These data are from the phosphorylase *a* native crystal. For clarity not all data points have been plotted.

If $k_1 + k_3$ is equal to k_2, the equation reduces to

$$I(t,s) = \exp(-k_2 t) + k_1 t \exp(-k_2 t) \exp(-Ds^2) \tag{8}$$

The equation used earlier in Hendrickson[4] can be obtained by setting $k_2 = k_3$ (note the original paper has a sign error). The expression used by Fletterick *et al.*[2] and in Eq. (6) above is obtained by setting $k_3 = 0$ and redefining the disorder parameter D. Equation (6) above can be obtained from Eq. (7) by setting $k_1 = k_2$ while simultaneously setting $k_3 = 0$. The analysis suggests the native molecules are first partially destroyed at one rate and then those are susceptible to complete destruction at yet another rate. No single process in this formulation will completely destroy a native molecule.

[4] W. A. Hendrickson, W. E. Love, and J. Karle, *J. Mol. Biol.* **74,** 331 (1973).

Hendrickson has tested the above models[3] using the radiation-damaged intensity data for myoglobin from Blake and Phillips.[1] The least-squares fits were made to the data sets for six different models, including the Blake and Phillips model with arbitrary time dependence. Hendrickson concluded that all of the models worked quite well to rationalize the data at moderate radiation damage, which is up to 25% loss of intensity. A second conclusion was that no rate model would adequately correct the radiation damage at 72% loss in intensity. However, at high levels of damage, up to 47% loss in intensity, the damage could still be described quite well by a simple rate model. This is an important conclusion from this study and the expression which seems to work well is a model in which k_3 is 0. This is Eq. (6) above.

Radiation damage corrections have traditionally been made in protein crystallography by periodic monitoring of a few reflections. It is recommended that 10 or preferably more reflections be constantly monitored (at about every one-tenth to one-twentieth of total data collection time) during the intensity measurement with careful attention to the distribution of those reflections in scattering angles. In order to use any of these models accurately it is essential to monitor the time dependency of the reflections over a range of scattering angles (e.g., $5° < 2\theta < 30°$) in order to provide maximum information on the disordering parameter D. The conclusion from these studies is that corrections can be made to much greater levels of radiation damage than are normally acceptable. One caveat with this correction paradigm is that the various reflections chosen to be monitored may not be fairly representative of the total radiation damage undergone by the crystal during data collection. Individual reflections will be found which will not be corrected; indeed, some reflections can increase in intensity after radiation damage.

Monochromators

The use of monochromators for X-ray data collection on diffractometers is relatively common in small-molecule studies. However, relatively few laboratories use monochromators for data measurement from single crystals of proteins or nucleic acids. The difficulty in setting up the monochromator system and making the appropriate adjustments as well as the change in scattering profile in the plane perpendicular to the incident and diffracted beam can be drawbacks. If a monochromator is used, then considerable attention must be given to measuring the background and determining peak shape as a function of position in the reciprocal lattice. This reduces the usefulness of simplified procedures such as the ω step scan[5] and data collection in which only the intensity about $\pm(0.1\text{–}0.2°)$ about the peak maximum is measured.

The unpopularity of the monochromator in data collection is surprising since the use of a monochromator has two positive features. While slightly diminishing the incident intensity, the monochromator has a pronounced effect at diminishing the background scattering. If however, a diffractometer is properly configured for data collection from large unit cells with attention to guard slits and windows[5] and if the collimated X-ray beam entry and exit slits are as close as possible to the crystal, then this is not a significant advantage over using filtered radiation. A second advantage is that monochromatized X rays are almost always less damaging to the protein crystal than unfiltered or nickel foil-filtered X rays. Stroud has reported a 7-fold increase in the useful life of trypsin crystals on using a graphite monochromator versus using nickel-filtered radiation.[6] A partial explanation for some of this increase in lifetime is certainly the lessened intensity of the direct beam. Michael James' laboratory has routinely used lower intensities (40 kV by 20 mA) for data collection on serine proteases.[7] It was found that running the X-ray tube at lower power also produces significantly lower radiation damage, though not the factor of 7 reported by Stroud.

Background

As with all scientific measurements, the signal-to-background ratio is to be maximized in order to obtain the most information from the experiment. For protein data collection using the diffractometer, it is essential to give extreme attention to lowering the background since the scattered intensity is very weak due to the typically large unit cell volume.

The most important determinant of the background in most experiments, excluding the flow cell technique,[5] is the glass capillary tube which houses the protein crystal. It is perhaps surprising that the scattered background of radiation is less from the protein and mother liquor. Stroud and collaborators have addressed this issue experimentally and quantified the effects.[8,9] Their conclusion is that the glass capillary should be kept

[5] H. W. Wyckoff, M. Doscher, D. Tsernoglou, T. Inagami, D. M. Johnson, and F. M. Richards, *J. Mol. Biol.* **27,** 563 (1967).

[6] R. M. Stroud, L. M. Kay, and R. E. Dickerson, *J. Mol. Biol.* **83,** 185 (1974).

[7] G. D. Brayer, L. T. Delbaere, and M. N. James, *J. Mol. Biol.* **124,** 241 (1978).

[8] M. Krieger, J. L. Chambers, G. G. Christoph, R. M. Stroud, and B. L. Trus, *Acta Crystallogr., Sect. A* **A30,** 740 (1974).

[9] M. Krieger and R. M. Stroud, *Acta Crystallogr., Sect. A* **A32,** 653 (1976).

uniformly bathed in the X-ray beam if possible. This can be accomplished by using a glass capillary which just allows the protein crystal to fit and fills the irradiated volume of the capillary as nearly as possible. (It should be noted that the glass capillaries which are available from most commercial sources are not low-scattering glass. They are constructed from sodium silicate and thus have a standard composition.)

A second important aspect of background reduction is to diminish air scatter from the irradiated volume on the diffractometer. Attention must be paid in particular to a guard aperture which should be positioned as close to the crystals as possible. The guard aperture is ideally only slightly larger than the diffracted beam at that point from the strongest reflection emanating from the crystal in an orientation which shows the largest cross section. In practice, a variable set of circular apertures are used and substituted according to size of crystal. It is important to keep this aperture as close to the crystal as possible since air can be eliminated immediately behind this aperture by enclosing the diffracted beam pathway in a helium-filled tube.[5] In phosphorylase *a*, the background was measured at the start of and at the end of data collection at one-half integral lattice points, both as a function ϕ and of 2θ. The total background counts were found to be the same before and after data collection, a result suggesting, as stated above, that the crystal makes little time-dependent contribution to the background even in the presence of radiation damage. Consequently, the backgrounds can be very well determined by counting for *long* times at the end of data collection, after the crystal has been considerably radiation damaged. This allows a very accurate determination of the background yet does not sacrifice valuable crystal life better employed for intensity data collection.

X-Ray Tube

If crystals are of a convenient size (0.3–1.0 mm), it is simplest to use a standard focus copper K_α high-intensity X-ray tube which provides a large uniform beam cross section. These tubes run reliably anywhere up to 5000 hr at 2000 W in favorable cases. Some laboratories have routinely run these tubes at 2250 W by increasing the milliamperes setting to 45 at a 50-kV accelerating voltage. In the author's experience, this practice has not significantly diminished the lifetime of the X-ray tube though extreme caution must be used if the diffractometer tube housing has not been rated at this power.

Several laboratories have diffractometers using 6- or 12-kW rotating anode sources with monochromators. These seem to work well where exceptional source brightness is required. These machines are very fast at data collection and when fitted with two-dimensional electronic detectors represent the state of the art in laboratory data measurement.

Strategies for X-Ray Data Collection on the Diffractometer

The strategy for data collection using the diffractometer is dependent on the type of data to be collected, whether they be data for the native protein, for heavy-atom derivatives of the native protein, or for difference Fourier maps. The measurement of diffraction data for native protein crystals is usually done in shells of resolution (intervals of 2θ).

Almost always, new data collection on a macromolecule is initiated at low limiting resolution such as 5, 6, or 7 Å. In measuring an initial three-dimensional data set, there is little to recommend which resolution between 7 and 5 Å is used as the limiting one. The reason for this is that these low-resolution images of usual macromolecular structures are approximately equivalent and the location of the heavy-atom sites can be usually accomplished at 7, 6, or 5 Å resolution. The advantage of measuring X-ray data at 7 Å resolution is that this limiting shell avoids the drop-off in intensity that usually occurs in approximately the 6 Å resolution region for proteins. Since there are 40% fewer reflections and they are more intense, rapid scans may be used. If crystal size or quality is severely limiting, it is quite appropriate to begin native and heavy-atom derivative collections at limiting resolution of 7 Å. If the lattice constants are unusually large so that there are a large number of measurements to be made, it is also advantageous to work at this lower resolution since the 7 Å sphere of resolution contains approximately one-third the number of reflections that are in the 5 Å resolution range.

There are, of course, advantages to working at a 5 Å resolution sphere. These are that the map indeed has a better appearance at this resolution and, second, that this larger set of data could be useful for solution of heavy-atom locations when there are many sites of substitutions. However, the use of direct methods at 7 Å resolution to find four equally substituted mercury positions was successful for a crystal form of yeast hexokinase B containing a 100,000-dalton asymmetric unit.[10]

Measurement of Weaker Reflections

Beyond 5 Å resolution, the data collection procedure should be altered to take into account the variation in intensity during the acquisition time. For measurements of heavy-atom derivative data all reflections should be

[10] W. F. Anderson, R. J. Fletterick, and T. A. Steitz, *J. Mol. Biol.* **86,** 261 (1974).

ideally measured with the same low percentage estimated standard deviation. This means that relatively little counting time should be given to strong reflections, but correspondingly, long counting times should be used to measure weaker reflections. This generality does not apply to the background data measurement since the backgrounds are, within a factor of 3 or so, approximately equal.

The obvious way of measuring reflections while taking into account their relative intensity is to do a quick scan (a few seconds) of the reflected intensity at the peak height, for example, and to use that information to determine how much counting time should be allocated to this reflection. Generally this works well in practice, provided most reflections are not weak. Precount scans are also useful to interrupt automatic peak search or scan software for those reflections which are too weak to be used to reliably reset the crystal parameters.

Using Lists of hkl for Data Measurement

An alternative to the above scheme which has advantages for measurement of data for difference Fourier analyses is to measure a complete data set at the working resolution for the native data and to process those data and sort the data set on uncorrected intensity. This sorted list can then be used to instruct the diffractometer software to measure diffraction data of a given shell of intensity. In order to keep the diffractometer from wildly wandering through reciprocal space (because the sorting procedure has discounted the lattice), it is necessary to sort the sorted list of intensities additionally on *h, k,* and *l.* Because counting times are relatively longer than slewing times for protein crystals, it has been found convenient to sort the sorted intensity list in blocks of as few as 100 reflections and to order the *hkl* values within those blocks in the usual fashion to minimize slewing time between reflections.

For the phosphorylase *a* data collection using diffractometer, the above scheme was used to prepare double-sorted lists at resolutions of 4.5, 3.5, 3, 2.5, 2.3, and 2.1 Å. These cutoff values in resolution were found to be convenient for data collection for the phosphorylase crystals which diffracted only to about the 2.1 Å sphere on a diffractometer equipped with 2000-W normal-focus sealed tube. (The heavy-atom data to 2.5 Å resolution were sufficient to produce a high-quality interpretable X-ray map, so phasing beyond that resolution was done by calculation.)

The measurement of anomalous data using the sorted list protocol proceeds as normal in that for each *hkl* generated within the double-sorted list an enantiomorphic reflection index is also generated. These enantiomorphs are interleaved within the data collection list so that a reflection and its Friedel mate are measured one after the other. Alternatively, if radiation decay is not a significant problem, the enantiomorphs could be measured after proceeding down one lattice line to minimize slewing time. Both of these procedures seem to give satisfactory results as judged by the merging *R* value on reflections which show a significant anomalous scattering contribution.[11]

Even with the above list it is, of course, impossible to increase the counting time sufficiently to measure the weakest reflections with the same percentage estimated standard deviations as of the strongest reflections. The compromise in counting time must be adjusted in order to obtain sufficient quality data to solve the heavy-atom position.

The use of list-directed data can be quite efficient for data collection. In the case of glycogen phosphorylase *a*, a complete 2.5 Å resolution data set (26 crystals) for a heavy atom which proved useful in solving the structure was obtained in 30 days from start to finish. This included collecting the entire set of anomalous data, scaling data, and, of course, the computing time for scaling and merging (about 30,000 reflections in all). The 30-day time period also included a few data sets which were measured but not usable due to high merging *R* factors.

Scaling

Multiple crystals are required for most data sets measured by single-counter diffractometry. Block scaling is often necessary to remove systematic errors but it is always preferable to scale partial data sets from separate crystals and merge them before scaling two complete data sets. It is important to note that the 26 data sets referred to above could be merged without block scaling. The data were carefully monitored for systematic error. The last 50–100 reflections measured at the end of data collection for one crystal were remeasured at the beginning of data collection for the successive crystal. Merging statistics for these reflections accurately represent the worst cases in systematic errors since they reflect variations due to crystal mounting, radiation decay, etc. Scaling is conveniently done by measuring the same set of 50 strong reflections, distributed in 2θ, at the start of data collection for each crystal. In the case of phosphorylase *a*, the 50 reflections were chosen pairwise on the basis of possible enantiomorphic discrimination. The criterion for acceptance of a new data set pairwise was that on the average, the pairwise differences within statistical bounds from the new data set are comparable to the differences of the previous data sets and are of the same sign.[11] The

[11] J. Sygusch and R. J. Fletterick, unpublished results.

data set was rejected if these criteria were not met. Additional scaling data in a plane (such as *hk*0 in phosphorylase *a*) are useful to assess the meaning of the merging *R* values. The only limitation of the above procedure would be the availability of a very large number of high-quality crystals. It should be noted that the same amount of data could be measured from two to four crystals using a two-dimensional area detector.

Difference Fouriers

Measurement of data for difference Fouriers can proceed quite readily using the list concept developed above. Though it is not necessary to measure anomolous pairs or symmetry mates for difference Fouriers, this should be done routinely to eliminate systematic errors in the intensity measurements. If a rough Fourier map is required merely to assay binding then it is not essential to measure duplicate reflections.

Further, it is not essential to use a complete set of data to a given resolution for difference Fourier measurements. The reason for this is that a large number of reflections can still be used to produce the image in the difference Fourier and series truncation effects are not a problem. At 4.5 Å resolution, there are approximately 6000 reflections for the glycogen phosphorylase crystals discussed above. It was the usual procedure to measure only partial lists, in this case, 2800 reflections which were the strongest in intensity, as this number could be conveniently collected in a 16-hr period from moderately sized crystals. This fragmentation of the data into strong and weak lists does not negatively influence the quality of the difference Fourier electron density maps. It is thus possible to measure a set of difference Fourier coefficients in 12–16 hr and produce data of sufficient quality to resolve the three chemical groups of AMP, for example.

Similarly, the same scheme can be used at higher resolution. Only about half or a third of the data to 2.8 Å need be used to generate a high-quality difference map. The intensity-sorted list is particularly useful for this. The intensity list used for the native data collection works perfectly well for the measurement of difference Fouriers since the relative intensity change is usually small. It is, however, advisable to modify the list during the initial stages of a structural analysis to account for the phasing errors as well. Thus the intensity list which is double sorted as described above should be further sorted to take into account the figure of merit associated with the phase for each given reflection. In this way, even when the overall figure of merit is not particularly good, it is possible to produce an enhanced difference Fourier map by using only reflections which are intense, well phased, and uniformly distributed in resolution.

Using these kinds of procedures with list-directed intensity measurement and careful attention to absorption and decay corrections, it is entirely feasible to collect full sets of diffraction data from as many as hundreds of crystals and to merge them quite successfully into a single data set of block scaling.

[26] On the Design of Diffractometers to Measure a Number of Reflections Simultaneously

By P. J. ARTYMIUK and D. C. PHILLIPS

Introduction

On a conventional diffractometer only one reflection can be measured at a time even though the conditions for reflection may be satisfied precisely for two, or approximately for many, reflections. This chapter is concerned with the methods for increasing data collection rates on diffractometers by the use of linear arrays of conventional proportional counters or of linear position sensitive detectors so that more than one reflection may be measured in a scan. These methods may be divided into two classes:

1. Those that deliberately exploit the possibility of simultaneous or quasi-simultaneous measurement of explicit groups of reflections in a short scan of about the same width as that required for a single reflection.
2. Those in which the linear detector is placed so that it will eventually intercept the diffracted beams from all the unique reflections within a given level normal to the mounting axis in the course of a complete rotation of (for example) 90° or 180°.

This chapter is concerned with methods in the first class,[1-3] but a brief review is given of methods of the second kind in the final section. With the use of position-sensitive detectors an order of magnitude increase in the rates of data collection can be obtained using either type of method.

[1] D. C. Phillips, *J. Sci. Instrum.* **41,** 123 (1964).
[2] U. W. Arndt, A. C. T. North, and D. C. Phillips, *J. Sci. Instrum.* **41,** 421 (1964).
[3] D. W. Banner, P. R. Evans, D. J. Marsh, and D. C. Phillips, *J. Appl. Crystallogr.* **10,** 45 (1977).

Principles of Simultaneous and Quasi-Simultaneous Measurement of Reflections

It is convenient to define the reciprocal lattice vector **S** in terms of cylindrical polar coordinates (ξ, ζ, and τ).[4] **S** is resolved into two components: ζ is parallel to the rotation axis; ξ is perpendicular to it; τ is the angle of the projection of **S** onto the plane normal to the rotation axis, with the direction of the incident beam being defined as $\tau = 0$. It is also useful to describe the direction of the diffracted beam by an azimuthal angle Υ normal to the plane containing the incident beam and the rotation axis.

Simultaneous Reflections

According to the familiar Ewald construction the conditions of reflection for a reflection *hkl* are satisfied when the corresponding reciprocal lattice point lies in the surface of the sphere of reflection. The conditions for this reflection remain satisfied if the crystal (and hence its reciprocal lattice) is rotated about the normal to the reflecting planes, i.e., about the reciprocal lattice vector **S.** During this rotation many of the other reciprocal lattice points pass through the surface of the sphere of reflection, and the corresponding reflections take place. Double reflections are well known to introduce errors into intensity measurements. These can be avoided by careful adjustment of the crystal orientation so as to satisfy the conditions for only one reflection at a time,[5] but this procedure is seldom attempted and indeed is impracticable in the study of complex crystals with large unit cells. Fortunately, the errors due to double reflection from such crystals seem usually to be small, and since they cannot be eliminated, the occurrence of simultaneous reflections can be exploited to speed up the collection of data.

The proportion of the other reflections within the whole of the limiting sphere which can be obtained simultaneously with any given reflection of reciprocal lattice vector **S** is a function of S and can be calculated easily from the volume of reciprocal space swept out by the surface of the sphere of reflection as it is rotated about **S.** This proportion, b, varies from 58.9% at $S = 0$ to zero at $S = 2$ according to the law $b = (3\pi/16)(1 - S^2/4)^{1/2}$, the radius of the Ewald sphere being taken to be unity. To any limited resolution a much greater proportion of the reflections can be simultaneously measured.

Within this limitation any two reciprocal lattice points can be brought into the sphere of reflection at the same time. However, since the origin, a

[4] M. J. Buerger, "X-ray Crystallography." Wiley, New York, 1942.

[5] E.g., M. Zocchi and A. Santoro, *Acta Crystallogr* **16,** A155 (1963).

third point, must always be in the surface of the sphere, it is not generally possible to satisfy the conditions for more than two points. This is, however, sometimes possible for reflections which are specially related to one another. One of the ways this can be achieved is most readily demonstrated by reference to inclination Weissenberg geometry.

If the crystal is mounted to rotate about a crystal axis which is normal to the other two axes and is thus coincident with a reciprocal lattice axis (say c^*), then any pair of reciprocal lattice points hkl_1 and hkl_2 which lie symmetrically on either side of the flat cone will lie simultaneously on the surface of the Ewald sphere.[1] In Fig. 1 this is shown for two reflections, P and Q, which are symmetrically related by the flat cone (as are the equi-inclination and anti-equi-inclination levels). In general the levels have a blind region at their centers within which the reciprocal lattice points cannot be brought into the sphere of reflection. For levels at distance $z = \zeta_1 - \zeta/2 = (\zeta/2) - \zeta_2$ from the level at $\zeta/2$ in the flat-cone setting the radius of this blind region is $(1 - z^2)^{1/2} - [1 - (\zeta^2/4)]^{1/2}$.

Subject to this reservation, then, the reflections in any two levels can be made to appear simultaneously in pairs simply by setting the inclination of a Weissenberg-type instrument so that the two levels at ζ_1 and ζ_2

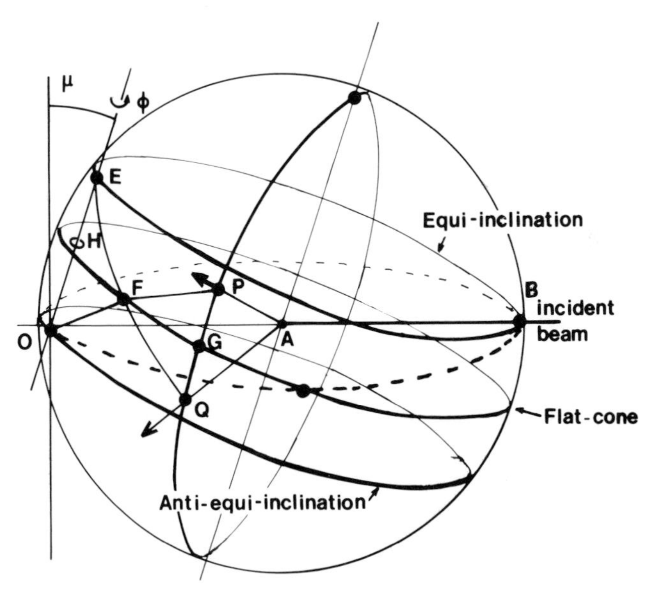

FIG. 1. Perspective drawing of the sphere of reflection showing inclination Weissenberg geometry with simultaneous reflections (reciprocal lattice points P and Q from general levels) symmetrically related by the flat-cone setting.

are symmetrically related by the flat cone, i.e., with

$$\sin \mu = (1/2)(\zeta_1 + \zeta_2)$$

The reflected beams lie in a plane containing the axis of rotation and they are inclined to that axis at angles $90° \pm \nu$, where

$$\sin \nu = (1/2)(\zeta_1 - \zeta_2)$$

Weissenberg Settings for Quasi-Simultaneous Reflections

When a crystal has large cell dimensions and two reflections are made to appear simultaneously, many other reflections appear quasi-simultaneously. The principles of quasi-simultaneous measurement of reflections were first described by Phillips[1] and were later extended by Banner *et al.*[3] (see p. 403).

The integrated intensities of reflections are usually measured by rotating the crystal through the reflecting position while the total reflected intensity is accumulated in a detector system. The rocking range must be great enough for all parts of the crystal to reflect from all parts of the X-ray tube focus and this depends, for example, on the geometry of the diffractometer, the mosaic structure of the crystal, and the range of wavelengths that have to be taken into account. In many investigations the required range is about 1° and this can be increased to take account of any error in setting, so long as the resolution of individual reflections is not impaired. When the spacing between reciprocal lattice levels is small enough, reflections from a number of them can be made to appear quasi-simultaneously during the same oscillation, even though the conditions for simultaneity are not satisfied.

General Settings. Figure 2a shows an elevation of the sphere of reflection with a crystal set for the upper level at ζ to be recorded in the flat-cone setting. Two general levels are shown at distances Z and $Z + z$ from the flat-cone level. The corresponding circles of reflection are shown in Fig. 2b, which also shows the points at which they are intersected by reciprocal lattice points at distance ξ from the axis of rotation and the directions of the corresponding reflected beams. The difference $\Delta\varphi$ in setting angle φ for reflections with the same ξ values in the levels $\zeta + Z$ and $\zeta + Z + z$ is easily calculated[1]:
for small values of $\Delta\varphi$,

$$\Delta\varphi = -(z^2 + 2Zz)/[2(1 - \zeta^2)^{1/2}\xi \sin \varphi] \quad (1)$$

where

$$\cos \varphi = (\xi^2 - \zeta^2 + Z^2)/[2(1 - \zeta^2)^{1/2}\xi]$$

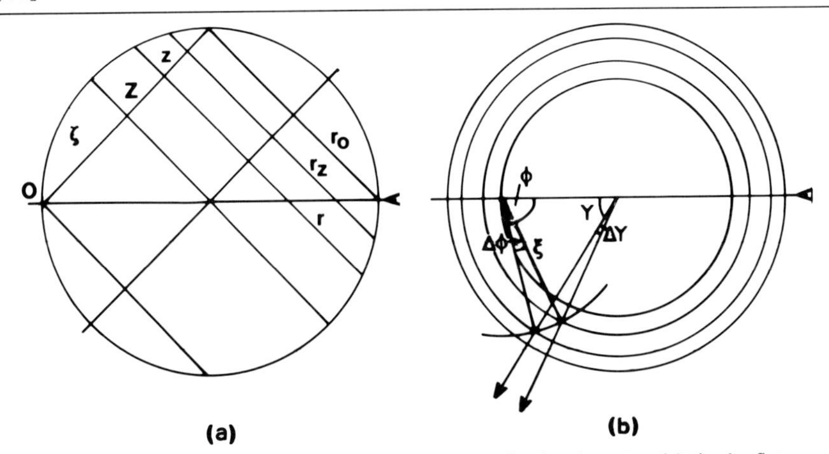

FIG. 2. (a) Elevation of the sphere of reflection showing levels at ζ and 2ζ in the flat-cone and equi-inclination settings, respectively, and general levels at $\zeta + Z$ and $\zeta + Z + z$. (b) Corresponding circles of reflection with the directions of reflected beams corresponding to reciprocal lattice points at ξ in the general levels.

Flat-Cone Setting. Differentiating Eq. (1) above shows that $\Delta\varphi$ changes least with z when $z = -Z$, i.e., near the flat-cone setting where the radii of the circles of reflection are most nearly constant. Near the flat cone when $Z = 0$,

$$z^2 = -[4\xi^2 - (\xi^2 + \zeta^2)^2]^{1/2}\,\Delta\varphi \qquad (2)$$

The region of reciprocal space within which $\Delta\varphi \leq 0.2°$, for example, is much more extensive in the flat-cone setting than in the equi-inclination setting and it varies must less with ζ and ξ.[1] There is, however, a blind region which increases with ζ, the minimum observable ξ being $1 - (1 - \zeta^2)^{1/2}$.

The acceptable values of z, that is, the separations of levels, in which reflections can be measured at the same time as those in the flat-cone setting are rather larger than Eq. (2) would suggest, since the criteria for $\Delta\varphi$ can be relaxed slightly: Eq. (1) shows that $\Delta\varphi$ is always negative with respect to the flat-cone level; if φ is set, therefore, at $+0.2°$ (for example) from its correct setting for the flat-cone level, a value of $\Delta\varphi$ of $-0.4°$ may be tolerated since the reflections in the flat-cone and neighboring levels would each be displaced by only 0.2° from the center of the rocking range. The region of reciprocal space that can be usefully measured becomes restricted when z becomes greater than (for example) 0.04 (equivalent to real spacings of less than 40 Å). The additional limitation in the minimum values of ξ for reflections that can be measured quasi-simultaneously corresponds to an effective increase in the radius of the blind region.

The difference in the values of Υ for the different reflections (i.e., ΔΥ, the equatorial component of the counter setting error) is given by

$$\Delta\Upsilon = [(\xi^2 + \zeta^2)z^2]/\{2[4\xi^2 - (\xi^2 + \zeta^2)^2]^{1/2}\}$$

with $\Delta\Upsilon_{min} = \xi z^2/[2(1 - \zeta^2)^{1/2}]$ for constant z and ζ.

This difference essentially imposes a restriction on the maximum resolution that can be obtained without requiring different positions of the counters for different levels. The maximum value of ΔΥ which can be tolerated is governed by the width of the detector apertures and the restriction is only important at high ξ.

Quasi-Simultaneous Measurement of Reflections on an Inclination Diffractometer

General. The linear diffractometer[6] was the first diffractometer to be converted to measure reflections quasi-simultaneously.[2] This machine was designed to be used with the flat-cone geometry and therefore the only modification required was the replacement of the single counter with a linear array of counters, and their associated electronics. At first three counters were used, but later this was increased to five.

The proportional counters were mounted a fixed distance u apart, and the crystal-to-counter distance D made variable. This made it possible to set the reciprocal lattice spacing of the axis the crystal was mounted about (a^*) by varying D:

$$a^* = u/D$$

In practice machine limits meant that $0.015 < a^* < 0.034$ r.l.u. The line of counters was parallel to the crystal rotation axis. It was found that ΔΥ was so small ($\Upsilon_{max} = 64°$) that there was no need to make special provision for it.

In order that five reflections could be measured at once with this instrument the crystal had to be mounted to rotate about a reciprocal lattice axis which was perpendicular to a principal reciprocal lattice plane (for monoclinic crystals the unique axis had to be used; triclinic crystals presented severe problems).

The instrument was set so that the central counter measured reflections in the flat cone (i.e., with $\sin \mu = \zeta$ for this level) and the length of the counter arm was adjusted to $D = u/a^*$ cm, fine adjustment being done using half-slits. Reflections from the five levels could then be measured.

The major limitation with this instrument was the existence of the blind region, which meant that the crystal had to be remounted to collect a

[6] U. W. Arndt and D. C. Phillips, *Acta Crystallogr.* **14,** 807 (1961).

complete data set. A further problem is that, because of the setting error of the diffractometer, reflections at lower than 8 Å has to be collected in single-counter mode or on a four-circle diffractometer.

Quasi-Simultaneous Measurement of Reflections On a Five-Circle Diffractometer

A diffractometer with equatorial geometry offers greater precision and flexibility of operation than does an inclination diffractometer. Because of the degrees of freedom allowed by the φ, χ, and ω circles a given reflection can be measured over a wide range of rotations, ψ, about its reciprocal lattice vector. Conventionally, some restraint is placed on ε, where

$$\varepsilon = \omega - \theta$$

In the single-counter symmetrical-A setting, for example, the restraint is $\varepsilon = 0$; in the flat-cone geometry $\varepsilon = \theta$. However by relaxing the restraint on ε it is possible to obtain a much more versatile multicounter diffractometer than is possible with an inclination machine.

The required modification to convert a conventional four-circle diffractometer into a five-circle machine is the addition of a fifth axis on the counter arm lying radially in the equatorial plane. This axis is called the σ axis: it carries a linear array of $2n + 1$ counters with an effective separation u, subtending an angle of η radians at the crystal when the crystal-to-counter aperture distance is D. For a crystal mounted about a reciprocal axis with spacings a^*.

$$u/D = \eta = a^* \tag{3}$$

With an array of proportional counters it is easiest to achieve the correct η by fixing the separation u, and varying the D (as was done on the linear diffractometer, above). If a position-sensitive detector is used, D can be fixed and the effective u can be defined by software. An upper limit is placed on u and D, however, by the requirement that all diffracted intensity should be collected by the counter.

Setting Geometry and Scans on the Five-Circle Diffractometer

The diffractometer is designed to measure $2n + 1$ reflections in reciprocal lattice lines parallel to a reciprocal lattice axis (the collection axis) as nearly simultaneously as possible. The operations needed to do this can be considered in three stages. First, the reciprocal lattice point corresponding to the central reflection is made to lie on the Ewald sphere by choice of the crystal setting angles φ, χ, and ω; second, the line of reciprocal lattice points is set in an appropriate relation to the sphere of reflection by rotation through an angle ψ about the vector to the central point; and third, the reflections are measured by rotating the crystal about an axis which will usually, but not necessarily, be an instrumental axis.

It is possible to scan the reflection using the flat-cone geometry φ scan (equivalent to the linear diffractometer in the flat-cone setting) in which the line of reflections to be measured lies tangential to the Ewald sphere: this setting is given by $\varepsilon = \theta$ for a crystal mounted with the collection axis along the diffractometer φ axis. The method has the disadvantage that, since the rotation axis φ makes an angle ρ with the reciprocal lattice vector of the central reflection, the peak width and therefore the required scan width are approximately proportional to $1/\sin \rho$, i.e., the peak width varies sharply throughout reciprocal space.

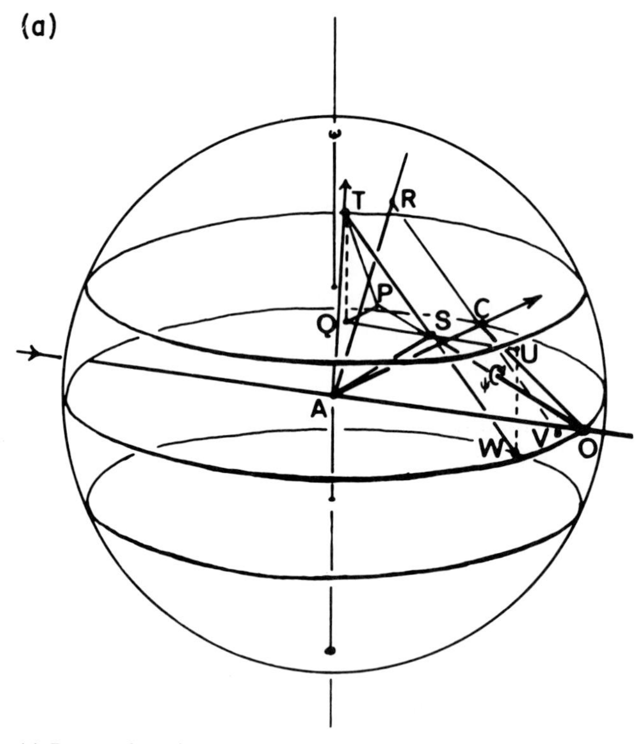

FIG. 3. (a) Perspective view and (b) equatorial projection of the sphere of reflection, showing the reciprocal lattice point corresponding to the central reflection passing through the sphere at C, and those corresponding to outer reflections passing through the sphere at T and W, Q and U are respectively the projections of T and W onto the equatorial plane of the instrument.

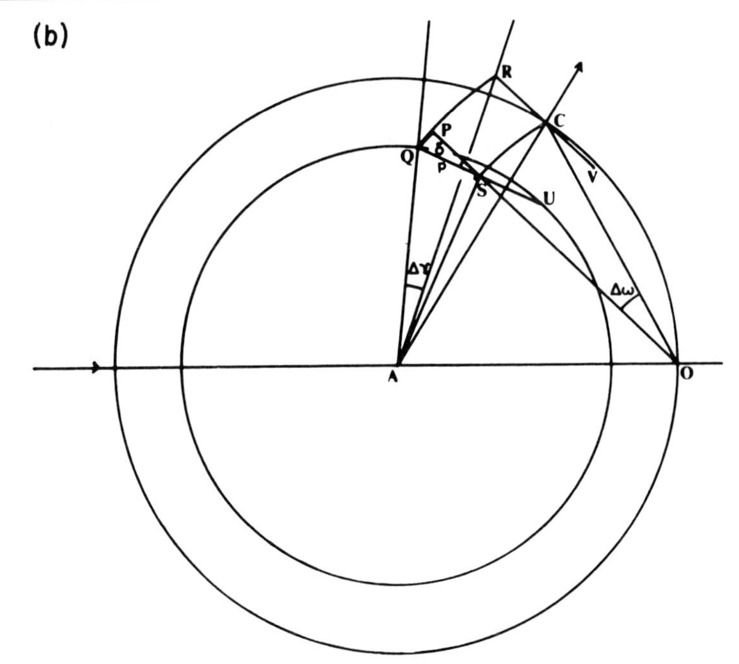

FIG. 3. (*continued*)

For the central reflection, the optimum scanning axis is ω since this axis is perpendicular to the reciprocal lattice vector (Fig. 3); $\Delta\omega$, the angle between the peak of the central reflection and the peak of the outer reflection, is always smaller than, or equal to, the corresponding $\Delta\varphi$ ($\Delta\omega \simeq \Delta\varphi/d^*$).[1] Moreover, with ω scans the peak width is much more constant over reciprocal space than with φ scans.

The Optimum Setting

However, if the ω scan is used, reflections symmetrically disposed about the central reflection no longer pass through the sphere simultaneously, particularly at low values of ρ. Therefore the ω scan seems preferable to the φ scan but the flat-cone setting is not optimum in the sense of minimizing $\Delta\omega$. The line of reflections to be measured is tangential to the sphere of reflection, but when the crystal is rotated about the ω axis the outer reflections do not cut the sphere at the same value of ω (Fig. 3). Their perpendicular distances to the sphere are the same, but in general their radii from the ω axis are different. However, by rotation by an angle ψ about the scattering vector for the central reflection, one outer reflection may be moved closer to the sphere and the other further away, such that when rotated about the ω axis they cut the sphere simultaneously. This reduces the ω rotation required to bring all the reflections through the sphere and is called the optimum setting: it exploits the degrees of freedom made available using φ, χ, and ω with the restraint on ε removed. The value of $\Delta\omega$ varies with θ:

$$\Delta\omega = t^2/(4 \sin \theta \cos \theta) \tag{4}$$

for small values of $\Delta\omega$, where t is the reciprocal spacing between the center and the outer reflection. At very low angles $\Delta\omega$ will be larger than the peak width of $2n + 1$ separate individual reflections: in that case there is no advantage in measuring more than a single reflection at a time. In data collection a check is made that $\Delta\omega$ is greater than this (or a lower) maximum value: if it is, the reflections in the quintuplet are measured individually in single-counter symmetrical-A mode. A similar check is made on ΔY to ensure that the outer reflections fall entirely in the detector: this has never proved a limiting condition since ΔY is always very small and may be approximated by

$$\Delta Y = (1/2)d^{*2}\,\Delta\omega = (1/2)t^2 \tan \theta$$

A value of ψ defines a set of diffractometer angles φ, χ, and ω, but this setting may not be usable, either because the χ circle hits the X-ray tube or the counter arm, or because an X-ray beam is obstructed by part of the χ circle. The range of accessible ψ values is smallest when ρ is near 90°, i.e., reflections for which χ is near zero. This is no problem for a crystal with the collection axis along the φ axis, but a crystal not in this special mounting will have a cone of reflections close to $\rho = 90°$ which cannot be measured in the multicounter mode.

For reflections close to the collection axis it is not possible to find a setting which brings the outer reflections to be measured through the sphere of reflection simultaneously. This region of reciprocal space is given by

$$\sin \rho' \leq (d^{*2} - t^2)/[2d^*(1 - t^2)^{1/2})]$$

where d^* is the reciprocal lattice vector length for the central reflection, and ρ' is the angle between the collection axis and the reciprocal lattice vector for the central reflection (Fig. 4). This blind region is very similar to that for the flat-cone setting, which is given by $\sin \rho' \leq d^*/2$, since t is generally small. Within this region it is still possible to find the setting which minimizes $\Delta\omega$. This is given by $\psi = \pm 90°$, i.e., the collection axis

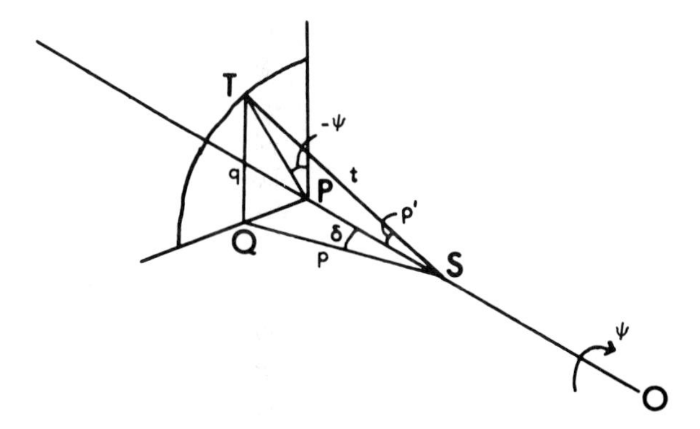

FIG. 4. Perspective view of part of Fig. 3.

lies in the equatorial plane. Expressions for $\Delta\omega$ and $\Delta\Upsilon$ in this condition are [3]

$$\Delta\omega = [t^2 + 2t\sin(\theta - \rho')]/[r(4 - r^2)^{1/2}]$$
$$\Delta\Upsilon \simeq r^2\,\Delta\omega/2$$

where r is the length of OT (Fig. 4). The value for $\Delta\omega$ must be calculated for each reflection in the line to be measured, since the central reflection is not necessarily the first to cut the sphere of reflection.

Figure 5 shows the regions of reciprocal space measurable with $\Delta\omega \leq 0.4°$ for reciprocal axis spacings of 0.01, 0.02, and 0.03 along the collection axis. Only a small volume of reciprocal space is made accessible by relaxing the criterion of simultaneous measurement of the outer reflections.

Use of the Optimum Setting to Measure Reflections in the Blind Region

Except in certain high-symmetry space groups, it is necessary to measure the inaccessible reflections close to the mounting axis in another setting. The simplest way of doing this is to use the central counter in conventional symmetrical-A geometry (this is always necessary for very low-angle reflections, because of the large $\Delta\omega$). For high-resolution data collection, however, this is not a very satisfactory solution: there are a large number of data in the blind region and multicounter data collection is therefore very desirable. In the flat-cone setting it is necessary to remount the crystal (or mount another crystal) about another axis, or to mount the crystal on a right-angled bracket to bring another axis along the φ axis.

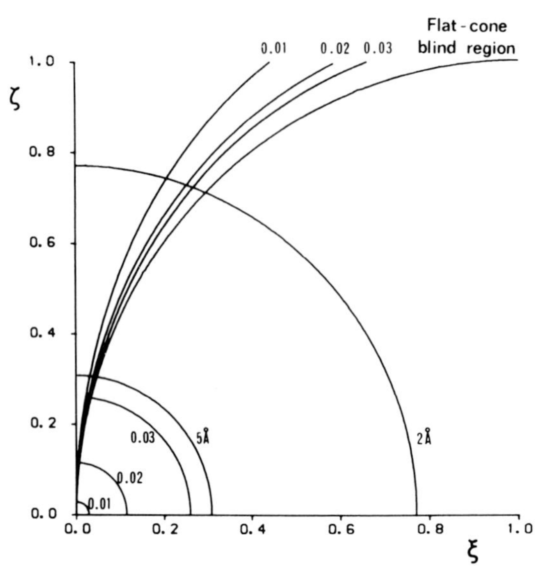

FIG. 5. Accessible regions of reciprocal space with $\Delta\omega \leq 0.4°$ for five counters and reciprocal lattice spacings of 0.01, 0.02, and 0.03 r.l.u. (ζ and ξ are the cylindrical polar coordinates about the collection axis).

Use of the optimum setting on the five-circle diffractometer makes these expedients unnecessary. Since the derivation of the multicounter optimum setting allows any mounting of the crystal, the missing reflections can be measured about another collection axis without remounting the crystal. If discrete counters are being used rather than position-sensitive detectors, the counter box has to be moved to a new position corresponding to the new reciprocal lattice cell dimension. There is little problem in setting the angles, as settings corresponding to a wide range of ψ values are physically accessible for reflections close to the mounting axis.

Calculation of Setting Angles

Figure 3a and b show the outermost reflections T and W projected on the equatorial plane at Q and U passing simultaneously through the sphere of reflection. **OS** is the reciprocal lattice vector for the central reflection lying in the equatorial plane. The possible settings of the diffractometer may be characterized by the clockwise rotation ψ about the vector **OS**; to calculate the setting angles it is necessary to find the value of ψ which brings the outer reflections T and W onto the sphere together. The zero of ψ is defined as corresponding to when the collection vector **ST** lies in the vertical plane through **OS**. The setting angle for this $\psi = 0$ setting may be calculated in a way similar to that described by Busing and Levy.[7]

It can be shown[3] that

$$\cos^2 \psi = \frac{(\sin^2 \rho' - m^2)}{\sin^2 \rho' (1 - m^2)}$$

where $m = (d^{*2} - t^2)/[2d^*(1 - t^2)^{1/2}]$. With $\psi = 0$ defined as the collection vector in the vertical plane it is only necessary to consider ψ in the range $-90° \leq \psi \leq +90°$ since the sphere is symmetrical to reflection in the equatorial plane. From ψ, the diffractometer angles φ, χ, and ω may be calculated.[7]

Counter Settings

The most convenient way to calculate σ is to consider rotation of the collection vector $\mathbf{v}_0$ into the 2θ frame of reference[7]:

$$\mathbf{v}_{0\ 2\theta} = \mathbf{\Theta}^{\mathrm{T}}\mathbf{R}\mathbf{v}_{0\ \varphi}$$

where $\mathbf{\Theta}$ is the rotation matrix corresponding to the θ axis, and $\mathbf{R}$ is the instrument angle matrix for the diffractometer angles, φ, χ, and ε. The value of σ may then be calculated from the two components of $\mathbf{v}_{0\ 2\theta}$ in the plane perpendicular to the diffracted beam.

Lorentz Factor

The inverse Lorentz factor for the outer counter T (Fig. 3) is given by

$$L^{-1} = (1/2)(4s^2 - r^4)^{1/2}$$

where $r = OT$ and $s = OQ$.

The Five-Circle Diffractometer: Instrumentation

The instrument in use at the Laboratory of Molecular Biophysics, Oxford University has five counters separated from each other by $u = 9.5$ mm, and D can vary over the range 25–100 cm. Therefore, from Eq. (3), this allows collection axis lengths of 41–163 Å using Cu K_α radiation. Full details of the diffractometer are given elsewhere.[3] It is a Hilger and Watts Y230 four-circle machine in which the standard counter arm has been replaced by a beam with a level surface extending from 20 to 100 cm from the crystal (see Fig. 6). The σ circle is mounted on the beam and may be clamped at any distance from the crystal with its axis horizontal (leveling screws are provided to adjust its height and tilt). The circle carries a

[7] W. R. Busing and H. A. Levy, *Acta Crystallogr.* **14,** 807 (1967).

FIG. 6. A view of the five-circle diffractometer showing the counter box on the extended θ arm. The helium tube has been removed for the photograph.

box with five proportional counters and five circular apertures, each aperture having half-slits to aid alignment. The limiting apertures may vary from 2 to 5.5 mm. The σ axis can be aligned optically to lie in the horizontal plane and to point at the common intersection of the other axes. The beam is sufficiently rigid for the vertical height of the σ axis to vary less than 0.02 mm between extreme positions. A tube filled with helium is normally placed between the counters and the crystal. This is made of plastic water pipe, closed with Mylar windows, and may be adjusted to any desired length. The diffractometer is controlled by an LSI 11/03 minicomputer which runs a FORTRAN program.[8] Reflection intensities are integrated using a five-counter modification of the centroid method of Tickle[9] which, allowing for $\Delta\omega$, adds up the ordinates of the five scans to produce a combined average peak profile of the five reflections with a higher probability of correctly finding the peak center. Indexed intensities and backgrounds are output to magnetic tape, together with the complete scan profiles for later analysis by a profile-fitting method,[10] if desired. A backup copy of the data is written to floppy disk.

[8] Program, written by W. C. A. Pulford and S. J. Oatley, is an adaptation of a four-circle control program written by P. R. Evans.

[9] I. J. Tickle, *Acta Crystallogr, Sect. B* **B31,** 329 (1975).

[10] S. J. Oatley and S. G. French, *Acta Crystallogr., Sect. A* **A38,** 537 (1982).

Data Collection Strategy on the Five-Circle Diffractometer

In general, the strategy of data collection on the five-circle diffractometer is very similar to that on a four-circle machine: backgrounds are minimized through the careful choice of collimators, guard rings, helium tubes, and counter apertures; standard reflections are measured at frequent intervals to monitor crystal slippage and radiation damage; Bijvoet pairs of reflections (quintuplets of them on the five-circle) are measured close together in time and at similar absorptions; absorption curves are measured before and after data collection. There are however certain special considerations:

1. *Scan width.* Because the reflections in a quintuplet are observed quasi-simultaneously, rather than absolutely simultaneously, the background of each reflection is observed while the peaks of other reflections in the quintuplet ($\Delta\omega$ away) are being measured in other counters (Fig. 7). The total scan, therefore, need be no wider than the peak width + $\Delta\omega$ + provision for missetting. Except at very low angle this need be no wider than a scan through the peak and background of a single reflection on a single counter machine.

2. *Counter–counter scales.* If an array of proportional counters is used, the reflections in different levels will be measured in different counters, whose efficiencies may be slightly different. Moreover, the path lengths of the diffracted beams to the outer counters will be slightly longer than those to the inner counters. The scale factors between counters are therefore determined in a separate run at the end of the main data collection: overlapping sets of five reflections (if there are five counters) are measured by stepping parallel to the collection axis, so that each reflection is measured sequentially in all five counters. These data are separated into five batches, according to the counter they have been measured in, and the batches are then scaled together.[11] The relative scale factors found between counters are usually in the range of 0.985–1.015 : 1.0. Each reflection in the main data set is then multiplied by the appropriate scale factor. However, as a special precaution it is normal to ensure that when collecting data with anomalous scattering, after a quintuplet has been measured the quintuplet of Bijvoet reflections is measured very shortly afterwards, with each reflection being measured in the same counter as its equivalent had been.

3. *Collection of data in the blind region.* As described above, it is not necessary to remount the crystal but merely to redefine the collection axis and to change the crystal-to-counter distance to be appropriate to the new collection axis cell dimension.

[11] G. C. Fox and K. C. Holmes, *Acta Crystallogr.* **20,** 886 (1966).

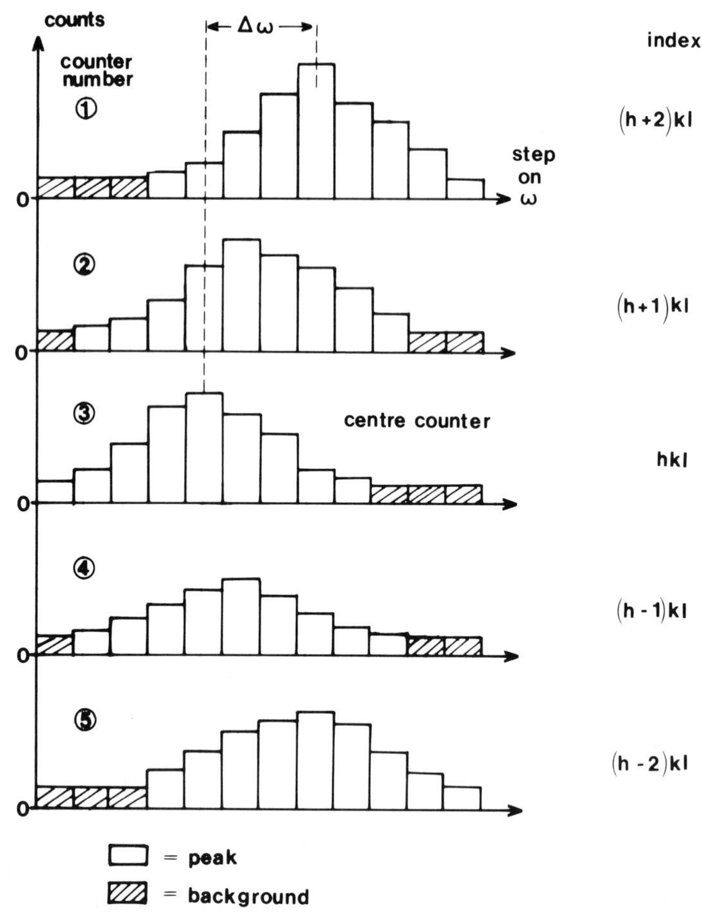

FIG. 7. Schematic diagram of scan profiles of a quintuplet of reflections. The peak in the center counter (3) leads the peaks in the outer counters (1 and 5) by $\Delta\omega$. Thus while the peaks of the outer reflections are being scanned the backgrounds of the center reflections are observed, and vice versa.

Efficiency of Data Collection

The five-circle diffractometer is a powerful machine for measuring diffracted intensities. The data compare well with data measured with a single counter on this and on other diffractometers. We measure typically about 200–250 reflections an hour, roughly five times the rate attained on a four-circle diffractometer in similar cases. It is clear however that by using a position-sensitive detector instead of discrete counters, data collection rates of at least twice this could be obtained.

Maximum Number of Quasi-Simultaneous Reflections for $\Delta\omega \leq 0.4°$

Resolution limit (Å)	Length of axis (Å)				
	50	75	100	150	250
6	3	7	9	13	15
3	5	9	13	17	21
2	7	11	15	21	27

Use of Position-Sensitive Detectors on a Five-Circle Diffractometer

The proportion of reciprocal space that can be explored depends on the length of the collection axis and the maximum value of $\Delta\omega$ that can be tolerated. The table shows the maximum number, N, of counters that can be used with $\Delta\omega \leq 0.4°$ for various axial lengths at different maximum values of sin θ.

It is clear that more than five channels could be usefully employed even with a collection axis length of 50 Å. This fact can be best exploited using a position-sensitive detector (PSD) in place of the proportional counters. PSDs are quantum counters which have been developed for both X-ray and neutron work.[12,13] An incident particle produces pulses at either end of a resistive central anode wire; the relative amplitudes of these pulses are a function of the distance along the wire at which the particle has been absorbed and so the assembly acts as a one-dimensional area detector. Use of a PSD also has the additional advantages of allowing a fixed crystal-to-detector distance and also permitting the measurement of backgrounds between diffracted beams.

A five-circle diffractometer using a linear PSD mounted on the σ circle has been built at the Medical Research Council Laboratory of Molecular Biology, Cambridge.[14] A modified θ arm carries a 10-cm-long PSD at a fixed distance of 40 cm from the crystal; apertures are defined by software, and the possibility of measuring backgrounds in the channels between apertures is being explored. It is expected that it may be possible to measure up to 20 reflections quasi-simultaneously in proteins with very large real cell dimensions.

[12] W. R. Kuhlmann, K. H. Lauterjung, B. Schimmer, and K. Sistemich *Nucl. Instrum. Methods* **40,** 118 (1966).

[13] C. J. Borkowski and M. K. Kopp, *Rev. Sci. Instrum.* **39,** 1515 (1968).

[14] F. Mallett and T. Bailey, unpublished.

"Adventitious" Quasi-Simultaneous Data Collection Methods

The second class of data collection method referred to in the introduction, are those methods in which a linear counter is placed where it will eventually intercept all the diffracted beams from a reciprocal lattice level during the course of a long (perhaps 90° or 180°) rotation of the crystal. In general, they are closely related to photographic methods (Weissenberg or oscillation camera) in their philosophy of collecting the data first and indexing and analyzing them afterwards, rather than by performing short scans around the positions of reflections of known index. Because the reciprocal lattice distances between reflections that happen to be measured at about the same time are not constant, discrete counters with "dead spaces" between them cannot easily be used and linear position-sensitive detectors have usually been employed.

There is, however, an exception to this: a parallel data collection system for the simultaneous measurement of single-crystal X-ray intensities has been described by Hamilton and co-workers[15,16] which uses an array of 128 discrete, liquid nitrogen-cooled, silicon oxide-passivated, diffused junction detectors, each with its own separate counting chain. These are arranged linearly around an annulus of 180°, this assembly being used, in effect, to occupy the place of the layer line slit in a Weissenberg camera. By rotating the crystal through 360° in steps of a few hundredths of a degree at a time, and writing the counts from the 128 channels out to disk at each stage for subsequent indexing and integration, a reciprocal lattice layer can be accumulated. Since several reflections are likely to be in or close to the reflecting position at any one time, it was estimated[15] that an order of magnitude enhancement of data collection could be obtained.

Methods Using Position-Sensitive Detectors (PSDs)

As has been observed above, a linear PSD can be mounted on the σ circle of a five-circle diffractometer, giving considerable advantages over an array of discrete counters. However they have been more widely used in both X-ray and neutron work in the alternative, less mechanically complex "adventitious" methods.

Cain *et al.*[17] describe a neutron position-sensitive detector system in

[15] R. Thomas, W. C. Hamilton, J. B. Godel, H. W. Kraner, V. Radeka, D. Stephani, G. Dimmler, W. Michaelson, and M. Kelly, *Acta Crystallogr., Sect. A* **A25**, S69 (1969).

[16] W. C. Hamilton, *Science* **169**, 133 (1970).

[17] J. E. Cain, J. C. Norvell, and B. P. Schoenborn, *in* "Neutron Scattering for the Analysis of Biological Structures" (B. P. Schoenborn, ed.), Rep. BNL-50453, VIII-43. Brookhaven Nat. Labo. Upton, New York, 1976.

which data are collected in normal beam geometry with crystal rotation, using a four-circle diffractometer. The detector (in fact, three PSDs connected linearly in series) is horizontally mounted on the detector arm and is positioned to intercept the diffracted beams within a particular level of reciprocal space normal to the mounting axis, as the crystal is rotated. However, to measure upper reciprocal lattice layers it is necessary both to change ν (the angle between the reflected beam and the plane perpendicular to the rotation axis) by physically raising the counter above the equatorial plane, and to increase the width of the slits in front of the detector (parallel to its length) to allow for the fact that in this setting (χ fixed at 0°, scanning on $\omega \equiv \varphi$) the cones of reflections for higher levels intercept the linear detector in hyperbolical and not linear loci. In addition to the inconvenience of this procedure, backgrounds are increased because of the wider slits.

A more elegant method of using a four-circle diffractometer and a PSD to measure reflections within a level has been described by Wlodawer, Santoro and co-workers.[18,19] They use flat-cone geometry so that the center of the sphere of reflection lies in the reciprocal lattice layer being measured. The diffracted beams from that level therefore lie in a plane, and the paths of all of them can be intercepted by a linear detector placed in the same plane. The detector is mounted on the 2θ arm of a four-circle diffractometer in a vertical plane which includes the diffractometer ω axis. Thus, the arm defines the μ angle, and the φ, χ, and ω axes of the diffractometer can be used to rotate the crystal around an axis perpendicular to the plane defined by the detector and the crystal. As noted above the major disadvantage of the flat-cone setting is the existence of a blind region. If the rotation around the zone axis had to be performed by a single circle it would be necessary to remount the crystal to measure the blind region reflections. However, use of the three independent circles of a four-circle diffractometer means that measurements may be made about more than one axis with a single mounting of the crystal, and the problem is thereby circumvented.

Data collection is performed by positioning the detector arm at $2\theta = \nu$ and then rotating the crystal through the (for instance) 90° required to observe all the unique reflections in the level. Channel profiles (1024) from the PSD are written out every few hundredths of a degree: they can either be written out to a large disk on the controlling computer for later analysis, or the peaks can be indexed and integrated concurrently if their positions have been predicted beforehand.

[18] E. Prince, A. Wlodawer, and A. Santoro, *J. Appl. Crystallogr.* **11,** 173 (1978).

[19] A. Wlodawer, L. Sjölin, and A. Santoro, *J. Appl. Crystallogr.* **15,** 79 (1982).

[27] Multiwire Area X-Ray Diffractometers

By RONALD HAMLIN

The multiwire proportional counter is at this writing the only type of two-dimensional position-sensitive X-ray detector capable of collecting diffraction data accurate enough for solution of new protein structures. The first diffractometer system to use this type of detector (the Mark I diffractometer system) was assembled at the University of California, San Diego and has collected the data used to solve four new protein structures. Similar diffractometer systems using a single thin, flat multiwire counter are now being constructed in several other laboratories around the world, and several of these should routinely be collecting good diffraction data from protein and perhaps even virus crystals by 1986. A table describing some of these other systems is included later in this chapter.

The next step in the evolution of area diffractometer systems based on the multiwire proportional counter is more complete coverage of the solid angle of the diffraction pattern—more complete than the 10–40% coverage possible with one flat multiwire counter. The phenomenon called "parallax" makes it impractical to intercept the whole diffraction pattern with one flat, xenon-filled multiwire counter. Two strategies for dealing with parallax are now being pursued. One strategy involves adding a spherical drift region to the front of a flat multiwire counter and a detector using this idea will be described. The other strategy, one being pursued by the author, involves building an array of flat detectors arranged to approximate a section of the surface of a sphere. The array of flat detectors gives more flexibility in crystal-to-detector distance and distributes the dead time over many detectors, thereby allowing the full array to have a high counting rate capacity even using only medium speed (2 μsec) position readout circuits for each individual detector.

Introduction

Crystals of large biological molecules such as proteins are gradually disordered by the incident X-ray beam during a diffraction experiment. In a time as short as a few hours they may lose as much as 15% of their diffracting power and be unsuitable for further use. Crystals of these large molecules of course have large unit cells—edge lengths are usually in the range from 50 to several hundred Angstroms—and so there are typically from tens of thousands to hundreds of thousands of relatively weak reflection intensities to measure. At any one crystal setting there will be from tens to hundreds of simultaneous reflections and so the obvious way to collect data most efficiently from this type of crystal would be to measure all the simultaneous reflections at the same time so that none of the diffracted radiation would be wasted. In the ideal case every diffracted X-ray photon would be detected and correctly assigned an *hkl* index so that the most possible information about the crystal structure would be obtained from each crystal during the finite amount of time it is usable.

Conventional data collection methods fall far short of this ideal. The standard single-crystal diffractometer is usually equipped with a scintillation counter or single-wire proportional counter which efficiently counts the photons in a diffracted beam but measures only one beam at a time, wasting all the other simultaneous ones—wasting, that is, from 90 to over 99% of the diffracted radiation from a crystal of a large biological molecule. Film, on the other hand, does intercept most or even all of the diffracted radiation from a large-molecule crystal, but film has poor usable detection efficiency. Film measurements suffer from a number of problems including film fog, local saturation, and inaccuracies in optical scanning and in film-to-film scaling. For these and other reasons film measurements of diffracted beam intensities come nowhere near yielding the accuracy which would be predicted purely from "counting statistics" based on the number of X-ray photons which have to be absorbed to darken a spot to a certain optical density. Clearly the detector which is needed for large-molecule crystallographic work would have the efficient, accurate, inherently linear, photon-counting ability of a scintillation counter. It would at the same time be able to intercept at least a large part of the solid angle of the diffraction pattern in the position-sensitive way film does, to allow measurement of many of the simultaneous reflections present at any given crystal setting.

The multiwire proportional counter is the first type of X-ray detector which has successfully met these requirements for photon counting and position-sensitive coverage of the solid angle, although other technologies such as television and solid-state detector arrays are being worked on. The discussion here will begin with a description of the prototype multiwire counter detector which has been used for more than 7 years in the Mark I area X-ray diffractometer system in San Diego. Also included here is a short description of the other similar flat, multiwire counters now being built for incorporation into diffractometer systems in other laboratories. The discussion will then turn to the next important phase in the development of multiwire counter area diffractometers, namely more complete coverage of the solid angle of the diffraction pattern, and will examine two ways to deal with an inevitable effect called "parallax" which arises with use of flat multiwire counters because the gas layer which absorbs the X-rays has finite thickness.

The Mark I Area Diffractometer System

The area detector used in the Mark I diffractometer system is a xenon-filled multiwire proportional counter and is an outgrowth of a technology first developed for use in high-energy particle physics in the late 1960s and early 1970s by Gorges Charpak and others at the CERN accelerator facility on the French–Swiss border.[1,2] The delay lines used in the two-dimensional position readout scheme in the Mark I detector are of the type developed, also in the early 1970s, by Victor Perez-Mendez and colleagues at the University of California, Berkeley.[3] The xenon gas handling system, the X, Y readout electronics, and the crystallographic software were developed in our laboratory at the University of California, San Diego beginning about 1972.[4] Serious protein crystallographic data collection began to be possible with the Mark I system in about 1975[5] and the first new protein structure solved with Mark I data, dihydrofolate reductase (*Escherichia coli*), was reported in 1977.[6] Since then three more new protein structures have been solved with Mark I data[7–9] and several others are in progress. The data collection software was described in 1978[10] and final, detailed characterization studies of the Mark I detector were reported in 1981.[11]

[1] G. Charpak, R. Bouclier, T. Bressani, J. Favier, and C. Zupancic, *Nucl. Instrum. Methods* **62,** 262 (1968).

[2] G. Charpak, *Annu. Rev. Nucl. Sci.* **20,** (1970). This paper is very valuable. The proportional counter discussion starts on p. 213.

[3] R. Grove, V. Perez-Mendez, and J. Sperinde, *Nucl. Instrum. Methods* **106,** 407 (1973).

[4] C. Cork, D. Fehr, R. Hamlin, W. Vernon, and Ng. H. Xuong, *J. Appl. Crystallogr.* **7,** 319 (1973).

[5] C. Cork, R. Hamlin, W. Vernon, and Ng. H. Xuong, *Acta Crystallogr., Sect. A* **A31,** 702 (1975).

[6] D. Matthews, R. Alden, J. Bolin, S. Freer, R. Hamlin, Ng. H. Xuong, J. Kraut, M. Poe, M. Williams, and K. Hoogsteen, *Science* **197,** 452 (1977).

[7] T. Poulos, S. Freer, R. Alden, Ng. H. Xuong, S. Edwards, R. Hamlin, and J. Kraut, *J. Biol. Chem.* **253,** 3730 (1978).

[8] D. Matthews, R. Alden, J. Bolin, D. Filman, S. Freer, R. Hamlin, W. Hol, R. Kisliuk, E. Pastore, L. Plante, Ng. H. Xuong, and J. Kraut, *J. Biol. Chem.* **253,** 6946 (1978).

[9] P. Weber, R. Bartsch, M. Cusanovich, R. Hamlin, A. Howard, S. Jordan, M. Kamen, T. Meyer, D. Weatherford, Ng. H. Xuong, and R. Salemme, *Nature* (*London*) **286,** 302 (1980).

[10] Ng. H. Xuong, S. Freer, R. Hamlin, C. Nielsen, and W. Vernon, *Acta Crystallogr., Sect. A* **A34,** 289 (1978).

[11] R. Hamlin, C. Cork, A. Howard, C. Nielsen, W. Vernon, D. Matthews, and Ng. H. Xuong, *J. Appl. Crystallogr.* **14,** 85 (1981).

Principles of Operation

Figure 1 is a drawing of the Mark I diffractometer system showing the quarter circle goniostat, the multiwire counter detector (the flat square box), and the detector positioning geometry. Copper X-rays from a standard fine-focus tube (not shown) are collimated by the collimater shown and the beam is directed at a protein crystal mounted in a glass capillary on the ϕ spindle. Some of the simultaneously occurring diffracted beams (dashed lines) are shown penetrating the 30 × 30-cm beryllium window of the detector. The detector can be placed a distance D from the crystal as indicated, and the track the detector is mounted on can be rotated around the ω axis by an angle θ_c with respect to the incident X-ray beam. During data collection the detector is locked in position and only the crystal orientation is changed by stepped rotation about the ω axis. A helium box can be placed between the crystal and the detector window as indicated in the inset. The front and rear windows of the helium box are of acetate 25 μm (0.001 in.) thick and helium is flowed through the box slowly. This helium box eliminates most of the 1% per centimeter attenuation which would occur if the diffracted beams had to traverse the relatively large crystal-to-detector distances (typically 30–60 cm) in air.

In cross section the Mark I detector appears as shown in Fig. 2. The 30 × 30-cm beryllium window (a cathode) is 1 mm thick, the anode wires are stainless steel, 20 μm in diameter, and the wires in the back electrode plane (also a cathode) are stainless steel, 50 μm in diameter. The volume

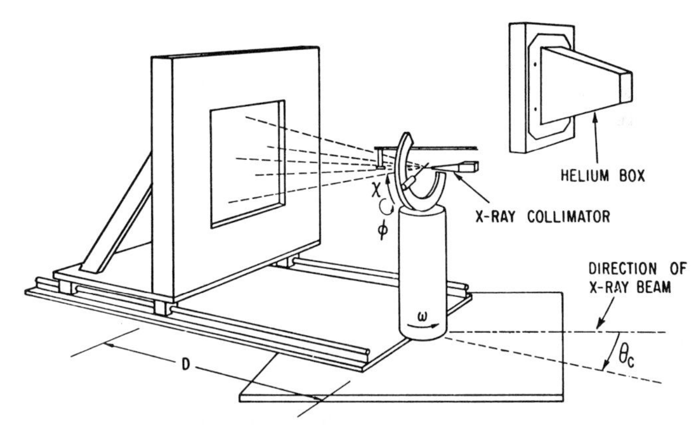

FIG. 1. Geometry of the Mark I area X-ray diffractometer system showing the multiwire proportional counter detector (flat box) positioned on a set of rails which can be rotated by an angle θ_c with respect to the incident X-ray beam. The detector can be moved along the track to place it a distance D from a protein crystal mounted in a glass capillary on the ϕ spindle of a quarter-circle goniostat. A typical helium box is shown in inset drawing.

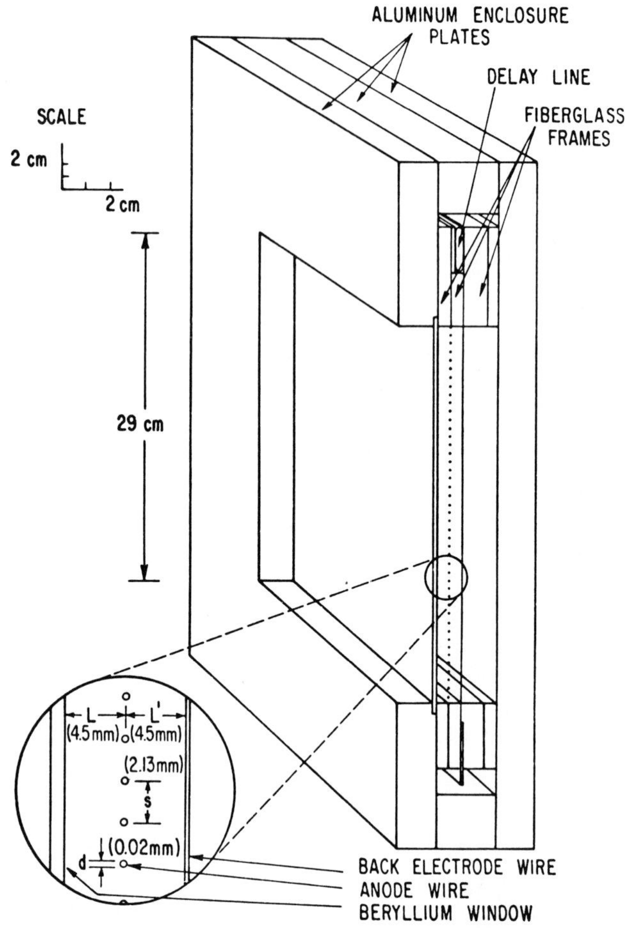

FIG. 2. Cross-sectional drawing of the Mark I multiwire proportional counter detector. The beryllium window is one of the three planar electrodes of the detector. The other two electrodes are plane arrays of parallal wires at distances of L and $L + L'$ from the beryllium window.

inside the detector is filled with a 90% xenon/10% carbon dioxide mixture at a pressure of 1 atm. The gas pressure is carefully regulated to keep it equal to the prevailing barometric pressure so the window will stay flat. This careful pressure regulation is required because the beryllium window is one of the electrodes and it must stay flat to maintain a uniform electric field at the anode wire surfaces. A simple bubble-in/bubble-out xenon gas delivery system is used to regulate the gas pressure inside the detector well enough to keep the window flat. The spacing of the wires in the anode plane is 2.13 mm (12 to the inch) and the spacing between the wires in the back cathode plane is half that. The spacing between electrode planes, L and L' in Fig. 2, is 4.5 mm in each case. Both cathodes (the beryllium window and the back wire plane) are held at ground (0 V) and the wires of the anode plane are held at +2.7 kV. Each wire in the anode plane acts independently as a proportional counter in much the same way as does the single anode wire in a standard single-wire proportional counter.

An X-ray photon which enters through the beryllium window is usually absorbed in the xenon gas (a small fraction do pass through the gas and get absorbed in the back plate of the detector). This absorption of an 8-keV X-ray photon in the "active" region between the window and the back wire plane takes place rather suddenly once it starts and all the photon's energy is quickly converted to ionization in the gas in a small region about 100 μm across. The few hundred electrons in this primary ionization then drift to the nearest anode wire where an ionization avalanche of 10,000 to 1,000,000 as many ion pairs results (depending strongly on the exact value of the electrical potential applied to the anode wires). The motion of the charged particles in this avalanche (chiefly the motion of the heavy positive ions *away* from the anode wire) causes a negative-going pulse on the anode wire and positive-going pulses on a few of the nearest wires in the back (cathode) wire plane.[2]

If the three electrode planes were removed from the detector and spread apart somewhat they would appear as shown in Fig. 3. The locations of the wirewound delay lines used for the X, Y position readout are shown in this figure too. The construction of these delay lines is described in Grove *et al.*[3] as mentioned above. The time required for an electrical pulse to travel the entire length of either delay line is 2 μsec. Each wire in each of the wire planes is soldered to a separate copper strip on one of two printed circuit boards (one board for each wire plane) and one delay line is clamped against (but electrically insulated from) each of these printed circuit boards. An electrical pulse from one of the wires is capacitively transferred into the appropriate delay line at the point where the copper strip for that wire presses against the delay line, and the pulse travels to the end of the delay line in a time proportional to the distance to the end of the delay line. Measuring this delay time thus gives the distance.

Besides the simplicity of this readout scheme there is a second advantage to use of the delay lines. Pulses are induced on several of the wires of the back wire plane (a cathode plane) not just one wire as is the case for the anode plane. These pulses are summed by the delay line for this wire plane into one smooth pulse. The pulses do not need to be handled separately through separate amplifier/discriminator channels as would be required using the high-energy physics style "wire-by-wire" readout scheme, and so the complexity of the readout electronics for the cathode wire plane is dramatically reduced. And, smooth position information to

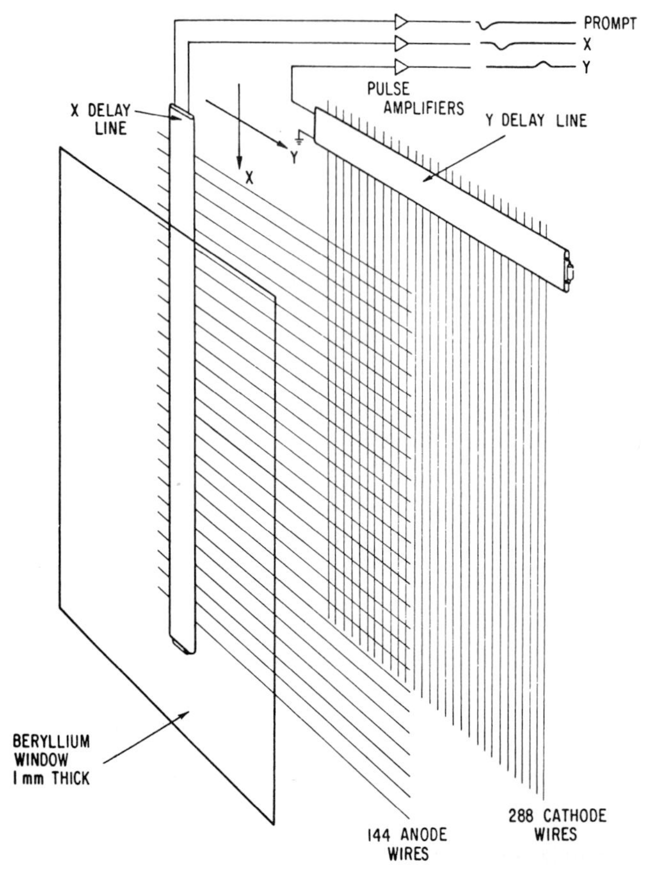

FIG. 3. Expanded view of the three electrodes in the multiwire proportional counter (separation between the three electrode planes is not here shown to scale). The positions of the two delay lines used in the X, Y readout scheme are also shown.

considerably better than the 1-mm wire spacing is obtained in the Y direction.

The maximum delay time for the delay lines used in the Mark I detector is 2 μsec. To measure the delay time of the pulses from these delay lines one needs a third pulse defining time zero, called the "prompt" pulse. This is obtained by sensing a small current which flows in the ground foil of the anode delay line immediately after an X-ray photon is detected anywhere on the detector surface. Two electronic clock waveforms (67 MHz and 130 MHz square waves) are started at the leading edge of "prompt" as shown in Fig. 4. Two binary counters—one for X and one for Y—count the rising edges in these clock waveforms until the arrival of

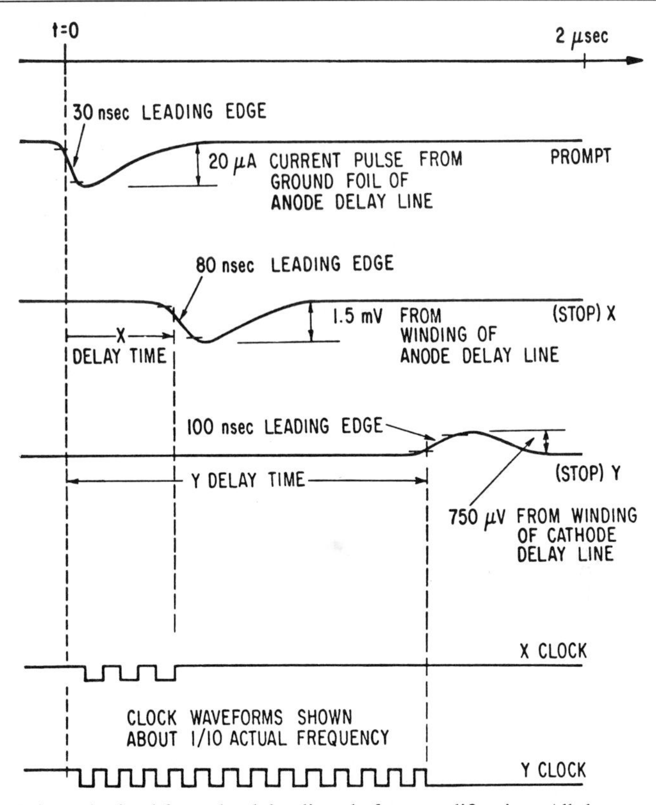

FIG. 4. Pulses obtained from the delay lines before amplification. All three are amplified to about 1 V before being sent into a constant fraction discriminator. Two counters are started at $t = 0$ and each is stopped when the corresponding delay pulse arrives at the end of its delay line.

the delay pulse from the corresponding delay line. The resulting numbers in the X and Y counters are interpreted to be the X, Y location of the detected X-ray photon. The range of X values is 0–127[11a] and the range of Y values is 0–255. The detector area is thus electronically partitioned into a 128×256 array of pixels each approximately 2×1 mm. A block of computer memory is used to record the X rays detected by this detector, one 16-bit word for each of the 32,768 pixels. The numbers in this memory are interpreted as a two-dimensional histogram called the "count histogram." After the detector has been exposed to X rays for a given exposure time the count histogram will contain a record of how many X rays were detected in each pixel on the detector surface during that time.

[11a] It is much simpler if each anode wire corresponds to one row of pixels. Position interpolation between the anode wires is not straightforward as it is for cathode wires.

Important Characteristics

For use in protein crystallographic work the detector must be able to spatially resolve neighboring diffracted beams so their intensities can be independently measured. To measure the sharpness of the spatial resolution of the Mark I detector a beam of X rays from an iron-55 foil source collimated down to about a 600-μm diameter was directed into the detector in a location on its surface chosen to be centered on 1 pixel. The counts put into the count histogram during a 5-min exposure (in the area around the chosen pixel) are shown in Fig. 5 along with a perspective

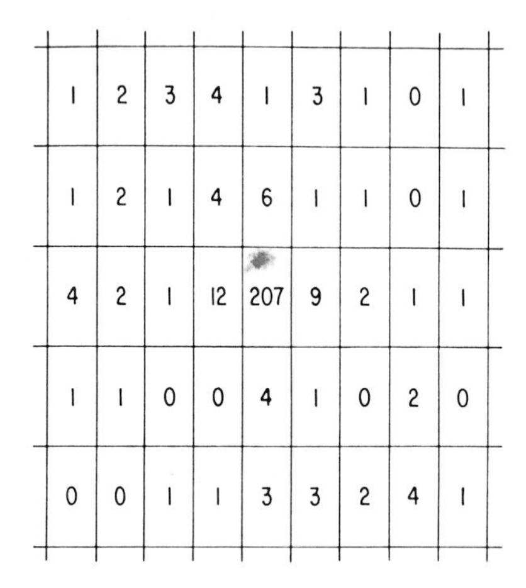

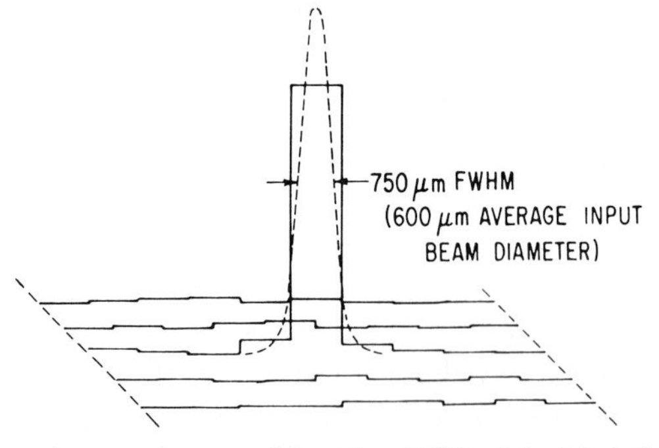

FIG. 5. The approximate point spread function (PSF) of the Mark I multiwire counter detector. The PSF is the response of the detector to a very narrow, "point-like" input beam of X rays. A rough estimate of the FWHM of this distribution is 750 μm as shown. This width is thus an estimate of the "spatial resolution" in the Y direction. The spatial resolution in the X direction is 2 mm, the anode wire spacing. The spatial resolution along Y has been reduced to about 500 μm in the newer, Mark II-type detectors mentioned below.

drawing of these counts. This distribution of counts is a rough measure of the point spread function, (PSF), a concept borrowed from optical imaging theory describing the response of a detector to a very narrow, essentially point-like input beam. The spatial resolution expressed as the full width at half-maximum (FWHM) of the point spread function is thus about 750 μm in the Y dimension as indicated in Fig. 5. In the X direction the spatial resolution is essentially the 2-mm anode wire spacing.

To demonstrate the effect this spatial resolution has in a crystallographic setting, a diffracted beam from a 0.7-mm crystal of the protein elastase was directed toward the detector, normal to its surface, with a crystal-to-detector distance of 38 cm and counts were accumulated in the count histogram for 30 sec. The counts accumulated in the histogram are shown in Fig. 6. The counts from the diffracted beam are almost all within the 3×5-pixel summing block normally used during data collection (indicated by dark box). Normal practice is to move the detector far enough away from the crystal so that the 3×5-cell summing blocks for neighboring simultaneous reflections are separated by at least one row or one column of cells (pixels). A safe center-to-center spacing for reflections at the detector surface is thus about 6 mm in Y or 8 mm in X. This complete separation of neighboring spots is required to allow the data extraction software to be relatively straightforward. Software separation of partially overlapping spots is in theory possible but would in practice be a nightmare because of the relatively coarse sampling of the spot shapes and because of the large number of reflections involved. The detector is simply placed a distance D away from the crystal sufficient to give the required spot separation. Larger unit cells require proportionally larger crystal-to-detector distances. A unit cell with 100 Å edges (and no major systematic absences) would, for example, require a crystal-to-detector distance of about 50 cm.

It is important to note here that the spatial extent of the reflection in Fig. 6 is at least a factor of 2 times broader than the point spread function shown in Fig. 5. This indicates that most of the apparent size of the reflection is crystallographically real, that is, due to main beam divergence and crystal mosaicity as described in Blundell and Johnson[12] and not due to limited spatial resolution of the detector. It would theoretically be possible to improve the spatial resolution of this type of flat xenon-filled multiwire counter (for radiation incident normal to the surface) by almost a factor of 3 to reach the limits of the gas physics involved,[13] but "parallax

[12] T. L. Blundell and L. N. Johnson, "Protein Crystallography." p. 254. Academic Press, New York, 1976.

[13] F. Sauli, Invited paper given at the Wire Chamber Conference, Vienna, Austria, 14–16 February 1978. Contact F. Sauli, CERN, Geneva, Switzerland.

81	84	78	74	89	83	69	104	101	81
77	88	88	93	181	140	105	83	102	83
73	63	97	193	4013	2640	196	106	79	84
79	71	80	84	87	77	102	73	73	75
65	63	88	74	78	76	78	82	74	80

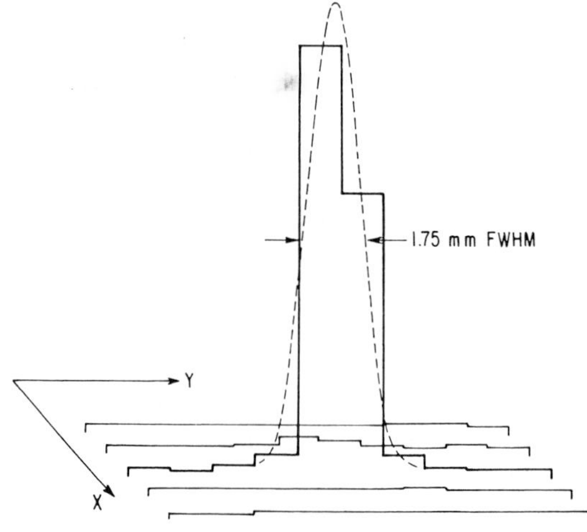

FIG. 6. Distribution of counts recorded in the count histogram of a diffracted beam from a 0.7-mm elastase crystal ($P2_12_12$, a = 52 Å, b = 58 Å, c = 75 Å) 380 mm from the detector, an appropriate data collection distance for this unit cell. Note that the reflection shown here is considerably broader than the distribution from the small test beam shown in Fig. 5. Thus most of the width of the reflection shown is the actual width of the diffracted beam at this crystal-to-detector distance.

broadening'' in the flat 9-mm thickness of the absorbing layer of the xenon gas would render any further improvements in the present spatial resolution of little use for most protein crystallographic work. Parallax is an important problem for multiwire counters and will be discussed in a separate section below.

A second important characteristic of a two-dimensional position-sensitive detector is the spatial linearity of its positional response (geometric linearity). It is important that this response be linear so that it is simple to locate the counts from a given reflection in a predicted place in the count histogram. Extensive studies were conducted with a computer controlled X, Y stage which could accurately position a collimated iron-55 X-ray source anywhere on the detector surface. It was found that the position sensitivity of the Mark I detector matched the intersections of an ideal linear grid to within 2 mm in both X and Y over the whole detector surface. This small amount of position error is stable over times of 6 months or more. It has been carefully mapped and is now routinely being compensated for in the data collection software.

A third important characteristic of the Mark I detector is its quantum detection efficiency, the fraction of the X-ray photons striking the active area of the detector which are actually counted by the detector. It is important to have the highest quantum detection efficiency possible to keep from wasting diffracted radiation from protein crystals which do not last very long in the X-ray beam. Ideally, as mentioned above, every X-ray photon diffracted from the crystal would be detected and included in the intensity measurement of the correct *hkl* reflection. An estimate of the quantum detection efficiency of the Mark I detector is 47%, a number which is the product of the 76% transmission factor of the 1-mm-thick beryllium window for 8-keV copper X-rays and the approximately 63% absorption of X rays at this energy in the 9-mm-thick volume of gas between the two cathodes. This 47% quantum detection efficiency is uniform to within 2% of that value over the entire detector surface and no corrections for small local variations are presently being applied. It is important to note that the value of the quantum detection efficiency is dependent also on the incident angle of the detected X-radiation.

The expression describing the probability for absorption in each material is the usual $P = 1 - \exp(-kx)$ where k is an empirical constant describing the absorbing power of the material for X rays of a given energy and x is the path length in the material. In the Mark I detector, by sheer chance, the extra absorption in the beryllium window for tilted incident beams (because of the longer path traversed in the beryllium) is nearly compensated for by the longer absorbing path in the xenon gas. But the correction, even though small, is being applied to the crystallographic data anyway since it is so easy to compute.

A fourth important detector characteristic is dead time. Besides efficiently detecting diffracted radiation, a position-sensitive X-ray detector and its attendant readout electronics must be able to compute X, Y addresses and transfer them to the histogram memory quickly enough to avoid serious dead times losses. Low dead time loss is important in protein crystallographic work, as is high quantum detection efficiency, to keep from wasting the limited amount of diffracted radiation available from the crystals. The fraction of detected X-ray photons lost to detector dead time for the Mark I detector is plotted vs the detection rate in photons per second in Fig. 7. The raw detection rate plotted on the horizontal axis is simply the "prompt" rate, and the fraction of those events which are *not* successfully transferred to the count histogram is plotted in the vertical direction. This plot covers the entire range encountered so far in protein crystallographic work on the Mark I diffractometer (using a Phillips-type fine-focus X-ray tube at 1200 W). Only rarely does the raw detection rate exceed 20,000 X-ray photons per second (and only in situations involving exceptionally large crystals or strongly scattering mounts such as flow cells) and so usually the dead time loss fraction is 10% or below. This loss fraction is nearly constant during data collection runs, first because more than 70% of the detected radiation is the nearly constant incoherent scatter from crystal and wet capillary mount, and second because the average detected flux from the many simultaneous reflections being observed is also nearly constant as the crystal is rotated during data

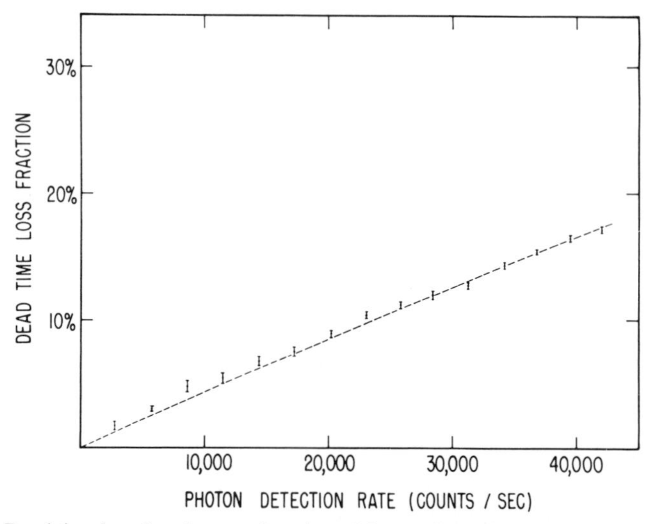

FIG. 7. Dead time loss fraction as a function of the number of X-ray photons detected per second over the entire detector surface. The range of detection rates shown here is the full range encountered so far in protein crystallographic work on the Mark I diffractometer using a Phillips-type fine-focus tube at 1200 W. Detection rates of up to 80,000 photons/sec have been observed for each detector of the newer Mark II diffractometer system which incorporates an Elliott rotating anode X-ray generator. But this newer system, as described below, is intended for use with larger unit cells where the detectors are placed farther away from crystals and the crystals diffract more weakly. In this setting typical detection rates are 20,000 photons/sec per detector and the fraction of these lost to dead time is under 10%.

collection. The small variations in count rate and hence the small variations in dead time loss fraction that do occur affect all reflections being observed on the detector surface at any given time by an equal amount (the largest local count rates are 1000 times smaller than those which would cause local saturation effects in the gas), and so are corrected for simply by rescaling the 32,000 numbers comprising each electronic picture by a factor of

$$F = 1/(1 - f)$$

where $f = f(R)$ is the loss fraction modeled as

$$f(R) = 1 - \exp[-R(2T_1 + T_2)]$$

and where R is the raw photon detection rate (the prompt rate). T_1 is the maximum delay time of a delay line (the factor of 2 is required because two photons detected in this time must both be rejected). T_1 is about 2 μsec using the present delay lines. T_2 is the time required to transfer an X, Y address to the histogram memory (no factor of 2 is required here since only the second detected photon must be rejected during this time and not the one whose X, Y address is known and being transferred). T_2 is of the order of a few hundred nanoseconds with the present hookup of the histogram memory.

The raw detection rate, R, is deduced in software based on the known exposure time, the total number of counts in the 32K histogram comprising the electronic picture, and a look-up table inversion of the expression

$$C = (1 - f)R = R \exp[-R(2T_1 + T_2)]$$

where C is the counts per second which were actually being transferred to the histogram memory during the exposure (total counts in the histogram divided by exposure time in seconds). R, T_1, and T_2 are as defined above. A look-up table inversion of this expression gives $R = R(C)$. Since C can easily be computed by the software after each exposure, R can be found by interpolating the look-up table $R(C)$. Then f can be determined using the expression for f given above and the correction factor F can be computed using the expression for F given above. The advantage of this dead time correction scheme is that no separate hardware monitoring of R, the raw photon detection rate, is required.

For the above dead time computation to be accurate, the position readout electronics must be able to read out all of the detector surface which is absorbing radiation. Any border area near the edges of the detector window, for example, outside the area which corresponds to the 128 × 256 array of pixels should be mechanically masked off to prevent absorption of X rays there. Allowing these "out-of-bounds" events into the detector would merely cause extra (and unwanted) dead time and require a more complex dead time model.

Data Collection with the Mark I Diffractometer

The method used for initial setting of protein crystals is similar to the method used with an oscillation camera. It is important, however, that the crystal orientation be set to within about 0.2° to be within the reliable capture range of the present crystal orientation refinement program. Setting is done using a television display of the count histogram as shown in Fig. 8. Good photographs of this display are difficult because of the strong contrast but it is possible to see even in this overexposed photograph that the display consists of small rectangles each representing one of the 2×1-mm pixels on the detector surface. The diffraction pattern is from the protein lysozyme with D about 30 cm and $\theta_c = 0$. The X-ray exposure time used for these "line up" shots is typically 20 sec and the display appears on the TV screen about 1 sec after the exposure is done.

FIG. 8. A photograph of the TV display of the counts recorded in the count histogram with a 20-sec exposure from a lysozyme crystal with $\theta_c = 0$ and $D = 300$ mm (as defined in Fig. 1). The spatial extent of the spots is exaggerated due to local overexposure in this photograph. This TV display and this detector position are used during initial setting of crystals. These "electronic photographs" can each be obtained in a few seconds and the setting method is similar to that used with an oscillation camera. (The photograph shown here is actually from the Mark II system discussed below. The detectors in this newer system are similar to the Mark I detector but have somewhat better spatial resolution and are free of the artifacts the Mark I detector has.)

For actual data collection the detector is positioned at a distance D that is large enough to resolve neighboring simultaneous reflections, as described in the spatial resolution discussion above, and then it is rotated out to a θ_c as shown in Fig. 1, appropriate for intercepting the desired part of the radiation diffracted from the crystal. If the square detector surface is imagined to be projected on a flat "film plane" normal to the main X-ray beam, the surface is seen to intercept a piece of the diffraction pattern such as the one represented by the large trapezoid in Fig. 9. For data collection from protein crystals with larger unit cells the detector has to be moved farther away from the crystal to allow good separation of the neighboring simultaneous reflections. It may then be necessary, because of the reduced solid angle coverage, to collect high-resolution data first (since the high-resolution reflections weaken fastest as the crystal disorders) and then move the detector to a smaller θ_c to collect the low-resolution data. These two detector positions are indicated by by the two smaller trapezoids in Fig. 9. The small overlap region shown is for cross-scaling of the data collected at the two different detector positions.

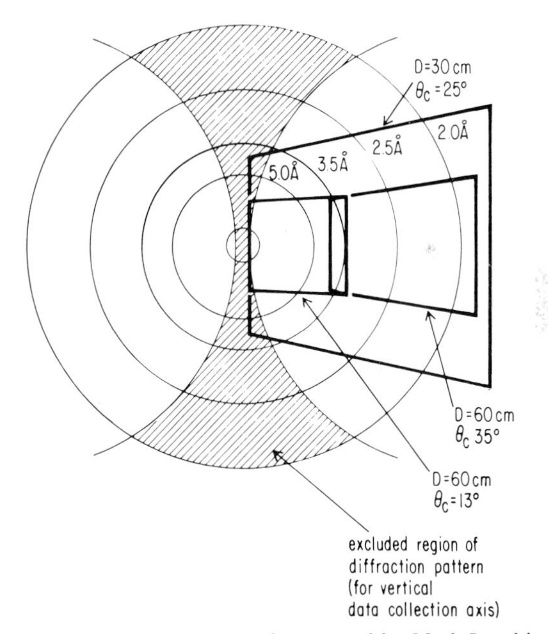

FIG. 9. Solid angle in diffraction pattern intercepted by Mark I multiwire proportional counter detector. Circles of constant Bragg plane spacing have the same meaning as they would have in an oscillation film. Shaded region is not used because the Lorentz correction is too difficult there (data collection axis is assumed to be vertical).

Once the crystal orientation has been set and the detector has been moved to a suitable position and locked in place, a list of reflection predictions is computed based on the parameters of the crystal lattice, the detector position, and the angle range through which the crystal orientation will be stepped during the data run. This prediction list contains all the reflections which will come onto the Ewald sphere and fall somewhere on the detector surface during the run. The reflections are listed in the order they will come onto the sphere as the crystal is rotated about the vertical data collection axis (either ω or else ϕ when χ is set at zero). Contained also in the list is the X, Y location on the detector surface where each reflection is expected to be detected.

The basic operation performed during actual data collection consists of stepping the crystal orientation through a small angle range such as 0.08° while counts are accumulated in the count histogram for a time such as 30 sec. A series of hundreds of these electronic "films" are exposed as a large section of the volume of the reciprocal lattice is swept through the Ewald sphere. This data collection scheme is essentially the rotation method except that each electronic film is exposed for only a very small range of crystal rotation angle. This small angle range would be impractical in work with real film since so many films would have to be processed, but is no hardship when done electronically. The advantage of the small angle range per electronic film is that counts from a particular location on the detector are used to compute a reflection intensity only during the few angle increments when the reflection is actually on the sphere. These counts are not mixed with the purely background scatter which would fall on that location during the rest of a standard (larger) oscillation angle such as 2°. Weak reflections can thus be measured which would probably be lost in the background on a standard oscillation film.[13a] Typically a reflection shows up above background in four or five successive electronic films. The partial observations of each reflection are sorted together on a disk file and the background corrected integrated intensity is computed as soon as the reflection has finished passing through the sphere. The computer running data collection is thus extracting the raw data from the

[13a] The weak reflections are also measured better because a multiwire counter does not have the inherent "film fog" type of background that further obscures weak reflections on X-ray film. With no incident X-radiation the inherent background in the San Diego detectors (from occasional cosmic rays) is less than 0.001 counts/sec per pixel so that the 15-cell summing block used for a reflection (see dark rectangle in upper part of Fig. 6) will on the average get less than one count from this source in a typical exposure time of 40 sec, while the typical number of counts accumulated for a *reflection* in that time is in the range 100–60,000.

count histogram and computing the integrated reflection intensities continuously during data collection.

Generalizations about the data collection speed of Mark I and the data quality compared to data collected with film or with a standard diffractometer are difficult because there are so many special circumstances to take into account in each case. It has usually been found that at least an order of magnitude more data can be collected from each protein crystal before it has disordered too much than can be collected with a standard single-crystal diffractometer or with film. It is not uncommon for the rate of measurement of reflection intensities to be 2000 or more per hour. And typical intensity R factors for agreement between symmetry-related reflections are in the 4–7% range, depending strongly of course on crystal quality and durability. Data from crystals that diffract especially well such those of lysozyme and elastase ordinarily have intensity R factors in the 2.5–5% range. The most important measure of the data, though, is that it has been used to solve four new protein structures.[6–9]

Nine other X-ray diffractometer systems incorporating a single, flat multiwire proportional counter as a position-sensitive detector are being assembled in various places in the world. Table I[14–20] is a synopsis of a few of the characteristics of each of the systems. The interested reader is encouraged to consult the sources given for more details. Information presented in the row "Number of protein crystal types" is almost certainly out of date even at this writing. Several of these other systems may be collecting crystallographic data accurate enough for solution of new protein structures by 1986. A major bottleneck in many of these systems is writing the detailed software required to carefully correct the reflection intensities for the local background in the presence of the observed geo-

[14] R. P. Phizackerley, C. Cork, R. C. Hamlin, C. P. Nielsen, Ng. H. Xuong, and V. Perez-Mendez, *Nucl. Instrum. Methods* **172,** 393 (1980).

[15] A. Gabriel and F. D'Auvergne, *Nucl. Instrum. Methods* **152,** 191 (1978).

[16] J. R. Helliwell, T. J. Greenhough, P. Carr, P. R. Moore, A. J. Thompson, G. Hughes, M. M. Przybylski, P. A. Ridley, J. E. Bateman, J. F. Connolly, and R. Stephenson, *Acta Crystallogr., Sect. A* **A37,** Suppl. C-316, 16.4-01.

[17] K. Mase, H. Hashizume, and Y. Iitaka, *Acta Crystallogr. Sect. A* **A37,** Suppl. C-316, 16.3-05.

[18] T. D. Mokulskaya, S. V. Kuzev, G. E. Myshko, A. A. Khrenov, M. A. Mokulski, Z. D. Dobrokhotova, A. Ya. Volodenkov, V. P. Rubanov, N. A. Ryanzina, B. I. Shitikov, S. E. Baru, A. G. Khabakhpashev, and V. A. Sidorov, *J. Appl. Crystallogr.* **14,** 33 (1981).

[19] S. E. Baru, G. I. Proviz, G. A. Savinov, V. A. Sidorov, A. G. Khabakhpashev, B. N. Shuvalov, and V. A. Yakovlev, *Nucl. Instrum. Methods* **152,** 209 (1978).

[20] Yu. S. Anisimov, S. P. Chernenko, A. B. Ivanov, V. D. Peshekhonov, S. A. Rozhnyatovskaya, Yu. V. Zanevsky, D. M. Kheiker, L. F. Malakhova, and A. N. Popov, *Nucl. Instrum. Methods* **179,** 503 (1981).

TABLE I

AREA DIFFRACTOMETER SYSTEMS USING SINGLE FLAT MULTIWIRE PROPORTIONAL COUNTERS[a]

	U. of California, San Diego	U. of Virginia	SSRL/ Stanford	Xenotronics/ Supper	EMBL, Hamburg, FRG	SRS, Darsbury, UK	U. of Tokyo	Moscow, USSR	Novosibirsk, USSR	Dubna, USSR
Method of readout	Delay line	Delay line	Delay line	Current division	Delay line	Delay line	Delay line	Wire by wire	Wire by wire	Delay line
Pixel array	128 × 256	128 × 256	128 × 256	Est. 300 × 300	Est. 50 × 200	Est. 200 × 200	64 × 64	64 × 64	128 × 128	350 × 200
Count rate for 20% loss fraction[b] (in kHz)	50	130	130	33	180	Est. 150	125	130	400	167
Detection gas	Xe	Ar or Xe	Ar or Xe	Xe	Ar or Xe	Xe	Ar	Xe	Ar or Xe	Xe
Number of protein crystal types data collected from[c]	20	4	4	2	1	1	1 full set[d]	1	?	?
New protein structures solved with data	4	0	0	0	0	0	?	0	0	0
References	11	[e]	14		15	16	17	18	19	20

[a] This table is a sampling and is not complete or very up-to-date. The references cited, however, can be used as a starting place for further investigation of the literature.

[b] Count rate determined from $R = f/2T, f = 0.2$, where R is the detection rate in photons per second, T is dead time of detector, and f is the dead time loss fraction. Count rate for 20% loss due to dead time is a useful parameter for protein crystallographers since loss fractions much above this are wasteful of the limited amount of diffracted radiation obtainable with a crystal easily disordered by X-ray exposure.

[c] Includes partial sets of test data.

[d] L-Asparaginase; full 3 Å data; 200,000 reflections. $R_{sym} = 4–7\%$.

[e] Contact Prof. Stanley Sobottka, Department of Physics, University of Virginia, Charlottesville, Virginia 22901.

metric nonlinearity and the only modest spatial resolution. The data collection software developed for use with either the Mark I or Mark II detector systems in San Diego has been more than 10 years in development and has grown to over 20,000 lines of "C" code. This software will be described in some detail in the following chapter in this volume [28]. Merely adapting oscillation film software packages for use with multiwire counters has not proved to be very productive since the problems encountered using film are different from those encountered using multiwire counters, particularly problems having to do with spatial resolution, spot shape, and X-ray background correction.

Parallax

It is evident in Fig. 9 that when the detector is placed 60 cm from the crystal it intercepts about 15% of the usable solid angle in the 2 Å cone. For protein work even at this relatively large crystal-to-detector distance the Mark I detector is still a big improvement over the detector normally used on a single-crystal diffractometer because it allows measurement of 30 or more simultaneous reflections at the same time and continuous measurement of the background scatter everywhere else on its surface (good *local* background measurements are crucial for measuring weak reflections). For a protein that diffracts to high resolution, however, most of the simultaneous reflections at any given crystal setting still will not fall on the Mark I detector surface, and so most of the limited amount of information available from a single protein crystal is still being lost. What is really needed is a detector which intercepts the whole diffraction pattern (excluding, perhaps, the shaded central region in Fig. 9 where the Lorentz correction is too difficult) so that none of the usable diffracted radiation is wasted. If one were to build one large, flat multiwire proportional counter, however, and use it to intercept the whole 2 Å cone of diffracted radiation one would encounter a severe practical difficulty collecting crystallographic data: parallax.

Parallax degrades the spatial resolution of the 1-cm-thick gas-filled detector as the incident angle of the detected radiation is tilted away from the local perpendicular to the detector surface. The reflections with Bragg plane spacing around 2 Å would be incident on the detector surface at an angle of about 45°. As X-ray photons were absorbed from this beam the primary electrons produced would drift over to the anode wires in a direction perpendicular to the anode plane, as shown in the lower part of Fig. 10, and clearly would not stay confined to the cylinder of volume defining the incident beam as they do for the normally incident beam shown in the upper part of Fig. 10. This spatial smearing of the primary

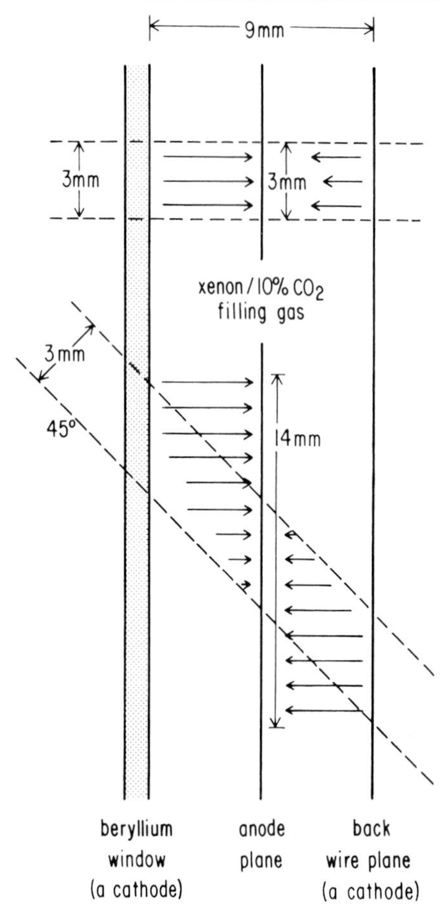

FIG. 10. Parallax as it manifests itself in flat multiwire proportional counters. The top pair of dashed lines indicate a 3-mm beam of X rays incident normal to the beryllium entrance window. No parallax smearing of the primary ionization results in this case. When the incident beam is tilted as shown in the lower part of the drawing, however, the primary ionization smears out along the anode plane as shown, resulting in a corresponding reduction of the detectors's spatial resolution.

electrons is the way parallax manifests itself in flat multiwire counters. The thicker the absorbing layer of gas the worse the smearing.[20a] The degradation of spatial resolution for X-ray beams incident at 45° would

[20a] Increasing the pressure of the gas inside the detector to between 5 and 10 atm would allow good absorption of copper X rays in a 1- to 2-mm-thick layer of xenon gas and would thereby give reduced parallax. But there is considerable mechanical difficulty building a large-area entrance window transparent to 8-keV X rays yet strong enough to hold this pressure safely. Explosion shielding would be required and the dust from an exploded beryllium window would be a serious health hazard.

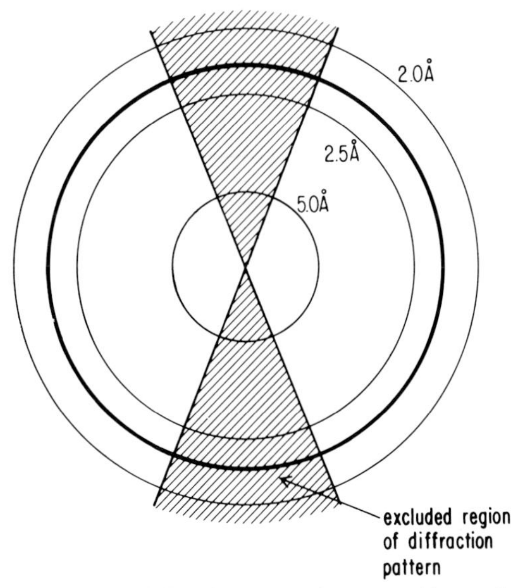

FIG. 11. Circular region of the diffraction pattern intercepted by a spherical drift chamber (dark circle) for 8-keV copper radiation. Reflections in the shaded region of solid angle are not measured—assuming a vertical crystal rotation axis—because the Lorentz correction is too difficult and the reflection profiles are too broad. This is roughly the same excluded region as the one shown in Fig. 9 for the San Diego Mark I system.

clearly be so severe that it would sometimes cause the counts belonging to neighboring simultaneous reflections to overlap in the count histogram to the extent that they would be nearly impossible to separate even with elaborate software. Intercepting the whole 2 Å cone with a single flat multiwire counter would thus not be very practical.

In 1974 G. Charpak *et al.* announced the development of a detector called a spherical drift chamber[21] which was designed to eliminate the effects of parallax in this kind of crystallographic situation. Their early prototype had an acceptance cone of only 50° but in 1977 they reported a new, larger spherical drift chamber with an angular acceptance cone of over 80° This detector could thus intercept the whole diffraction pattern out to 2.2 Å (see Fig. 11) for work with 8-keV copper X-radiation.[22] Figure 12 is a cross-sectional drawing of this detector.

A 50 × 50-cm flat multiwire proportional counter is fitted with a spherical drift region which has two electrodes of its own: one is an

[21] G. Charpak, Z. Hajduk, A. Jeavons, R. Stubbs, and R. Kahn, *Nucl. Instrum. Methods* **122,** 307.

[22] G. Charpak, C. Demierre, R. Kahn, J. C. Santiard, and F. Sauli, *Nucl. Instrum. Methods* **141,** 449 (1977).

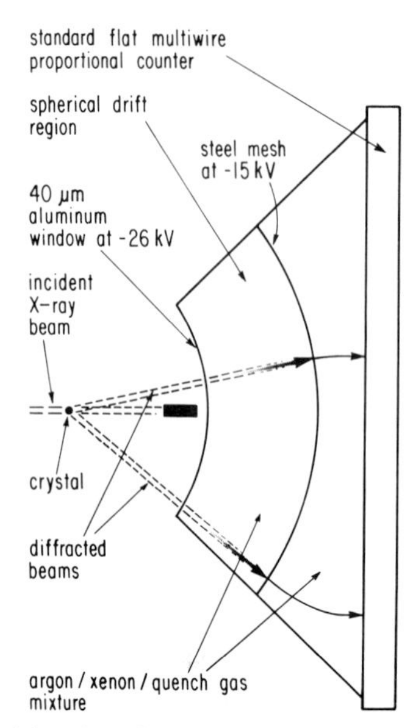

FIG. 12. Cross-sectional drawing of the spherical drift chamber. Diffracted X-ray beams from the protein crystal pass through the thin, spherical aluminum entrance window and are absorbed in the gas mixture in the spherical drift region as shown. The beams are completely absorbed in the spherical drift region and the resulting primary ionization (shaded arrows) travels in the same direction as the radiation in each incident beam was traveling, thus avoiding parallax. The crystal must, of course, be very accurately positioned at the geometric "focus" of the spherical window for this scheme to work well.

aluminum entrance window 40 μm thick and the other (10 cm behind it) is a stainless-steel mesh. Each of these electrodes is the shape of a section of a sphere centered at the intended crystal location. These two spherical electrodes and the three standard wire plane electrodes in the flat multiwire counter section are all in a gas-tight container filled with a mixture of argon, xenon, and a quench gas such as isobutane or carbon dioxide. Diffracted beams from the crystal enter the spherical drift region through the thin spherical aluminum window as shown and are absorbed completely in the 10-cm spherical drift region. A strong electric field is maintained between the two electrodes such that the electric field lines are radial and directed toward the crystal location, and thus exactly match the paths of the entering diffracted beams everywhere in the spherical drift region. Primary electrons produced in the gas as these diffracted beams are absorbed thus travel in the *same direction* as the absorbed X-ray photons which caused them. If the detector is very precisely built and the crystal location is carefully adjusted to be at the natural focal point of the detector, there will then be no smearing out of the primary ionization with respect to the incident diffracted beams and parallax is elegantly avoided. To complete the detection process, the primary electrons pass through the steel mesh and proceed to predictable locations in the fairly standard flat multiwire proportional counter as shown, and their X, Y locations are then sensed by two-dimensional position encoding electronics. These electronics were built for part of the active area and tested at CERN in Geneva[22] and a full set was built for a second, nearly identical detector at MIT.[23] Higher speed X, Y position readout electronics have since been developed for a third similar detector at LURE near Paris for work with synchrotron radiation from the DCI storage ring.[24]

The count rate for 12% dead time loss for the latter detector is reported as 370 kHz. The flat part of this detector is partitioned into a 480×240 array of pixels each about 1×2 mm, and reflections as projected onto this flat region from the spherical drift region (see Fig. 12) must be separated by 5–6 mm to be spatially resolved. Since the crystal-to-detector distance is fixed, this required minimum separation distance for neighboring simultaneous reflections translates directly to a minimum angular separation and limits the maximum crystal unit cell size to about 92 Å (assuming no major systematic absences) for 8-keV radiation. Longer wavelength X rays could be selected at this synchrotron installation and this would help separate the diffracted beams, but then the 2.2 Å minimum Bragg plane spacing would be increased accordingly (thus work with larger unit cells could only be done at lower resolution) and the softer radiation would not penetrate the aluminum entrance window as well. Detection efficiency would be reduced (steeply as a function of energy) below the 58% now reported for 8-keV radiation.

For protein crystallographic work with unit cell edge lengths up to about 90 Å, however, this detector will undoubtedly be useful and new protein structures may be solved using data collected with it. Some test data have been collected from crystals of several small proteins already, including lysozyme B, cytochrome c_7 and gramicidin A. The major disadvantage of this detector, however, is its fixed geometry. It cannot be moved farther away from the crystal to get the increased angular resolution needed for work with larger unit cells (the way a thin, flat multiwire counter can) without the introduction of severe parallax in the 10-cm-thick spherical drift region. Larger spherical drift chambers could perhaps be built and could be used to collect diffraction data from crystals with

[23] C. Bolon, J. Crawford, M. Deutsch, and G. Quigley, *IEEE Trans. Nucl. Sci.* **NS-28,** No. 1 (1981).

[24] R. Kahn, R. Fourme, B. Caudron, R. Bosshard, R. Benoit, R. Bouclier, G. Charpak, J. C. Santiard, and F. Sauli, *Nucl. Instrum. Methods* **172,** 337 (1980).

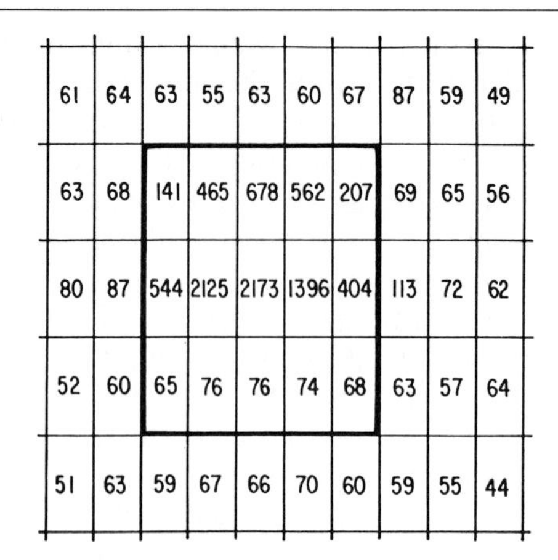

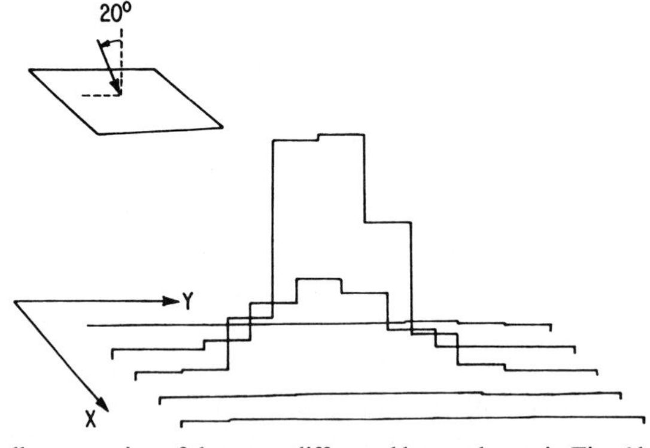

FIG. 13. Parallax smearing of the same diffracted beam shown in Fig. 6 but with this beam tilted 20° away from the normal to the detector. This amount of tilt is about the maximum allowable under ordinary circumstances using the Mark I diffractometer. Nearly all the counts belonging to the reflection are still contained in the 3 × 5-cell summing block normally used during data collection (dark outline above).

appropriately, larger unit cells, but they would be very difficult to build precisely enough and hence would be very expensive.

The Mark II Area Diffractometer, A Multidetector Array

For the Mark I detector whose active absorbing thickness is 9 mm (the volume of xenon/10% CO_2 gas mixture between the front window and the back wire plane), radiation incident in a direction tilted only 20° from the normal to the window surface is detected with only a moderate amount of broadening due to parallax as is demonstrated in Fig. 13. This figure shows the count distribution from the same reflection used in Fig. 6 but with the detector tilted 20° with respect to this diffracted beam. The counts actually belonging to the reflection are still well contained in the standard 3 × 5-pixel summing block normally used during data collection. With the detector placed at $D = 30$ cm as shown in Fig. 9, and intercepting diffracted radiation near the beam stop shadow on the left and extending out beyond 2 Å on the right, the range of entrance angles of intercepted diffracted beams is only about ±23° with respect to a perpendicular to the detector surface. The moderate amount of parallax broadening which results near the extreme left and right sides of the detector surface causes no trouble since this is essentially the same situation as shown in Fig. 13. Parallax is thus not a serious problem for flat detectors as long as the incident diffracted beams are not tilted more than ~25° away from a perpendicular to the detector window.

A simple, flexible way to intercept nearly all of the solid angle in the 2 Å cone, therefore, without encountering severe parallax is to use an array of relatively small flat multiwire counters positioned to approximate a section of the surface of a sphere centered at the crystal. When the detectors must be moved back to spatially resolve diffracted beams from larger unit cells, the detectors can be positioned to approximate a section of the surface of an appropriately larger-radius sphere, and still keep parallax effects to a minimum. Solid angle coverage can thus be flexibly traded for better angular resolution as required for the particular problem and the disadvantages of fixed-radius spherical detectors are thus avoided.

Such an array system, called Mark II, is being assembled in San Diego by the author. The first two detectors of this array are now in routine operation and are shown in Fig. 14 correctly positioned for X-ray data collection from a protein crystal having 180 Å unit cell edges (crystal-to-detector distance about 90 cm). The frame around the window of each Mark II detector has two narrow sides and two wider sides as shown in Fig. 14. When the detectors are positioned as shown, there is only a 2.5-cm-wide strip of "dead" area (the width of one of the narrow sides) between the active detector surfaces. Thus the penalty for dividing up the solid angle among several detectors in an array is minimized; little solid angle is lost between detector windows. Also shown in Fig. 14 is the full circle goniostat and the rotating anode X-ray generator fitted with a graphite monochromator. The monochromator is especially important for work with protein crystals since it removes the softer "white" radiation that would contribute nothing to the measured diffracted intensities and is believed to cause considerable damage to the crystal (because it is

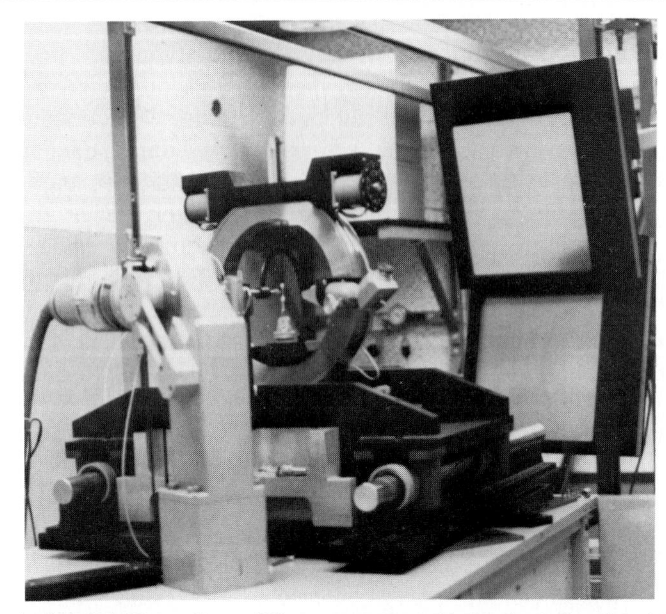

FIG. 14. The Mark II area X-ray diffractometer assembled at the University of California, San Diego by the author. In the foreground are the rotating anode X-ray tube and a full-circle goniostat. In the background are the first two of the new Mark II-type position-sensitive X-ray detectors which will be part of a multidetector array. As of July 1, 1983, this instrument with these first two detectors became available to outside users as a research resource funded by NIH.

strongly absorbed). A block diagram of the Mark II diffractometer system as it is presently configured is shown in Fig. 15. The incident beam monitor will be used to detect momentary "arc-overs" in the rotating anode tube, and the beam intensity information could eventually be incorporated into data scaling to compensate for fluctuations in incident beam intensity which occur then. With the installation of ferrofluidic seals in the rotating anode tube, however, the arc-over problem has very nearly vanished and so no compensation in the data scaling seems to be required.

A low-temperature device is also available on Mark II now. A stream of cold dry nitrogen can be delivered coaxially along the capillary using a set of the jointed Syntex (Nicolet) glassware. The present range of crystal temperatures possible is from room temperature down to about $-40°$ and the servoed temperature control has been measured to be stable to within about $\pm 0.1°$. Crystal temperatures of 5° down to about $-10°$ are often useful for increasing crystal lifetime in the X-ray beam. Temperatures

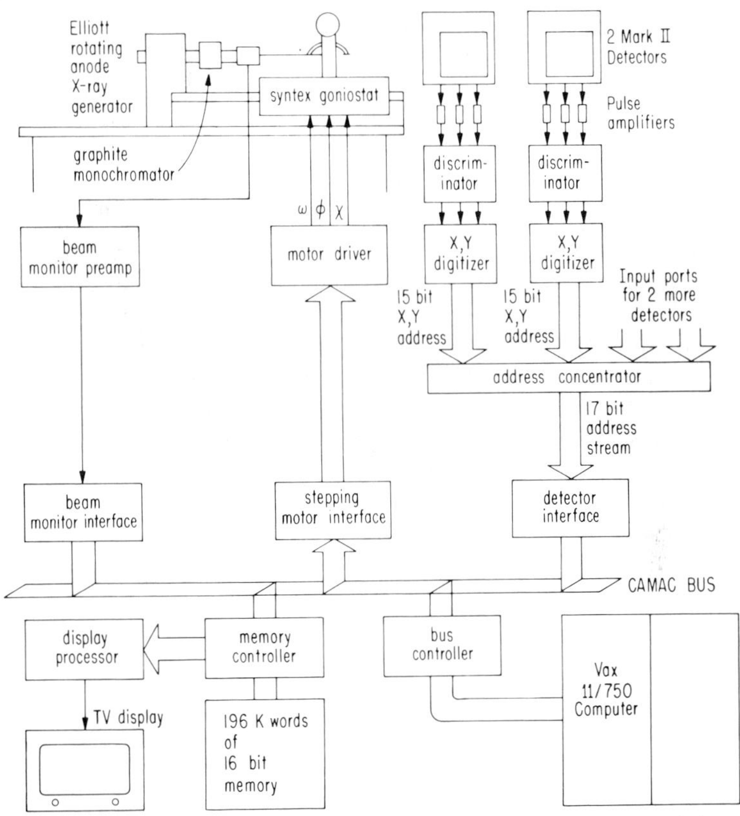

FIG. 15. A block diagram of the Mark II area X-ray diffractometer as it is presently configured. For some data collection projects the computing power of the VAX 750 is more than half-saturated computing background-corrected integrated intensities as fast as the corresponding electronic pictures are accumulated. The planned upgrading to four detectors may require either installation of a bigger computer or direct buffering of raw electronic pictures on a large disk system (500 Mbyte or larger) for Ethernet transfer to another computer.

below $-10°$ can be used, when suitable cryosolvents are known, to do special studies which attempt to "freeze" the protein molecules in some intermediate state long enough (e.g., 4–5 hr) to collect a full set of medium- or even high-resolution data. It may thus be possible using low-temperature techniques and the full data collection speed of Mark II to take crystallographic "snapshots" of several of the steps an enzyme molecule goes through during a particular chemical reaction. The cytochrome *c* peroxidase studies listed in Table III below represent one of our early efforts to do this kind of work. In this case the investigators are attempting to see the intermediate "compound I" state of the cytochrome *c* peroxidase molecule.

General Characteristics

The Mark II detectors are based on the technology borrowed from high-energy physics for use in the Mark I detector. Key parts of the Mark I detector (delay lines and wire planes) were built for us by Victor Perez-Mendez at University of California, Berkeley but the Mark II detectors are now built in our own laboratory at University of California, San Diego. The Mark II detectors use 2-μsec delay lines and wire plane electrodes similar to those of Mark I but, as mentioned above, the shape of the detector enclosure has been changed somewhat compared to the Mark I detector to minimize "dead" solid angle in detector arrays. The Mark II detectors have a somewhat improved spatial resolution in the Y direction (as defined in Fig. 5), about 0.5 mm compared to the 0.75 reported for Mark I. The geometric linearity is improved, too, with maximum position discrepancies compared to an ideal pixel grid of about 1 mm instead of the 2 mm reported for Mark I. The X-ray window is made of a food-packaging material which is more transparent to soft X rays, more uniform in thickness, flatter, very much cheaper, and much safer than the beryllium used for the Mark I detector window. The increased window transparency together with a 20% increase in the thickness of the xenon gas volume and considerably better gas seals to keep contaminants out result in a 60% detection efficiency compared to the rather optimistic 47% reported for the Mark I detector. This detection efficiency is uniform to within 1% of its value over the surface of the Mark II detectors instead of within the 2% reported for Mark I. Since the Mark II detectors use the same kind of 2-μsec delay line position readout scheme, their dead time loss behavior is essentially the same as that given for the Mark I detector in Fig. 7. The more flexible window material used in the Mark II detectors requires better pressure regulation to keep the window from bowing in response to barometric pressure changes (a situation which could cause output pulse-size variations). The pressure regulation is now done statically using a soft, flexible expansion chamber made of the same food-packaging material the front windows are made of. Since material in the window is stretched tightly on the front window frame, the required small volume changes occur in the soft expansion chamber and the window stays flat. There are now no moving parts in the xenon plumbing system; it is simple, gas-tight, and reliable. Also there is no wasteful flow of xenon into the atmosphere as there is with the Mark I bubble-in/bubble-out pressure regulation scheme; hence there is no replacement bottle of xenon required every few months as there is with Mark I. The first two Mark II detectors have been very reliable so far. The older of the two detectors has been operating for more than 2 years without breakdown.

Data Collected with Two Mark II Detectors

Routine crystallographic data collection with the two-detector version of Mark II began in February, 1983, and a number of research groups including our own (Xuong) scheduled time on it. Table II is a brief sum-

TABLE II
SUMMARY OF DATA COLLECTED DURING A 1-MONTH PERIOD

Date	Group	Protein	No. of observations	No. of reflections	Resolution range (Å)	Intensity *R* factor (%)
3/11/83 to 3/15/83	Kraut, UCSD	Dihydrofolate reductase (chicken liver), native	39,000	18,000	4–1.6	2.9
3/17/83 to 3/19/83	Xuong, UCSD	Phospholipase (cobra), Hg derivative	126,000	22,000	Inf.–2.4	7.8
3/21/83 to 3/23/83	Kraut, UCSD	Dihydrofolate reductase (chicken liver), native	39,000	18,000	4–1.6	3.1
3/24/83 to 3/26/83	Fletterick, UCSF	Phosphorylase *a* + maltopentaose	59,000	21,000	Inf.–3.5	6.9
3/27/83 to 3/28/83	Fletterick, UCSF	Phosphorylase *a* + maltopentaose + AMP	65,000	20,000	Inf.–2.8	4.9
3/28/83 to 3/29/83	Fletterick, UCSF	Phosphorylase *a* + maltopentaose + ATP	114,000	23,000	Inf.–2.8	7.6
4/6/83 to 4/8/83	T. Steitz, Yale	DNA polymerase	84,000	21,000	Inf.–3.0	5.7
4/9/83 to 4/10/83	T. Steitz, Yale	DNA polymerase with substrate	81,000	28,000	Inf.–2.6	3.5

mary of the data collected in a mouth-long period shortly after the two detector version of Mark II became operational. Since only 3 weeks of the month were actually used to collect the data in Table II it is believed that it will be routinely possible to collect about one million reflection observations a month with this version of Mark II (if the required steady stream of data collectors with good crystals can be found).

In certain favorable cases the data collection rate can be much higher than this. When crystals with large unit cells diffract especially well so that the exposure time per electronic picture can be as low as 25 or 30 sec, data collection rates over 100,000 observations per day can be achieved. Data from native and heavy-atom derivative crystals of histidine carboxylase ($I4_12_21$, $a = b = 222$, $c = 108$) were collected at a rate of 180,000 observations per day, a rate which is computed on the basis of calendar time and includes the overhead operations which must be performed occasionally (once or twice a day) during data collection. These data scaled with an intensity R factor of 4–5% and the investigators have since informed us that the Patterson maps clearly show the heavy-atom sites and the structure solution is proceeding well. It is expected that the data collection rate will increase by about another factor of 2 (at least for large-unit-cell problems) when Mark II is expanded to a four-detector machine soon (see below).

Besides the histidine carboxylase data (M. Hackert, University of Texas, Austin) a considerable amount of crystallographic data of comparable quality have been collected on Mark II (Table III) since the period described in Table II. An interesting comparison can be made with the single-crystal diffractometer (Nonius, CAD-4) on the basis of the DNA polymerase data (in $P4_1$ $a = b = 107$ Å, $c = 90$ Å). Diligent effort had been

TABLE III
DATA COLLECTED ON MARK II

Protein	Group
Ribulose-bisphosphate carboxylase	D. Eisenberg, University of California, Los Angles
DNA polymerase	T. Steitz, Yale University
Elongation factor—TU	F. Jurnak, University of California, Riverside
Nucleosome (chicken erythrocyte)	G. Bunick, Oak Ridge National Laboratory
Cytochrome *c* peroxidase ("compound I" at −25°)	N. Xuong and J. Kraut, University of California, San Diego
Aspartate carbamoyltransferase	W. Lipscomb, Harvard University

given to collecting native and heavy-atom derivative data on a single-crystal diffractometer with specially modified software to maximize data collection speed. Due to constraints from crystal lifetime and diffracting power and the inevitable errors from scaling together the data from many heavy-atom derivative crystals, the practical resolution limit of this data was about 3.5 Å. The density in the multiple isomorphous replacement (MIR) map was only poorly connected and chain tracing was going slowly. After an exploratory expedition to Mark II in April, 1983, the investigators decided to come back to San Diego with some of their best crystals and recollect all their native and heavy-atom derivative data on Mark II. Using only one or two crystals of each type they collected a heavily replicated set of 2.9 Å data. They have found that the MIR map based on this new data is dramatically better than the one based on the single-crystal diffractometer data. The main chain is now well connected and readily traceable and structure solution is proceeding rapidly.

A careful comparison of data quality between the Mark II system and X-ray film has not been done yet. It was found that for most protein work the older Mark I system was able to collect more data per crystal by a factor of at least 10 than the oscillation film method could, with an accuracy that was somewhat better (giving 5% *R* factors on intensity when film gave 8%). The Mark II system, with its monochromatized rotating anode X-ray source (300-μm focus) instead of the fixed-tube source Mark I has, and its present two detectors instead of the one Mark I has, appears to be giving an increase of at least a factor of 4 in data collection speed compared to Mark I. But since the hotter beam (even though it is now monochromatized) probably disorders the crystals somewhat faster than the Mark I beam does, the diffraction information extracted per crystal may at the moment be only be a factor of 3 more than it is for Mark I. The inference here, therefore, is that the Mark II with two detectors is getting about 30 times more information from most protein crystals than the traditional oscillation film method would. And of course there is no film to scan afterwards, because the data are already in numerical form, first as counted photons and then as integrated intensities computed while data collection is taking place. It must be noted here that the inference about information per crystal just drawn does not hold for cases in which the unit cells are very large, having ~300 Å or longer cell edges. In these cases the detectors have to be moved rather far away from the crystal (about 150 cm) to give the required 6- to 8-mm separation of nearest neighbor diffracted beams at the detector surfaces (although finer beam collimation could reduce this requirement somewhat). Much less solid angle of the diffraction pattern is then intercepted than is the case in film work, and much of the advantage of the present Mark II detectors is traded way. Efficient work with unit cells over 300 Å awaits more and/or larger area detectors of this type.

Mark II Array Characteristics

Two more detectors will soon be added to Mark II (i.e., there will then be a four-detector array). The two new detectors will usually be positioned similarly to the first two shown in Fig. 14 except that they will intercept the other side of the diffraction pattern. Each of the four detectors will be positioned exactly facing the crystal so that a ray from the center of each window and perpendicular to it will *intersect* the crystal (or nearly so; there is an angular tolerance of a few degrees here). When these detectors are positioned about 45 cm from the crystal they will intercept the four regions in the diffraction pattern shown in Fig. 16 The array configured as shown in this figure will be able to collect diffraction data from protein crystals with unit cell edge lengths up to about 90 Å (or

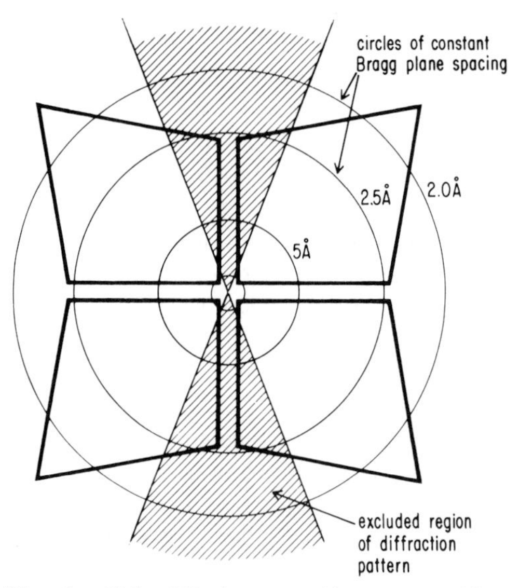

FIG. 16. The solid angle which will be intercepted by an array of four of the Mark II area X-ray detectors. The crystal-to-detector distance implied (50 cm) is appropriate for work with crystals having 100 Å unit cells edges. Each detector is tilted to face the crystal (to minimize parallax), and therefore the square detector surfaces undergo perspective distortion as shown when they are projected onto a flat "film plane" perpendicular to the main X-ray beam.

appropriately longer if there are major systematic absences in the space group and the reflection centers are separated by the required 6–8 mm at the detector surface). It should be noted here that the solid angle covered by this configuration of the four-detector array is nearly the same as that shown in Fig. 11 for the spherical drift chamber (also for 90 Å unit cell edges). The array, however, has several important advantages.

The flat detectors are much cheaper and easier to build and the size of the data collection system can be expanded to a modular fashion as the need arises and funds become available. The X, Y position readout channel for each detector operates independently of the others and so the detector dead times are independent. For 12% dead time loss the four-detector array can operate at a photon detection rate of 100 kHz even with relatively slow 2-μsec delay line position readout similar to that in the Mark I detector (25 kHz point in Fig. 7). This rate is only about a factor of 3 times slower than the rate quoted for the LURE synchrotron spherical drift chamber, yet the readout electronics are dramatically simpler and cheaper. The most important advantage of the array, however, is that it can be reconfigured for data collection from crystals with larger unit cells (while the spherical drift chamber, of course, cannot).

Figure 17 shows the regions of the diffraction pattern that would be intercepted by the four detectors positioned to approximate a section of the surface of a sphere 85 cm from the crystal, as would be appropriate for data collection from a crystal with 170 Å unit cell edges. Some solid angle coverage has been traded for finer angular resolution. Figure 18 shows the two-step data collection procedure which would be used to collect data from a crystal with 250 Å unit cell edges. High-resolution data would usually be collected first since the high-resolution data would ordinarily decay fastest as the crystal disorders in the beam. Low-resolution data would then be collected using relatively short exposure times since the low-resolution diffraction is usually strong even after the high-resolution diffraction was weakened considerably.

A pyramid-shaped helium-filled box is of course required between the crystal and each of the detectors to eliminate the 1% per centimeter attenuation of copper radiation in air. A design has now been developed which allows the front windows (3.5 cm^2) of these helium-filled boxes to be placed very close together, about 9 cm from the crystal with only about 0.2 cm of "dead" area between them. Thus even when the detectors are in "close-packed' arrays the "noses" of their individual helium boxes will not get in each other's way. A set of four of these fixed-length boxes will be required for each of four or more crystal-to-detector distances (the length ratio between sets will be 1.4) and the boxes do promise to be something of a storage problem.

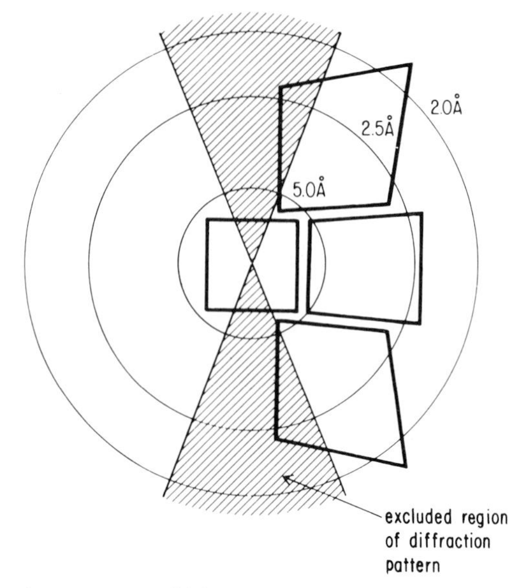

FIG. 17. A four-detector array which would be appropriate for X-ray data collection from a protein crystal with 170 Å unit cell edges. The detectors are positioned to approximate a section of the surface of a sphere centered at the crystal and are each at a distance of about 90 cm from the crystal. Perspective distortion is again evident when the square detector surfaces are projected onto a flat "film plane" perpendicular to the main X-ray beam.

Conclusion

The flat xenon-filled multiwire proportional counter has been found well suited for collecting accurate diffracted X-ray intensity data from the easily disordered crystals of proteins. It allows at least an order of magnitude more data to be collected from each crystal than can be done with traditional film or diffractometer methods, and often allows a complete, heavily replicated set of 2.5 Å data to be collected from one crystal. Such data have then been used to solve more than a dozen new protein structures. No other two-dimensional X-ray detector technology is yet able to do this. Understandably a number of other single-detector multiwire counter diffractometer systems are now being built in various places in the world and some of them were outlined in Table I.

Parallax prevents full coverage of the 2 Å cone with one flat multiwire counter. The spherical drift chamber is a modified multiwire counter designed to defeat parallax in a 2.2 Å cone (at 8 keV) but its fixed crystal-to-detector distance puts severe restrictions on its use for data collection from crystals with large unit cells. An array of four or more flat detectors can cope with parallax in a simpler, cheaper, and much more flexible way

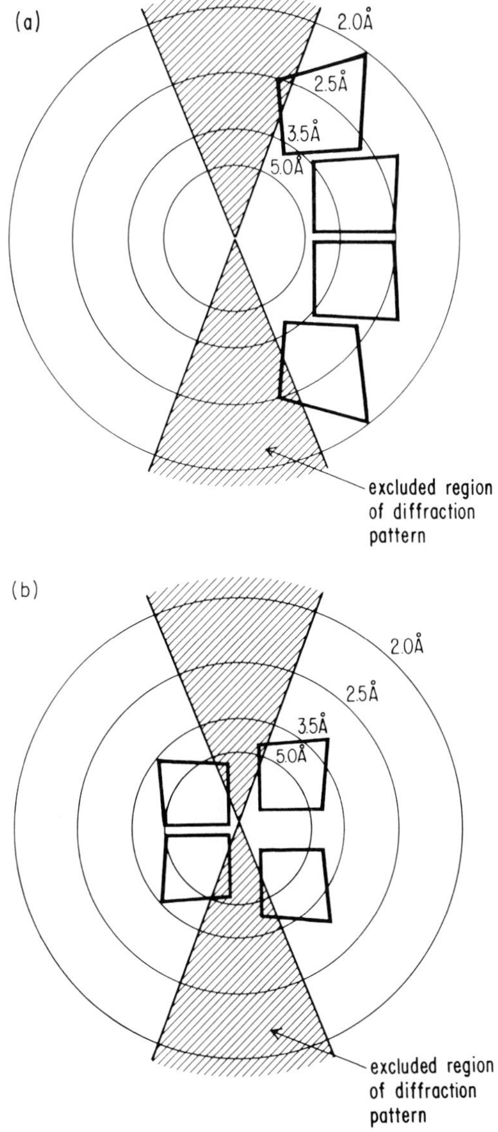

FIG. 18. Two arrangements of a four-detector array which would be used to collect X-ray intensity data from a protein crystal with 250 Å unit cell edge lengths. (a) shows the array arrangement which would be used to collect the high-resolution data (data which should be collected first while the crystal still diffracts to high resolution) and (b) shows the array arrangement which would be used to collect the low-resolution data. The small amount of overlap of the two configurations with respect to each other will be used for cross-scaling of the data.

for unit cells of arbitrary size by allowing a trade-off between angular resolution of the diffracted beams and solid angle coverage of the diffraction pattern. Because of the division of dead times mentioned above such an array will be able to handle photon detection rates approaching those encountered in at least some synchrotron work, even with the relatively slow, 2-μsec delay line position readout being used now.

Acknowledgment

This work is now being supported by the National Institute of Health under General Medical Grant GM 20102 and under Research Resource Grant RR 09144. As a Research Resource we are making about 40% of the machine time on Mark II available to outside users on a service basis (no co-authorship required) and another 30% or so of the machine time available on a collaborative basis for data collection problems requiring special equipment or special participation by local personnel. If you are interested in using Mark II and feel you have a data collection problem really requiring use of this machine submit a two-page summary to Professor Xuong or the author (Department of Physics B-019, UC San Diego, San Diego, California 92093) for use by the oversight committee who will recommend a priority for your project to us.

[28] Software for a Diffractometer with Multiwire Area Detector

By A. J. HOWARD, C. NIELSEN, and NG. H. XUONG

Introduction

As shown in the previous article,[1] the use of the multiwire proportional chamber as an X-ray area detector can increase the rate of data collection for protein crystals by at least an order of magnitude.

One could collect intensity data with the multiwire area detector (MAD) diffractometer by using the standard rotation method,[2] but this would not take full advantage of the capabilities of the detector. Therefore we have devised an improved technique called the "Electronic Stationary Picture" method.[3]

[1] R. Hamlin, this volume [27].

[2] U. W. Arndt, J. N. Champness, R. P. Phizackerley, and A. J. Wonacott, *J. Appl. Crystallogr.* **6,** 457 (1973).

[3] Ng. H. Xuong, S. T. Freer, R. Hamlin, C. Nielsen, and W. Vernon, *Acta Crystallogr., Sect. A* **A34,** 289 (1978).

In this method of data collection, reflection intensities are extracted from a sequence of electronic pictures, each of which is exposed while the crystal is stationary (but see later). Between successive pictures the crystal is rotated about a fixed axis by a constant angle of approximately 0.06°. This is to be contrasted with standard rotation methods[4] in which the crystal is continuously oscillated by a few degrees during the entire time the detector is recording the diffraction pattern. The "Electronic Stationary Picture" technique has three principal advantages. First, since all data are measured with the crystal stationary, reflection overlap is minimized. Second, since intensity data are extracted from a definite X, Y location of the detector *only* during the interval when the diffracted beam is actually present, the peak-to-background counting ratio is maximized; by contrast in crystal rotation methods counts are accumulated even when the Bragg conditions are not satisfied. Third, one can use the count rate measured at this X, Y location before (or after) the arrival of the diffracted beam to obtain a precise estimate of the background. These advantages are particularly significant for data collection from crystals with large unit cells, which produce large numbers of densely packed weak reflections.

Since the profiles of some reflections can be quite sharp, a stepping of, for example, 0.06° between frames would not allow the exact estimate of their intensities. Therefore we have slightly changed the procedure so that each frame will be the sum of six subframes of equal but smaller exposure time in which the crystal will rotate by 0.01° between successive subframes.

Our method is similar to the step scanning technique[5] used in conventional diffractometry except that a large number of simultaneously diffracting reflections are measured together. Because the integrated intensity of each reflection is extracted from several consecutive pictures, the advance prediction of its scanning angle and spatial location is required; therefore the data extraction algorithm is necessarily complex.

In the following, we discuss in detail the computer programs which drive the area detector system. The programs are written in the "C" language and run under the UNIX[5a] operating system on a Digital Equipment Corporation VAX 11/750 computer. After outlining the organization of the software, we shall describe the programs used in the alignment of the crystal, the precise determination of the crystal orientation and pa-

[4] U. W. Arndt and A. J. Wonacott, "The Rotation Method in Crystallography." North-Holland Publ., Amsterdam, 1977.

[5] U. W. Arndt and B. T. M. Willis, "Single Crystal Diffractometry." Cambridge Univ. Press, London and New York, 1966.

[5a] UNIX is a trademark of Western Electric Company.

rameters, the data gathering, and the data output and scaling. Finally, we shall present a strategy for data collection and discuss the results obtained so far.

Program Organization

Area detector data collection involves a number of complex, interrelated tasks: crystal alignment and refinement, prediction of reflection locations, collection of intensity data, and data transfer. We have developed an organizational scheme behind the programs controlling these operations so that data can flow readily from one task to the next, and so that user intervention and input is minimized. The software comprising the University of California, San Diego (UCSD) system involves the programs listed in Table I, together with a library of files containing crystallographic information for previously studied crystals and for the current crystal. Each file contains the cell parameters, orientation parameters, current goniostat position, space group information, protein name and crystal type, scanning angle and stepsize, starting and ending goniostat positions of the current data run, and so forth. As these values change in the course of data collection, the file for the current crystal is continually updated.

The program package is designed to facilitate the input of adjustable parameters necessary to collect data. This operation is largely accomplished with the program "TABLET." This program enables the user to edit all values necessary to set up a data collection run by typing the desired values in clearly delineated slots in data tablets. The UCSD system uses two tablets. One describes the overall properties of the crystal

TABLE I
COMPUTER PROGRAMS FOR DATA COLLECTION

Program	Function
TABLET	Data Entry Tablet: for general crystal information and for a specific data run
ALIGN	Crystal Alignment: interactive program for alignment
PREDICT	Reflection Prediction: precomputes reflection positions
DATAC	Data Collection: collect data, computes intensities
PARAM	Parameter Refinement: refines cell parameters, orientation angles, and detector position
BGNCI	Generates an initial background picture
SCALE1	Scaling data of different runs together
REJECT	Identifying and deleting faulty measurements
EABCOR	Ellipsoidal-surface approximation for absorption correction

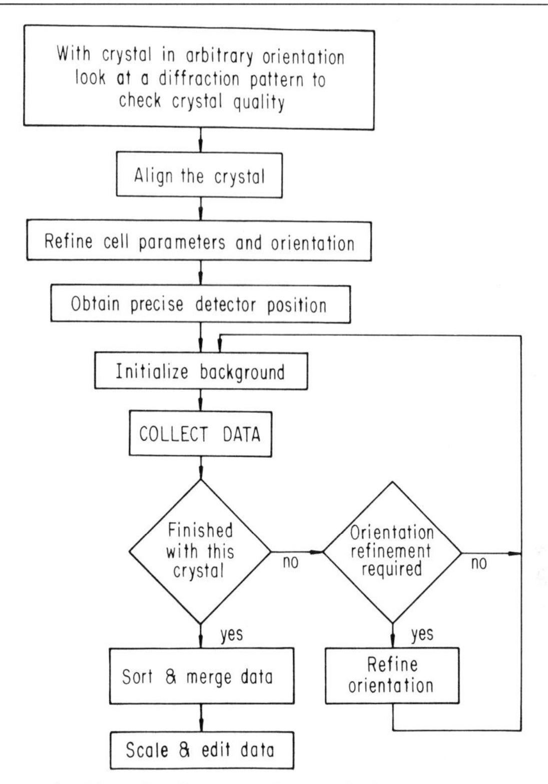

FIG. 1. Steps required in collecting crystallographic data on the multiwire area detector diffractometer.

such as its name, an identifier code, its cell parameters, and the alignment angles. The other includes parameters describing a data collection run such as the initial and final goniostat positions, the exposure time per picture, and the stepsize between frames. The program minimizes the number of entries the user must explicitly change from one run to the next by automatically computing some parameters as functions of others; for example, the resolution limit of a data run is a function of detector position, so this limit is simply computed for the user. If it is necessary to override the computed values of any parameters there is a facility for doing so.

The execution and sequencing of the programs listed in Table I proceed as described in Fig. 1. In almost all situations, the user simply has to type the name of the program to run; the programs are designed to automatically find the correct input files and generate output files, and to initialize the hardware, if required. As with the tablets above, it is possible to override the defaults if desired.

Crystal Examination and Alignment

The first task the computer can help with is the initial examination of the diffraction pattern. It is usually possible to obtain a still "photograph" of a diffraction pattern in less than 1 min using an area detector. The single still is enough to give the user a general notion of how well his crystal diffracts, and it will often allow him to make approximate reciprocal space axis assignments. The computer's contribution to this task is to direct the displays of the digitized electronic picture on a video screen.[1] This display provides the user with a rapid, qualitative view of reciprocal space. The UCSD system allows the storage of two images simultaneously; these can be displayed on a television screen individually or overlaid upon one another. The alignment program ALIGN permits the user to take electronic pictures, move motors, and perform some simple calculations based on spot positions. It allows for the collection and display of electronic oscillation and pseudo-precession photographs as well as still pictures; these "pooled" displays can be used in the same way that small-angle precessions are used in film data collection for lining up principal axes. The software permits motion of the crystal around the conventional goniostat axes ω, χ, and ϕ, and it also permits two more complex motions: rotation around the direct beam direction, and rotation around an (horizontal) axis perpendicular to the direct beam and the goniostat axis. These motions are sufficiently general to permit crystal alignment even under adverse circumstances (as when a crystal is mounted in a flow cell) without the need to adjust the goniometer head arcs. The calculated features in the program permit estimates of cell axis lengths and directions based on the positions of spots along a row.

The next task is crystal alignment. Usually the crystal parameters and its space group symmetry have been determined previously, but the crystal still needs to be aligned along a known orientation. If one knows the relationship between the morphology of the crystal and the orientation of the unit cell axis, then the alignment procedure is rather simple, and resembles that used with a rotation camera.

But many crystals do not show clear morphology, and for these crystals a partially automated method has been in use for several years at UCSD. In this method, the initial goniostat angles are set at $\omega = 0.0°$, $\chi = 90.0°$, and $\phi = 0.0°$. The crystal is then rotated about the ϕ axis, until a zero circle appears in the diffraction pattern. This zero circle is then brought under the incident beam and a small-angle precession picture taken. The measurement of the spacing between the spots on a row will show whether it is parallel to a principal axis. If this is not the case, then one continues the search along the ϕ axis. In most cases, one of the rows visible in the small-angle precession picture will be along a principal axis. This axis will then be set vertical or horizontal and the crystal rotated about the axis to reveal a principal zone, often with symmetry in the diffraction pattern. The user can then make small adjustments in the goniostat angles to arrive at an orientation which is correct to within 0.25° in every angle.

Refinement of the Orientation of Crystal

In principle a suitably sophisticated alignment program could provide a knowledge of crystal orientation which is sufficiently precise that data collection could begin as soon as alignment is complete. It has, however, proven more practical to separate crystal alignment from a refinement procedure, in which several parameters are made more precise by minimizing differences between some observed and calculated quantities. The parameters which must be made precise fall into two categories: unit cell lengths and angles, and crystal orientational parameters.

The observable quantities used in the crystal parameter refinement are the scanning-angle centroids of some reflections measured in a brief "refinement" data run. The scanning-angle centroids prove to be more effective in refinement than the positions of the reflections on the detector face, because they can be precisely determined and depend only on the crystal parameters, whereas the reflections' positions are sensitive to the detector's own parameters and its nonlinearities. The crystal refinement involves minimization of the quantity

$$R_\omega = \sqrt{\frac{1}{n}\sum_{j=1}^{n}(\omega_{oj} - \omega_{cj})^2}$$

where ω_{oj} is the observed scanning-angle centroid for the reflection and ω_{cj} is the computed centroid.

The "refinement" data runs involve prediction and collection of Bragg reflections much as in a full-fledged data collection run (see below). On the UCSD system a relatively primitive background correction is applied during a refinement data run: the background is estimated from the average count on the detector "free" cells surrounding the reflection, that is, the neighboring cells not included in another reflection's summing area. Otherwise, the method of intensity data measurement is as described below. The data are collected in two runs of roughly 2° of rotation each but with a separation of at least 60° between the two runs.

All the reflections with (intensity/standard deviation) ratios above a user-set minimum are used in the refinement. The computed centroids are functions of the cell parameters (a, b, c, α, β, γ) and the crystal orientation, parameterized as (ω_A, χ_A, ϕ_A), the goniostat setting at which the crystal axes lie along the laboratory axes. The dependence of ω_{cj} on these parameters is nonlinear, so either a linear least-square approximation or a nonlinear refinement is required. The latter approach has been adopted in a fairly simple way: the parameters are arranged in the order (ω_A, χ_A, ϕ_A, a, b, c, α, β, γ) and one parameter at a time is varied to obtain a minimal value of R_ω for the collected data. Then the next parameter in the list is varied to again optimize R_ω, and so on to the end of the list. To minimize the likelihood of refining into a local minimum higher than (and distant from) the global minimum, the parameter are then varied in reverse order, again minimizing for each parameter. If a unit cell parameter is constrained by space group symmetry to be a constant or to be equal to another parameter, the refinement simply skips that parameter in the minimization. Thus in monoclinic space group, seven parameters are refined; in cubic space group only four.

A successful refinement yields $R_\omega \cong 0.01°$; thus the refined parameters bring the observed and calculated scanning-angle centroids to within 0.01° of one another. The refined parameters will be used to predict reflections for a full-fledged data run; the reflections collected in that run will have well-centered scanning-angle profiles.

Data Acquisition Programs

Once the crystal parameters are known, data acquisition can begin. This procedure consists of two principal steps: (1) Precalculating the scanning angle and detector positions of the reflections which will appear during the data collection run. This is done by the PREDICT program. (2) Actual gathering of the data and estimating the integrated intensities of the predicted reflections. This is done by the DATAC program.

In the following we shall describe these two programs in more detail. In addition, we shall present a discussion of a new method of background estimation and initialization which is well suited to the area detector system.

Reflection Prediction

In order to collect and reduce data on-line it is necessary to precalculate the positions on the detector face and the scanning angles at which Bragg reflections will appear. A method has been developed for determining which reflections will be in diffracting position during a data run without solving the detailed geometry problem for each *hkl* triplet. This method is a modification of a technique developed by George Reeke for oscillation photography[6]; it works in this context because data collection on the detector involves rotation of the crystal about a fixed axis (ω) while χ and ϕ are held fixed.

In the detector version, the index of the crystallographic axis closest to the direct-beam direction (x) is varied slowest, the index of the axis closest to the rotation axis (z) is varied fastest, and the remaining axis is varied intermediately. Thus if a crystal is oriented with **a*** closest to z and **c*** closest to x, then h will be varied slowest, k next slowest, and l fastest. The limits on h (in this example) are determined from reciprocal space geometry and the position of the detector; then for each h the limits of k appropriate to that h are determined, and for each hk pair one or two ranges of permissible l values are obtained, depending on how the Ewald sphere intersects the resolution-limit sphere at the extrema of the data run. The list of trial hkl values is only slightly longer than the list of reflections which ultimately get predicted, because the restrictions on h, k, and l are based on both the "scanning range" and the detector position. Only the hkl values generated by this "range-limiting" algorithm are considered in the geometrical construction which follows.

A reflection **h** will have a scattering vector $\mathbf{s} = \lambda A\mathbf{h}$, with

$$A = \begin{pmatrix} a_x^* & a_y^* & a_z^* \\ b_x^* & b_y^* & b_z^* \\ c_x^* & c_y^* & c_z^* \end{pmatrix}$$

where a_x^*, a_y^*, a_z^*, etc., are the projections of the reciprocal-space unit cell vector $\mathbf{a}^*$, $\mathbf{b}^*$, $\mathbf{c}^*$ on the laboratory axes. Initially we calculate this matrix at the fixed χ and ϕ values of the data run and at $\omega = 0$.

We express **s** in cyclindrical polar coordinates (ζ, ϕ, ξ), with

$$\begin{aligned} \zeta &= s_x \\ \phi &= \tan^{-1}(s_z/s_y) \\ \xi &= (s_y^2 + s_z^2)^{1/2} \end{aligned}$$

As shown in ref. 2, a reflection (hkl) will cross the Ewald sphere at two setting angles

$$\omega_1 = (\pi/2) - \phi - \beta \quad (\text{modulo } 2\pi) \tag{1}$$

and

$$\omega_2 = (\pi/2) - \phi + \beta \quad (\text{modulo } 2\pi) \tag{2}$$

where β is defined by

$$\cos\beta = (\zeta^2 + \xi^2)/2\xi \tag{3}$$

[6] G. Reeke, *Am. Crystallogr. Assoc. Program Abstr.* **7**(1), and 14 (1979).

Sometimes owing to goniostat misalignment, the rotation axis is tilted so that it deviates from the normal to the direct beam by an angle μ. Then the setting angles can still be calculated by Eqs. (1) and (2) but with

$$\cos \beta = (\xi^2 + \zeta^2 + 2\zeta \sin \mu)/(2\xi) \tag{4}$$

If the chamber is positioned at a distance D from the crystal and its normal is set at an angle θ_c from the incident beam then the coordinates of the diffracted rays on the chamber[3] are:

$$X_1 = \frac{(\zeta + \sin \mu)D}{[1 - (\zeta + \sin \mu)^2]^{1/2} \cos(\Gamma - \theta_c)} - D \tan \mu + X_0 \tag{5}$$

$$Y_1 = D \tan(\Gamma - \theta_c) + D \tan \theta_c + Y_0$$

and

$$X_2 = \frac{(\zeta + \sin \mu)D}{[1 - (\zeta + \sin \mu)^2]^{1/2} \cos(-\Gamma - \theta_c)} - D \tan \mu + X_0 \tag{6}$$

$$Y_2 = D \tan(-\Gamma - \theta_c) + D \tan \theta_c + Y_0$$

where

$$\cos \Gamma = \frac{\cos^2 \mu + \xi^2 - [1 - (\zeta + \sin \mu)^2]}{2 \cos \mu \, [1 - (\zeta + \sin \mu)^2]^{1/2}}$$

and (X_0, Y_0) are the chamber coordinates of the undiffracted beam.

Reflection prediction requires first the generation of a trial list of *hkl* triplets which satisfy the condition of the "range-limiting" algorithm described above.[6] For each trial *hkl* triplet the program computes ω_1 and ω_2 as defined above to see if either falls within the range of data collection. The reflection (X, Y) position is also computed to make sure the reflection will actually hit the active area of the detector. Any reflection which passes both tests is output with its indices, its scanning-angle value and detector position, and a calculated correction factor

$$C = \eta / Lp$$

where η is a path absorption correction, L is the Lorentz factor, and p is the output-beam polarization factor. The factor C is to be multiplied by the raw integrated intensity and its standard deviation in data collection to yield corrected values. Formulas for the elements of C are given in Chap. 9 and Appendix I of ref. 7. After all the trial *hkl* triplets have been checked, the output reflections are sorted into increasing scanning-angle order and output to disk.

[7] A. Howard, Ph.D. Dissertation, University of California, San Diego (1981).

The use of the "range-limiting" algorithm, which has only been a part of the UCSD package since early 1982, has resulted in a dramatic reduction in computer time. For small scanning-angle ranges (1°), predictions can be computed in a few seconds; as the range increases the computer tends to increase proportionally. The prediction algorithm is now coded as a callable subroutine. Thus any program which needs the ability to predict reflections can do so. At present, all programs which use predictions can do their own predictions. This includes the programs ALIGN, PREDICT, BGNCI, and DATAC.

Before starting the actual data collection, it is useful to check the reflection predictions against actual diffraction patterns. A stationary picture is taken at a given goniostat position (ω, χ, ϕ). The χ and ϕ angles are those for the proposed data collection run. We use the ability of ALIGN to perform its own predictions to predict ± 2.5 frames in the scan angle ω about the current position. We can then superimpose a set of small crosses placed at the predicted chamber coordinates. This superposition can be viewed on the video display, giving the operator a method of checking the quality of the prediction. A bright reflection must be accompanied by a cross but a cross can appear at a very weak reflection and does not show up on the video display of the diffraction pattern. This display is very useful in making the final adjustment of X_0, Y_0, and D, which characterize the position of the detector.

Data Collection

The program which directs the counting of diffracted photons and the integration of the reflection intensities is the heart of an area detector software package. It must receive the counts, move the scanning motor, match collected intensity with predicted reflections, sum the intensities, correct for background, and output the results to an appropriate storage medium.

At least two general schemes for data collection can be envisioned. The first, described by Arndt,[8] closely resembles screenless oscillation photography. In it, the crystal is rotated continuously about an axis, and the intensity of a reflection is computed as the total number of counts received at the appropriate position on the detector as the reflection passes through the Ewald sphere, minus a background estimate. The accumulated intensity would be determined separately for each reflection as it finishes its transit through the sphere's surface, so no discrete "pic-

[8] U. W. Arndt and A. R. Faruqi, "The Rotation Method in Crystallography," Chapter 15. North-Holland Publ., Amsterdam, 1977.

tures'' are taken; rather, each ''cell'' of the detector surface is sampled and zeroed out at the appropriate moments, independent of all the other cells.

At UCSD we use the other technique, which involves the collection and readout of discrete electronic ''pictures.''[3] Each picture is a digitized image of the counts on the detector, and is read out at regular intervals as an array of count values, one value for each 32,768 cells of the detector surface. The picture is formed with the goniostat rotating through an (0.05°–0.12°) angle about the fixed rotation axis. The next picture is taken at the ''next'' position along the scanning range, which is the 0.05°–0.12° range that begins where the previous one ended. A reflection's intensity is measured by summing counts at its detector position over all the pictures in which it diffracts. As in the oscillation technique, a background estimate must be subtracted from the raw intensity (see next section).

The readout of discrete pictures simplifies data handling, and because explicit profile elements are measured, some angular information is retained. Also, the program performs the intensity summation using only the profile elements which actually contribute to the total; this improves the signal-to-noise ratio.

A Bragg reflection extends over a finite angular width, so its projection onto a detector surface is not infinitesimally small. Therefore, counts for a single reflection actually extend over several cells of the detector face, and the reflection summation must involve all those cells in each of the pictures in which the reflection appears. The exact nature of the summing ''block'' will depend on the resolution of the detector and on how precisely the crystal parameters and the detector position are known. In our system we use a 3×5 block of cells representing an area on the detector surface of 6.4×5.9 mm.[1]

Computing integrated intensities is relatively straightforward. In Arndt's oscillation technique, the integrated intensity is nothing more than the background-corrected count over the summing area. With the UCSD technique, the counts over a group of cells in a single picture are summed to give a raw intensity in that picture. This value, termed the raw profile-element intensity, is corrected for dead time loss.[1] Meanwhile, the background in the cells which make up the reflection is estimated by either local averaging or image updating as described below. The difference between the raw intensity and the background is the corrected profile-element intensity for the picture. A section (three to seven pictures' worth depending on the intensity of the reflection) of the nine-picture profile is taken to be the useful portion of the profile, and the intensities in that portion are summed to yield a raw integrated intensity.

The raw intensity can be altered on-line with a number of correction factors. Lorentz, polarization, and air absorption corrections are applied at this point. As mentioned above, the magnitude of several of these corrections can be precalculated as part of the reflection prediction operation, and need only be multiplied by the raw intensity during data acquisition.

An estimated standard deviation should be computed along with each intensity measurement. It is useful to include a "floor" term as an additive term in the variance, so that the estimated standard deviation for a reflection of intensity I and background B becomes

$$\sigma = [\text{var}(I + B) + \text{var}(B) + (QI)^2]^{1/2}$$

since the raw intensity $(I + B)$ and the background B are the independently measured quantities. We assume the raw intensity obeys Poisson statistics so that $\text{var}(I + B) = I + B$; $\text{var}(B) = B/16$ because the background is effectively summed over 16 pictures (see below) and therefore is much better measured than $(I + B)$. The constant Q which scales the "floor" term to the counting statistics variance may vary from system to system; at UCSD $Q = 0.025$ has proven satisfactory. Thus

$$\sigma = [I + (17/16)B + (0.025I)^2]^{1/2}$$

Some reflections should not be integrated at all. There is enough redundancy in a typical area detector experiment that it is probably better to throw out potentially spurious observations than to try to incorporate them in with the rest of the data. Table II lists some of the reasons for on-line rejection of reflections. The computer must keep track of rejected reflections so that the detector addresses involved can be excluded from the background-updating operation.

As data collection proceeds, the software may be able to improve the centering of the reflection profiles as it goes along. If reflections are predicted concurrently, the computer can periodically refine the cell parame-

TABLE II
CRITERIA FOR REJECTION OF REFLECTIONS

Reason for rejection	Method of identification
Too close to detector border	Predicted (X, Y) location
Too close to beam stop	Predicted (X, Y) location
Lorentz factor too large or varying too rapidly	Predicted Lp factor
Quantum efficiency nonuniformity near reflection	Predicted (X, Y) location and current list of nonuniformities
Miscentered profiles	Examination of profile
Overlap with other reflections	Predicted (X, Y) location and setting angles

ters and orientation matrix of the crystal based on already observed reflections and apply the updated parameters in the subsequent prediction. Even when reflections are predicted ahead of time, as in the current implementation at UCSD, an overall shift in crystal orientation can be corrected if the shift moves all the profile centroids in the same direction. The software keeps track of the mean offset to adjust the upper limit of predicted scanning angle that will be associated with each subsequent picture. This dynamic recentering has proved successful at UCSD; profiles stay centered when the crystal is slowly moving or the crystal parameters were slightly wrong to begin with.

Background Estimation and Initialization

As with any other data collection technique, area detector collection requires careful correction for background. The magnitude of the background relative to the Bragg intensity is lower with area detector data than with film data because background counts are accumulated only while the reflection is in diffracting position, but for weak reflections the background can still be a considerable percentage of the total intensity.

A straightforward approach to background estimation is to sum the counts in a region surrounding the reflection's location on the detector face. Thus, if a reflection appears at a location X, Y the background is taken as the average of the counts in the detector locations or cells inside a rectangular box centered on X and Y and with dimension ΔX and ΔY but excluding the cells which are parts of this reflection or others.

This technique is similar to that used in most film data reduction, and it has the advantage that the background for a reflection is entirely determinable from the detector image currently under consideration. Due to geometrical nonlinearity of the detector[1] and the low statistics of the background counts, this method yields unsatisfactory background values. At UCSD it is only used in "refinement" data collection where an exact estimation of the integrated intensity of the reflection is not required.

A different technique for background estimation is used for actual data collection. In this "image-updating" technique a background image, containing estimated background counts on all 32,768 points on the detector face, is maintained. To improve statistics, actual counts in 16 consecutive pictures are stored. This image is updated point by point after each picture is taken except in cells used in the summing areas of reflections. Thus, if in a single cell, the "old" background image value as of picture n is B_n and the value in the data image is D_n, then the new background estimate for that cell is

$$B_{n+1} = (15/16)B_n + D_n$$

This is a straightforward updating algorithm with a correlation length of 16. Such an cumulative technique requires a starting point, and on the UCSD system this is provided by a background initialization program (BGNCI).

The initialization program directs the collection of 16 pictures at goniostat positions near the beginning of the data collection range. The background at each cell is taken as the average of the counts over the 16 pictures except where reflections appear. Any cell which is part of a reflection in a specific picture has its background estimated by an average over nearby cells in that picture. This operation requires foreknowledge of the reflections which will appear in a run, so it must be done after reflection prediction.

Data Output and Reduction

The output file from the data collection (DATAC) program includes for each reflection its indices, the integrated intensity, the estimated standard deviation, the predicted and observed chamber coordinates, the scanning-angle centroid, and the nine individual measurements that form the "profile" of the reflection. The "profiles" are displayed on a video terminal as they are collected, allowing the operator to verify the quality of the current data run.

The data reduction operations are performed after collection: sorting, merging, scaling, identification and elimination of faulty data, and crystal absorption correction. Sorting and merging the collected reflections involve putting the reflections in *hkl* order (or *khl* order in monoclinic spacegroups) and gathering together all instances of a reflection and its symmetry equivalents so that they follow one another in the output list. The details of the scanning-angle profile are not saved in the merged file, but the X, Y and ω values are, so that an off-line absorption correction program can use that information.

Scaling in the most general sense allows reflection collected under differing conditions to be properly comparable. Thus, scaling should be able to correct for differences in exposure time, extent of crystal decay, input beam intensity, and crystal absorption. The UCSD scaling program operates by grouping reflections into scanning angle ranges or "shifts," each with a defined width in ω (about 5° in range). The integrated intensity measurement I_{ij} of each reflection in shift i is compared to measurements made in other shifts, and a scale constant g_i for the shift is obtained by minimizing the differences between the I_{ij} values and the associated average intensities Y_j. The expression to minimize is

$$\sum_{ij}[w_{ij}(g_i Y_j - I_{ij})]^2$$

Here w_{ij} is the weight associated with I_{ij}; it is $[\sigma(I_{ij})]^{-1}$ if reflection j appears in shift i and is zero otherwise. As shown by Monahan *et al.*,[9] this is a linear least-squares problem for the determination of either Y_j or g_i and is suitable for iterative refinement. If on the kth iteration $g_i = g_i^{(k)}$ then the solution of the least-squares problem for the average intensity is

$$Y_j^{(k)} = \sum_i w_{ij}^2 g_i^{(k)} I_{ij} \Big/ \sum_i (w_{ij} I_{ij})^2$$

and the solution for g_i on the next iteration is

$$g_i^{(k+1)} = \sum_j w_{ij}^2 Y_j^{(k)} I_{ij} \Big/ \sum_j (w_{ij} I_{ij})^2$$

Initial values for the shift-scale factors g_i are set to unity but they converge to within 1% of their final values after five or six iterations.

A typical shift at UCSD contains about 500 observations, of which 350 have symmetry equivalents in the data set and can contribute to the scaling. (A typical data set contains 100–300 of such shifts.) The copious redundancy with which reflections are observed allows the scale factors to be quite well determined for area detector data. As indices of the quality of the data we define symmetry R factors

$$R_{\text{rms}} = \left\{ \sum_{ij} [w_{ij}(g_i Y_i - I_{ij})]^2 \Big/ \sum_{ij} (w_{ij} I_{ij})^2 \right\}^{1/2}$$

$$R_{\text{av}} = \sum_{ij} \left| g_i Y_j - I_{ij} \right| \varepsilon_{ij} \Big/ \sum_{ij} \varepsilon_{ij} I_{ij}$$

where $\varepsilon_{ij} = 1$ if reflection j appears in shift i and $\varepsilon_{ij} = 0$ otherwise. For a typical protein data set R_{rms} is between 0.03 and 0.08, and R_{av} is usually smaller than R_{rms}.

The scale factors g_i computed in the scaling program are used to help search for and delete "outliers," i.e., erroneous measurements. A program searches through the data set and finds the worst agreements among symmetry-related reflections according to the criteria

$$\begin{aligned} D_1 &= U_{ij} - Y_j \qquad (U_{ij} = I_{ij}/g_i) \\ D_2 &= (1 - Y_j/U_{ij})^2 \\ D_3 &= (U_{ij} - Y_j)/\sigma(U_{ij}) \\ D_4 &= (U_{ij} - Y_j)/\sigma(Y_j) \end{aligned}$$

[9] J. E. Monahan, M. Schiffer, and J. P. Schiffer, *Acta Crystallogr.* **22**, 322 (1967).

Usually the worst 200 observations according to each criterion are deleted from the data set.

The shift-scaling technique applies one scale factor to all the reflections striking the detector in each scanning-angle range. This will not correct for changes due to crystal absorption in the outgoing beam direction, since all the reflections diffracting over a scanning range are assigned the same scale factor regardless of resolution or output beam direction. An approach to solving this problem has been taken at UCSD. Crystal absorption and other factors which effect intensity over the detector face are paramaterized according to

$$K_{ij} = \sum_{m=1}^{3} \sum_{n=1}^{3} a_m a_n A_{mn}$$

where a_m and a_n are the direction cosines of the output beam for reflection ij relative to axes fixed on the crystal and A is a symmetric 3×3 matrix of refinable parameters. Determining a_m and a_n requires knowledge of an observation's ω, X, and Y values, which is why those are saved. There is a separate A matrix for each shift i, and its elements are determined by least-squares minimization of

$$\sum_{ij} [w_{ij}(K_{ij} I_{ij} - Y_j)]^2$$

This problem is linear and an iterative refinement of the matrices A can be performed. Since the matrix A describes a three-dimensional ellipsoid, this technique is termed "ellipsoidal absorption correction," and is an extension of the "local scaling" concept of Matthews and Czerwinski[10] by an algorithm developed by Hendrickson (personal communication). The use of the ellipsoidal absorption connection generally improves the symmetry R factor of a data set but only slightly; consequently this operation is frequently omitted.

Once the reflections in the data set have been merged, scaled, edited, and (perhaps) absorption corrected, the averaged intensities Y_j can be computed and output. At this point area detector data is functionally indistinguishable from film or diffractometer data, and can be used as structure factor inputs to Fourier or Patterson syntheses.

Data Collection Procedure and Results

In this section the symbols ϕ, χ, and ω designate conventional setting angles for a three-circle goniostat, 2θ is the Bragg angle, and θ_c is the angle

[10] B. W. Matthews and E. W. Czerwinski, *Acta Crystallogr., Sect. A* **A31,** 480 (1975).

between the chamber normal and the incident beam. A data collection run consists of a consecutive series of electronic pictures with ω advanced by a specific amount between each picture and with every picture in the series taken at the same χ and ϕ setting angles and exposure time.

The most convenient goniostat setting is with $\chi = 0°$, since at this angle it is possible to avoid obscuring the detector by the χ circle for an entire run by keeping $\omega = 0°$ and stepping along ϕ. Because it is necessary to discard reflections from reciprocal lattice points that lie near the rotation axis, other runs with different χ settings may be required to obtain a complete set of data.

The crystal-to-detector distance is chosen so that the detector is as close as possible to the crystal without producing overlapping reflections. Since with our present chamber all counts that fall within a 6 × 6-mm area contribute to the intensity of a particular reflection, adjacent reflections should not produce spots closer together than 6 mm. Thus, for a typical crystal with 100 Å unit cell axes, the crystal-to-detector distance is set at 50 cm. At this distance, the chamber, which has a detection area of 25 × 25 cm, is set at $\theta_c = 20°$ in order to measure data with Bragg spacing from 20 to 2.6 Å. To collect higher or lower resolution data, the chamber may be set at larger or smaller θ_c.

The data collection procedure is straightforward, as can be seen in Fig. 1. The crystal is mounted on the goniostat and roughly aligned, usually with a principal axis along the X-ray beam, with the help of program ALIGN. A small set of data is collected with program DATAC and the unit cell parameters and initial setting angles are refined by program PARAM. For each data collection run, setting angles, detector coordinates, and Lorentz polarization corrections are computed for all reflections by programs PREDICT. Predicted scanning angles and detector coordinates are checked using ALIGN program (see section on Reflection Prediction) and the initial background picture is generated by program BGNCI. Intensity data are then measured automatically by program DATAC. Operator intervention is required only to start the next run. The system is easy to use and an operator can be trained in 2 days.

Quality of the Data

During the last few years, we have measured approximately three million reflection intensities on seven proteins. The reproducibility of these data is given in Table III together with pertinent crystal parameters. These projects are done in collaboration with several research groups.

Dihydrofolate reductase from Escherichia coli. (Work done in collaboration with Prof. Kraut's group from UCSD.) For 2.5 Å resolution data

TABLE III
REPRODUCIBILITY OF MAD INTENSITY DATA

Protein	Space group	Cell dimension (Å) a	b	c	Number of intensity measurements	R_{sym} (%)
Dihydrofolate reductase (*E. coli*)	$P6_1$	93	93	73	350,000	4
Dihydrofolate reductase (*L. casei*)	$P6_1$	72	72	92	200,000	5
Cytochrome *c* peroxidase	$P2_12_12_1$	107	51	77	180,000	7
Cytochrome *c*′	$P2_12_12_1$	56	72	76	500,000	6
Elastase	$P2_12_12_1$	51	58	75	400,000	4
RuBPCase	I422	149	149	137	700,000	6
B-Phycoerythrin	R3	189	189	60	220,000	6

we were able to measure in 6 days approximately 80,000 observations of 12,000 reflections on one crystal before the average intensity dropped by 15%. With a Hilger and Watts diffractometer it was possible to obtain only 6000 observations with a crystal of the same size. Reproducibility of the data was the same for both instruments. Intensity measurements were also collected for two heavy-atom derivatives. An electron density map phased using these derivatives was readily interpretable and a model of the protein, with two 17,000-dalton molecules in the asymmetric unit, has been constructed.[11] We have also obtained intensities to 1.7 Å resolution on the parent crystal; the structure has been refined to *R* factor of 15% at that resolution.[12]

Dihydrofolate reductase from Lactobacillus casei. (Work done in collaboration with Prof. Kraut's group at UCSD.) We have measured 2.4 Å intensity data on the parent and one heavy-atom derivative. An electron density map phased on this single isomorphous replacement is even more easily interpretable than the map from the *E. coli* reductase, and a model of this protein has been built.[13] Native data has been collected out to 1.7 Å

[11] D. A. Matthews, R. A. Alden, J. T. Bolin, S. T. Freer, R. Hamlin, Ng. H. Xuong, J. Kraut, M. Poe, M. Williams, and K. Hoogstein, *Science* **197,** 452 (1978).

[12] D. A. Matthews, R. A. Alden, J. T. Bolin, D. J. Filman, S. T. Freer, R. Hamlin, W. Hol, R. L. Kislink, E. J. Pastore, L. T. Plante, Ng. H. Xuong, and J. Kraut *J. Biol. Chem.* **253,** 6946 (1978).

[13] J. T. Bolin, D. J. Filman, D. A. Matthews, R. Hamlin, and J. Kraut, *J. Biol. Chem.* **257,** 13650 (1982).

resolution and the structure has been refined to an R factor of 15% at that resolution.[12]

Cytochrome c peroxidase. (Work done in collaboration with Prof. Kraut's group at UCSD.) These crystals are frequently twinned and we usually have to mount at least five crystals before finding one of sufficient quality for data collection. Even with these crystals the intensity profiles of reflections in some orientations are fairly broad and sometimes split into two peaks. This probably accounts for the relatively high R_{sym} of 7%. We have measured intensities to 2.6 Å for the parent protein and two heavy-atom derivatives. The electron density map was somewhat noisy, but a model has been successfully built.[14] Native data to 1.8 Å resolution have been collected, and protein structure refinement is in progress.

Cytochrome c'. (Work done in collaboration with Prof. Salemme's group from the University of Arizona.) During a period of 4 weeks, we were able to make more than 500,000 intensity measurements on this type of crystal. For the parent crystal we have made about 100,000 intensity measurements of 26,000 reflections, with a full set of data out of 2.2 Å resolution and with many reflections out to 1.6 Å resolution. R_{sym} for the native data is about 6%. We have also collected data to 2.5 Å resolution for five heavy-atom derivatives with about 80,000 observations of 17,000 reflections for each. R_{sym} varies from 6.5 to 7.8%. Only two heavy-atom derivatives were useful [$K_2H_gI_4$ and $UO_2(NO_3)_2$]. The resulting electron density maps permitted construction of a model of the active dimer. Results from a 2.5 Å resolution structure have been published.[15] The structure refinement with data to 1.6 Å resolution is now in progress.

Elastase. In low-temperature studies of elastase, we have collected about 400,000 reflections intensities with crystals under different conditions.[16] In particular, we have succeeded in cooling a native crystal to −73° and collecting a set of data to 1.6 Å with an R_{sym} of 2.9. We have refined this structure to an R factor of 18%.

Ribulose bisphosphate carboxylase/oxygenase (*RuBPCase*). In a collaboration with Prof. Eisenberg's group from the University of California, Los Angeles, we have collected more than 700,000 reflections intensities on RuBPCase crystals.[17] A multiple isomorphous replacement (MIR)

[14] T. L. Poulos, S. T. Freer, R. A. Alden, Ng. H. Xuong, S. L. Edwards, R. C. Hamlin, and J. Kraut, *J. Biol. Chem.* **253,** 3730 (1978).

[15] P. C. Weber, R. G. Bartsch, M. A. Cusanovich, R. Hamlin, A. Howard, S. R. Jordan, M. D. Kamen, T. E. Meyer, D. W. Weatherford, Ng. H. Xuong, and F. R. Salemme, *Nature* (*London*) **286,** 302 (1980).

[16] Ng. H. Xuong, C. Cork, R. Hamlin, A. Howard, B. Katz, P. Kuttner, and C. Nielsen, *Acta Crystallogr., Sect. A* **A37,** 551 (1981).

[17] W. W. Smith, S. W. Suh, D. Eisenberg, R. Hamlin, A. Howard, C. Nielsen, and Ng. H. Xuong, *Acta Crystallogr., Sect. A.* **A37,** S38 (1981).

electron density map has been calculated to 3.0 Å resolution based on three derivatives. It is being improved by density modification.[18]

B-Phycoerythrin.[19] In collaboration with Dr. R. Sweet and F. Tsui from the University of California, Los Angeles we have measured more than 220,000 reflection intensities on native crystals and three heavy-atom derivatives. Difference Patterson maps clearly reveal the positions of the heavy atoms. Phase refinement and density modification are now in progress.

Compound 1 of cytochrome c peroxidase. (Work done in collaboration with Prof. Kraut's group.) In addition to very rapid data collection, the UCSD multiwire system can be used to study relatively short-lived phenomena. An example of such a problem is the determination of the structure of the ES complex of cytochrome *c* peroxidase (CCP) called "compound 1."[20] On addition of hydroperoxides such as H_2O_2, crystals of cytochrome *c* peroxidase are converted to those of complex ES. The half-life of the spontaneous conversion of the ES complex to CCP is 4 hr at 23°. CCP is crystallized in 30% methylpentanediol (MPD) and the crystals can be cooled to −15°. Visible spectra of a compound 1 crystal taken after 7-hr X-ray exposure at −15° indicate that at least 90% of the protein molecules are still in the compound 1 state.

Therefore, our data collection strategy involved the collection of about 70° of data per crystal, using seven crystals to obtain a complete data set to 2.7 Å. For each crystal, we followed the following procedures, designed to maximize the amount of data per crystal.

Each crystal can be mounted so that its axis orientation is approximately known. Fifteen minutes are spent obtaining a rough, visual alignment. We then commence the process of taking data, using an ω step of 0.1° per picture, each picture being exposed for 30 sec. Instead of being analyzed on-line, the images are simply stored on the computer's disk. When 700 pictures have been taken, accounting for 70° of data, another 45 min are spent obtaining a more precise visual alignment. The crystal is then removed, dissolved, and tested for the amount of compound 1. The images stored on disk are used to obtain precise lattice and orientation parameters, and the data collection program is then run, obtaining its pictures also from the stored disk images.

We have collected 52,800 observations of about 14,000 unique reflections using the technique described above. These reflections form a unique data set to 2.7 Å with a scaling R factor of 6.5% on intensity. Our

[18] R. W. Schevitz, A. D. Podjarny, M. Zwick, J. J. Hughes, and P. B. Sigler, *Acta Crystallogr., Sect. A* **A37,** 669 (1981).

[19] F. C. Tsui, Ph.D. Dissertation, University of California, Los Angeles (1983).

[20] T. Yonetani, B. Chance, and S. Kajawara, *J. Biol. Chem.* **241,** 2981 (1966).

next step will be to obtain a complete very high-resolution data set of the native CCP crystal at −15° to enable an accurate comparison with the compound 1 crystals.

This method of data collection is an exciting development, making possible experiments completely impractical using conventional crystallographic techniques. With the addition of more disk storage for our computer, we will be able to do such projects much more easily in the future.

In summary, we have devised and written a complete software package for a method of data collection for a diffractometer with an area detector. The system is easy to use and has provided high-quality reflection intensity data.

Acknowledgments

We would like to thank Drs. J. Kraut, W. Vernon, R. C. Hamlin, and C. W. Cork for their participation in the project and their advice. We thank Drs. J. T. Bolin, S. L. Edwards, B. A. Katz, P. G. Kuttner, and K. W. Volz, and Mr. D. H. Anderson for their help and patience as they collected and reduced the many data sets required in debugging the software. We thank J. M. Stronski, D. L. Sullivan, and W. A. Weeks for their technical assistance.

This project was funded by grants from the National Institute of Health (RR 01644, GM 20102) and from the National Science Foundation (PCM81-00302).

[29] Television Area Detector Diffractometers

By U. W. Arndt

Introduction

Photographic film is an excellent recording medium for X-ray diffraction patterns, especially when these are produced by X rays of wavelengths longer than about 1.5 Å. It does, however, suffer from three principal defects: it can record only where and not when an X-ray photon has arrived; it gives only a relative and not an absolute measure of X-ray intensity; and although the blackening (optical density) of an exposed film is fairly accurately proportional to the incident X-ray intensity, the quantity which is actually measured, by means of a microdensitometer, is the amount of light transmitted through the film, and this is an exponential function of the density and thus of the incident X-ray intensity.

The first shortcoming greatly hampers the interpretation of, for example, screenless oscillation photographs in which the exact estimation of "partiality" often presents a considerable problem (see this volume,

Harrison [19], and Rossmann [20]). More seriously, the signal-to-noise ratio, that is, the reflection-to-background ratio, of single-crystal photographs is high: a typical oscillation photograph corresponds to a crystal rotation of, perhaps, 2°, during all of which background radiation reaches the film, while the angular width of any one individual reflection may be less than 0.5°. If it were possible to obscure any given portion of the film at all times except during the recording of a reflection at that point the signal-to-background ratio would be four times better. An electronic area detector effectively provides the function of such hypothetical shutters. The sharper the reflections, the more important is this electronic shuttering.

The variation of the absolute sensitivity of film from one sheet to another, even with the most carefully controlled developing conditions, leads to the need to scale the data recorded on different films, with a consequent loss of accuracy.

Because of the logarithmic relation between the light transmitted through a blackened film and the incident X-ray intensity, the film must be sampled on a very fine grid of raster points: typically a single diffraction spot is sampled at more than a hundred points in a microdensitometer. This greatly lengthens the microdensitometry of films. With steeply sloping spot profiles it also reduces measuring accuracy because of the so-called Wooster effect[1] which expresses the fact that the sum of logarithms is not equal to the logarithm of the sum. The spatial resolution of electronic area detectors is much poorer than that of X-ray film. The diffraction pattern can be sampled at best on a grid of 512 × 512 picture elements instead of on a normally employed microdensitometer grid of at least 2000 × 2000 points. Because the measured quantity with an electronic detector is a much more nearly linear function of X-ray exposure, the required spatial resolution is lower.

A further disadvantage of film is that it requires a certain minimum exposure to produce a measurable pattern.

The necessary exposure time with an area detector is determined solely by the desired precision of the counting statistics. It thus becomes possible to record a number of complete low-precision data sets which can then be combined to give an adequate final precision. During the same total exposure time it is usual to collect only one photographic data set. The advantages of the former procedure are obvious when the specimen crystal is suffering progressive radiation damage during exposure.

The use of wavelengths shorter than 1 Å increases the amount of diffraction information which can be collected for a given amount of

[1] W. A. Wooster, *Acta Crystallogr.* **17**, 878 (1964).

radiation damage[2] and reduces errors due to specimen absorption. In addition, we can expect increasing use of anomalous dispersion phasing methods as a result of the availability of tunable synchrotron radiation sources[3–7]; these methods usually require shorter X-ray wavelengths.

Area Detectors

An area detector suitable for recording the diffraction pattern from single crystals with large unit cells must have the following properties.

1. It should subtend a large solid angle at the specimen crystal in order to record as much of the pattern as possible at one setting.
2. It must have a good spatial resolution so that it can separate neighboring reflections in a dense pattern.
3. It must have a high detective quantum efficiency[8,9] (DQE), that is, the precision of intensity measurements must be limited principally by the counting statistics of the number of *incident* photons, in order to compete with photographic film as an X-ray detector. This requirement presupposes that the detector has a high absorption efficiency.
4. The absorption and detection of the incident X rays should be in a thin layer so that the resolution is not degraded for obliquely incident X rays. (If the detector surface is spherical with the specimen crystal at the center, this requirement can be relaxed; however, such detectors do not allow the crystal-to-detector distance to be varied, as may be desirable in order to achieve a suitable compromise between requirements 1 and 2.)
5. The detector should be capable of dealing with a high counting rate integrated over the whole of its area without large counting losses; a very high local count-rate capability is less important for biological crystals. (In the diffraction pattern from a typical protein crystal with a 100 Å unit cell a medium-strength reflection gives rise to fewer than 10 counts/sec, but since some 90% of the total diffracted intensity is in the background,

[2] U. W. Arndt, *J. Appl. Crystallogr.* **17,** 118 (1984).

[3] W. Hoppe and U. Jakubowski, *in* "Anomalous Scattering" (S. Ramseshan and S. C. Abrahams, eds.), p. 437. Munksgaard, Copenhagen, 1975.

[4] J. C. Phillips, A. Wlodawer, M. M. Yevitz, and K. O. Hodgson, *Proc. Natl. Acad. Sci. U.S.A.* **73,** 128 (1976).

[5] J. C. Phillips, A. Wlodawer, J. M. Goodfellow, K. D. Watenpaugh, L. C. Sieker, L. J. Jensen, and K. O. Hodgson, *Acta Crystallogr., Sect. A* **A33,** 445 (1977).

[6] U. W. Arndt, *Nucl. Instrum. Methods* **152,** 307 (1978).

[7] U. W. Arndt, T. J. Greenhough, J. R. Helliwell, J. A. K. Howard, S. A. Rule, and A. W. Thompson, *Nature (London)* **298,** 835 (1982).

[8] R. C. Jones, *Photogr. Sci. Eng.* **2,** 57 (1958).

[9] P. Fellgett, *Mon. Not. R. Astron. Soc.* **118,** 224 (1958).

the integrated counting rate may exceed 10^5 counts/sec. These values are for a rotating anode X-ray source; with storage-ring synchrotron radiation sources they have to be multiplied by two or three powers of 10.)

6. The detector should be capable of being moved relative to the specimen since it will frequently be impossible to record the diffraction pattern at one setting.

This list does not include such obviously desirable qualities as economy and long life. It also does not include stability with time and temperature, linearity, and uniformity of response, since these qualities are implicit in the need for a high detective quantum efficiency. These requirements can largely be met by an area detector based on a television camera. The essential components of such a system are shown in Fig. 1; a large variety of different configurations is possible, depending upon the way in which the different functions are combined in individual components. Economic feasibility requires that standard commercially available devices be used where possible. Some of the available alternatives have

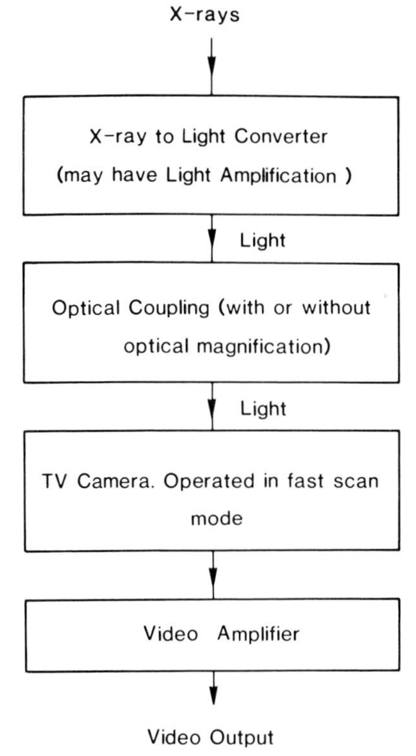

FIG. 1. Component parts of an X-ray television area detector.

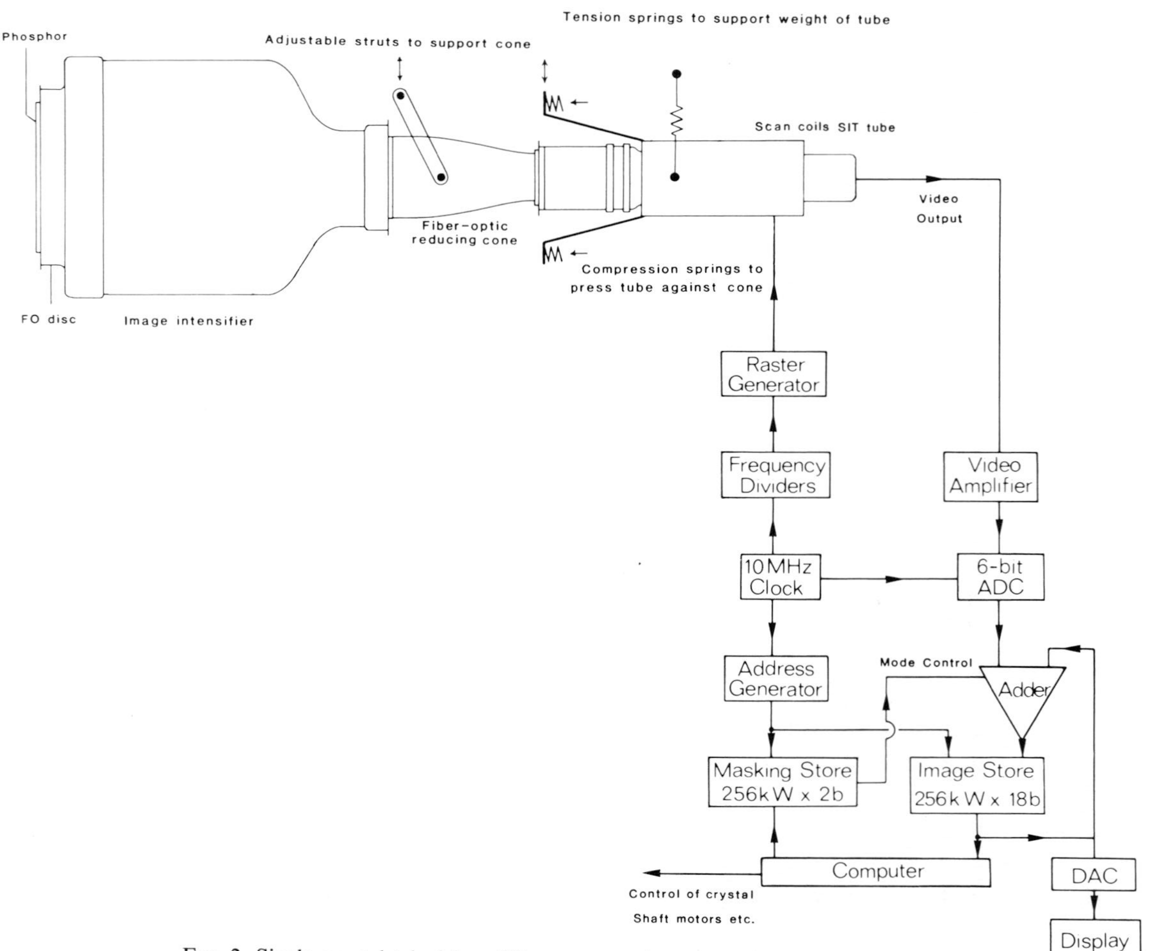

FIG. 2. Single-crystal television diffractometer: detector and its associated circuits.

been discussed elsewhere[10,11]; here we shall examine the present form of a system which has evolved over a number of years[12–14] and which is shown in Fig. 2.

The Phosphor

The phosphor layer must fulfill several requirements which are partly in conflict: it must have a high X-ray absorption; it must be as thin as possible in the interests of high spatial resolution and low parallax; it must have a high fluorescence conversion efficiency; it must be stable; it must be uniform and it must have a short decay time constant, that is, no appreciable afterglow. The best compromise is offered by polycrystalline layers of gadolinium oxysulfide or of zinc sulfide.[10] The former has a higher absorption coefficient for 8-keV X rays (Cu K_α), the latter a somewhat better fluorescent conversion efficiency. In the interests of uniformity the layer thickness should be many times the grain diameter, but the light output of polycrystalline phosphors is reduced as the particle size is diminished; the smallest useful grain size of the two phosphors is about 5 μm. The X-ray absorption is an exponential function of the layer thickness; inevitable small variations in thickness become progressively less important as the absorption approaches 100%.

One X-ray photon produces between 250 and 500 visible light photons from the phosphor. It is essential to collect the highest possible fraction of these on the first photocathode of the detector since any information loss at the input cannot be recovered by subsequent light amplification. In practice, therefore, the phosphor must be fiber optically coupled to the photocathode. The input image intensifier must, therefore, have a fiber optics entrance window with the photocathode directly on the inside: the phosphor layer is deposited directly on the outside of the window, or more conveniently on a separate fiber optics disk which is then brought into optical contact with the window. A separate disk may be either planar or concavo-planar.

In a diffraction experiment the scattered X rays diverge from the specimen and the size of the diffraction pattern can thus be scaled up or down within limits by changing the specimen-to-detector distance. In protein crystallography, however, the range within which this can be done is limited, especially as regards scaling down: in photographic meth-

[10] U. W. Arndt, *Nucl. Instrum. Methods* **201,** 13 (1982).
[11] S. M. Gruner, J. R. Milch, and G. T. Reynolds, *Nucl. Instrum. Methods* **195,** 287 (1982).
[12] U. W. Arndt and D. J. Gilmore, *Adv. Electron. Electron Phys.* **52,** 209 (1979).
[13] U. W. Arndt and D. J. Gilmore, *J. Appl. Crystallogr.* **12,** 1 (1979).
[14] U. W. Arndt and D. J. Thomas, *Nucl. Instrum. Methods* **201,** 21 (1982).

ods the practice of using 125 × 125-mm films has become almost invariable because it is well suited to the size of the specimen (determined by optimum dimensions from the point of view of absorption), available X-ray tube focal spot sizes and convenient collimator dimensions.[15]

The maximum input size of television detectors is about 56 × 56 mm. This is the square inscribed in a circle 80 mm in diameter, which is the diameter of the largest fiber optics image intensifiers which are commercially available. It might be possible to couple the phosphor to the image intensifier by reducing fiber optics, but the cost of such large-diameter couplers is at present prohibitive. A television diffractometer thus requires shorter crystal-to-detector distances and smaller collimator and crystal sizes than are customary in X-ray cameras and this scaling down leads to a deterioration in the diffraction-spot-to-background intensity ratio. However, this is more than compensated by the electronic shuttering which was referred to in the introduction.

The Image Intensifier

The sensitivity even of so-called low-light-level television camera tubes is not sufficient to detect individual X-ray photons and some degree of image intensification is, therefore, necessary. In the arrangement shown in Fig. 2 this is provided by a 2:1 demagnifying diode-type electrostatically focused image intensifier.[16] In this type the light gain is due to the acceleration of the photoelectrons by about 15 kV before they strike the output phosphor of the image intensifer where they give rise to an intensified output image. This type of device, known as a first-generation image intensifier,[17,18] has a photon gain of about 100. Higher gains (~1000) are obtained with second or third generation image intensifiers, which incorporate a microchannel plate (MCP) electron multiplier.[17,19,20] However, these devices tend to have a slightly lower spatial resolution than the diode type; they are also less readily available with fiber optics input and output faceplates.

The image intensifer is coupled to the television camera by means of a reducing fiber optics cone which matches its 40-mm output diameter to

[15] U. W. Arndt and R. M. Sweet, *in* "The Rotation Method in Crystallography" (U. W. Arndt and A. J. Wonacott, eds.), Chapter 5. North-Holland Publ., Amsterdam, 1977.

[16] VLI-116. Varian LSE, Palo Alto, California.

[17] "Image Intensifiers," Tech. Inf. No. 39. Mullard Ltd., London.

[18] P. R. Collings, R. R. Beyer, J. S. Kalafut, and G. W. Goetze, *Adv. Electron. Electron Phys.* **28A,** 105 (1969).

[19] B. W. Manley, A. Guest, and R. T. Holmshaw, *Adv. Electron. Electron Phys.* **28A,** 471 (1969).

[20] G. Eschard and J. Graf, *Adv. Electron. Electron Phys.* **28A,** 499 (1969).

the 16-mm input diameter of the camera tube.[21] The cone also serves the purpose of a long-leakage-path electrical insulator: the output of the intensifier is at a potential of +15 kV and the input of the camera tube at a potential of −9 kV; vacuum fiber optics windows should not be operated with an appreciable electrical potential across them.

The various fiber optics surfaces are coupled with optical grease or oil. They must be strictly parallel to one another and it is, therefore, necessary to suspend the fiber optics cone and the camera tube on a system of springs to counteract gravitational torque and to force them together by means of axial springs. The optical axes of the various devices do not coincide exactly with their geometrical axes, and small translational adjustments are provided to center the images.

The Television Camera Tube

At present the most suitable high-sensitivity camera tube is the silicon intensifier target (SIT) tube.[22] This tube consists of a photoemissive cathode followed by an intensifier section (electron-optically similar to the image intensifier discussed above). The accelerated electrons produce a charge image on a silicon diode array target which is scanned in a raster by an electron beam.

The signal current is the difference between the incident electron beam and the current needed to neutralize the charge image. The raster has 625 lines which are swept out in 40 msec, in accordance with normal European television standards. Some time is required for line and frame flyback; so that the image consists of 512 active lines. Each line lasts 64 μsec; the flyback occupies 12.8 μsec and the active part of the line 51.2 μsec.

The SIT tube has a high sensitivity, a linear transfer curve (i.e., a γ of unity), and a stability limited only by that of the associated electronic circuits. The spatial resolution is determined by the spacing of the discrete silicon diodes of the target and is thus better for larger diameter (but more expensive) SIT tubes.[23] However, the resolution of a relatively inexpensive 25-mm diameter tube is reasonably well matched to those of a polycrystalline X-ray phosphor and of the image intensifier.

A disadvantage of the tube is the disortion of the image and the nonuniformity of its response across its area. (The response of the detector at the edges is about 50% down on that at the center of the field of view;

[21] Galileo Electro-Optics Corp.

[22] V. J. Santilli and G. B. Conger, *Adv. Electron. Electron Phys.* **33A,** 219 (1972).

[23] SIT Tubes, also referred to as E.B.S. Tubes, are available from several manufacturers with photocathode diameters of 16, 25, 40 and 80 mm.

vignetting in the fiber optics coupling contribute to this factor.) Before a television area detector can be used for quantitative measurements the distortion and response must be carefully calibrated. Fortunately the performance is stable in time and calibration is necessary at infrequent intervals only.

It can be expected that vacuum camera tubes will eventually be replaced by solid-state television cameras on single silicon chips.[11,24] The charge-coupled device (CCD) cameras look the most promising; present-day versions have a smaller image area and a lower spatial resolution than 25-mm SIT tubes. They are also less sensitive and therefore, require the use of external image intensifiers with a greater light gain and a higher demagnification. The compactness and complete positional stability of solid-state cameras suggest that they will lend themselves to being stacked side by side with each device looking at part of the field of view, with a consequent gain in spatial resolution.

It should be noted that silicon devices are directly sensitive to X rays and to accelerated electrons. In principle, therefore, they can be used without some of the processes shown in Fig. 1 which involve intermediate conversions to light. In practice, however, they are damaged by exposure to energetic radiation, and with standard products the continued use of intermediate light images will probably remain necessary.

The main advantage of these devices lies in their good signal-to-noise ratio, especially when operated at low temperatures. Consequently, detectors based on CCD cameras should have a higher DQE, especially at high X-ray intensities.[24] At present, their principal disadvantage is that it is difficult to obtain chips which are completely free of defective elements.

Electronic Circuits

The storage target of a television camera tube sums the photons which arrive during one frame period of 40 msec, that is, between successive discharges of the target by the scanning electron beam. This period is not sufficient to give adequate counting statistics, nor, since the maximum charge which can be stored in the target is determined by its electrical capacity, an adequate dynamic range. A large number of frames must thus be summed. As shown in Fig. 2, this summation is carried out by converting the amplified video signal to digital quantities in an analog-to-digital converter (ADC)[25] and then adding the corresponding readings from each picture point digitally in a large semiconductor store. During this proce-

[24] N. M. Allinson, *Nucl. Instrum. Methods* **201**, 53 (1982).
[25] TRW, LSI Products Type TDC1007J.

dure much of the random electrical noise originating in the camera tube and the video amplifier is integrated out. The process is thus one of the recovery of a periodic signal (the repeatedly scanned X-ray pattern) from noise. Any other periodic error signal which may arise in the scanning circuits or elsewhere is, of course, also enhanced, and extreme care in the design of these circuits is necessary to eliminate such signals.

The sampling rate of the ADC determines the number of picture elements (pixels) into which the image is split up. In Fig. 1 this sampling rate is shown as 10 MHz and each of the 512 active (i.e., nonflyback) line periods is thus sampled 512 times in the active line period of 51.2 μsec. The store then needs $512 \times 512 = 262{,}144$ locations and it must be capable of a read–modify–write operation, the addition of the current intensity to that previously accumulated, in 100 nsec. The store thus needs to be specially constructed since normal computer-type memories cannot operate at this speed.

The store is 18 bits wide. With the new 8-bit ADC (256 intensity levels) a minimum of 10^{10} frames (40 secs) can thus be summed before an overflow can occur.

The passage of new information into the store is conveniently controlled by a 2-bit wide masking register which is also 256K words long and which can be loaded from the controlling computer. It allows the separate selection at any pixel of one of four functions: addition (for the summing of successive frames), subtraction (of background), recirculation (of the accumulated data), and clearing.

The store is composed of dynamic metal-oxide-semiconductor (MOS) random-access integrated circuits. Any individual pixel can be directly addressed by the controlling computer to read out the accumulated intensity. The two least significant bits of the data word are rejected. The dynamic range in the integrated image can thus be $2^{16}:1$ or 65,536 : 1 after the maximum number of summations.

Properties of the Detector

Sensitivity. The sensitivity of the detector can be varied within wide limits by varying the high tension applied to the image intensifier and to the SIT tube. At maximum sensitivity the maximum video signal (= ADC output of 2^8) is produced by 3 8-keV quanta per picture element per frame period. At minimum sensitivity this figure becomes about 3000 quanta. The minimum and maximum peak counting rates are thus 75 and 75,000 photons pixel^{-1} sec^{-1} (with a frame period of 40 msec). It should be noted that the detector has no deadtime and, therefore, no counting losses as such at any counting rate up to the maximum peak rates. In particular, the performance at one point of the detector is not affected by the intensities incident on any other part. In practice one works with the maximum sensitivity that avoids saturation of the detector since the detective quantum efficiency (DQE) decreases at low sensitivities.[13] One can, therefore, use the best compromise between DQE and dynamic range.

The sensitivity, unfortunately, is not uniform over the area of the detector. There is a smooth long-range variation of about 50% between the center and the edges of the active area due to vignetting and to variations in the photocathodes of the image and camera tubes. This can be corrected easily using a low-order polynomial correction function. Superimposed on this is a small number of blemishes in the targets which are flagged for the rejection of any reflections which fall on them, and an irregular variation of about 1.5% due to the phosphor graininess. This last variation can be corrected with a look-up table with an entry for every pixel. The seriousness of the effect is reduced by a factor of about 4, since an average diffraction spot occupies about 16 pixels.

Stability. The detector and its associated circuits reach thermal equilibrium within a few hours after switching on. Thereafter, the position of the image remains constant to ± 0.25 pixel over a 24-hr period, and the positional response becomes stable to about 0.3% over a similar period. Spatial distortion and response correction measurements are, therefore, only necessary at infrequent intervals. The stability is checked with the help of four radioactive light sources[26] mounted on the input faceplate of the detector. These produce four patches of light at the corners of the image, the intensities and positions of which are monitored from time to time.

Spatial Resolution. The resolution of the system is best expressed in terms of the line spread function (LSF).[27] This is the Fourier transform of the system modulation transfer function (MTF), which in turn is the product of the MTFs of the individual components. The LSF, which is approximately Gaussian in shape, has been found to have decayed to 1% of its peak value at a distance of ± 2.5 pixels of that peak. The closest practicable separation of neighboring spots is then about 6 pixels, the collimation being so chosen that the spots occupy an area of about 4×4 pixels with 2 pixels of "background" between the spots. The bandwidth of the video amplifer must be matched to the resolution: the above requirements imply that the 3-dB point on the gain versus frequency curve occur at 5 MHz.

It will be noted that this resolution allows the recording of about 85

[26] Saunders-Roe Ltd., Beta Light.

[27] See, for example, J. C. Dainty and R. Shaw, "Image Science." Academic Press, New York, 1974.

orders parallel to the edge of the detector surface and that at this spatial separation some errors may occur in very weak reflections in the immediate neighborhood of and on the slopes of very strong reflections.

Detective Quantum Efficiency. The DQE has been measured to be ~50% at a sensitivity setting of 0.5 quanta per digitization level. The way in which the DQE can be expected to vary with sensitivity has been discussed elsewhere.[13] There the "noise" due to digitization round-off errors was included: these errors are much smaller with the present 8-bit ADC and can be eliminated completely by the addition of pseudo-random noise at the input of the ADC.[28]

The Diffractometer

So far we have discussed the bare area detector. We must now consider its use for data collection from single crystals. The detector is best mounted on the (strengthened) counter arm of a single-crystal diffractometer which employs an Eulerian cradle or κ bracket[29] for mounting the crystal. The discussion of the spatial resolution of the detector given above shows that the entire pattern can be recorded with a single symmetrical setting of the detector only when no more than ±40 orders need to be recorded (e.g., 100 Å unit cell to to 2.5 Å resolution). Note that with a 56-mm square detecting surface the size of a pixel is 110 μm and our 4 × 4-pixel spot actually measures 440 × 440 μm on the detector. In the above example, when using Cu K_α radiation, the crystal-to-detector distance must be 28 mm. If, on the other hand, a 200 Å unit cell is to be measured to a resolution of 2.5 Å orders 0 to +80 can be recorded by moving the detector back to a distance of 68 mm and swinging the counter arm to θ_{max}, i.e., to 22.6°, so that the direct beam is in line with the edge, instead of the center of the detector. Extension of the data set in the vertical plane requires that the crystal be moved using ϕ and χ movements to bring the appropriate part of the reciprocal lattice into the equatorial zone (just as in a normal single-counter diffractometer the reflections are brought into the equatorial plane for measurement).

The collimation of the diffractometer is designed so as to reduce to a minimum the cross-fire of the X-ray beam incident on the specimen; the mosaic spread of native proteins is frequently less than 0.1° and the signal-to-background ratio of the pattern is greatly improved with a near-parallel primary beam. (It is the good collimation at synchrotron radiation beam lines which is largely responsible for the high-quality diffraction patterns

[28] D. J. Thomas, *Nucl. Instrum. Methods* **201,** 31 (1982).

[29] Enraf-Nonius. N. V., Delft CAD-4 Diffractometer.

which are being recorded there.) The diffractometer is, therefore, used with a microfocus X-ray tube[30] which has a 150-μm square foreshortened focus. The illuminated area of the crystal is about 250 × 250 μm so as to produce a 350-μm spot size at the detector. An even more parallel beam can be produced by using a pair of Franks mirrors.[31] The situation is analogous to that when single-crystal patterns from very large unit cells are recorded photographically.[15] An incidental benefit of a television detector is that its use during the setting of the Franks mirrors takes all the difficulty out of this normally tedious task!

Strategy of Data Collection

The most obvious way of using an area detector diffractometer is to record a series of contiguous rotation patterns and to analyze these in much the same way as in oscillation photography. However, the angular range of each image must be matched to the reflecting range of the crystal to obtain the optimum signal-to-background ratio and this implies a rotation of only about 0.1° per image. Each 512 × 512-pixel image contains 0.5 Mbytes of information and a full 180° rotation of the crystal thus requires the storage of 900 Mbytes of information. Trial measurements have been made in this mode of operation, and have confirmed the high quality of the data obtainable with the diffractometer in that reproducibility and symmetry R factors are essentially limited only by counting statistics. Recently, routine use of the instrument has been improved by new real-time software which calculates the positions and angular ranges of all reflections in terms of the distorted detector coordinates as the crystal rotates. These coordinates must be supplied by the computer for each reciprocal lattice point just when it reaches the Ewald sphere. Before data collection can begin the orientation of the specimen must be determined. This is done by a variation of the normal technique in oscillation photography which employs two still photographs at 90°.[32] In the present case small ω angle rotations are used, and the center of gravity in ω of a number of selected indexed reflections is determined precisely by recording their intensity profiles.

The crystal is now rotated at a uniform rate about the vertical ω axis, the masking store being used to set measuring boxes for each reflection at the calculated positions and of the appropriate calculated dimensions in X, Y, and ω.

[30] Telefunken-AEG F-50.

[31] A. Franks, *Proc. Phys. Soc., London* **368,** 1054 (1955).

[32] A. J. Wonacott, *in* "The Rotation Method in Crystallography" (U. W. Arndt and A. J. Wonacott, eds.), Chapter 7. North-Holland Publ., Amsterdam, 1977.

After a full ω rotation of 180° (or less, depending on crystal symmetry) the crystal is reorientated using the ϕ and χ rotations in order to fill in the "cusp" or blind region by a limited rotation about a second axis.[32] The rotation, of course, is still about the vertical axis.

In practice, better results can be expected by collecting a number of complete data sets at different orientations so chosen that the blind regions do not coincide, and then summing the low-precision data for each reflection from the individual sets in the manner suggested above.

Thomas[33,34] has developed algorithms which generate the coordinates in real time even at the rates of data collection (and therefore rotation rates) to be expected with synchrotron sources. This procedure is much to be preferred to their generation before data collection starts because it allows the continuous refinement of a multidimensional matrix as measurements proceed in the light of the centers of gravity determined for individual reflections. The quantities to be refined are the lattice parameters of the specimen and its orientation, as well as the crystal-to-detector distance, the orthogonality of the detector surface to the incident beam, and the position of the origin of the detector coordinate system relative to the direct beam.

However, the behavior of the detector and all its associated circuitry is now completely characterized and the essential parts of the software system have been tested. The television diffractometer is capable of achieving an accuracy in protein data collection rather better than that obtainable by photographic methods, at a rate about 30 times faster and in a manner which will require very much less operator interference. It has been shown that the detector is usable over a considerable range of X-ray wavelengths and that it can be set to cope with a very large range of X-ray intensities. It is constructed in modular form from standard image intensifiers and TV camera tubes used in a standard way; it will thus be possible to take full advantage of future improvements in the design and performance of these components.

Acknowledgments

The television diffractometer owes much to the work of many collaborators and colleagues, especially D. J. Gilmore, S. H. Boutle, F. Walker, D. J. Thomas, and R. L. Tooze and to ideas contributed by the team at Enraf-Nonius, Delft, especially J. van Bekkum and N. v.d. Putten, who are engaged on producing a fully engineered version of the instrument from our laboratory prototype. I am particularly grateful for the continued support of the Medical Research Council in what has proved to be a very long development.

[33] D. J. Thomas, *Nucl. Instrum. Methods* **201,** 27 (1982)

[34] D. J. Thomas, Ph.D. Thesis, Cambridge University (1982).

[30] A General Purpose, Computer-Configurable Television Area Detector for X-Ray Diffraction Applications

By Kenneth Kalata

During the last two decades developments in broadcast and closed-circuit television and low-light-level image intensifiers have increased the sensitivity of television cameras to the single photon level and have made high-resolution television equipment widely available. Ongoing advances in this technology will further increase the resolution and performance of television systems in the next few years. Coincident with these developments has been an increasing interest in and need for area detectors for the collection of two-dimensional images in a wide range of research problems. Over the last 15 years television detectors which provide high resolution, accuracy, quantum efficiently, and dynamic range have been built for such applications as astronomy,[1–6] X-ray diffraction,[7–10] electron microscopy,[8] and fluorescence spectroscopy.[11] A more detailed list of references can be found in references 12–16, and references 14 and 15

[1] B. L. Morgan and D. McMullan, eds., "Advances in Electronics and Electron Physics," Vol. 52. Academic Press, New York, 1980.

[2] B. L. Morgan, D. McMullan, and R. W. Airey, eds., "Advances in Electronics and Electron Physics," Vol. 40, Parts A and B. Academic Press, New York, 1976.

[3] M. Duchesne and G. Lelievre, eds., "Astronomical Applications of Image Detectors with Linear Response," IAU Colloq. No. 40. Meudon Observatory, Paris, 1976.

[4] C. de Jager and N. Nieuwenhuijzen, eds., "Image Processing Techniques in Astronomy." Reidel Publ., Dordrecht, Netherlands, 1975.

[5] J. W. Glaspey and G. A. H. Walker, eds., "Astronomical Observations with Television-Type Sensors." University of British Columbia, Vancouver, 1973.

[6] J. C. Geary and D. W. Latham, eds., "Solid State Imagers for Astronomy," Vol. 290. Soc. Photo-Opt. Instrum. Eng., Bellingham, Washington, 1981.

[7] U. W. Arndt and D. J. Gilmore, *J. Appl. Crystallogr.* **12,** 1 (1979).

[8] K.-H. Hermann, D. Krahl, and H.-P. Rust, *in* "Electron Microscopy at Molecular Dimensions," (W. Baumeister and W. Vogell, eds.), p. 186. Springer-Verlag, Berlin and New York, 1980.

[9] U. W. Arndt and D. J. Thomas, *Nucl. Instrum. Methods* **201,** 21 (1982) (also see p. 27).

[10] J. R. Milch, S. M. Gruner, and G. T. Reynolds, *Nuc. Instrum. Methods.* **201,** 43 (1982).

[11] D. W. Johnson, J. A. Gladden, J. B. Callis, and G. D. Christian, *Rev. Sci. Instrum.* **50,** 117 (1979).

[12] K. Kalata, *Proc. Soc. Photo-Opt. Instrum. Eng.* **331,** 69 (1982).

[13] K. Kalata, *IEEE Trans. Nucl. Sci.* **NS-28,** 852 (1981).

[14] S. M. Gruner, J. R. Milch, and G. T. Reynolds, *Nucl. Instrum. Methods* **195,** 287 (1982).

[15] U. W. Arndt, *in* "The Rotation Method in Crystallography" (U. W. Arndt and A. J. Wonacott, eds.), p. 245. North-Holland Publ., Amsterdam, 1977.

[16] K. Kalata, *Trans. Am. Crystallogr. Assoc.* **18,** 69 (1982).

also provide a good general discussion of television X-ray detectors. These detectors incorporate an S-T, SIT, Plumbicon, or SEC vidicon tube or a CCD array, which, in many systems, is coupled to an image intensifier with fiber optics or a lens system. Most recent detectors use a SIT (silicon intensifier target vidicon) tube or CCD (charge-coupled device) array because of their high sensitivity, dynamic range, and accuracy. The image integration and readout with television detectors are accomplished in one of three ways. The image can be integrated on the cooled target of the vidicon tube or CCD array for a period ranging from a fraction of a second to several hours and then read out and digitized in a single slow scan, or the television camera can be continuously scanned (usually at standard TV rates) and the image integrated in a large digital memory. In the latter mode the video signal can be digitized at each pixel and the digitized amplitudes added to the image accumulated in memory on previous frames, or single photon (or particle) events can be detected, their locations computed, and the corresponding memory locations incremented. The vidicon tubes and image intensifiers are sensitive to visible light; X-ray images can be acquired by using a thin phosphor or scintillator to convert each incident X ray into several hundred visible light photons. CCD arrays have been usually used with visible light but they are also directly sensitive to X rays.[17-20] Additional studies and improvements in radiation hardening are still needed before X-ray-sensitive CCDs can be used for routine data collection. Because of their small active area (approximately 12 × 9 mm for the larger standard devices), the present lack of a standard chip optimized for sensitivity to 8-keV X rays, and their fixed gain, these devices should be presently considered only for those applications for which their small physical size is essential. However, CCDs coupled with reducing fiber optics to an image intensifier and an X-ray converter should prove to be of great use to X-ray diffraction applications in a few years. Television detectors can also be used for neutron diffraction if they are lens coupled either to a neutron converter backed with a scintillator or phosphor, or to a neutron scintillator.

These television detectors (and other developed for such applications as diagnostic radiology) have been developed for specific applications, and their utility in other applications has usually been limited by the

[17] R. E. Griffiths, G. Polucci, A. Mak, S. Murray, and D. A. Schwartz, *Proc. Soc. Photo-Opt. Instrum. Eng.* **290,** 62 (1981).

[18] N. G. Loter, P. Burstein, A. Krieger, D. Ross, and D. Harrison, *Proc. Soc. Photo-Opt. Instrum. Eng.* **290,** 58 (1981).

[19] M. C. Peckerar, W. D. Baker, and D. J. Nagel, *J. Appl. Phys.* **48,** 2565 (1977).

[20] G. R. Riegler, R. A. Stern, K. Liewer, F. Vascelus, J. A. Nousek, and G. P. Garmire, *JPL Astrophys.* Reprint No. 38 (1982).

specific choices of image intensification and readout devices, formats, and image acquisition and integration modes. With few exceptions they have not been made generally available to other researchers. The time, effort, funding, and expertise needed to develop independently a television detector system has limited the application of this technology in the past. In principle, however, a single general purpose detector with an interchangeable front end and programmable readout, image processing, and integration electronics can combine essentially all of the capabilities of previous television detectors and can be made available at a reasonably low cost. It is this latter approach that the author has taken in developing a versatile system[12,13] which can be easily configured to meet the requirements of particular research programs. Two of the major applications for which this detector is currently being developed are X-ray crystallography and small-angle X-ray scattering, and others include electron microscopy and X-ray astrophysics. The detector system consists of an X-ray converter, an image intensifier, fiber optics couplers, and a SIT tube, all of which can be replaced by similar components of different sizes or characteristics, and computer-configured controlled-scan and processing electronics which generate a digital image which is integrated in a separate memory system or in a computer. The scan rates, gain, and the sizes of the image and the array into which it is digitized can be chosen to match the requirements of the particular application. The detector system can operate in the analog-to-digital converter (A/D) mode, in which the intensity each pixel is converted into an 8- to 14-bit digital word, or in the photon counting mode, in which the location of each incident event is determined to a small fraction of the analog resolution of the camera system. A block diagram of the system is shown in Fig. 1. The following is a description of the detector subsystems and the performance of their various configurations.

Camera System

The camera system incorporates an interchangeable, reconfigurable camera head, and a separate set of camera electronics which controls the camera readout. In its most general form the camera head consists of a conversion and image intensification section which is coupled by fiber optics to an SIT or other television tube, or to a CCD array. A thin phosphor or scintillator coupled to the image intensifier converts each incident X ray into a burst of several hundred photons with wavelengths of 4000–6000 Å, depending on the phosphor. Active areas from 4 mm square to over 100 mm in diameter can be obtained through the use of demagnifying or magnifying fiber optics to couple the image intensifier to the converter and the SIT tube, and through the use of reducing image

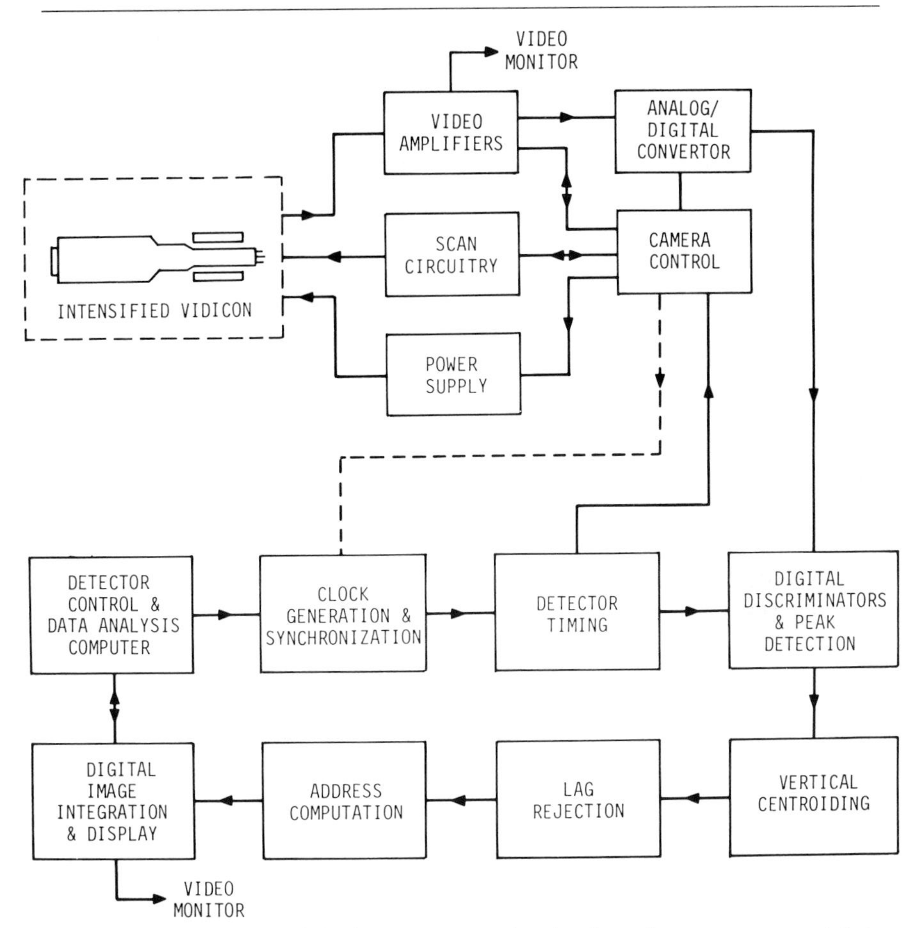

FIG. 1. Block diagram of the detector system showing the major components and their interconnection.

intensifiers. Drawings of two camera heads presently in use are shown in Fig. 2a and b. The original camera head, shown in Fig. 2a, consists of an interchangeable $La_2O_2S(Tb)$ (P-44) convertor, two 40-mm electrostatic image intensifiers, and a 16-mm SIT vidicon tube with associated yoke and preamplifier. In a second camera head, shown in Fig. 2b, the two diode image intensifiers are replaced by a single microchannel plate image intensifier which is coupled to the SIT tube with an interchangeable 2:1 fiber optics reducer or a 1:1 coupler. The second-generation image intensifier uses lower voltages than the pair of first-generation intensifiers, has a lower ion background (caused by the ionization of residual gas atoms), and has a higher gain so that it can be directly coupled to a CCD array.

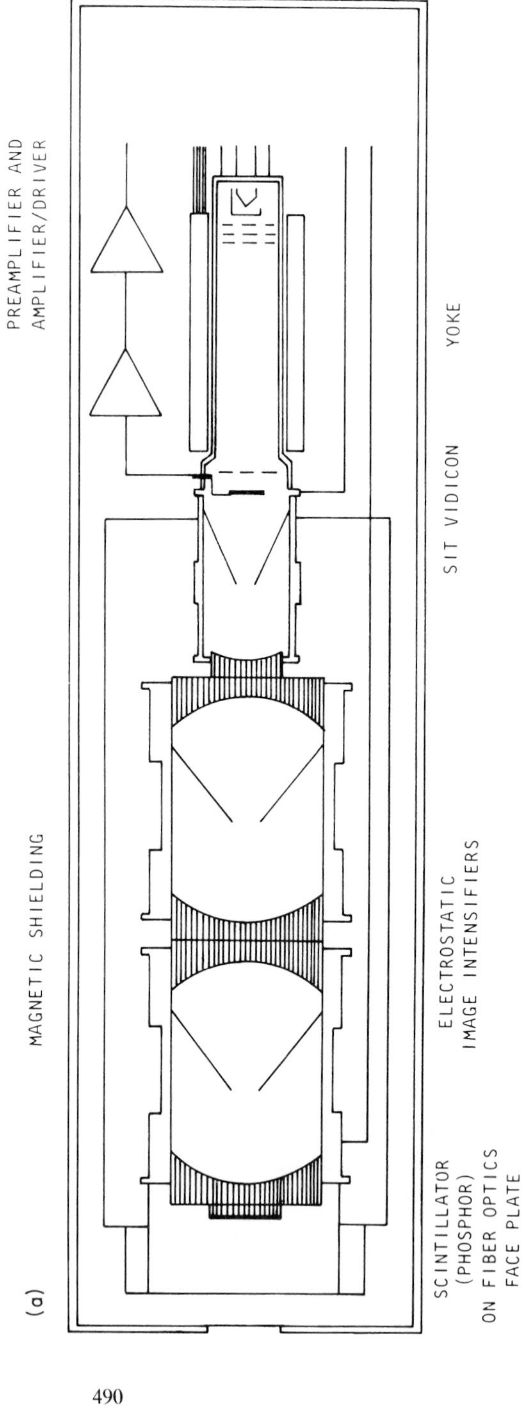
(a)
MAGNETIC SHIELDING
PREAMPLIFIER AND
AMPLIFIER/DRIVER
SCINTILLATOR
(PHOSPHOR)
ON FIBER OPTICS
FACE PLATE
ELECTROSTATIC
IMAGE INTENSIFIERS
SIT VIDICON
YOKE

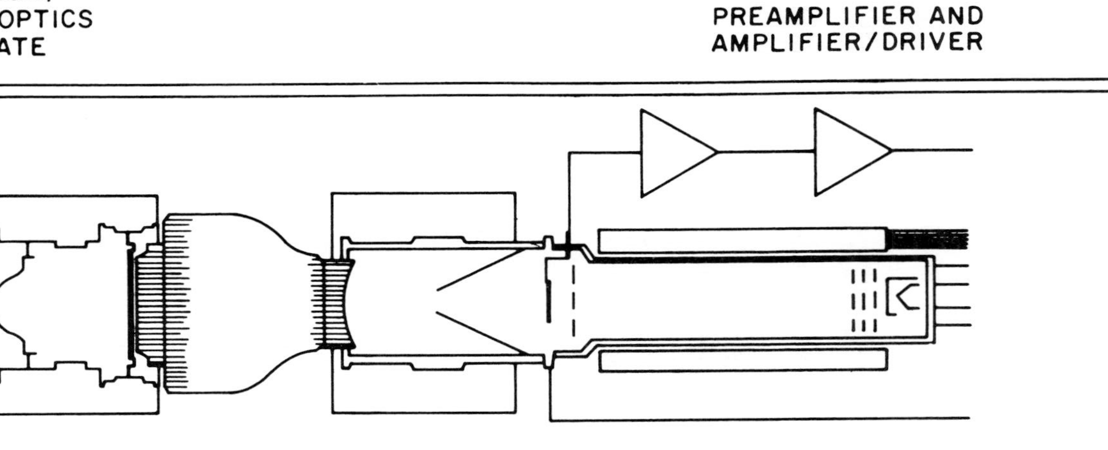

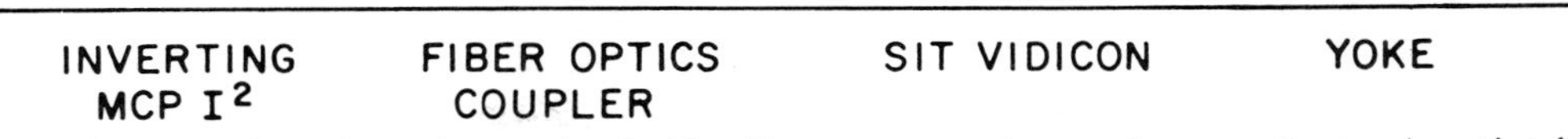

FIG. 2. Scale drawings of (a) a detector head with a 13-mm square active area incorporating two inverting diode image intensifiers and (b) a more recent detector head with a 25-mm diameter active area incorporating a higher gain microchannel plate image intensifier (MCP I^2) and demagnifying fiber optics.

With the 2:1 reducer the resolution of the intensified image is better matched to the resolution of the SIT tube so that the final spatial resolution of this camera head is only slightly less than that of the first even though the active area is twice as large. Another camera head which has just been completed incorporates two 40-mm proximity focused diode intensifiers and a 40-mm inverting diode intensifier coupled to the SIT tube with a 2:1 fiber optics taper. The first image intensifier has a low-noise bialkali photocathode. With any of the detector head the gain can be varied over five orders of magnitude so that rates from 10^0 to 10^{12} events/second can be accommodated.

The camera head can be modified to optimize it for a particular application. A proximity focused intensifier can be used, thereby eliminating the pincushion distortion found with inverting image intensifiers. Proximity focused and inverting microchannel plate image intensifiers are available with active areas from 18 to 75 mm in diameter. A larger active area can be obtained by coupling an 80/40 mm reducing image intensifier or a 75-mm proximity focused diode intensifier with a 2:1 or 3:1 fiber optics reducer to a smaller diameter intensifier. Reducing fiber optics can also be used to directly couple a phosphor screen up to 80 mm in diameter to a smaller image intensifier, but the coupling efficiency in this case is low (10%) so that this should typically be done only for brighter images or if a cooled bialkali photocathode is used to minimize thermionic background. For large active areas a higher resolution can be obtained with a 40- or 80-mm diameter SIT tube. The zoom feature of these tubes also allows the active area to be reduced electronically by up to a factor of 2. Gated image intensifiers or variations of a sweep tube can be used when a time resolution better than a frame time is needed.

The camera electronics generate the sweep currents and tube voltages needed to scan the SIT tube or any other standard TV tube, and process and digitize the analog video signal. The length and duration of each scan line and the number of scan lines/frame can be set under computer control. Any one or several rectangular portions of target can be scanned with up to 4095 scan lines. The focus and alignment currents, the black level, blanking, and the tube heater are also under computer control. The sweep, focus, alignment currents, and the black level are generated by high-resolution D/A converters for stability and ease of digital control. The camera can operate in a fast or slow scan mode or the image can be integrated on the target and read out in one or several frames. Since the scan parameters are not limited to standard closed-circuit TV values, they can be chosen to optimize spatial, intensity, or time resolution and the data rate. A differential amplifier, filter, and clamp circuit generates a video signal which is digitized each pixel by a 30-MHz 8-bit converter.

Other converters of up to 14 bits can also be accommodated. A video amplifier generates video and sync signals so that the analog image can be displayed on a video monitor.

The camera system has so far usually been operated with a 64-μsec line length and 238 active scan lines/frame so that a standard video monitor can be used to display the real-time image. However, the camera electronics also include an additional sweep card which can drive an $X-Y$ display or a modified video monitor at nonstandard rates and formats. Tests have been made at scan rates several times faster and slower than standard and with up to 1000 scan lines/frame. The camera electronics can also be used with a standard television camera or one modified to accept the horizontal and vertical reset signals generated by the camera electronics. The first camera head was scanned by the electronics from an inexpensive CCTV camera modified in this fashion. A CCD camera can similarly be used, and timing pulses are provided to directly drive a CCD array.

Scan Control and Image Processing Electronics

The clock and timing electronics generate the basic clock frequencies, timing pulses, and gating signals that control the operation of the camera electronics and image processing logic. The basic pixel and detector clocks can be set under computer control. The timing electronics generates the horizontal and vertical reset, sync, and blanking pulses used by the camera. Their widths and relative timing can be set by the computer to any value, permitting fast or slow scan operation or single frame integration and readout at whatever spatial and time resolution best matches the data. If an unmodified standard TV camera is used in place of the camera system previously described, its reset or composite sync pulses can be used to synchronize the system timing. In order to minimize the amount of data to be stored and processed, a gated window from 1 to 4095 pixels wide and 1 to 4094 lines high can be placed around the region of interest in the image. The most significant timing and status signals are made available to the associated computer so that it has the capability of controlling the detector operation on a line-by-line basis if real-time scan control is needed.

The intensity at each pixel is compared to upper and lower level values set by the control computer so that in the photon counting mode ion events, events outside the desired energy range, and electronic noise can be rejected. In the A/D mode these discriminators may also be used to prevent the accumulation of noise and ion events in the image being integrated in memory. This is accomplished by setting the digital value of the intensity to zero when it is outside of the discriminator settings. In the photon counting mode this circuitry also detects and flags the maxima of the intensity distribution along each scan line by comparing the intensity at each pixel with the value at the previous pixel. Except when the image consists of discrete point sources, the photon counting mode circuitry cannot accurately detect multiple events which occur in the same resolution element in a single frame time (typically 10–30 msec). Images in which the counting rate in the brightest pixel of interest exceeds several tens of events per second should usually be integrated in the A/D mode. A current addition to the discriminator circuitry allows the active area to be divided into an array of cells up to 128 × 128 elements in size. The lower level discriminator setting can be set independent in each cell. This accommodates the falloff in gain near the edge of inverting image intensifiers.

The image of a single event on the camera target may be crossed by several scan lines and may also be detected on the following frame. In the photon counting mode the vertical centroiding and lag rejection circuitry can be used to provide a single address and strobe pulse for each incident event and determine the location of its center to a fraction of its size. There are several different types of centroiding which may be selected to match the needs of the experiment. In one type of centroiding, the pattern of spot crossings on a number of adjacent lines is detected and the size of the spot is determined. Events which are too large or small in size, caused by ion events, α particles emitted from the fiber optics faceplate, or thermionic noise, can therefore be rejected. The flags in the last four lines are stored at a rate of up to 25 MHz, for a maximum of 2048 pixels across a scan line, and the lower level discriminator signal at each point on the previous eight lines is stored. Four lines after a flagged maximum the lower level signals in that column on the four preceding and succeeding lines are checked to determine if the spot intensity is above the lower level discriminator on an equal number of lines on each side. If there are an equal number of crossings on both sides of this fiducial line, then the flag on this line is at the center of the spot. If there is one additional crossing in the lines following this fiducial line than in the lines before it, the center of the spot lies one-half line down. This information is used in the address generation if there is enough memory to accommodate this effective doubling of the vertical resolution. If not, the effective resolution may be increased somewhat by storing in memory one-half count on the line before the center and one-half count on the line after it. The operation of this type of centroiding is shown in Fig. 3. Another type of centroiding operates by storing for one line the intensity at each pixel and the flags in that line at a rate of up to 25 HMz for a maximum of 2048

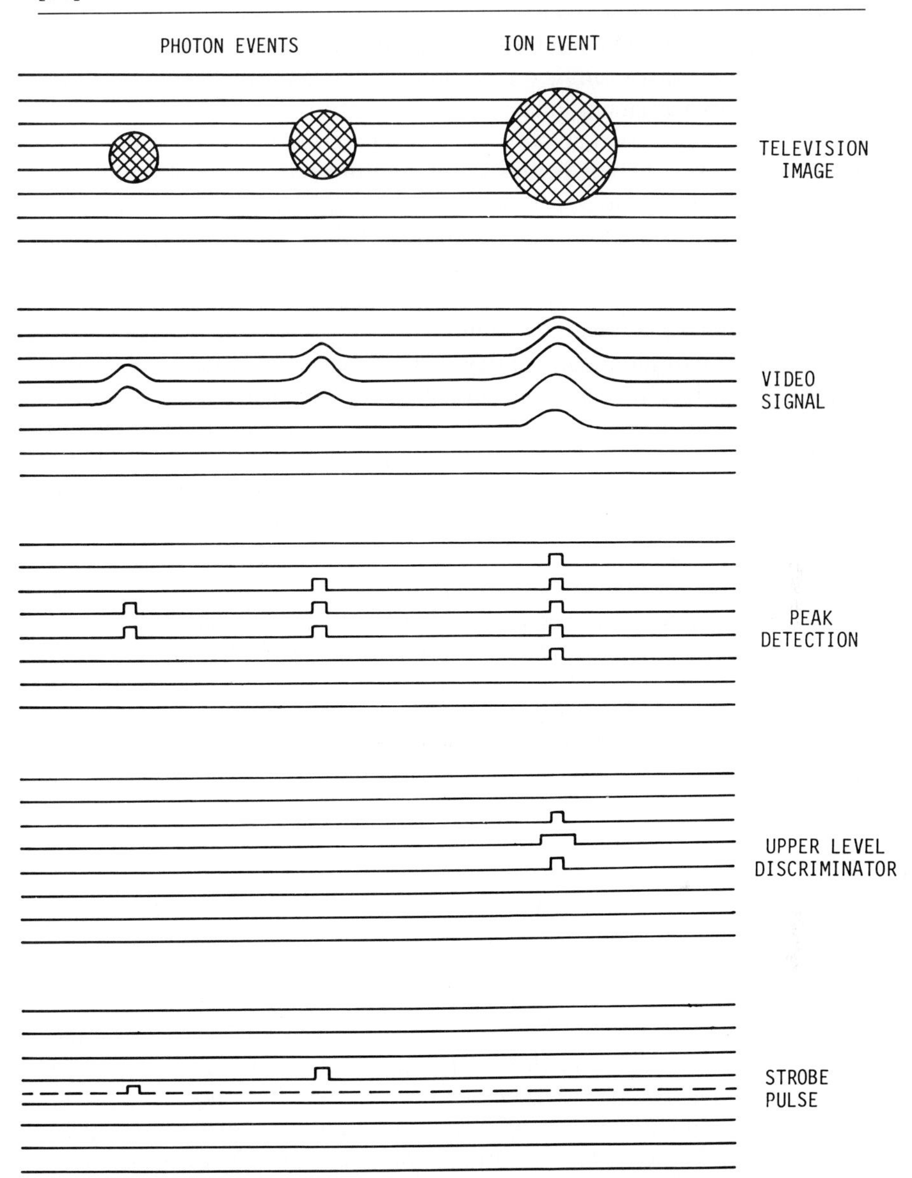

FIG. 3. Operation of one centroiding mode showing the detection of the peaks in the video signal on each scan line, rejection of events which are too bright or too large, and the determination of the centers of the patterns of event crossings.

pixels. The peak of the amplitude distribution in the spot is located when the intensity of the maximum on the fiducial line is greater than the intensities on the two adjacent lines.

In fast scan operation all of the charge on the camera target is not read out in a single frame. The lag rejection circuitry inhibits the detection of any residual signal as an event. The active area is divided into an array of elements up to 256 × 256 in size, each from 1 to 16 pixels wide and up to 16 lines high, into which the threshold information present in each frame is stored. An event is rejected if a signal in that element on the previous frame was above the lower threshold.

The address of an event occurring in a given pixel is set equal to the number of previous pixels in the gated portion of the active area that have been scanned since the beginning of the frame. An offset can be added to all addresses so that multiple images can be stored in memory. If an interlaced video image is digitized, corresponding pixels in alternate frames can be assigned the same address, pixels on even fields can be assigned even addresses and pixels on odd fields assigned odd addresses, or even and odd fields can be assigned successive memory blocks. The image should usually be digitized in an array which is comparable to or finer than resolution of the camera in order to avoid missing events or throwing away data. When a coarse resolution along one or both axes is desired, the image can be divided into an array of cells, with each cell up to 256 pixels wide and 256 lines high. The same address is assigned to all pixels in a given cell.

Memory System

The image can be integrated in a large, high-speed memory system, the memory of a computer system, or the target of the TV tube or CCD array. The computer can be used for integration in the photon counting made or slow scan A/D mode if the data rate does not exceed the computer's processing speed and if the number of pixels does not exceed the available memory capacity. However, when high resolution or high data rates are needed, a dedicated memory is required. The memory system can accommodate a work size of up to 24 bits and has a maximum capacity of 16 million words. One current memory has a capacity of 1024K × 24 bits. The memory system accepts the data word from the detector electronics and adds or subtracts it to the sum accumulated in that address on the previous frames. In the A/D mode this occurs every pixel. In the photon counting mode, the appropriate memory location is incremented when an event is detected. After integration the contents of the memory can be addressed by the computer and reset for the next run. The image accumulated in memory can also be displayed on a video monitor or *X-Y* display during and after integration. In order to accommodate the 5- to 10-MHz data rate when the A/D mode is operated at high resolution at standard TV line rates, four or eight sets of memory cards can be sequentially multiplexed. For operation in the photon counting mode or slow scan A/D mode only one set of memory cards is needed. When a large dynamic range is not needed, each word can be divided into two independent 12-bit pixels, thereby effectively doubling the memory capacity. Each memory location can be gated by its most significant bit, so as to allow gated regions of the image to be set and changed by the computer in real time or to inhibit overflow in bright channels in the A/D mode.

The integrated image or the digitized real-time image can be scaled and displayed during and after integration on a video monitor or *X-Y* display. The memory can also be used as a black and white or color graphics system by the computer. Four overlay channels are provided to display such information as cursors, windows, axes, and text. The 24-bit word can be scaled so that only the desired 16 bits are sent to the computer, or the full 24 bits can be sent as a 16- and 8-bit word or as two 12-bit words. The computer can access the memory during and after integration and can also read the data sent to the memory by the signal processing electronics. The memory system also serves as a convenient large scratch-pad memory for data analysis and can be interfaced to look like a portion of the computer memory.

System Calibration

Software has been developed to perform flat field and distortion corrections on the raw integrated images. These corrections are particularly necessary when inverting image intensifiers are used because of the pincushion distortion which can reach 8% per stage at the edges and because of the falloff in gain at the edges of the intensifier.[5,6] Some reducing image intensifiers have a pincushion distortion of less than 2% and correspondingly lower gain falloff. In the first camera head with two 20-mm inverting diode intensifiers with S-20 photocathodes, the lower level discriminator can be set to reject the thermionic noise while rejecting less than 20% of the 8-keV X rays at the center, but because of the radial falloff in gain, as many as 90% of the X rays can be rejected at the extreme corners of the image. This problem can be alleviated by using proximity focused image intensifiers which have no pincushion distortion or gain falloff, by using an image intensifier with a low-noise photocathode so that a smaller lower level discriminator setting can be used, or by using a 64 × 64 matrix of lower level discriminator setting so that an appropriate setting can be used for each point on the image.

The distortion correction program operates by moving radially inward each pixel by an amount approximately proportional to its radius and by reducing its size by a similar amount. The corrected radius is $R_c = R/(1 + aR + bR^2)$, where the quadratic term is small and can be ignored in many cases. The intensity in each uncorrected pixel is divided up among as many as 4 pixels in the corrected image, depending upon where the corrected radius falls relative to the grid of pixels in the corrected image. This division reduces the spatial resolution in the corrected image by a fraction of a pixel, so that ideally a pixel size smaller than the inherent resolution of the detector should be selected. The existing program reduces the residual distortion to less than 1/2% of the image size.

Phosphor Efficiencies and Characteristics

Measurements of the conversion efficiencies of a number of phosphors have been made for X rays between 0.7 and 5.9 keV. Also measured were the effects of overcoating and optical attenuation and dead layers in the phosphor. A much more complete and detailed description of these measurements and modeling of the results can be found in reference 21 or can be obtained from the authors.

The X-ray conversion efficiencies of 14 phosphors were measured at 5.9 keV by irradiating phosphor-coated glass disks with a calibrated iron-55 source and measuring the number of photoelectrons which were subsequently produced in the bialkali photocathode of a photomultiplier tube coupled to the disks. The number of photoelectrons produced by each X ray for the most efficient phosphors are as follows: P45, 14; P22B, 8; LaOBr, 8; P11, 8; P44, 6; BaFC1, 5; P43, 4; P24, 3. These numbers would be higher with an S-20 photocathode.

The phosphors used were deposited by settling onto a glass disk with potassium silicate used as a binder. Details of the optimized techniques and formulations used can be obtained from the authors. Approximately half the photons produced escape through the front surface of the disk. The efficiency can be increased by overcoating the phosphor layer with a thin (~15 μm) sprayed polyurethane overcoating. The efficiency is decreased if the phosphor screen is directly aluminized but is increased if a thin aluminized polypropylene window is placed over the disk. It can also

[21] J. H. Chappell and S. S. Murray, *Nucl. Instrum. Methods* **221,** 159 (1984).

be increased if an overcoating sufficiently thick (>30 μm) to leave a flat surface is subsequently aluminized.

Detector Performance

The spatial resolution of the detector depends on the active area, the operating mode, and the quantization of the image, as well as the resolution of the SIT tube and the image intensifier. The spatial linearity (distortion) and the detector background are mostly determined by the characteristics of the image intensifier. For a given application, the SIT tube or CCD array, the image intensifier, and the coupling fiber optics can be chosen to optimize the detector performance. For most applications, however, a reasonably low-cost camera head will provide sufficiently high performance.

The spatial resolution of the 25-mm detector head is approximately 100 μm full width, half-maximum (FWHM) in the A/D mode. Because of the centroiding logic the inherent resolution in the photon counting mode is about 20×30 μm. The photon counting mode resolution can still be significantly improved by improving the phosphor screen and reducing electronic noise and pickup. Larger pixels can be chosen when this fine a resolution is not needed. The spatial resolution of the 13-mm square detector head is similar to that of the 25-mm detector head, even though its active area is smaller, because the event spot size is not reduced by demagnifying fiber optics and because of the lower resolution of the inexpensive camera electronics used with a smaller detector head. The highest overall performance is obtained when the image intensifier resolution is matched to that of the SIT tube with 2 : 1 or 3 : 1 reducing fiber optics. For small active areas the highest resolution is achieved by coupling the image to a larger image intensifier with magnifying fiber optics and then coupling the image intensifier to the SIT tube with reducing fiber optics. When active areas larger than 25 mm are obtained by coupling a large image intensifier to a 16-mm SIT tube or CCD array, the resolution scales somewhat more slowly than the active area. For larger active areas a 40 or 80 mm SIT tube can provide approximately 50% higher resolution than the combination of a 16-mm SIT tube and reducing fiber optics. When coupled to a large diameter image intensifier, CCD arrays with 1000×1000 elements will be able to provide almost twice the resolution that can be obtained with a 16-mm SIT tube, but they will not be commercially available for several years.

In the photon counting mode with 8-keV X rays the detector noise is about 10–30 counts/sec over the entire active area. This noise can be greatly reduced with a slightly improved phosphor screen, by cooling the first photocathode, or by using an image intensifier with a lower noise photocathode. Images with a dynamic range of $10^4:1$ have been obtained in the photon counting mode, and larger dynamic ranges are possible. In the A/D mode the dynamic range for very bright images is limited by the quantization of the A/D converter to 512:1 for a 9-bit converter, but when the maximum rate in any resolution element is less than about 5000 events/sec the A/D mode can be used to detect individual events (although with a lower spatial resolution than in the photon counting mode) and can provide a dynamic range comparable to what can be obtained in the photon counting mode.

The maximum data rate for which the photon counting mode can be used depends on the intensity distribution in the image to be acquired. In the photon counting mode pileup occurs when two or more events occur within a given camera resolution element in the interval between two successive readouts of that point (one frame time). A correction factor can be calculated from the instantaneous counting rate at each point if the pileup is not too large. If the incident flux is uniformly distributed over the active area, images with rates of up to 10^5 sec^{-1} can be integrated with a pileup of less than 10%. In an image such as a crystal diffraction pattern, in which a small number of small spots are visible at any one time, single events can be counted at a maximum rate of only a few hundred counts per second or less. More diffuse patterns, such as those from solution scattering, can be integrated at rates of up to about 10^4 sec^{-1}. The pileup can be reduced by decreasing the frame time either by scanning each line faster or by reducing the number of scan lines per frame (usually by scanning a smaller portion of the target). It can also be reduced by scanning a crystal at several degrees per second so that any one spot is in reflecting position for only a short time each cycle. In general, however, brighter reflections should be acquired in the A/D mode at a lower gain or by reducing the X-ray flux. The weaker reflections can then be acquired either in the A/D mode, or in the photon counting mode when higher resolution and greater statistical precision are needed. Because of the low blooming of just about all television tubes and CCD arrays, the presence of extremely bright, totally saturated portions of the image does not significantly affect the data from faint portions of the image more than a few pixels away.

Most of the images collected so far have been obtained with the 13-mm detector head and its modest resolution camera electronics. The 25-mm detector head provides a higher resolution and a better quality image, but the computer to which it connected has a very limited display capability at this time. The following images were therefore taken with the 13-mm detector head. A 50-mm square detector head will be completed before the publication of this book and the 25-mm microchannel plate head is now being operated in the cooled mode. The performance of these detector heads will be significantly greater than that of the original detector head since their active areas and spatial resolutions are better matched to the size of the diffraction patterns and the focused beam size of 200 μm. Detailed information on the performance of these detector heads can be obtained from the author.

The central portion of the oscillation pattern from a polyoma capsid crystal scanned over a range of 5.0° around the [110] axis at 0.2°/sec is shown in Fig. 4. The image was integrated in the photon-counting mode in a 240 × 205 array for 10 min at 30 kV and 35 mA. A rotating anode X-ray

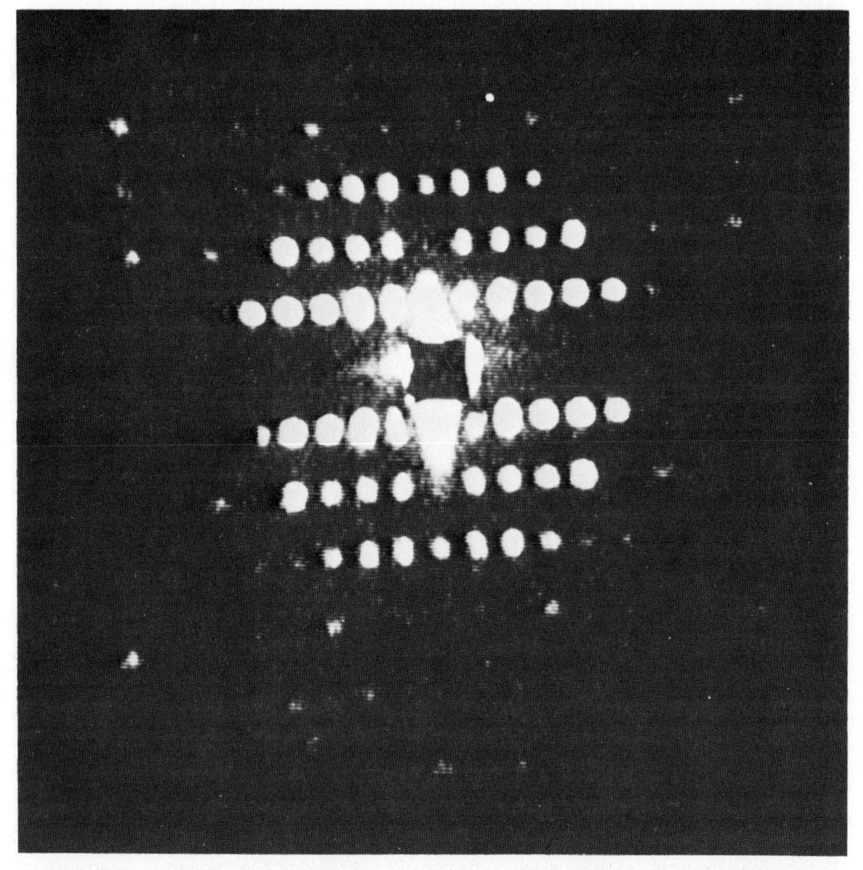

FIG. 4. An oscillation pattern from a polyoma capsid crystal collected in the photon counting mode in a 10-min integration at 30 kV and 35 mA. The brightest spots are saturated in this exposure.

machine was used with a double-mirror (Franks) camera to produce a 200-μm focal spot. The brightest spots are totally saturated at this incident flux but the fainter spots are accurately recorded. A pattern obtained in a 10-min integration with the same crystal scanned over a range of 1.5° is shown in Fig. 5. For the spots in reflecting position the signal-to-background ratio is higher than in the previous pattern and fainter reflections are visible. Measurements with similar patterns have shown that the detector can be used in a manner similar to film to collect an oscillation pattern with better statistics and higher accuracy 10–20 times faster than film. Reflections not visible on a 12-hr exposure on film can be accurately measured in a 1-hr integration with the detector. However, area detectors are most effectively used to acquire diffraction patterns when the data are taken as a series of stills or oscillations over a small angular range. This minimizes the background and allows the background to be measured for each reflection at angles adjacent to those where the spots are in reflecting

FIG. 5. An oscillation pattern similar to that shown in Fig. 4 except that the crystal was scanned over a range of 1.5° instead of 4.95°.

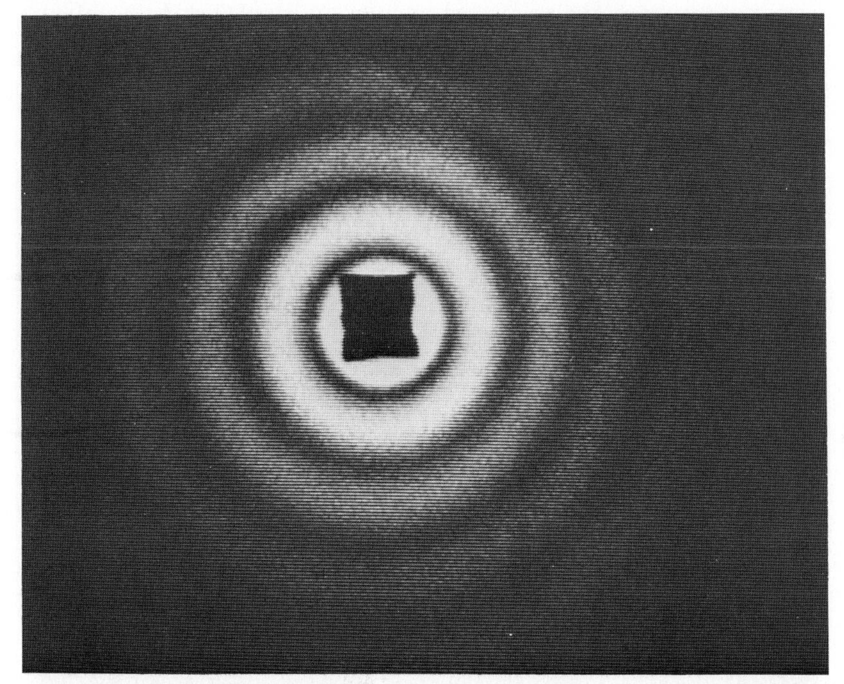

FIG. 6. The X-ray diffraction pattern of a solution of cowpea mosaic virus integrated in the photon counting mode for 30 min at 30 kV and 30 mA. The intensity is displayed on a linear scale so that only a small portion of the dynamic range of over 1000 : 1 is visible. The bright portions of the image are set to a level of 256.

position. Software to accomplish this has been developed by Ng. H. Xuong,[22] U. W. Arndt, and M. Deutsch and is available directly or through commercial sources.

An X-ray diffraction pattern taken in the photon counting mode of a solution of cowpea mosaic virus is shown in Fig. 6. The image was integrated for 30 min at 30 kV and 30 mA in a 360 × 200 array. The counting rate was about 2000 events/sec so that pileup was neligible. The dynamic range of the image is about 5000 : 1; only a small portion of this dynamic range can be seen in this one picture. The visible light image of a test pattern digitized in the A/D mode in a 390 × 238 array at an event rate of about 3×10^5 sec^{-1} is shown in Fig. 7. A single frame of date (1/60 sec) is shown although since the data were sent directly to the computer, they were actually sampled on a number of successive frames. The detector

[22] Ng. H. Xuong, S. T. Freer, R. Hamlin, C. Nielsen, and W. Vernon, *Acta Crystallogr., Sect. A* **A34,** 289 (1978).

FIG. 7. One frame of an image digitized in the A/D mode and illuminated with visible light with an event rate of about 3×10^5/sec.

memory can now be multiplexed and images have recently been integrated in the A/D mode in real time. The images shown here have not been corrected for the detector spatial efficiency or pincushion distortion in the image intensifier. Calibration programs to do this have recently been implemented. A corrected X-ray diffraction pattern of a tobacco mosaic virus liquid crystal taken in the A/D mode is shown in Fig. 8. The same pattern taken in the photon counting mode is shown in Fig. 9. The larger detector heads will have a much higher spatial uniformity in addition to their significantly higher fractional resolution.

System Configuration

A number of factors should be considered by researchers who are interested in configuring and implementing a television detector system. Because of the effort and expense involved in designing and building television detector electronics, researchers should whenever possible ei-

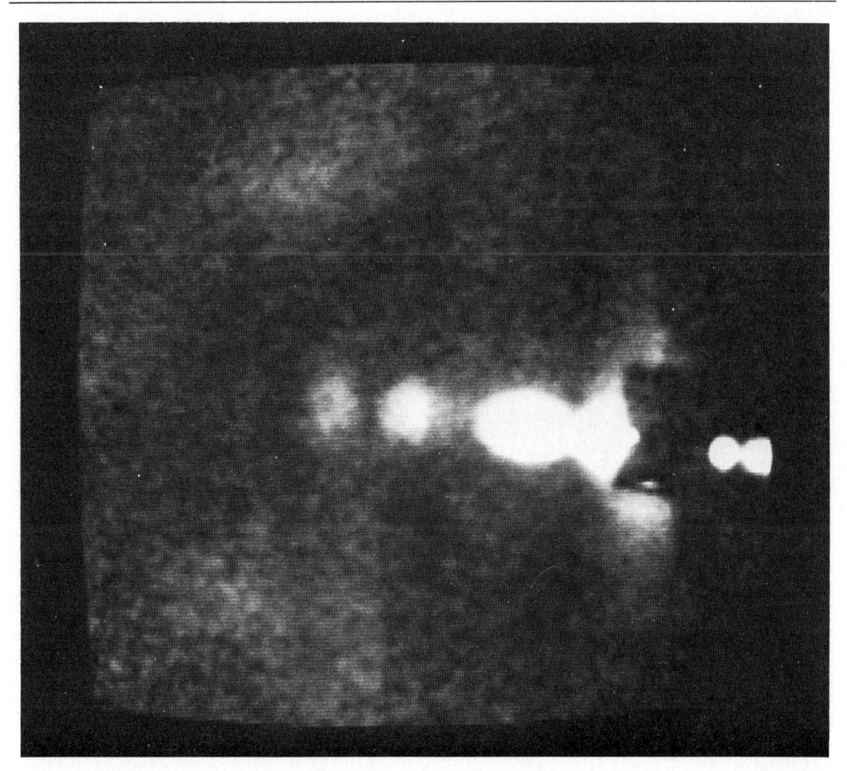

FIG. 8. A distortion-corrected 60-min X-ray diffraction pattern of a tobacco mosaic virus liquid crystal taken in the A/D mode.

ther purchase a commercial system, have most or all of the electronics built by a laboratory which has previously built a working television detector system, or build a copy of an existing system, making modifications and improvements if necessary. The camera head, however, can be built by many researchers at a considerable savings in cost, and even if it is purchased its components should be specified to the extent possible to provide the optimal performance in a particular application at the lowest cost. The costs of image intensifiers and television tubes vary widely and particular performance characteristics can often be specified for a small additional cost.

Electrostatic image intensifiers with fiber optics input and output windows should be used in almost all applications because they can be directly coupled to the phosphor converter and the television tube. Magnetically focused image intensifiers can provide somewhat higher performance, but they have to be lens coupled to a magnetically shielded

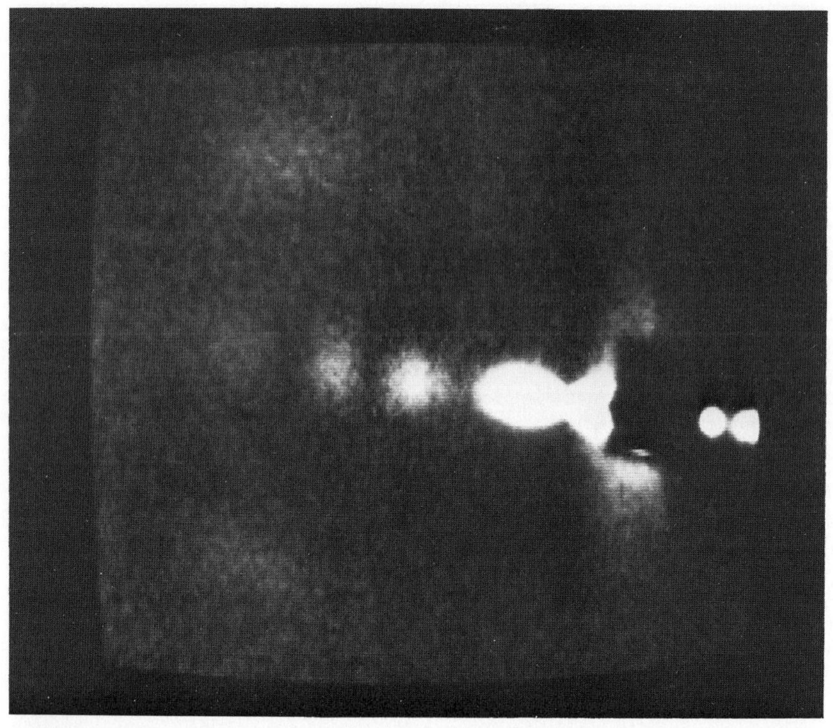

FIG. 9. A diffraction pattern similar to Fig. 8 but taken in the photon counting mode.

television tube, their gain cannot be easily varied, and their costs are considerably higher than electrostatic image intensifiers. The electrostatic intensifiers are available in inverting and in proximity focused versions. The lower gain intensifiers (diode tubes) have a maximum light gain of 50–100 per stage, while the higher gain tubes incorporate a microchannel plate in front of the output phosphor screen and have a maximum gain of between 10,000 and 50,000. Varo Electron Devices (Garland, Texas), ITT Electro-Optical Products Div. (Fort Wayne, Indiana), Thomson CSF (Paris, France), and Mullard (Mitcham, England) offer a broad line of image intensifiers, Varian LSE Div. (Palo Alto, California) offers an 80/40 mm reducing zoom inverting diode intensifier, and Proxitronic Funk GmbH (Darmstadt, West Germany) offers one- and two-stage 25-mm diode proximity focused intensifiers. These intensifiers should be ordered with external power supplies which can be well shielded or mounted outside the camera head since the high-voltage power supplied built into standard intensifiers operate at about 15 kHz and generate pickup in the camera preamplifier. For the highest accuracy a separate, well-regulated high-voltage supply should be used. Inverting intensifiers have pincushion distortion which reaches about 8% per stage at edge of the intensifier. This distortion can be corrected in software but where this is inconvenient, proximity focused intensifiers can be used. One or more small bright spots or other blemishes are often found in the outer portion of the intensified image so that it should be specified that the intensifier have no significant blemishes within the area viewed by the television tube. If a SIT tube or other intensified camera tube is used, the image intensifier high voltage can be floated on the SIT voltage or a high-voltage resistor divider can provide both the intensifier and SIT voltages. However, this leads to voltages of −20 to −30 kV on the front end of the detector and is inconvenient when microchannel plate intensifiers, which use multiple voltages, are used. A better solution is to operate the photocathode of the intensifier at a lower voltage and drop 15–20 kV across the fiber optics coupler between the image intensifier and the SIT tube. This coupler should be at least 25 mm thick, coated on both faces with thin tin oxide layers, and specified to withstand this voltage. The high voltages at the output of intensifier and at the photocathode of the SIT tube should be applied to the mating conducting coatings, with high-voltage wires connected to the beveled edges of the fiber optics with a dab of conducting epoxy. All mating faces should be coupled with a thin layer of immersion oil to match the indices of refraction.

While standard image intensifiers can be used in many applications, the manufacturers can provide several options which will improve their performance for X-ray applications. The aluminized phosphor screen in proximity focused diode image intensifiers should be blackened to prevent the diffuse reflection of incident light not absorbed by the photocathode, which would otherwise produce a faint halo several millimeters in size around bright regions of the image. For most applications, these intensifiers should also be ordered with a photocathode-to-phosphor spacing sufficient to withstand at least a 12-kV potential. This permits operation at a higher gain, and at lower gains the electrostatic field, and therefore the dark noise, is lower. Except where very high-resolution television tubes are used, the resulting decrease in the resolution of the camera system as a whole is negligible. Tubes rated for operation at 6 kV can be used if an additional intensification stage is incorporated when high gain is needed. In general, it should be specified that there be no emission point defects in the active area. For low-flux operations a low-noise photocathode should be specified unless the image intensifiers are to be cooled below 0°. This is especially important if microchannel plate image intensifiers are used, because of the long exponential tail in the pulse height distribution of the microchannel plate. A standard microchannel plate intensifier at room temperature has over 10^3 thermionic events/sec which have amplitudes which are as large as those produced by an average 8-keV X-ray. A standard low-noise intensifier has a dark noise (EBI) of $0.5–1 \times 10^{-11}$ lumens/cm^2 at room temperature (which corresponds to approximately 50,000 thermionic events/cm^2-sec), but ITT (Fort Wayne) and Proxitronics Funk can provide image intensifiers with bialkali photocathodes with under 1000 thermionic events/cm^2-sec. In general, the thermionic noise is reduced by a factor of 2 for every 4–6° that the photocathode is cooled and the S-20 noise is reduced to about 1500 events/cm^2-sec at 0°. Ongoing developments in image intensifiers with bialkali photocathodes should eventually provide even lower noise performance. These photocathodes make it much easier to distinguish single X rays from thermionic events.

In the photon counting mode, the image intensifier should typically be at least as large as the desired active area. In the A/D mode, the phosphor screen can be coupled to a smaller, less expensive image intensifier with 2:1 reducing fiber optics if a low-noise photocathode is used and single X rays do not have to be accurately measured.

The lowest grade 16-mm (diagonal) SIT tube from RCA costs $1500, while the B and A grades cost $2500 and $4000, respectively. The lowest grade tubes typically have one or two defects at the very edge of the target but provide satisfactory performance for most applications. A 40-mm SIT tube costs between $20,000 and $35,000 depending on grade, while an 80-mm tube costs between $30,000 and $80,000 except for certain possibly acceptable rejects. SIT tubes that are 16 mm are also available from most TV tube manufacturers. Unpotted tubes should be specified since potted tubes have a glass faceplate over the fiber optics input. Other television tubes are available for costs between $1000 and $30,000, but SIT (or S-T) tubes have been used in most television detectors because of their high performance and sensitivity. Of the other tubes, only Plumbicon and SEC vidicon tubes have been used to any significant extent in research applications. All of the previously mentioned companies are very helpful in providing advice and special configurations, and should be contacted for further information.

A set of programmable, high-resolution camera electronics is available through the author. A somewhat less versatile set of camera electronics is available from Sierra Scientific (Mountain View, California). If less accuracy and larger drifts are acceptable and the detector is to be operated at standard TV rates, the electronics of a silicon target TV camera (manufactured by RCA and others) can be used to scan and read out a SIT tube. Most of these cameras can be easily modified to accept external horizontal and vertical reset signals from the detector electronics, a modification which provides higher resolution than syncing the detector electronics to the camera.

If the detector is connected to a dedicated computer with at least 128K–256K words of memory the image can be integrated in the computer memory in the photon counting or slow scan A/D modes, if the data rate does not exceed the processing speed of the computer. Many applications, however, require the high speed, large capacity, display capabilities, and convenience of a separate memory system. Four or eight memory boards with capacities up to 2 Mbyte per board should be multiplexed to accommodate the data rate in the A/D mode when the camera is scanned at standard TV line rates. Only one or two high capacity memory boards are needed if only the photon counting mode is used or if the camera is scanned at one-quarter or one-eighth of the standard TV line rate when the A/D mode is used. Memory boards incorporating high-speed static RAM chips can be used in the high-speed A/D without multiplexing. However, their costs are at least several times higher than dynamic memory boards.

Detector Availability

Researchers can obtain a television detector system for X-ray diffraction applications in one of several ways. A complete system for crystallographic work can be obtained from Enraf-Nonius. It includes an X-ray generator, a diffractometer, an 80-mm detector operating in the A/D mode on a 512 × 512 array, and a minicomputer with crystallography software. A slow scan SIT system (OMA 2 SIT) which operates in the A/D mode is available from EG&G Princeton Applied Research. This system can digitize the 12-mm square active area on an array which is 512 pixels wide by 1 to 256 pixels high. While the entire height of the target is digitized, a smaller portion of the width of the target can be digitized by reducing the number of 25-μm-wide pixels along a scan line ("track"). The data is not integrated in a separate memory, so that for diffraction applications the image would typically be integrated on the SIT target for a short period of time and then read out in a single scan about 5–10 sec long.

Photometrics Ltd. (Tuscon, Arizona) makes a cooled, slow scan CCD camera which could be lens coupled to a second-generation image intensifier with a phosphor converter. The camera system incorporates an RCA CCD which can be read out in a 320 × 256 or 320 × 512 array at rates up to 200 kHz. The data can be binned along both axes to create larger pixels and readout of an arbitrary rectangular portion of the CCD can be selected. A shutter must be used during readout when the array is scanned in a 320 × 512 array. Fairchild and RCA can provide CCD cameras with fiber optics windows which the user could directly couple to a phosphor screen or cooled image intensifier.

Several companies make video digitizers which can be used to digitize and integrate images from a user supplied TV camera system operating at standard scan rates. These systems can be configured to provide a 512 × 512 × 16-bit A/D image. MCI/LINK (Palo Alto, California) make an image processor which can be connected to an LSI 11/23 computer. It is also capable of integrating slow scan images. Interactive Video Systems (Bedford, Massachusetts) provides a stand-alone system with an interface to a separate computer. Imaging Technology (Woburn, Massachusetts) provides a set of cards which can be plugged into a Q-bus or Multi-Bus computer, but the integration capability is not particularly straightforward and requires some care in implementing the system software.

The detector system described in this chapter is available through the author in a variety of compatible versions which can be easily tailored for particular applications. Schematics and descriptions are also available to those who would like to build a copy of a similar system.

Conclusion

Television area detectors can be used to collect high-resolution X-ray diffraction data over a wide range of fluxes in applications in which the significant portion of the diffraction pattern is typically less than about 80 mm in diameter. They are particularly useful with focused X-ray beams or synchrotron radiation. A wide range of system configurations are possible. Some of the more important practical details and trade-offs have been discussed here. The author and other researchers in the field can provide further information on these and other significant considerations.

[31] Experimental Neutron Protein Crystallography

By B. P. SCHOENBORN

Introduction

During the past 25 years or so, great advances have been made in X-ray protein crystallography, and, as seen in this volume, hundreds of protein structures have been determined. Neutron-scattering studies can

TABLE I
NEUTRON CROSS SECTIONS AND SCATTERING LENGTHS FOR ELEMENTS OCCURRING IN BIOLOGICAL SYSTEMS

	Cross section (10^{-24} cm^2)		Mass absorption coefficient, μ/p (cm^2/g)		Scattering length (10^{-12} cm)	
Element	Total	Coherent	Neutron	X ray (1.54 Å)	Neutron	X ray ($\sin\theta = 0$)
H	81.5	1.8	0.11	—	−0.38	0.28
D	7.6	5.4	0.0001	—	0.65	0.28
C	5.5	5.5	0.0002	6	0.66	1.7
N	11.4	11.0	0.048	9	0.94	2.0
O	4.2	4.2	0.0	13	0.58	2.3
Na	3.4	1.6	0.008	31	0.36	3.11
Mg	3.7	3.6	0.001	40	0.53	3.4
P	3.6	3.5	0.002	73	0.51	4.2
S	1.2	1.2	0.006	91	0.28	4.5
Cl	15.2	12.3	0.3	103	0.96	4.8
Ca	3.2	3.0	0.004	170	0.46	5.6
Mn	2.0	1.6	0.083	284	−0.36	7.0
Fe	11.8	11.4	0.015	325	0.95	7.3
Ni	18.0	13.4	0.03	50	1.0	7.9
Zn	4.2	4.2	0.006	60	0.56	8.3
Ge	9.0	8.8	0.01	70	0.84	9.0

expand these investigations considerably by exploiting the large and significantly different scattering factors of hydrogen and deuterium. In neutron diffraction analysis of single crystals of proteins, it is possible to determine the positions of hydrogen atoms—atoms that play a dominant role in enzymatic function but cannot be localized in electron density maps.

Neutron diffraction also facilitates the differentiation of hydrogen from deuterium and of nitrogen from carbon and oxygen (Table I). The large scattering factors of hydrogen and deuterium make it possible to study hydrogen exchange and to localize water of hydration and determine its coordination.

Neutron Diffraction

Neutrons were discovered by Chadwick in 1932,[1] based on the Joliot-Curie studies of α-irradiation of beryllium. The flux from these very weak

[1] J. Chadwick, *Proc. R. Soc. London, Ser. A* **136,** 692 (1932).

sources was slowly increased by using more intense α emitters, but only with the advent of the fission reactor in 1942 could intense enough neutron beams be produced for even very simple diffraction studies.[2] Development of better and specialized "beam reactors" rapidly increased the neutron flux (polychromatic) from 10^5 neutrons/cm^2-sec to 10^{16} neutrons/cm^2-sec, which now makes possible even the study of proteins[3] containing over 3000 atoms. But the most intense neutron beams are still about 10^5 less intense than common X-ray sources. This handicap can, however, be overcome with efficient data collection techniques using linear or two-dimensional position-sensitive detectors as described below.

To a large degree, neutron diffraction is very similar to X-ray diffraction, subject to the same space group conditions, Bragg's law, and Lorentz, but not polarization, conditions. In the following discussion the differences, rather than the similarities, will be stressed.

Generally, neutron scattering factors do not increase with increased atomic number as X-ray scattering factors do, but show marked variation even for isotopes of the same element (Table I). For protein crystallography, the most important isotopic difference in scattering factor is found for hydrogen and deuterium with -0.374×10^{-12} cm/atom for the former and 0.667×10^{-12} cm/atom for deuterium. The negative scattering factor for hydrogen is caused by a nuclear resonance phenomenon which retards the phase of the scattered neutron by 360° instead of the usual 180°. In Fourier maps, atoms with negative scattering factors will show up as negative density levels. Inspection of Table I shows that hydrogen is the only negative scatterer of importance to protein crystallography. In addition to the coherent elastic scattering that gives rise to Bragg peaks for crystalline samples and therefore contains the static structural information, a number of other neutron interactions are of some importance to protein crystallography.[4] Hydrogen atoms and some other elements (isotopes) that possess unpaired nuclear spins produce strong incoherent scattering that results in a large increase in background scattering. This incoherent scattering does not produce any interference effects and therefore does not produce any diffraction peaks, but produces a uniform background radiation. Free hydrogen atoms (H gas) have a particularly large incoherent cross section of 80 barns with only a 2-barn cross section for coherent scattering. In proteins where H atoms are mostly bound, the incoherent cross section reduces to ~40 barn. [A barn is a unit of cross section which is a number that denotes the effective area presented by a

[2] R. M. Brugger, *Phys. Today* **21,** 23 (1968).

[3] B. P. Schoenborn, *Nature (London)* **224,** 143 (1969).

[4] B. P. Schoenborn and A. C. Nunes, *Annu. Rev. Biophys. Bioeng.* **1,** 529 (1972).

nucleus for an interaction process (scattering or absorption). For example, a nucleus with a 1-barn cross section placed in a neutron beam of 1 cm^2 in area will scatter 10^{-24} neutrons of the beam. For a glossary of common terms used in neutron diffraction, see Schoenborn and Nunes.[4]]

Generally, absorption factors for neutrons are considerably smaller than those found for X rays, permitting the use of large crystals to assure a large diffraction peak. The electrically neutral neutron does not produce free radicals as X rays do and, therefore, does not cause significant radiation damage, thus enabling the collection of all diffraction data from one crystal.[5] This is a great advantage considering the tedium of growing crystals, especially large ones. The use of only one crystal also reduces scaling problems and eliminates perturbations due to differences in crystals.

Hydrogen–Deuterium Exchange

The ability to distinguish between the presence or absence of hydrogen or deuterium adds a major dimension to neutron structural studies and has been used to determine the protonation state of histidines in myoglobin[5–7] and in trypsin.[8] H–D exchange studies can be used to measure the conformational stability of proteins. Although it is difficult from the present studies to draw detailed conclusions about the localized H–D exchange, there is clear evidence that less H–D exchange takes place in hydrogen-bonded peptides than in peptides not hydrogen bonded. In the CO-myoglobin[9] study, the occupancy for deuterium in a hydrogen-bond configuration is 35% compared to 58% in a nonbonded arrangement. In order to draw accurate conclusions on H–D exchange, the error in the observed Fourier density map has to be assessed. This presents three problems:

1. The observed peak heights (weights Z_i) are a measure of the hydrogen–deuterium exchange with $Z_i = 6.7 - 3.7(1 - X)$ where X is the fraction of D present. For an average error of <2 fermi units (F) below and above the zero level, one calculates that the fractional D occupancies (X) lying between 0.16 and 0.54 are undetermined.

[5] B. P. Schoenborn, *Cold Spring Harbor Symp. Quant. Biol.* **36,** 569 (1972).
[6] S. E. V. Phillips and B. P. Schoenborn, *Nature (London)* **242,** 81 (1981).
[7] J. C. Norvell, A. C. Nunes, and B. P. Schoenborn, *Science* **190,** 568 (1975).
[8] A. A. Kossiakoff and S. A. Spencer, *Biochemistry* **20,** 6462 (1981).
[9] J. C. Hanson and B. P. Schoenborn, *J. Mol. Biol.* **153,** 117 (1981).

2. The neutron weights and temperature factors of H and D atoms are difficult to determine because of their high correlation with those of the atoms to which they are bonded.

3. The average error observed in the Fourier map depends on the accuracy of the observed reflection magnitude and the derivation of accurate phases. Accurate phases not only depend on the precise locations of atoms, but include contributions from the solvent space. This is particularly important in cases in which H_2O has been replaced by D_2O. D_2O has a scattering density of 0.6 F/Å^3 compared to -0.06 F/Å^3 for H_2O and contributes significantly to the low-order reflections.

The use of D_2O as a solvent in neutron protein crystallography not only allows the determination of H–D exchange, but increases the visibility of water molecules in Fourier maps.[5] In a Fourier map derived from neutron data, a water molecule appears about three times as strong as in the equivalent electron density map. Observations on the CO-myoglobin structure[9] showed only 40 water molecules with high occupancy. The effect of salt and pH on the role of bound water is, however, still unclear and might explain some of the large differences that have been observed in different proteins.

It also should be noted that the scattering length of nitrogen (9.4 F) is significantly larger than that for oxygen (5.8 F) and carbon (6.6 F) allowing correct positioning of side chains such as histidines.

Another remarkable feature of neutron density maps is the large peak observed for N^{ζ} of lysine in the deuterated case and the absence of any peak in the H case. The presence of the ND_3 feature is a valuable aid in tracing lysine side chains and assessing the hydrogen bonding of this N.

The Neutron Protein Diffractometer

For diffraction studies, neutrons must have wavelengths of the order of 1 Å. The fast, highly energetic neutrons produced by fission of ^{235}U are exposed to a moderator such as D_2O or graphite, and through collision lose energy until they reach thermal equilibrium. Thermal neutrons possess a Maxwellian distribution with a flux maximum at ~1 Å for a moderator at a temperature of ~40°.

Neutrons are extracted from the reactor by evacuated beam tubes into an external shield that houses a monochromator (graphite or copper, etc.) and serves as beam stop for fast neutrons and γ radiation. The crystal orienter is a three-circle Eulerian cradle with a massive arm to support a tower for the detector support (Fig. 1a). The control arrangement of the whole data acquisition system is shown in Fig. 1b. The system is designed in modular fashion to permit use of any subsystem without affecting the

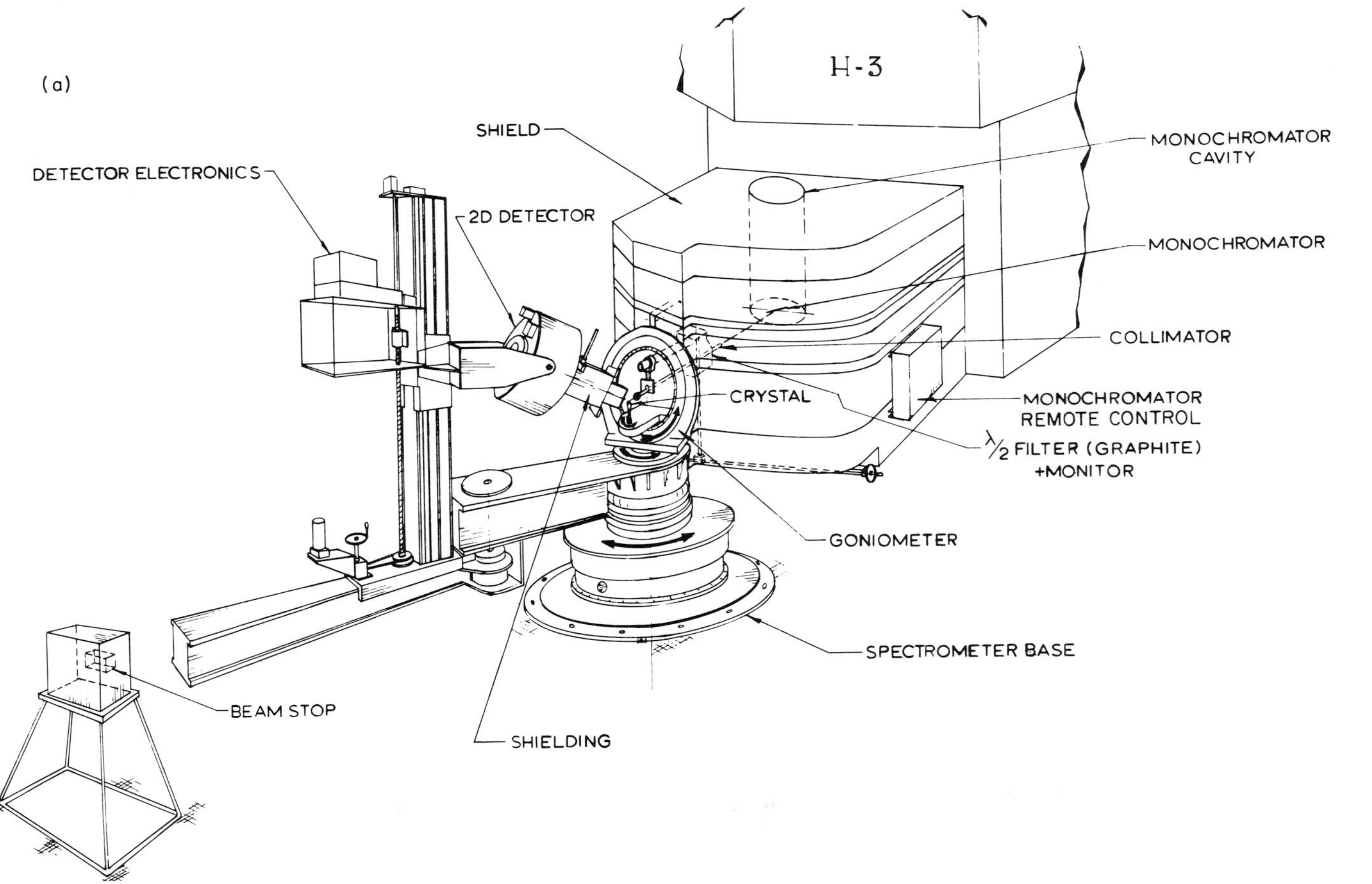

FIG. 1. (a) Schematic drawing of the protein crystallography station at the Brookhaven National Laboratory High Flux Beam Reactor (HFBR). (b) Schematic diagram of the electronic control and data acquisition system.

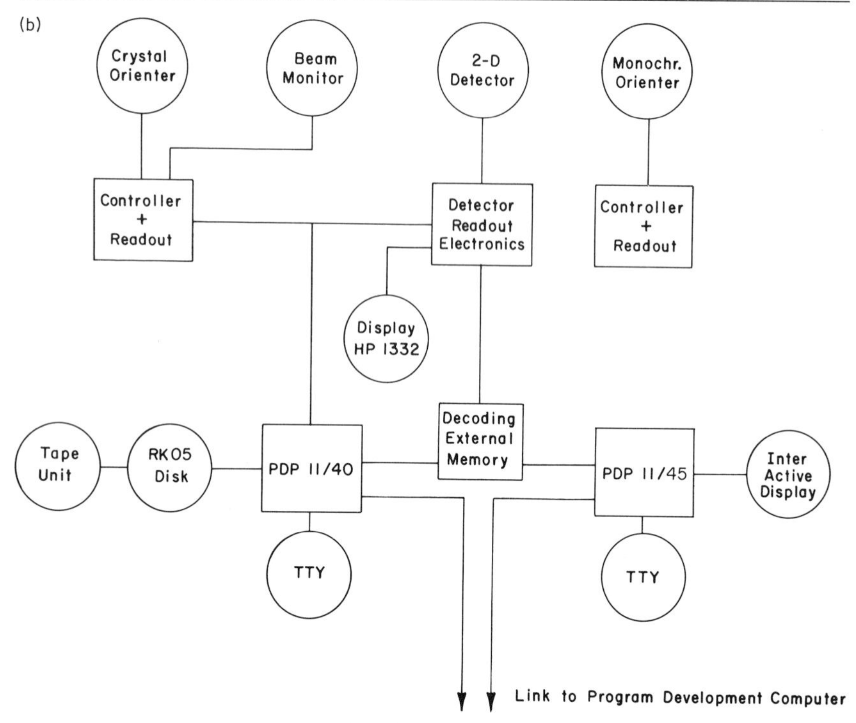

FIG. 1. (*continued*)

whole control environment. This allows easy servicing of the individual modules and gradual extension of the system by simply adding new detectors. The direct addressing external memory interface will permit up to four detectors of 32,000 elements each. The spectrometer utilizes stepping motors and encoders for accurate positioning, while the monochromator orienter relies on stepping motors and pulse counting for positioning. Both devices are controlled by microprocessors connected via standard serial links.

An analog storage oscilloscope connected to the two-dimensional detector readout allows immediate observation of strong diffraction peaks for initial detector and crystal alignment. The on-line interactive display system has direct access to the external memory and permits inspection of raw and processed data for accurate detector calibration, crystal alignment, and data monitoring during collection. The main computer (PDP 11/40) carries out the crystallographic calculations and controls the input and output of the system. Raw reflection data and preprocessed background substracted intensities are written to tape for further processing as described below.

The crystals are mounted in such a way that their most densely populated reciprocal lattice planes are horizontal to coincide with the high-resolution direction of the two-dimensional counter, with the crystal rotation axis parallel to the vertical y axis and perpendicular to the beam. The crystals are then rotated in small $\Delta\omega$ steps ($\sim$0.05°) similar to the rotation method. For a given reflection, every ω step results in the 2θ–y profile, with the reflection centered at $2\theta_{hkl}$ and ω_{hkl}. The data profile is stored as a three-dimensional array in memory. This three-dimensional array is summed along the vertical y direction to produce an ω–2θ projection (Fig. 2). Change in the shape of reflections along the y direction is small and monotonic. The use of a two-dimensional rather than a three-dimensional

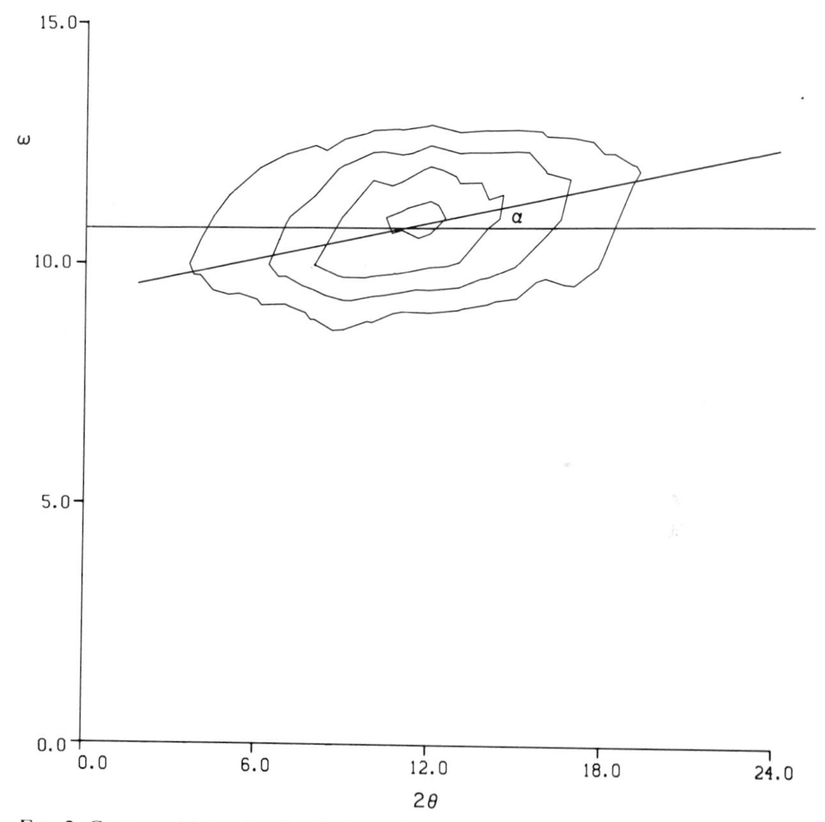

FIG. 2. Contoured intensity distribution of a neutron reflection. The vertical coordinate axis is the crystal rotation (ω) with 0.07° per step. The horizontal coordinate axis is the 2θ direction with 0.07° per step. The major reflection axis with the inclination angle α is indicated and is derived from the second moments of the density distribution. The reflection has been integrated in the vertical detector direction.

reflection representation simplifies the discussion somewhat and reduces the computational effort which is particularly desirable for on-line calculations. The reflection within this array is approximately ellipsoidal and can be characterized by an inclination angle α and major and minor axes. The orientation and shape of such depicted reflections vary as a function of *hkl*. To obtain integrated intensities with an accurate subtracted background, the shape and orientation of the reflection have to be known.

Observations of reflection profiles from proteins and other crystals showed that the shape of neutron reflections is different and more varied than shapes observed in X-ray diffraction analysis. Profile fitting techniques, as developed for the X-ray oscillation and diffractometry techniques,[10] were thus not useful. A simple integration scheme using a rectangular box in ω-2θ as developed by Cain *et al.*[11] was used for the early data processing of protein data. A profile fitting and filtering technique was described by Spencer and Kossiakoff,[12] particularly for the detection of very weak reflections. Sjolin and Wlodawer[13] used this method in the evaluation of ribonuclease data collected on the National Bureau of Standards linear position-sensitive detector system. These profile fitting techniques are lengthy processes, not easily suitable for on-line data processing procedures, and they tend to overestimate very weak reflections. The shape of a neutron reflection is strongly dependent on the wavelength bandwidth $\Delta\lambda$ and the counter resolution. These parameters, as discussed below, are much larger for neutrons than for X rays and do not permit the use of reflection orientation parameters developed for the X-ray case. A number of basic analyses of neutron diffraction profiles have, however, been published, but these descriptions of the resolution function do not treat the conditions encountered with position-sensitive detectors.[14,15]

Position-Sensitive Detectors

Large and efficient area detectors are needed to collect diffraction data in a reasonable time since counting time per diffraction peak can take several minutes. Many possibilities suggest themselves[16,17] as the method

[10] G. C. Ford, *J. Appl. Crystallogr.* **7,** 555 (1974).

[11] J. E. Cain, J. C. Norvell, and B. P. Schoenborn, *Brookhaven Symp. Biol.* **27,** VIII-43 (1975).

[12] S. Spencer and A. Kossiakoff, *J. Appl. Crystallogr.* **13,** 563 (1980).

[13] L. Sjölin and A. Wlodawer, *Acta Crystallogr., Sect. A* **A37,** 594 (1981).

[14] M. J. Cooper and R. Nathans, *Acta Crystallogr., Sect. A* **A24,** 619 (1968).

[15] S. A. Werner, *Acta Crystallogr., Sect. A* **A27,** 665 (1971).

[16] J. B. Davidson, *Brookhaven Symp. Biol.* **27,** VIII-3 (1975).

[17] U. W. Arndt and D. J. Gilmore, *Brookhaven Symp. Biol.* **27,** VIII-16 (1975).

TABLE II
NEUTRON–CHARGED PARTICLE REACTIONS SUITABLE FOR THERMAL NEUTRON DETECTION[15]

	Reaction Q value	Thermal cross section (barns)
(a)	$^{10}B + n \rightarrow {}^{7}Li + \alpha + 2.79$ MeV (6.1%)	3840 ± 11
	$\searrow {}^{7}Li^{*} + \alpha + 2.31$ MeV (93.9%)	
	$\searrow {}^{7}Li + \gamma + 0.478$ MeV	
(b)	$^{6}Li + n \rightarrow {}^{3}H + \alpha + 4.786$ MeV	936 ± 6
(c)	$^{3}He + n \rightarrow {}^{3}H + p + 0.764$ MeV	5327 ± 10

best suited to fulfill this need. Gas discharge detection of neutron-induced charged-particle reactions is, however, considered to be the method of choice under existing technology for the detection of thermal neutrons.[18] Among the possible types of gas discharge counters, practical considerations such as γ-ray discrimination, count-rate capacity, stability, resolution, and efficiency led to the selection of the proportional system.

The nonionizing neutron cannot be as easily detected as X rays. In order to count neutrons, they have first to be converted to an ionizing radiation by reaction with nuclei such as ^{10}B or ^{3}He (Table II).[19] Typical particle reactions for X rays and neutrons are depicted in Fig. 3. The area detectors at Brookhaven National Laboratory use ^{3}He for detection and Argon or C_3H_8 as a stopping gas. These gases allow operation at high pressures (high efficiency) with relatively low voltages (~3 kV), good gas gain (multiplication), and high thermal neutron cross section. The detectors used at the High Flux Beam Reactor use the resistive wire charge division method[19,20] although a number of other techniques are available such as delay line position sensing[21] or diffusive RC lines.[22]

The method of charge division is shown in Fig. 4 and is particularly suited to high-pressure proportional chambers with slow signal rise times. The basic characteristics of such position-sensitive detectors are given in Table III. One of the most serious handicaps, particularly for protein crystallography, of these position-sensitive detectors is the resolution of

[18] J. Alberi, J. Fischer, V. Radeka, L. C. Rogers, and B. P. Schoenborn, *IEEE Trans. Nucl. Sci.* **NS-22**(1), 255 (1975).

[19] J. L. Alberi, *Brookhaven Symp. Biol.* **27,** VIII-24 (1975).

[20] J. L. Alberi and V. Radeka, *IEEE Trans Nucl. Sci.* **NS-23,** 251 (1976).

[21] V. Radeka, *IEEE Trans. Nucl. Sci.* **NS-21** (1), 51 (1974).

[22] C. J. Borkowski and M. K. Kopp, *Rev. Sci. Instrum.* **39,** 1515 (1968).

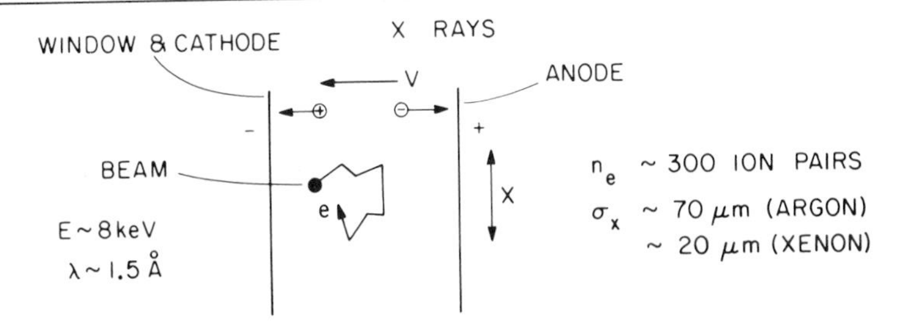

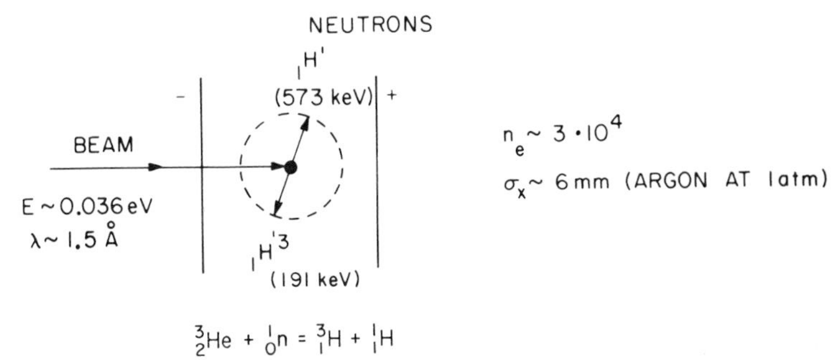

FIG. 3. Schematic representation of the events associated with the detection of X rays and thermal neutrons. For neutrons, the gas amplification as described by the ion pair formation (n_e) is large; but the range (σ_x) of the resultant particles is also large ~2.7 mm at 4 atm argon, limiting the resolution of such detectors.

TABLE III
CHARACTERISTICS OF TWO-DIMENSIONAL POSITION-SENSITIVE DETECTOR

Characteristic	Value
Neutron detection efficiency at 1.5 Å wavelength	70%
Position resolution	2.7 mm
Positional uniformity of detection efficiency	3%
Integral linearity	1%
Differential linearity	0.1%/% of full scale
Event resolving time	1 μsec
Event processing time	3 μsec
γ-Ray sensitivity	2.7×10^{-6} detected/incident γ
Sensitive area	18 × 18 cm
Gas filling composition (absolute pressure)	6 atm ^3He
	4 atm Ar
	0.5 atm CO_2

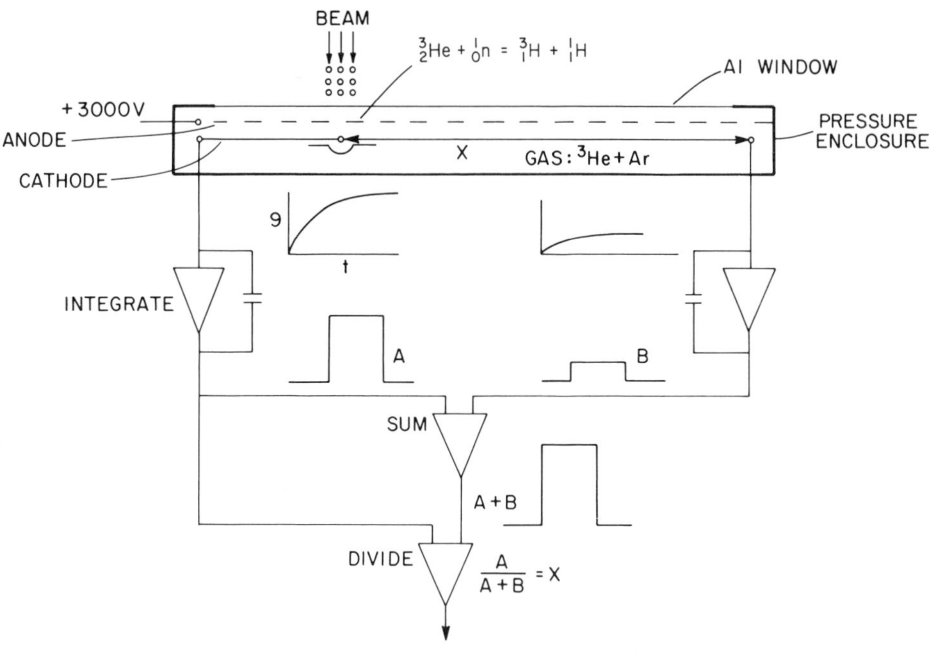

FIG. 4. Schematic presentation of the charge division method for a linear detector.

~3 mm full width half-maximum (FWHM). Recently, Radeka *et al.*[23] have, however, developed a new readout system with a very low avalanche gain coupled with a precision centroid finding filter. The use of these new detectors with a resolution of ~1.5 mm will reduce reflection overlap. For the same basic diffraction conditions, this will reduce the crystal to detector distance and increase the number of simultaneously collected reflections, significantly reducing data collection time.

The A Shape of a Neutron Reflection

The shape of reflections presented in the ω-2θ coordinate system (Fig. 2) is dependent on (1) the beam divergence, (2) the crystal mosaic, (3) the crystal size, (4) the crystal to detector distance, (5) the detector resolution, and (6) the wavelength spread. In the X-ray case as recently described by Mathieson[24] only the first four conditions are important since

[23] R. A. Boie, J. Fischer, Y. Inagaki, F. C. Merritt, H. Okuno, and V. Radeka, *Nucl. Instrum. Methods* **200,** 533 (1982).

[24] A. McL. Mathieson, *Acta Crystallogr.* **A38,** 378 (1982).

the X-ray emission lines (Cu K_{α}) normally used are very sharp and the resolution of X-ray position-sensitive detectors (~0.1 mm) or film is small compared to the neutron case. The effect of each of these parameters on the spot shape will briefly be described to highlight the impact of these parameters on the design of an experimental station and on the shape of reflections as presented in references 24–26.

The Influence of Beam Divergence and Crystal Mosaicity. Most natural compounds, particularly proteins, do not form perfect crystals but grow with many different crystalline regions that are slightly inclined relative to each other. For proteins, this so-called mosaic spread (μ) is in the order of a tenth of a degree. For mosaic crystals, reflections will therefore be observed at the given Bragg angle $2\theta_{hkl}$ for a finite crystal rotation width, with $\Delta\omega = \mu$ as depicted in Fig. 5a. (In this case, only the effect of the mosaic distribution is considered.) The intensity distribution within this rotation ($\Delta\omega$) width is often Gaussian. Even in a well-collimated beam, there is, however, a finite beam divergence. This beam divergence $\Delta\theta$ in effect changes the origin of the diffraction by $\Delta\theta$ so that a reflection is observed from $2\theta + \Delta\theta$ to $2\theta - \Delta\theta$ as the crystal is rotated by $\Delta\omega = \Delta\theta$. This beam divergence produces in the ω–2θ coordinate system a reflection that is inclined at 45° (Fig. 5b). Both effects of beam divergence and crystal mosaic occur simultaneously, producing a convolution of the two diffraction effects. The intensity distribution within each profile depends on the particular conditions; the mosaic often produces a Gaussian distribution, while the beam divergence is often trapezoidal. Figure 6a shows that such a convolution produces an ellipsoidal shape with an inclination angle between 45° and 90° depending on the relative magnitude of the two variables ($\Delta\theta$ and μ).

The Influence of Crystal Size and Counter Resolution. The physical size (width) of a crystal produces a reflection smearing that in the ω–2θ space results in a smearing of the 2θ direction; in essence, the origin of diffraction is translated over the width of the crystal. In practice, this results in the further convolution of the spot shape with a rectangle. If the crystal is asymmetric, the effective width of this function will vary with the crystal rotation ω. Protein crystals used in neutron diffraction experiments are often as large as the detector resolution (~3 mm), and produce significant reflection broadening (Fig. 6b).

The effect of the counter resolution is similar to the crystal size effect and smears the reflection; with a typical counter resolution of ~3 mm FWHM this has a significant effect on the size of a spot. The detector

[25] B. P. Schoenborn, *Acta Crystallogr.* **A39,** 315 (1983).

[26] B. P. Schoenborn, *Acta Crystallogr., Sect. A* **A37,** 16.X-08, C 313 (1981).

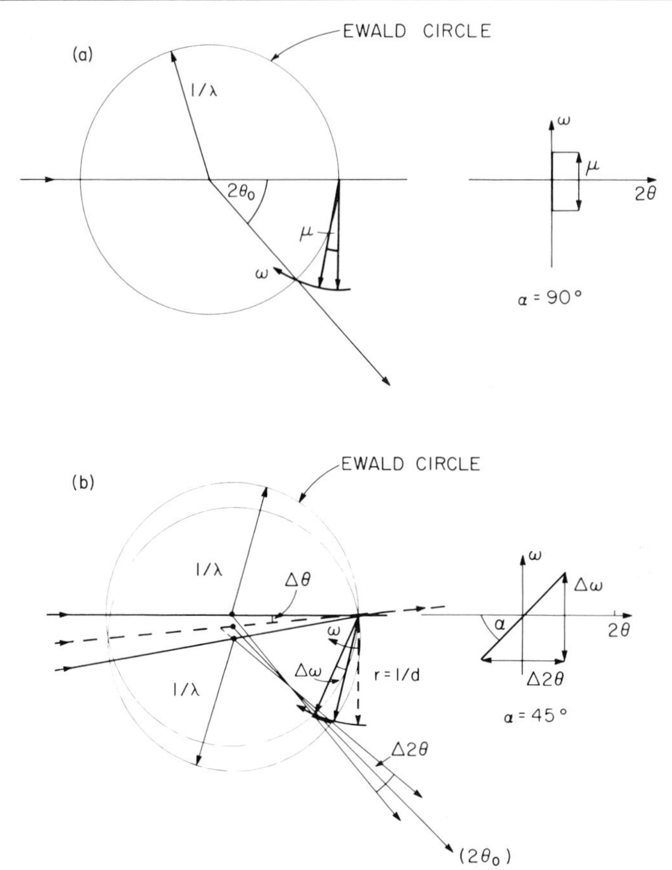

FIG. 5. (a) Ewald sphere construction describing the effect of the crystal mosaic for a given wavelength λ. The reciprocal lattice vectors distributed within angular width μ produce diffraction at the Bragg angle $2\theta_0$ over the rotational width $\Delta\omega = \mu$. (b) Ewald sphere construction describing the effect of beam divergence for a given μ on the diffraction process. The reciprocal lattice vector $\nu = 1/d$ is in diffraction condition for the rotational width $\Delta\omega$ (heavy line) producing diffraction over the angular width $\Delta 2\theta = \Delta\omega$ centered on the Bragg angle $2\theta_0$. On the right side, the resulting intensity distribution (heavy line) is mapped as a function of the crystal rotation ω and the diffraction angle 2θ.

smearing is Gaussian; the numerically convoluted probabilities of the above discussed effects are shown in Fig. 6c. In the equatorial diffraction plane, the above effects are not dependent on $2(\sin\theta)/\lambda$.

The Influence of the Wavelength Bandwidth. The neutron source has a finite size producing radiation with a given divergence $\Delta\theta$; this beam divergence and the mosaic of the monochromator result in a diffracted

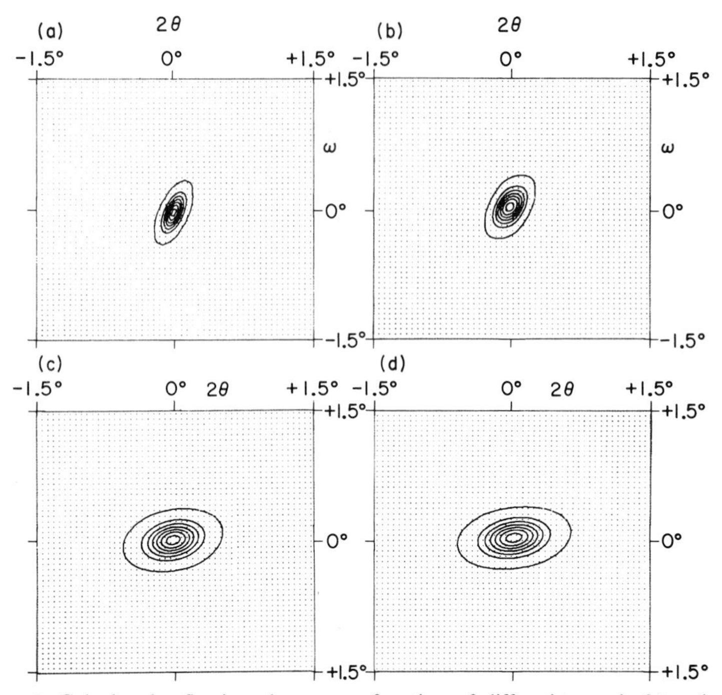

FIG. 6. Calculated reflection shape as a function of diffraction angle 2θ and crystal rotation ω showing the effects of (a) beam divergence and crystal mosaicity, (b) the added effect of the crystal size (2 mm), (c) the added effect of the detector resolution (3 mm), and (d) the effect of the wavelength bandwidth $\Delta\lambda$. Beam divergence = 0.1°, crystal mosaic = 0.1°, $\Delta\lambda$ = 0.01, r.l.u. = 0.3 Å^{-1}.

beam with a given wavelength bandwidth $\Delta\lambda$.[27] $\Delta\lambda$ is calculated from the differentiated Bragg equation and $\Delta\lambda = \lambda\ \Delta\theta \cot(\theta_M)$ where $\Delta\theta$ is the effective beam divergence and θ_M the Bragg angle of the monochromator. The effect of $\Delta\lambda$ on the diffraction condition is depicted by an Ewald[28] sphere construction (Fig. 7). In this case, the two limiting spheres have radii $r_1 = 1/(\lambda - \Delta\lambda)$ and $r_3 = 1/(\lambda + \Delta\lambda)$. The equatorial circles will intersect at two points, the origin of the reciprocal lattice at O and the point M, the so-called focusing position at the end of the reciprocal lattice vector ν_m of the monochromator spacing.[27] Note that the wavelength is correlated with the diffraction angle θ, larger wavelength corresponding

[27] U. Arndt and B. T. M. Willis, "Single Crystal Diffractometry," p. 208. Cambridge Univ. Press, London and New York, 1966.

[28] M. M. Woolfson, "X-Ray Crystallography," p. 63. Cambridge Univ. Press, London and New York, 1970.

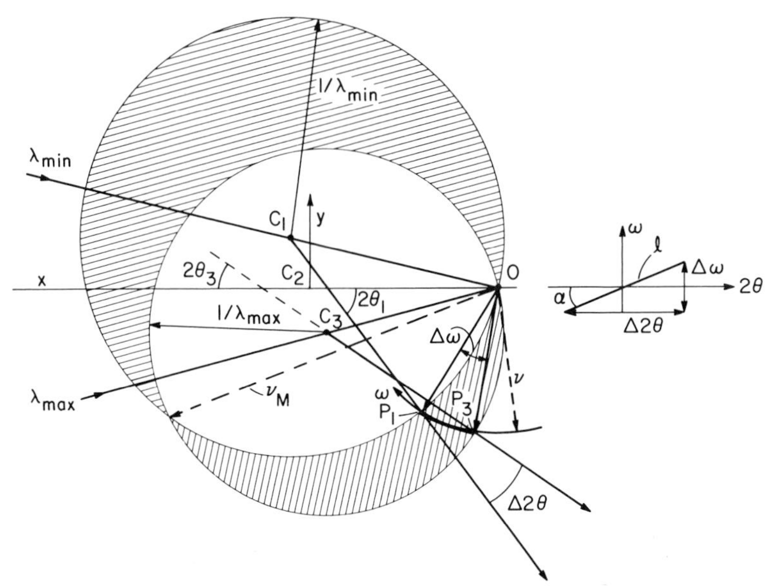

FIG. 7. Ewald sphere construction showing the width of the reflection due to $\Delta\lambda$, the wavelength bandwidth. The limiting spheres for the smallest (λ_{min}) and the largest wavelength (λ_{max}) are drawn. Diffraction occurs at any point in the shaded regions. The large shaded region at the top is due to diffraction in the "antiparallel" mode.[23] The shaded region on the bottom represents diffraction in the parallel or focusing mode. The vector length ν_M is given by the lattice spacing of the monochromator selecting the radiation. On rotation ω, the reciprocal lattice vector γ (of the sample) will be in reflecting position from P_1 to P_3 over the rotational width $\Delta\omega$ and will produce diffraction over an angle $\Delta 2\theta$. $\Delta\omega$ and $\Delta 2\theta$ vary as a function of reciprocal lattice length ν and layer line height θ (θ is perpendicular to the plane of the drawing). The diffraction process is mapped on the right side as a representative function of ω and 2θ. α is the angle formed between the diffraction line and the horizontal axis (2θ). The reflection length l is the length of the distribution at half-height measured in degrees.

to larger 2θ angles. The beam size and divergence result additionally to a smearing of every λ component about its mean distribution. In this construction, the required diffraction conditions are satisfied for all points in the shaded area. A crystal vector ν rotating perpendicular to the rotation axis ω located at O will produce diffraction from point P_1 to P_3 during the rotation $\Delta\omega$; the resulting rays will diverge at an angle $\Delta 2\theta$.

The effective widths $\Delta\omega$ and $\Delta 2\theta$ depend on the vector length ν. This diffraction behavior is again depicted in the ω-2θ space (Fig. 5) with a reflection of length l inclined at an angle α. The vector ν crosses the circle r_1 and r_3 and the intercept points ω_1 and ω_2 can be calculated from geometry. Note that the origins of the Ewald circles are not concentric. There-

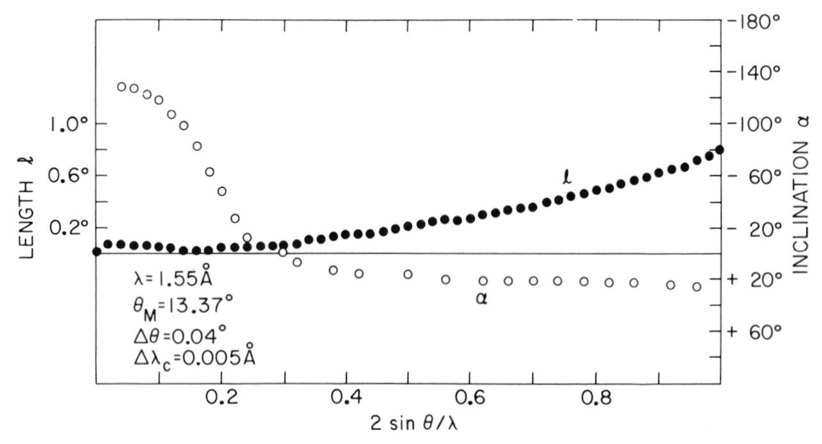

FIG. 8. The effect of $\Delta\lambda$ on the inclination α and the reflection length l in the ω-2θ space as a function of reciprocal lattice length $\nu = 2(\sin\theta)/\lambda$ for a graphite monochromator with $\theta_M = 13.4°$.

fore, in a numerical approach, the vector ν is rotated along ω and the distance (d) from the end of vector ν to the respective Ewald centers (C_1 or C_3) are calculated. Intercepts exist where $d_1 = r_1$ and $d_3 = r_3$. The origin of the coordinate system is at C_2, the center of the Ewald sphere for the average λ.

For given intercepts ω_1 and ω_3, the respective diffraction angles (2θ) are calculated. As a function of rotation ω, a reflection will be observed moving from $2\theta_3$ to $2\theta_1$ as depicted on the right side of Fig. 7. The center of a reflection occurs at the Bragg angle for the mean wavelength. In Fig. 8, the reflection length and orientation due to the effect of $\Delta\lambda$ only are presented as a function of reciprocal length ν. In this case, calculations were performed for a graphite monochromator set at $\theta_M = 13.4°$ for a wavelength of 1.55 Å and a beam divergence of 0.10° with $\Delta\lambda = 0.01$ Å.

Inspection of Fig. 7 shows that special conditions arise for $\nu = \nu_M$ and $\nu = 0.5\,\nu_M$:

$$\text{at} \quad \nu = 0.5\,\nu_M, \qquad \Delta 2\theta = 0° \quad \text{and} \quad \alpha = -90°$$
$$\text{at} \quad \nu = \nu_M, \qquad \Delta\omega = 0° \quad \text{and} \quad \alpha = 0°$$

The intensity distribution due to $\Delta\lambda$ over the reflection length l is assumed to be Gaussian. Since this intensity distribution only reflects the effect of $\Delta\lambda$, the effects due to beam divergence, crystal mosaic, crystal size, and counter resolution are numerically convoluted.[29] The resultant probability ellipsoid is then convoluted with the diffraction length l as evaluated for

[29] H. Lipson and C. A. Taylor, "Fourier Transforms and X-Ray Diffraction," p. 20. Bell, London, 1958.

the $\Delta\lambda$ effect. Figure 6 shows the efect of the successive convolutions while in Table IV shape variation as a function of reciprocal lattice length $\nu = 2(\sin\theta)/\lambda$ is tabulated.

To delineate the shape of the reflection for ease of integration, the spot shape is described by a minor and major ellipsoidal axis with the major axis inclined by an angle α towards the $\Delta 2\theta$ coordinate axis. These parameters are best calculated from the second moments of the theoretical intensity distribution (see reference 8). Comparison of such calculated spot shapes with experimental data from perfect germanium, postassium bromide, alumina and protein crystals gave good agreement as summarized in Table IV.

Data Collection and Processing Strategy

The crystal is mounted so that reciprocal lattice planes with the highest reflection density lie in the equatorial plane, which is also the direction that has the highest counter resolutions. The crystal is then rotated in discrete steps (~0.05°) through ω, which is the axis perpendicular to the beam. The (ω, 2θ) coordinates of reflections (*hkl*) are calculated from the crystal orientation matrix for reflections that fall on the detector. The unit cell dimensions and the diffraction geometry determine the spatial resolution and fix the crystal-to-detector distance which, for a medium-size protein, is 60–100 cm. With area detectors of 20 × 20 cm, the acceptance angle is ~15°, permitting observations of many simultaneous reflections depending on the overall 2θ and λ settings. For the duration of every $\Delta\omega$ step (~60 sec), the whole counter is mapped into a direct addressable external memory. At the end of every ω step and for every active reflection, a mapped region with height Δy and width $\Delta 2\theta$ is extracted and is stored separately as a function of ω. A reflection is represented, therefore, as a three-dimensional data array centered on $2\theta_{\mathrm{hkl}}$, y_{hkl}, ω_{hkl}. This array is stored on disk for further processing as described below.

The best peak-to-background ratio is achieved by delineating the reflection as a three-dimensional ellipsoid. In practice, however, this is very time consuming for an on-line integration. The reflection is, therefore, summed over Δy to produce a two-dimensional profile in ω-2θ which is then integrated after delineation of the reflection according to precalculated spot shape parameters as described above.

For strong reflections, the center of mass is determined after background subtraction to monitor the crystal alignment and the counter electronics. Note that a well-adjusted counter is, however, stable and linear over many months with a positional accuracy better than 0.5%. The orientation of strong reflection within this ω-2θ array is then determined to check on the precalculated orientation parameters.

TABLE IV

OBSERVED AND CALCULATED SPOT SHAPE PARAMETERS FOR SOME WELL-CHARACTERIZED TEST CRYSTALS

Crystal	Monochromator setting (°)	Beam divergence (°)	Crystal mosaic (°)	λ (Å)	ν [=2(sin θ)/λ]	Angle α (°)		Major axis length (°)		Minor axis length (°)	
						Calc.	Obs. (±3°)	Calc.	Obs. (±0.28°)	Calc.	Obs. (±0.07°)
KBr	13.4	0.05	0.15	1.5	0.35	2	0	0.5	0.6	0.3	0.4
					0.71	7	5	0.8	0.8	0.3	0.4
Germanium	13.4	0.08	0.001	1.5	0.30	6	4	1.0	0.9	0.3	0.2
					0.50	8	9	1.4	1.8	0.4	0.5
					0.59	16	12	1.8	1.3	0.3	0.3
Myoglobin	13.4	0.08	0.15	1.51	0.08	12	10	0.27	0.24	0.15	0.16
					0.12	7	8	0.22	0.20	0.15	0.16
					0.18	7	6	0.22	0.24	0.15	0.16
					0.24	6	6	0.21	0.24	0.15	0.16
					0.30	5	5	0.23	0.24	0.15	0.16
					0.40	4	5	0.24	0.24	0.15	0.16

The orientation parameters are then used to calculate the reflection figure for integration. The axial (major and minor) lengths are adjusted depending on the magnitude of the reflection, so that 99% of the statistically significant intensities lie within the delineated area. Areas outside of this region are used as background. The background information is accumulated for given regions of reciprocal lattice space to improve the signal-to-noise ratio. Background is, therefore, averaged for a number of reflections (~100). Reflections with background regions that show large deviations from their group average are eliminated, marked, and treated separately. The background area for one reflection is divided into eight separate regions. The summed background for each region is compared to the expected value as determined for reflections with similar indices. If any of the eight regions does not lie within the expected distribution (3σ) it is not used for the group averaged background. This procedure eliminates counter edge effects. If a reflection does not have at least five acceptable background regions, it is rejected. Since reflections collected at the beginning suffer from lack of background accumulation, a second cycle of integration is beneficial using the accumulated background from the whole data set.

Conclusion

The development of efficient high-resolution position-sensitive detectors[18,23] and the use of accurate peak integration techniques[7–9,11–13,21,25] make it possible to collect accurate neutron diffraction data from medium-size crystals (~1 mm^3) to determine all atoms including hydrogens, hydrogen–deuterium exchange, and water of hydration. The above-described neutron techniques have been used at Brookhaven National Laboratory in the analysis of a number of protein structures such as myoglobin,[3,5–7,9,30] gramicidin,[30] crambin,[31] plastocyanin.[32] With further improvements in focusing neutron beam optics, reactor power, and the use of multiple two-dimensional detectors, it is expected that data collection time will gradually be reduced from the present 2–3 months to ~2 weeks.

Acknowledgments

The author wishes to acknowledge the many contributions made to the development of the neutron protein crystallography station by J. Alberi, V. Radeka, G. Dimmler, E. Kelly, J. Cain, and E. Caruso. Research carried out at Brookhaven National Laboratory under the auspices of the United States Department of Energy.

[30] R. Koeppe and B. P. Schoenborn, *Biophys. J.* **45**, 503 (1984).
[31] M. Teeter, in preparation.
[32] M. Freeman, T. Garrett, and B. P. Schoenborn, in preparation.

[32] Automated Peak Fitting Procedure for Processing Data from an Area Detector and Its Application to the Neutron Structure of Trypsin

By ANTHONY A. KOSSIAKOFF and STEVEN A. SPENCER

Introduction

By virtue of its ability to locate hydrogen atoms experimentally, neutron diffraction has the potential to answer many questions about the structure of proteins and their interactions with substrates and solvent. To realize this potential, however, two inherent experimental limitations must be overcome. The major limitation is that the flux from current neutron sources is roughly five orders of magnitude less than the flux from conventional X-ray sources. The second limitation is the large isotropic background produced by biological materials as a consequence of the incoherent inelastic scattering of neutrons by hydrogen atoms in the sample. This incoherent scattering can be reduced by exchanging deuterium for hydrogen, but because many of the hydrogens cannot be exchanged, the desired coherent diffraction usually contains a large background component.

Together these experimental problems constitute a formidable barrier to the successful completion of a protein structure unless very large crystals are available. To circumvent the need for large crystals, considerable effort has been expended to develop more efficient instrumentation. One important development has been the introduction of position-sensitive detectors for neutrons.

Although many of the advantages of position-sensitive detectors for neutron and X-ray work have already been described by other authors[1-3] (and references therein), one which has not been sufficiently appreciated is that these detectors add a second dimension to the data, in the sense that each crystal rotation step is subdivided along the detector axis. The resulting two-dimensional array of intensities permits more accurate extraction of the peak counts from the background, because the boundary between peak and background can be more carefully defined.

[1] H. D. Bartunik and V. Jacobe, *in* "Proceedings of the Neutron Diffraction Conference," p. 430. Reactor Centrum Nederland, Petten, 1975.

[2] U. W. Arndt and A. R. Farugi, *in* "The Rotation Method in Crystallography" (U. W. Arndt and A. J. Wonacott, eds.), p. 219. North-Holland Publ., Amsterdam, 1977.

[3] Ng. H. Xuong, S. T. Freer, R. Hamlin, C. Nielsen, and W. Vernon, *Acta Crystallogr., Sect. A* **A34,** 289 (1978).

We have developed the set of algorithms described here to take advantage of properties of position-sensitive detectors. A more detailed discussion of the following material can be found in Spencer and Kossiakoff.[4]

Introduction to the Method

The method is based on the fact that, as shown in Fig. 1, the reflection boundaries are nearly elliptical in shape. This leads to a relatively simple set of mathematical expressions from which the size, shape, and orientation of the reflection boundaries can be determined solely from the information contained in each data array. One advantage of this approach is that no detailed analysis of the diffraction experiment is required.

The processing of a data array involves three basic operations: (1) background calculation, (2) determination of the reflection center, and (3) calculation of the ellipse orientation and axial lengths. An iterative procedure is used. First, an initial value is determined for each parameter, either by appropriate treatment of the data arrays (for the initial background, reflection center, and ellipse orientation) or from a look-up table (for the initial ellipse size and shape). These preliminary values are then refined to tailor the ellipse to fit the reflection boundary as closely as possible. Figure 2 contains a flow chart outlining these operations as performed in practice. A more detailed outline of the refinement of each parameter is given in the table.

Background Determination

Two passes through the data are used to determine and apply the backgrounds. In the first pass, the background for an individual reflection is determined by an iterative procedure using all elements in its data array not enclosed by the reflection boundary. Errors due to statistical differences among these individual backgrounds are then reduced by applying an averaging procedure, which yields a graph of average background versus position along the detector. In the second pass through the data, values from this background curve replace the individual backgrounds. A new curve is obtained for each level of data.

Since the size and position of the reflection ellipses vary from one data array to the next, it is not possible to determine the background values by simply assigning the same elements in each array as background. Instead, the backgrounds are determined from histograms, such as the one in Fig. 3, which shows the distribution of elements in a data array as a function of intensity.

[4] S. A. Spencer and A. A. Kossiakoff, *J. Appl. Crystallogr.* **13,** 563 (1980).

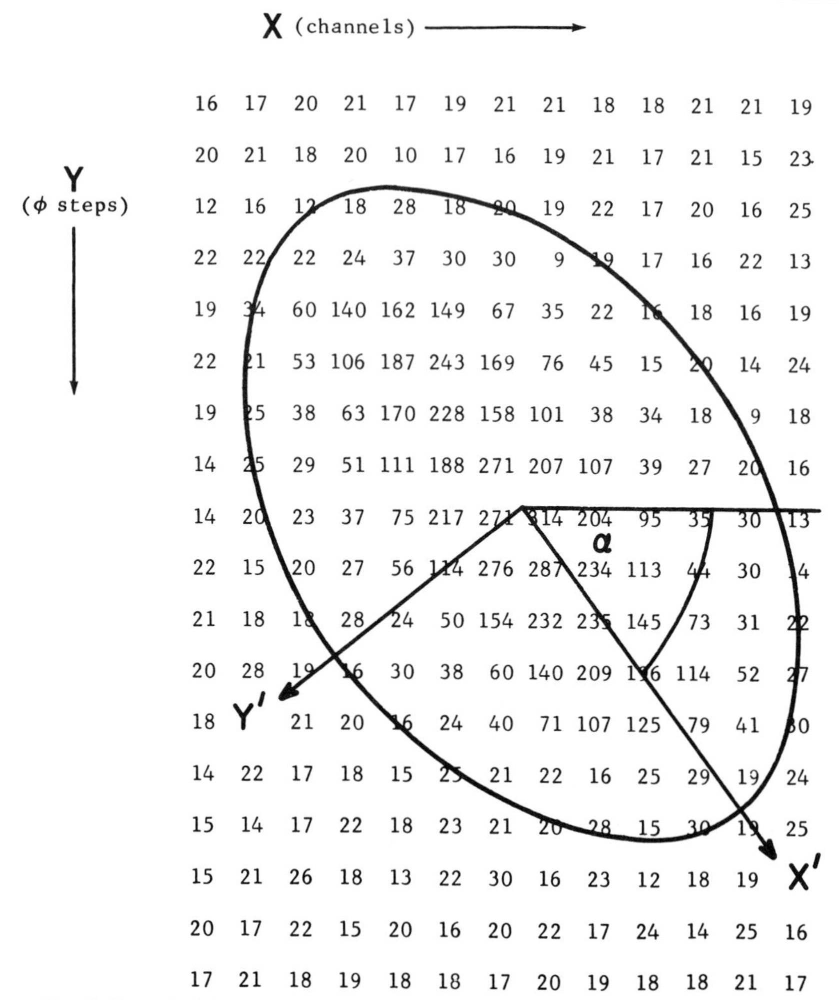

16	17	20	21	17	19	21	21	18	18	21	21	19
20	21	18	20	10	17	16	19	21	17	21	15	23
12	16	12	18	28	18	20	19	22	17	20	16	25
22	22	22	24	37	30	30	9	19	17	16	22	13
19	34	60	140	162	149	67	35	22	16	18	16	19
22	21	53	106	187	243	169	76	45	15	20	14	24
19	25	38	63	170	228	158	101	38	34	18	9	18
14	25	29	51	111	188	271	207	107	39	27	20	16
14	20	23	37	75	217	271	314	204	95	35	30	13
22	15	20	27	56	114	276	287	234	113	44	30	14
21	18	18	28	24	50	154	232	235	145	73	31	22
20	28	19	16	30	38	60	140	209	[illegible]	114	52	27
18	[illegible]	21	20	16	24	40	71	107	125	79	41	30
14	22	17	18	15	25	21	22	16	25	29	19	24
15	14	17	22	18	23	21	20	28	15	30	19	25
15	21	26	18	13	22	30	16	23	12	18	19	
20	17	22	15	20	16	20	22	17	24	14	25	16
17	21	18	19	18	18	17	20	19	18	18	21	17

FIG. 1. Smoothed data array for a moderately intense reflection. The calculated elliptical peak boundary and principal axes (X', Y') are shown. The principal axes originate at the center of mass of the peak and are rotated by the tilt angle, α, relative to the data array axes (X, Y).

These histograms are the sum of two distributions, one composed of elements from the peak and the other of elements from the background. The theoretical shape of the peak distribution depends on the details of the diffraction experiment and is not readily determined. However, the background distribution results from statistical fluctuations about the mean background value, and so the mean background intensity can be

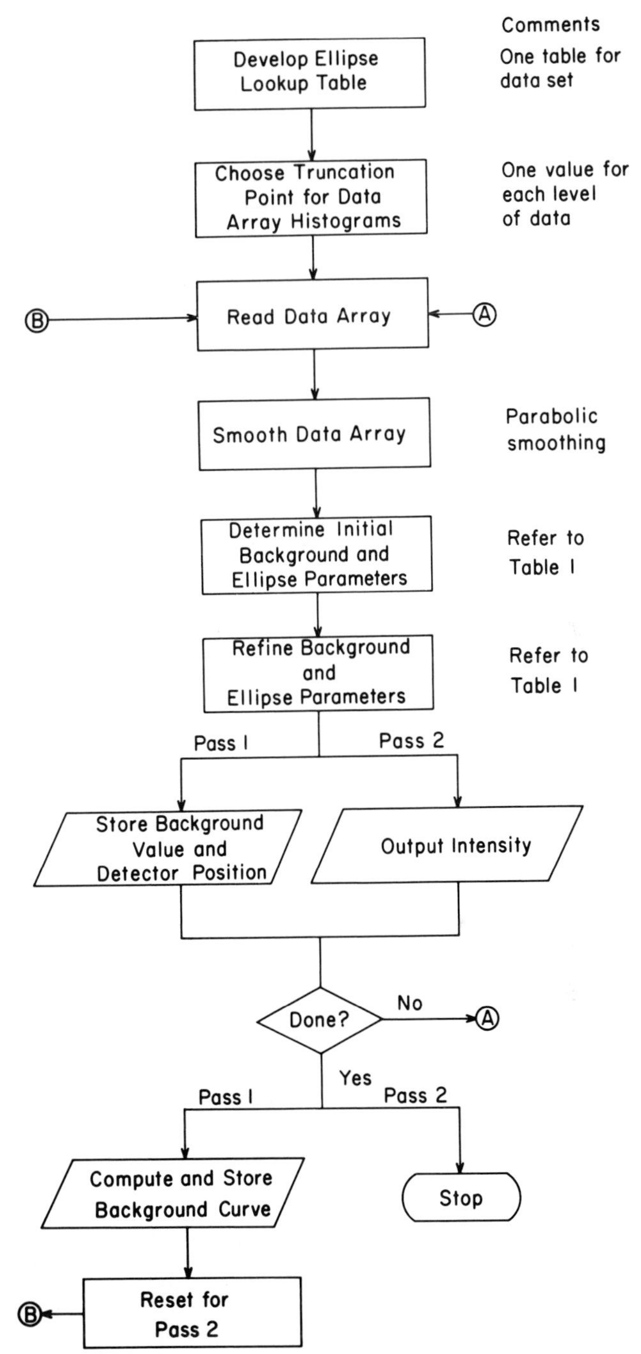

FIG. 2. Flow diagram for the ellipse method. Pass 1 is used to obtain a background curve, which is then used in Pass 2 for computation of the final reflection intensities. The individual steps are described in the text.

Refinement Procedures for Background, Peak Center, and Ellipse Parameters

	Background	Peak center	Ellipse tilt	Ellipse shape	Ellipse size	Intensity
Approach	Data array histogram	Calculate center of mass	Calculate momental ellipse	Calculate momental ellipse	Empirical fit to reflection	Summation
Initial value	Truncate histogram, fit Gaussian	Use σ array elements ≥ 0.5 the largest element	Use σ array elements ≥ 0.5 the largest element	Look-up table	Look-up table	
Refinement	Exclude data array elements within reflection ellipse from histogram, fit Gaussian	Use data array elements within reflection ellipse	Use data array elements within reflection ellipse	Calculate momental ellipse with data array elements within reflection ellipse	Expand size in steps	Sum data array elements within reflection ellipse, subtract background
Constraints	Pass 1 only, background curve used in Pass 2	Restrict center if intensity $>3\sigma$	Limit tilt to within $\pm 10°$ of theoretical value	Elliptical reflection boundary	Limit size if intensity $<3\sigma$	Stop refinement if intensity increases $<0.5\sigma$ or cycle 5

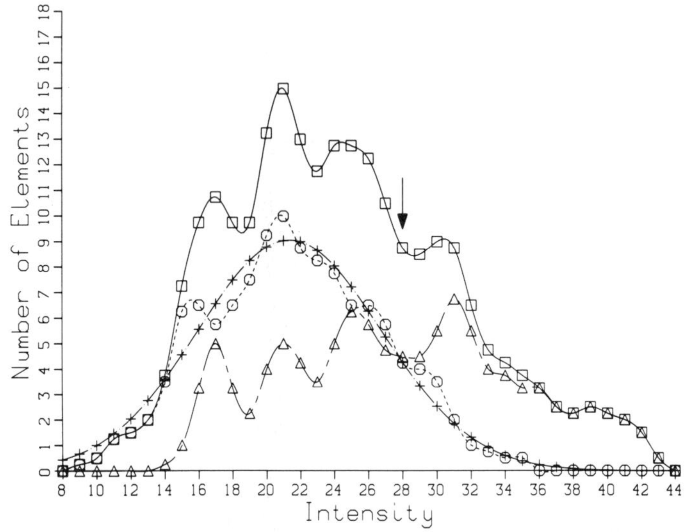

FIG. 3. Data array histogram for a relatively weak reflection. Intensity is in counts per array element. The portion of the total (peak + background) array histogram (□) to the left of the arrow was used to determine the initial background value (see text). The final background (○) and peak (△) distributions are shown, as is the least-squares fit of a Gaussian curve (+) to the background distribution. For an average data array, about one-half the array elements were background.

obtained from the least-squares fit of a Gaussian curve to the background distribution.

Only for very intense reflections are the peak and background distributions sufficiently separated to obtain the background distribution directly; so an iterative process is used to separate them. To initiate this process, a maximum expected value for the background is found by inspection of the data. The data array histogram is then truncated somewhat above this value, thereby reducing the contribution from the peak distribution. The exact truncation point is arbitrary, but must be greater than the mode of the background distribution (see Fig. 3). A Gaussian curve is then fitted to the truncated histogram and an initial background determined. This initial background is normally too large, since at this stage the contribution from the peak distribution has not been completely eliminated. However, it is a sufficiently accurate starting point for the refinement procedure.

For the first refinement cycle, a starting reflection ellipse (described in the Ellipse Parameters section) is placed on the data array at the peak center. All array elements contained within this ellipse are excluded from the array histogram, thereby further reducing the contribution of the peak distribution to the histogram. A Gaussian curve is fitted to this modified histogram and a more accurate background value obtained. With this new background value, new reflection ellipse parameters are calculated. During the next refinement cycle, the array elements within this new ellipse are similarly excluded from the histogram. In this way, as the reflection ellipse is refined to its final size, shape, and position, the background is also refined to its final value.

Once the individual backgrounds are determined, they are arranged as a function of their detector position and backgrounds from reflections in each group are then averaged. The resulting curve is used in the calculation of the final reflection intensities in the second pass through the data.

Determination of Ellipse Location

It has been our experience when dealing with reflections with poor signal-to-noise ratios that the probability of successfully extracting accurate intensity information from the data arrays is highly coupled with the ability to define the peak centers correctly. Ideally, they should occur at their precalculated positions or at least deviate from these positions in a systematic way so that accurate compensation can be made by applying an orientation matrix. Unfortunately, due to instabilities inherent in the experimental setup (crystal slippage, detector position drift, and so forth) neither of these conditions may hold true. It is therefore necessary to devise a way to determine the peak centers on a reflection-by-reflection basis. The method chosen is based on a pattern recognition technique known as a matched filter.

As used here, the matched filter is simply a means of correlating a group of array elements, thereby reducing the statistical variations found for each element alone and emphasizing the intensity differences between peak and background regions of the array. The filter incorporates information about the expected shape and orientation of the reflection, so elements whose relative positions correspond to this pattern are grouped together. This helps to discriminate against correlations not fitting the expected pattern, and considerably reduces the effect of random large background elements on the center determination.

The very simple matched filter shown in Fig. 4a is used to treat the data arrays. It consists of a nine-element array whose shape approximates an ellipse in the expected orientation. Experience has shown that the exact shape of the filter is not overly critical.

In use, the filter is placed in the position shown, the nine included elements are summed, and the background is subtracted. From counting statistics, the number of standard deviations (σ) above background is computed for this difference intensity and stored in a second array. The

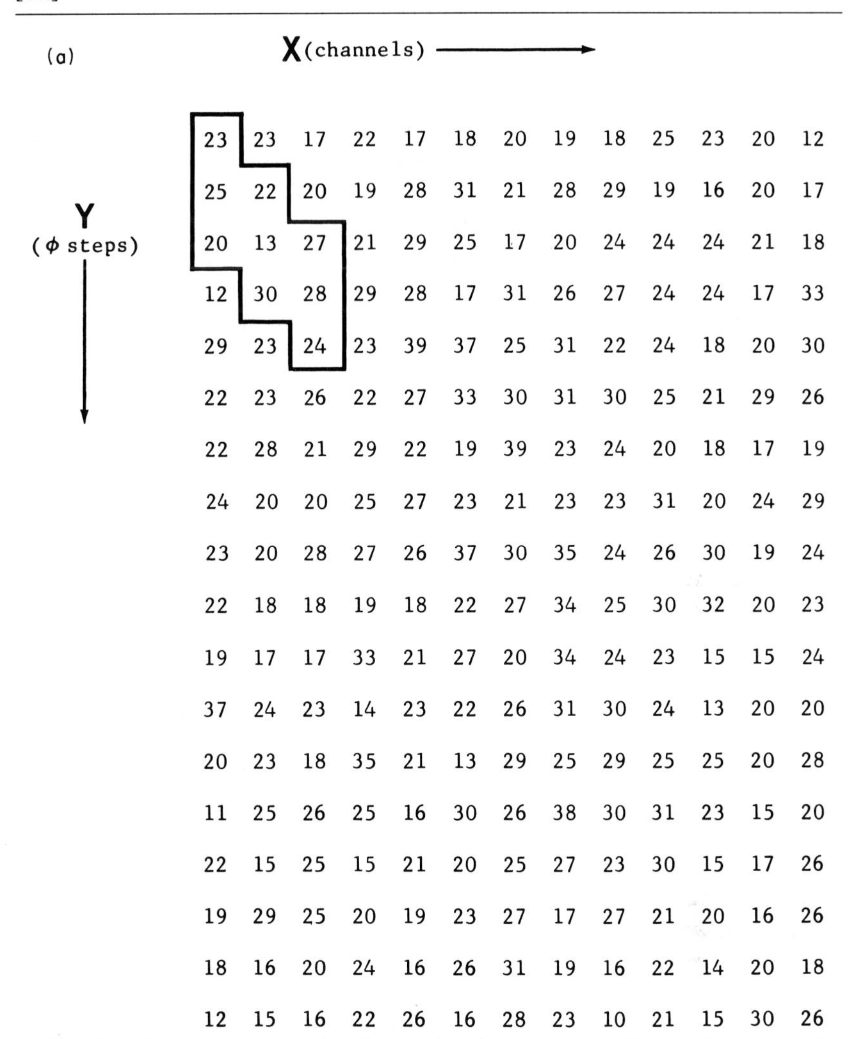

23	23	17	22	17	18	20	19	18	25	23	20	12
25	22	20	19	28	31	21	28	29	19	16	20	17
20	13	27	21	29	25	17	20	24	24	24	21	18
12	30	28	29	28	17	31	26	27	24	24	17	33
29	23	24	23	39	37	25	31	22	24	18	20	30
22	23	26	22	27	33	30	31	30	25	21	29	26
22	28	21	29	22	19	39	23	24	20	18	17	19
24	20	20	25	27	23	21	23	23	31	20	24	29
23	20	28	27	26	37	30	35	24	26	30	19	24
22	18	18	19	18	22	27	34	25	30	32	20	23
19	17	17	33	21	27	20	34	24	23	15	15	24
37	24	23	14	23	22	26	31	30	24	13	20	20
20	23	18	35	21	13	29	25	29	25	25	20	28
11	25	26	25	16	30	26	38	30	31	23	15	20
22	15	25	15	21	20	25	27	23	30	15	17	26
19	29	25	20	19	23	27	17	27	21	20	16	26
18	16	20	24	16	26	31	19	16	22	14	20	18
12	15	16	22	26	16	28	23	10	21	15	30	26

FIG. 4. (a) Data array for a weak reflection showing the matched filter used to process the arrays. In practice, the filter is initially placed in the indicated position, then moved stepwise along the horizontal and vertical directions. At each position, the enclosed elements are treated as described in the text, producing the σ array. (b) Three-dimensional graph of the σ array produced by filtering the data array in (a). The filter size prevents the first and last two rows and the first and last columns from being filtered. Negative values have been set to zero. The small peaks on the periphery of the main peak give an indication of the noise level.

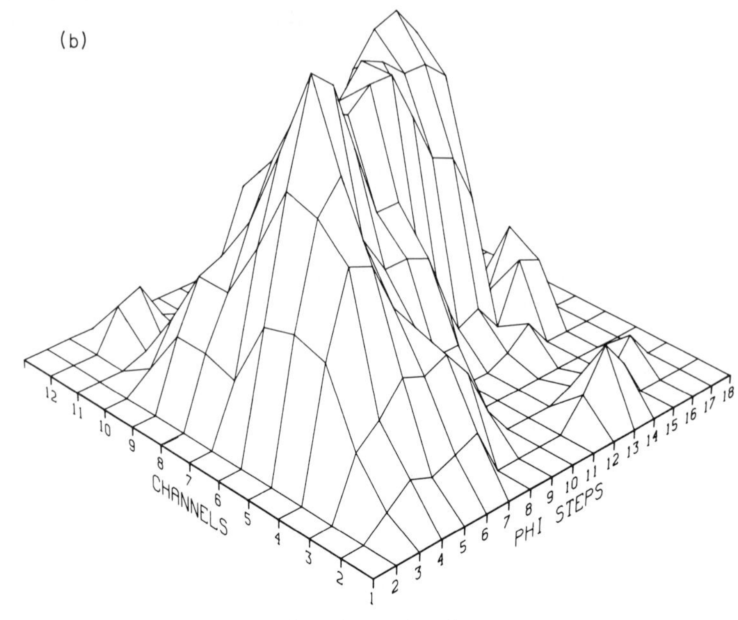

FIG. 4. (*continued*)

filter is then shifted down a row and the process repeated. Systematic movement of the filter both horizontally and vertically is continued until the whole array has been filtered.

The result of this filtering is a new "sigma array" (Fig. 4b) which contains the values of σ calculated at each position of the filter. Since the filter is designed to find elliptically shaped peaks, the σ array contains large values when the filter overlaps the peak and small or negative values when it is over the background region.

The initial peak center is determined from the σ array by computing the center of mass of all elements greater than half the magnitude of the largest element. During subsequent refinement of the size, shape, and position of the reflection ellipse, the center is recomputed as the center of mass of all data array elements (corrected for background) which are contained within the ellipse perimeter for that refinement cycle. Only small movements from the initial center were normally observed.

Ellipse Parameters

The shapes and orientations of the reflection ellipses are determined by borrowing the concept of the momental ellipse from classical mechanics. The momental ellipse for an object has the property that the length of any radius is inversely proportional to the radius of gyration of the object about that radius.[5] In the present case, this means that if the intensity distribution within the reflection is properly symmetric about the center of mass, then the corresponding momental ellipse will conform to the shape and orientation, though not the size, of the reflection ellipse. Hence, the reflection ellipse parameters can be determined by calculating the momental ellipse and scaling it to the correct size.

The equations for the momental ellipse are derived using centric moments, so the results are independent of the position of the reflection ellipse in the data array. The centric moments are calculated using the following expression:

$$M_{pq} = \sum_x \sum_y (x - \bar{x})^p (y - \bar{y})^q I_{xy} \quad (1)$$

where x, y is the peak center, I_{xy} is the background corrected intensity of the element at position x, y in the array, and the sums are over the same array elements as used for determining the peak center. A general form for the momental ellipse can then be written as follows [a derivation of Eqs. (2)–(6) may be found in Lambe[5]]

$$M_{02}x^2 + M_{20}y^2 - 2M_{11}xy = (M_{02}M_{20} - M_{11}^2)/M_{00} \quad (2)$$

This expression can be simplified by choosing a new set of axes for which the cross term disappears. These new axes, the principal axes of the ellipse, are related to the data array axes by a rotation α (Fig. 1), where

$$\tan 2\alpha = 2M_{11}/(M_{20} - M_{02}) \quad (3)$$

In this rotated coordinate system, the momental ellipse equation reduces to

$$M'_{02}x'^2 + M'_{20}y'^2 = M'_{20}M'_{02}/M_{00} \quad (4)$$

where

$$\begin{aligned} M'_{02} &= \tfrac{1}{2}(M_{02} + M_{20}) - \tfrac{1}{2}[(M_{20} - M_{02})^2 + 4M_{11}^2]^{1/2} \\ M'_{20} &= \tfrac{1}{2}(M_{02} + M_{20}) + \tfrac{1}{2}[(M_{20} - M_{02})^2 + 4M_{11}^2]^{1/2} \end{aligned} \quad (5)$$

From (4) the momental ellipse axes are found to be as follows:

$$\begin{aligned} &\text{Semimajor axis} \quad (M'_{20}/M_{00})^{1/2} \\ &\text{Semiminor axis} \quad (M'_{02}/M_{00})^{1/2} \end{aligned} \quad (6)$$

[5] C. B. Lambe, *in* "Applied Mathematics for Engineers and Scientists," p. 80. English Universities Press, London, 1958.

For reflections which are clearly above background, these calculations produce satisfactory values for the orientations and axial length ratios of the reflection ellipses. The momental ellipses then need only to be scaled to the correct size. Unfortunately, problems can arise when weaker reflections are tried. In particular, the scale factor does not remain constant, making it necessary to find a new value for each reflection; and the calculations themselves become unreliable for very weak reflections due to the small intensity differences between array elements from the peak and background.

To overcome these problems, the initial lengths for the reflection ellipse axes are chosen from a short look-up table, and then refined to their final values. The look-up table is designed to make the starting ellipse smaller than the correct reflection ellipse, so the starting ellipse can be refined by using the momental ellipse calculation to modify its shape while simultaneously increasing its overall size in small increments until the proper size is reached. It was found that in this way the best size and shape for the reflection ellipse are obtained. Also, abnormally small or misshapen ellipses for very weak reflections are avoided, since the look-up table provides a reasonable elliptical template for them.

The values in the look-up table are empirically derived from a small set of representative reflections by correlating peak height with final ellipse size; that is, a larger starting ellipse is chosen for the more intense reflections. It was found that a look-up table containing axial lengths for five different starting ellipses gave sufficient flexibility to cover the observed reflection ellipse sizes. Since the table values serve only as initial approximations, their precise magnitudes are not critical, as long as the starting ellipses are smaller than the correct reflection ellipses.

The size of the increment used to enlarge the ellipse during the refinement is chosen based upon the starting ellipse size and the desired number of refinement cycles to reach convergence. It was found that increasing the ellipse axes in increments equal to 30% of their initial lengths gave reasonable results. To allow the axial length ratio to vary from one refinement cycle to the next, the axes are not simply lengthened by fixed amounts. Instead, for each refinement cycle the major and minor axes for the corresponding momental ellipse are calculated from Eq. (6); then these axes are scaled to make the area of the ellipse the same as it would have been if the starting ellipse axes had been lengthened in fixed increments for the same number of cycles. This forces the area of the ellipse to increase for each cycle without restricting its shape.

The test for completion of the refinement process consists of summing the array elements inside the ellipse for the current cycle, subtracting the background, and comparing the resulting intensity to that obtained on the previous cycle. If the intensity increases less than 0.5σ over the previous value, the refinement is halted. A maximum of five refinement cycles is allowed.

For the weaker reflections, the ellipse tilt angle α must be constrained. This is done by restricting the angle calculated from the data array to be within $\pm 10°$ of the theoretical angle α_t. The 10° limit was chosen to allow some flexibility in ellipse orientation while preventing unreasonable excursions.

Treatment of Bias

A potential pitfall of any data reduction method which searches for regions of highest average intensity is that the resulting intensities may be biased toward positive values. The ellipse method was tested for this possibility by generating a set of data arrays containing random numbers normally distributed around a known mean, and then processing these arrays in the normal way to see whether the resulting intensities were evenly distributed around zero, indicating no bias, or around some positive value.

The results of this test indicated that two sources of bias existed. The major source of bias came from allowing an unrestricted choice of the ellipse center. This bias affected only the very weak reflections for which there was no detectable center, since this forced the centering algorithm to choose a random region of higher than average intensity as the center. The magnitude of this bias was reduced by restricting the centers of the weak reflections (those under 3σ) to be within three array elements of the average center for the stronger reflections in the same segment of data. The bias could have been totally eliminated by fixing the centers for these reflections at the average center, but this would have increased the risk of miscentering some reflection ellipses. The limit of three array elements was chosen as a reasonable compromise between these conflicting requirements.

A relatively small source of positive bias arose from choosing the final ellipse size based on the intensity above background of the enclosed array elements. Since the ellipse was allowed to expand until the intensity increased less than 0.5σ, this bias tended to be greatest for the weaker reflections, where the standard deviations for the peak and background were of comparable magnitudes. This bias was reduced by limiting the maximum ellipse size for reflections under 3σ. An alternative would have been to increase the intensity limit above 0.5σ, but this might have prevented the fringe elements of strong reflections from being included in the intensity calculation, thus biasing these intensities toward lower values.

The ellipse program was tested with the random data arrays after incorporating the limits on ellipse size and placement, and it was found that a positive bias of 0.4σ still existed. This bias was reduced to 0.1σ by preventing the ellipse centers from moving and totally eliminated by also using a fixed ellipse size. However, it was felt that the errors introduced by placing additional restrictions on the ellipse size and placement could be larger than any errors attributable to a bias of 0.4σ in the weaker reflections. Therefore, no attempt was made to reduce the bias below this level.

Comparison to Other Methods

To assess the merits of the ellipse method, a comparison was made between data processed as described above and the same data processed by some alternative techniques. The criteria for this comparison were the reproducibility of intensities for symmetry-equivalent reflections and the relative statistical significance of the data processed by each method. For this test, the *h*-5 and *h*-11 data levels were chosen as representative of the entire data set and were combined for presentation here.

Of the four methods used in this comparison, three relied upon various aspects of the ellipse technique, while the fourth used a scheme analogous to the standard ω scan. In Method 1, the complete ellipse treatment described in this chapter was used. Backgrounds for individual reflections were determined during the first pass through the ellipse fitting program, averaged as a function of detector position and reflection level, and then used together with the ellipse-fitting algorithm to determine final intensities during the second pass through the program.

Method 2 was an abbreviated version of Method 1, in which the reflection intensities were determined during a single pass through the program. The ellipse-fitting algorithm was used, but individual backgrounds replaced the averaged values used in Method 1.

In Method 3, backgrounds were obtained and averaged as in Method 1. However, instead of fitting an elliptical profile to the peak, the final intensity was obtained by integrating the whole data array and subtracting the background.

For Method 4, backgrounds were obtained by integrating the first and last two rows of the data array, while the remaining 14 rows were integrated to give the peak. This resembles a standard ω scan, in which backgrounds are taken several steps before and after the peak scan.

The relative statistical significance (expressed as number of σ above background) for data processed by each of the four methods is illustrated in Fig. 5. From these curves, it is apparent that Method 4 was the least

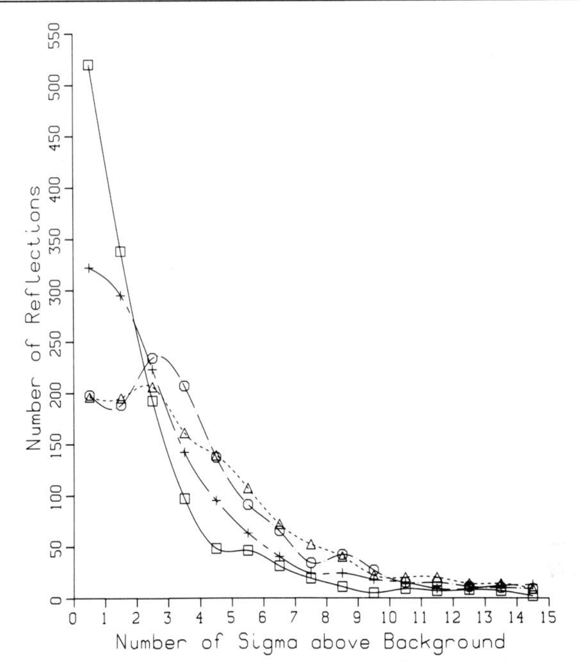

FIG. 5. Histograms showing the relative statistical significance of data processed by the four methods described in the text. Each point is the sum of all reflections within ±0.5 of the indicated value on the abscissa. (△) Method 1; (○) Method 2; (+) Method 3; (□) Method 4.

reliable technique, due to the relatively crude way in which the peaks and backgrounds were determined. Using either the averaged backgrounds (Method 3) or the elliptical reflection boundary (Method 2) substantially improved the results, although the latter produced the greater improvement. Surprisingly, combining both of these techniques (Method 1) produced only a small additional improvement compared to Method 2 alone, perhaps because errors in the backgrounds become less significant once the many extraneous background elements are excluded from the peak by the elliptical reflection boundary.

Figure 6 contains graphs of R_{sym} versus intensity for the symmetry-equivalent reflections. From these graphs, it can be seen that the background averaging and elliptical reflection boundaries affect mainly the small and intermediate size reflections (intensity groups 1–6), and that the four methods are essentially equivalent for large reflections (intensity group 7). For intermediate reflections (intensity groups 5 and 6), errors in the backgrounds seem to be the main factor affecting R_{sym}. Thus, Methods 1 and 3, which used the averaged backgrounds, produced significantly

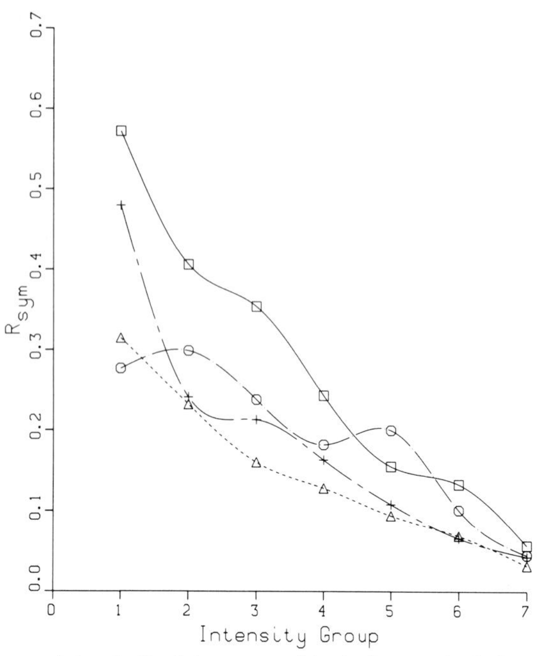

FIG. 6. R_{sym} versus intensity for data processed by the four methods described in the text.

$$R_{sym} = \sum_{hkl} \sum_{i=1}^{N} |I(hkl)_i - \langle I(hkl) \rangle| \Big/ \sum_{hkl} \sum_{i=1}^{N} |I(hkl)_i|$$

where the inner summation is over all measurements of symmetry-equivalent reflections. Each intensity group contains the same reflections for all four methods. Intensity groups were chosen, based on Method 2 intensities, as follows: group 1, 50–100 counts above background; 2, 100–150 counts; 3, 150–250 counts; 4, 250–350 counts; 5, 350–500 counts; 6, 500–1000 counts; 7, over 1000 counts. Symbols are the same as in Fig. 5.

lower values of R_{sym} than Methods 2 and 4, which used individually determined backgrounds. For smaller reflections (intensity groups 1–4), the elliptical reflection boundary assumes greater importance, and by using both the background averaging and the elliptical reflection boundary (Method 1), it is possible to reduce R_{sym} below the values achieved using either alone. These results, together with those presented in Fig. 5, demonstrate that the complete ellipse method (Method 1) is decidedly superior to the other techniques when dealing with small to intermediate reflections.

The reader should be aware of other recent efforts to improve data accuracy. Sjölin and Wlodawer[6] have devised a variant of the technique

[6] L. Sjölin and A. Wlodawer, *Acta Crystallogr., Sect. A* **A37,** 594 (1981).

described here which they call the "dynamic mask" procedure. Their method defines the size, shape, and orientation of a reflection peak strictly on a statistical basis, and makes no prior assumptions about the expected geometric properties of the peak based on the conditions of the diffraction experiment. The background intensity is obtained from a "universal" background curve which is collected simultaneously with the intensity data. The backgrounds which are applied are the average of many background observations taken in equivalent regions of reciprocal space, and thus are of high statistical quality. Schoenborn has recently developed a procedure which predicts the size and orientation of a reflection peak using measured crystal and beam characteristics. This procedure holds promise as a method to accommodate easy on-line data processing.[7]

Application of the Method

The results of the above comparison indicate that an improvement in both the statistical reliability and reproducibility of the data is attained with the ellipse procedure. However, the most appropriate evaluation of the data quality can be made by using the processed data for a crystallographic analysis. For this test, the 2.2 Å trypsin neutron data set is ideal due to the low average intensity of the diffraction data. Since there are rather substantial differences between X-ray and neutron methods in the application of refinement techniques and evaluation of phasing models, a brief description of the steps involved in structural analysis is presented in the following sections.

Starting Phasing Model

The initial phasing model was calculated by applying the appropriate neutron scattering lengths to the refined X-ray monoisopropylphosphoryl-trypsin coordinates of Chambers and Stroud.[8] The resulting structure factor calculation for the neutron data, excluding all hydrogen atoms and water molecules, gave an R of 0.304.

Considering that nearly one-half of the total atoms (hydrogen) in the structure were omitted from the structure factor calculation, this statistic indicated a strong correlation between the neutron amplitudes and the starting coordinate set. Perhaps more encouraging, however, was the amount of interpretable information contained in the resulting difference map.

[7] B. P. Schoenborn, *in* "Neutrons in Biology," Basic Life Sciences 27 (B. P. Schoenborn, ed.), p. 261. Plenum, New York.

[8] J. L. Chambers and R. M. Stroud, "Protein Data Bank." Brookhaven Natl. Lab., Upton, New York, 1977.

Addition of Hydrogen and Deuterium Atoms to the Model

Each residue in the structure was inspected on an individual basis using Fourier and Fourier difference maps. Where the residue was deemed "well ordered," the hydrogen atoms that were in stereochemically constrained positions were added to the model. If a region of the residue showed disorder (or misplacement), the hydrogens pertaining to that portion were omitted. Hydrogen temperature factors were assigned as $B + B^{1/2}$, where B is the temperature factor of the parent atom.

In a similar manner, deuterium atoms were searched for in difference maps by looking at individual residues which contained potential sites of exchange. The criterion for assigning a deuterium position was that there be significant positive density (one-half of the expected density) in the difference map at the predicted atomic center.

Refinement

Refinement of atomic coordinates was carried out using the constrained difference Fourier technique. After each cycle of coordinate shifts, the structure was rebuilt to ideal bond lengths and bond angles and to a lower global energy using a program developed by Hermans.[9] The calculated function which was minimized for each atom was

$$F_1 = K_x \Sigma \tfrac{1}{2}(X - X_0)^2 + K_l \Sigma \tfrac{1}{2}(l - l_0)^2 + K_\theta \Sigma \tfrac{1}{2}(\theta - \theta_0)^2 + K_\rho \Sigma \tfrac{1}{2}(\rho - \rho_0)^2 + K_e(-A/r^6 + B/r^{12} + C/r) + K_d \Sigma \tfrac{1}{2}E_b[1 - \cos n(\phi - \phi_0)]$$

where the energy contributions are derived from (a) approach to initial coordinates, (b) bond lengths, (c) bond angles, (d) fixed dihedral angles, (e) nonbonded energy, and (f) torsional potential.

Insofar as most close contact regions in a protein involve hydrogen–hydrogen interactions, the last two terms in the energy refinement, the nonbonded and torsional energies, are of special importance in neutron structure analysis. In situations in which the methyl rotors appeared to adopt high-energy conformations, the weight of the torsional energy parameter was set to rotate the group only 10°–15° toward its energetically preferred orientation. A rotation of this limited magnitude allowed the rotor to readjust back toward its initial position if subsequent difference maps continued to show a strong preference for the higher energy conformer.

The nonbonded energy function used in the refinement consisted of three terms. The first two terms are equivalent to the Lennard–Jones "6–12" potential, in which coefficients A and B represent the attractive and

[9] J. Hermans and J. E. McQueen, *Acta Crystallogr., Sect. A* **A30**, 730 (1974).

repulsive components of the interaction. The third term expresses the electrostatic interaction between the atoms calculated as the monopole potential between the partial charges.[10] When a group of atoms was found to be involved in hydrogen bonding, the coefficients A and B of the donor and acceptor atoms were adjusted to redefine the resulting potential energy surface. Rather than including a separate set of A and B values for each type of hydrogen bonding couple, an average value for A and B was selected and applied in all cases.

Difficulties Presented in Refinement

In any protein structure refinement there exists a number of inherent difficulties: the large number of atoms in the structure, the limited resolution, errors in the model, and so forth. However, neutron structures present several unique problems not encountered in X-ray analysis. These special problems predominantly arise from the close proximity of hydrogen atoms to their parent atoms, coupled with the effects of the negative scattering length of hydrogen atoms. Clearly, potential problems exist when the difference density generated from positional errors of one atom overlaps an adjacent atom site. This situation is further complicated by the fact that, because of this negative scattering length, an error in a hydrogen atom position is minimized by moving the atom down the difference density gradient, that is, opposite to the direction required for correcting parent atom positions.

The success of the curvature-gradient refinement technique is predicated on the assumption that since the measurement of the gradients is restricted to a small range around the position of the atom, the influence of the neighboring atoms is largely minimized. However, at the resolution to which most protein data are collected, there exists a real danger that gradient values can be adversely affected by the influences of proximate atoms. In viewing this problem, there was concern that at the resolution of our maps, productive refinement would be difficult, or perhaps even impossible, because of the short hydrogen–parent atom bond distances.

To evaluate the refinement capabilities of the technique, a test was devised in which all the coordinates in the model were perturbed by a varying but known amount from their ideal positions, and a difference synthesis was computed using a set of 2.2 Å data equivalent to the collected trypsin data. It was determined that, in general, dependable shifts could be obtained when the coordinate errors were less than 0.3 Å. (A dependable shift was defined as one which was in the right direction and

[10] D. Poland and H. A. Scheraga, *Biochemistry* **6,** 3791 (1967).

would likely converge to the proper atomic position in three cycles or less.) Also, it was found that if a parent atom (an atom with one or more hydrogens attached to it) was displaced by over 0.6 Å from its correct position, the effect of the neighboring hydrogens rendered the gradient at the parent atom center inaccurate. This problem was not investigated exhaustively. However, this test did give helpful general indications of what kind of response could be expected from the technique in certain defined situations, and it clearly emphasized the necessity that a neutron analysis begin with a well-determined X-ray model.

The following subsections discuss examples of specific refinement problems with brief explanations of how they were handled.

Constraining Hydrogen Atom Temperature Factors. When two or more hydrogen atoms were bound to the same parent atom, their individual temperature factors (B factors) were adjusted toward parity, since it is hard to rationalize a situation in a refined structure in which two stereochemically constrained hydrogens, bound to the same parent atom, would have substantially differing temperature factors. The adjustment was performed by calculating the average B of the atoms involved, and then readjusting the value of each B back to one-third the difference between the average and its original value. This scheme effectively reduces unwarranted discrepancies between structurally similar hydrogen atoms, but does allow some influence by the independently arrived at temperature factor value. It was found that moderate adjustments toward equalization, extended over several cycles, constituted a more effective procedure in arriving at the stabilized set of B values than one which simply sets them to the average value in a single step.

Refinement of Partial Occupancy Deuterium Atoms. Assigning a proper temperature factor to a deuterium atom proved to be quite difficult because, unlike an unexchanged hydrogen atom which can be treated as an extension of the parent atom, the true density for the deuterium is affected by both an occupancy and a temperature factor. There is no straightforward way to deal easily with this problem at the current resolution of the data (2.2 Å); accordingly, a reasonable guess for the temperature factor was made and it was allowed to converge to its desired value during refinement. At present, our refinement program adjusts only temperature factors. A more satisfactory approach would be to assign a temperature factor derived from that of the parent atom and refine the occupancy factor. The actual effect on the phasing model of the present mode of refinement is probably quite small, since the only atoms which suffer from this treatment are those of low occupancy.

Refinement of Water Molecules. Each deuterium atom of a water molecule (D_2O) has the potential to scatter at a magnitude comparable to the parent oxygen. It is therefore possible, in principle, to establish the coordination geometry of the well-ordered waters from the density features in a Fourier map. Very helpful in this respect is the knowledge of the oxygen atom position from the X-ray analysis. The oxygen can be subtracted from the map by a difference synthesis, in effect isolating the scattering density for the deuterium atoms alone.

Before water molecule refinement was initiated, a significant percentage of the largest peaks in the "oxygen removed" difference map could be attributed to these isolated deuterium atoms associated with D_2O molecules. The difference density was usually not characterized by two distinct lobes, but rather by a single broad asymmetric peak with the oxygen position located on the periphery. After studying several approaches to orienting the D_2O molecules using both empirical and energetic criteria, it became apparent that it was not realistic to pursue detailed conclusions of this nature at the current resolution. However, these attempts did indicate that at an extended resolution, the orientation of D_2O molecules should be readily obtainable from the maps.

Determination of the Protonation States of the Catalytic Residues

A primary goal of the trypsin analysis was to resolve a much debated mechanistic issue concerning the identification of the residue (Asp-102 or His-57) which functions as the catalytic group during the transition state of the hydrolysis reaction.[11,12] Many other physicochemical methods have been employed in attempts to resolve this issue, but the results have been contradictory. We felt that the most direct method to resolve the issue would be to locate directly the proton in the Asp-102, His-57 hydrogen bond using neutron diffraction.

Since a major objective of the study was to identify conclusively the group acting as the base, it was important that the interpretation of the proton positions in the catalytic site be unambiguous. Toward this end, three methods were devised to test the preferred location of proton H-1 (proton in the Asp–His H bond). In the first method, the deuterium (proton H-1 was exchanged for deuterium during the soaking procedure) was omitted from the model and the full structure refined through one cycle of coordinate shifts and model reidealization. This refinement cycle was run to eliminate any bias in the phases that may have been introduced by a prior placement of this atom in the model. Figure 7a shows the result of the difference synthesis calculated from these structure factors and is

[11] A. A. Kossiakoff and S. A. Spencer, *Nature (London)* **288,** 414 (1980).
[12] A. A. Kossiakoff and S. A. Spencer, *Biochemistry* **20,** 6462 (1981).

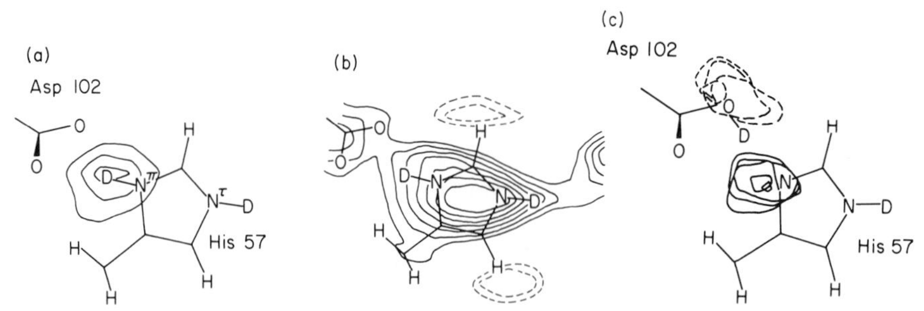

FIG. 7. (a) A difference map $[(F_o - F_c) \exp(i\phi_c)]$ calculated with only the deuterium (H-1) between the His-57 and Asp-102 side chains left out of the phases. The difference peak is approximately 5.5σ above the background level and shows the deuterium to be bound to the imidazole nitrogen. (b) A Fourier map computed with terms $(2F_o - F_c) \exp(i\phi_c)$. In this map, both of the catalytically important deuteriums (bound to the imidazole) were omitted from the phases. It is clear from this map that both are located on the imidazole. (c) A difference map in which the deuterium was placed by stereochemistry on atom $O^{\delta 2}$. The difference density peak clearly indicates that the preferred location of the deuterium is the imidazole of His-57.

interpreted as showing a distinct preference for the deuterium to reside on the imidazole.

In the second method, a deuterium atom was placed by stereochemistry on the nitrogen of the imidazole and its position refined for two cycles. Throughout the refinement, the deuterium atom exhibited stable refinement characteristics and a relatively low temperature factor, indicating that it was tightly bound to the parent N^{π} atom. The same process was repeated but with the deuterium atom placed in its alternative position on the $O^{\delta 2}$ of the Asp-102 side chain. In this case the outcome was quite different; the refinement was unstable with respect to both positional and thermal parameters and the deuterium atom resided along a distinct gradient, a clear indication of misplacement.

The third method involved the qualitative examination of two independent difference maps. For the first map, structure factors were calculated with the deuterium atom on the N^{π} of His-57. In agreement with the refinement results, a nearly featureless map was produced, signifying that this placement was in close accord with the data. In contrast, the placement of the deuterium on the $O^{\delta 2}$ of Asp-102 resulted in the interpretable difference map shown in Fig. 7c. The directionality indicated by the difference peak is distinctly toward the imidazole ring.

The results from each of these three techniques independently support the conclusion that the mechanistically important proton is coordinated to the imidazole of His-57. Taken together, we believe they offer compelling proof that, at physiological pH, the protonated species in the tetrahedral intermediate is the imidazole of His-57 rather than the carboxylate of Asp-102.

Conclusion

The trypsin analysis reported here was performed with data collected from a crystal over an order of magnitude smaller than any crystal previously used for neutron protein crystallography; yet the refinement statistics and the wealth of information obtained from the resulting maps clearly indicate that the data are of high quality. It has been shown that even at 2.2 Å the locations of hydrogen and deuterium atoms can be determined. This is a clear demonstration of the accuracy of the data obtained by the data reduction methods described above. The usability of crystals previously considered too small for neutron studies, coupled with the number of detailed structural features which can be evaluated at a resolution above 2.0 Å, should dramatically increase the number of protein systems which can be studied profitably by neutron diffraction.

Acknowledgments

This research was carried out at Brookhaven National Laboratory under the auspices of the U.S. Department of Energy. The authors wish to thank Drs. Barry Nelson and Alexander Kossiakoff for their valuable suggestions during the course of this work. We also wish to acknowledge our indebtedness to Drs. James Cain and Benno Schoenborn, who designed and developed the protein crystallographic station at the Brookhaven High Flux Beam Reactor.

[33] Neutron Diffraction: A Facility for Data Collection and Processing at the National Bureau of Standards Reactor

By Alexander Wlodawer

One of the important reasons for the relatively slow pace of the development of neutron protein crystallography is the low flux available from the neutron sources. The highest flux of 2×10^8 neutrons/cm^2-sec was reported in the sample position at the Institut Laue-Langevin high-flux reactor.[1] The flux estimates at the Brookhaven biology beam line were

[1] G. A. Bentley, E. D. Duee, S. A. Mason, and A. C. Nunes, *J. Chim. Phys.* **76,** 817 (1979).

5×10^7 neutrons/cm^2-sec,[2] while the flux obtained on a medium-flux reactor at the National Bureau of Standards (NBS) was only 6×10^6 neutrons/cm^2-sec.[3] While these numbers are not directly comparable since they depend on the neutron wavelength, beam divergence, monochromator type, etc., it is clear that they are much lower than available X-ray intensities. Phillips *et al.*[4] estimated 3×10^{10} photons/cm^2-sec as a flux generated by a sealed X-ray tube, 1.6×10^{11} photons/cm^2-sec obtained using a GX-6 rotating anode generator, and 1.2×10^{13} photons/cm^2-sec as a flux at the Stanford synchrotron source, which was in the early stages of development. The fluxes are even higher for the new synchrotron sources, and in any case the highest neutron flux is many orders of magnitude smaller than the available X-ray fluxes. For these reasons, the neutron data collection facilities have to be highly optimized in order to be useful at all. This was quite obvious during the design of a neutron diffractometer for the National Bureau of Standards reactor, where the neutron flux is lower by an order of magnitude than at the two high-flux facilities.

For reasons of technical difficulties and of cost, we chose to use a linear position-sensitive detector (PSD) in the construction of a diffractometer which could hasten the data collection by at least a factor of 10 over the standard four-circle diffractometer. Higher efficiency was possible with an area detector,[5] but such counters were still in the development stage, while linear PSDs for both X rays and neutrons were available commercially. Area detectors were also at least a factor of 10 more expensive than the linear PSDs.

The details of the design and construction of the NBS neutron diffractometer have been described by Prince *et al.*[6] and will not be repeated here. Only a general description of the instrument will be provided. It should be stressed that, while the NBS diffractometer was built specifically for neutron diffraction, a similar design (incorporating commercially available detectors) can be easily adapted for X-ray diffractometers as well.

In order to utilize the properties of the linear PSD, it is necessary to use a diffraction geometry which causes the diffracted beams for a large number of reflections to appear in a common plane. The most straightforward and efficient way to accomplish this objective is to use flat-cone geometry, which was one of the Weissenberg methods first studied and

[2] A. A. Kossiakoff and S. A. Spencer, *Biochemistry* **20,** 6462 (1981).

[3] A. Wlodawer, *Acta Crystallogr., Sect. B* **B36,** 1826 (1980).

[4] J. C. Phillips, A. Wlodawer, J. M. Goodfellow, K. D. Watenpaugh, L. C. Sieker, L. H. Jensen, and K. O. Hodgson, *Acta Crystallogr., Sect. A* **A33,** 445 (1977).

[5] U. W. Arndt, *Acta Crystallogr., Sect. B* **B24,** 1355 (1968).

[6] E. Prince, A. Wlodawer, and A. Santoro, *J. Appl. Crystallogr.* **11,** 173 (1978).

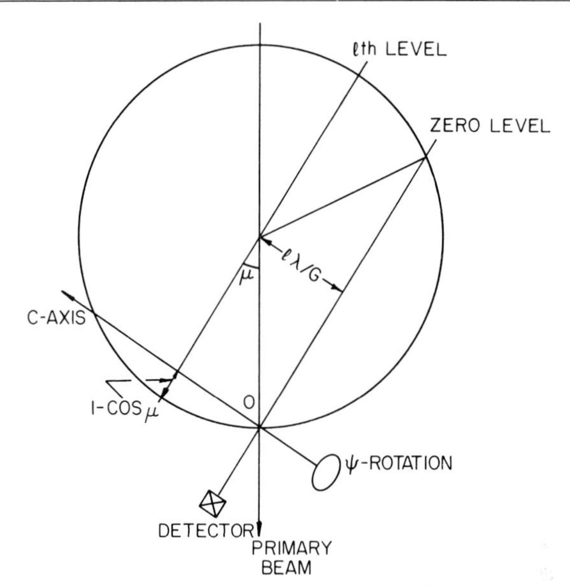

FIG. 1. Schematic representation of flat-cone geometry. The rotation axis in the figure is c, and the lth level is in the flat-cone position. [By permission of *J. Appl. Crystallogr.* **11,** 173 (1978).]

applied by Buerger.[7] With this method the crystal is mounted so that it may be rotated around an axis normal to a set of layers of the reciprocal lattice (see Fig. 1), for example, the c axis. If this axis makes an angle of $90° + \mu$ with the incident beam, where $\sin \mu = \lambda l/c$, the center of the sphere of reflection will lie in one of the reciprocal layers, and as the crystal is rotated about the c axis, the lattice points in that layer will intersect the reflection sphere on a great circle. If the detector lies in the plane parallel to the great circle passing through the center of the sample, it will intercept all the diffracted beams from that layer.

The flat-cone geometry, however, has two main disadvantages: (1) for upper levels ($\mu \neq 0°$), there is a circular region of reciprocal space of radius $1 - \cos \mu$ which is not accessible to the diffractometer and (2) if, because of crystal symmetry, the chosen zone axis is also a reciprocal lattice row, the flat-cone geometry, like the equi-inclination and normal-beam geometries, has the undesirable property of intrinsic simultaneous diffraction.[8,9] It has been shown[10] that, for certain crystals, simultaneous

[7] M. J. Buerger, "X-ray Crystallography," p. 301. Wiley, New York, 1942.

[8] R. D. Burbank, *Acta Crystallogr.* **19,** 957 (1965).

[9] A. Santoro and M. Zocchi, *Acta Crystallogr.* **21,** 293 (1966).

[10] W. H. Zachariasen, *Acta Crystallogr.* **18,** 705 (1965).

diffraction causes significant systematic errors in intensity measurements. On the other hand, in studies of complex crystals such as proteins, in which the PSD can be most efficiently used, the errors due to this effect usually seem to be small enough to be neglected.[11]

If the rotation around the zone axis is performed by a single circle (this axis is designated ω in the Weissenberg camera; it will be termed ψ here to avoid confusion with the ω axis of a four-circle diffractometer), a remounting of the crystal will be necessary to measure reflections in the blind region of the reciprocal lattice or to measure reflections particularly affected by simultaneous diffraction. If, however, the crystal rotation is performed by using the three independent angles of a four-circle diffractometer, measurements may be made about more than one axis with a single mounting of the crystal, and the disadvantages mentioned above may be circumvented. Of course, all such rotation axes must lie within a cone around the ψ axis, and this cone should have a semiangle of no more than about 45°.

For these reasons a PSD should be mounted on the detector arm of a four-circle diffractometer in a vertical plane which includes the diffractometer axis. The arm then determines the angle μ, and the ϕ, χ, and ω axes of the diffractometer can be used to rotate the crystal around an axis perpendicular to the plane defined by the detector and the crystal. The diffractometer settings necessary to bring successive levels into diffracting positions can be calculated using the formulas provided by Prince *et al.*[6] and Busing and Levy.[12]

The flat-cone neutron diffractometer has been constructed as a modification of an existing four-circle instrument (Fig. 2). A single counter was replaced on the 2θ arm by a vertically mounted, 1 m long, ^{3}He PSD (2.5 cm diameter). The counter shield (35 × 35 cm in cross section) and its support were the only items specially manufactured for this project. The detector can be pivoted around its center to depart from the vertical setting for up to 30°. With the detector vertical, the distance D between the sample and detector can be varied from 70 to 200 cm. Thus for an incident wavelength of 1.67 Å, D set to 100 cm at $\alpha = 90°$, and then α moved to 60°, data can be collected up to 2 Å resolution for the zero level (higher for upper levels). If needed, the maximum resolution can be increased by using neutrons with shorter wavelength, but this may lead to problems with resolving the reflections. The detector may also be raised in such a way that it will no longer intersect the horizontal plane (although some low-order reflections will become inaccessible in this configuration).

[11] D. C. Phillips, *J. Sci. Instrum.* **41,** 123 (1964).

[12] W. R. Busing and H. A. Levy, *Acta Crystallogr.* **22,** 457 (1967).

FIG. 2. Flat-cone neutron diffractometer installed at the National Bureau of Standards reactor.

The effective aperture of the detector can be changed by adjusting the two parallel cadmium masks on the front face of the shield. The width of the opening can be experimentally adjusted to minimize the background while leaving it wide enough to pass all neutrons belonging to the reflections from a level under investigation.

The electronic circuits have been described by Alberi *et al.*[13] The output from each end of the detector wire is fed into a delay-line-clipped

[13] J. L. Alberi, J. Fisher, V. Radeka, L. C. Rogers, and B. P. Schoenborn, *IEEE Trans. Nucl. Sci.* **NS-1,** 255 (1975).

linear amplifier through a charge-sensitive preamplifier. The output of one amplifier (the "numerator signal") is fed directly to a linear gate and stretch (LGAS) circuit which produces a flat-topped output signal with a height equal to the highest point of the amplifier output and a width of 9 μsec. The outputs of both amplifiers are fed to a summing circuit, and its output (the "denominator signal") is fed to another LGAS circuit. The outputs of the two LGAS circuits are then fed into a custom digital divider circuit, which consists of two fast analog-to-digital converters (ADC) and a processing unit. The resulting channel numbers are stored in the computer's memory. The uncertainty, due to electronic circuits, is ± 1 channel out of a total of 512. The computer used for instrumental control is a PDP 11/44 with 640K of core memory. This computer controls eight other spectrometers in addition to the flat-cone diffractometer and is used for program development. It also functions as a remote job-entry terminal to a large computer.

The 9-μsec output pulse width of the LGAS circuit accurately defines the dead time of the counting system. Counting-loss corrections, which amount to 1% with a count rate in all channels of 1 kHz and 10% if the count rate reaches 10 kHz, are made (in the first approximation) by a specialized circuit which stops the system clock when the counting chain is busy. So far, we have not experienced any difficulties caused by high count rates.

One of the advantages of this design for a neutron diffractometer equipped with a linear PSD is that it can be used in the four-circle mode without any hardware modifications. This can be done by simply disregarding all the data falling outside of a small number of channels spanning the horizontal plane. This option is advantageous for several reasons. First, the initial orientation matrix can be found and refined in a standard way. Second, the reflections falling in the blind region of the flat-cone geometry can be easily collected without any need for remounting the crystal. Third, selected reflections can be counted in both flat-cone and four-circle modes and can be used for testing the hardware and the software.

This instrument was indeed initially used in the four-circle mode to collect data extending to 2.8 Å from a crystal of ribonuclease A (space group $P2_1$, $a = 30.18$ Å, $b = 38.4$ Å, $c = 53.32$ Å, $\beta = 105.85°$). Each reflection was measured by an ω scan 1.6° wide in 64 steps, with 10 steps on each side used as background. Reflections to 3.2 Å resolution were each counted for 10 min, and reflections to 2.8 Å for 20 min. Two standard reflections were monitored with no change in intensity throughout data collection. A substantial number of Friedel pairs were measured yielding $R = 4.4\%$ upon merging. Data were corrected for absorption, but the influence of neutrons with half-wavelength was neglected. The number of observed reflections ($F > 2\sigma$) was 2773, which was 87% of all reflections at 2.8 Å resolution.[3]

Another data set to 2 Å resolution was collected with the diffractometer operated in flat-cone geometry. Each layer of reflections was collected by rotation of the crystal around the *a* axis in order to limit the number of necessary scans by measuring the msot densely populated planes. A complete 360° scan in ψ increments of 0.09° took 60 hr. Each general reflection was measured twice, and special reflections, four times. An example of a frame of the detector output is shown in Fig. 3.

With the instrument operating in the flat-cone mode, all data falling in the vicinity of each reciprocal lattice point have been saved on magnetic tape for future processing. Data collection with a linear PSD gives two-dimensional images of the reflections, with one dimension being the position on the detector and the other the scanning angle. For neutron data, the ratio of peak to background is usually poor due to incoherent scattering of hydrogen atoms present in the sample, and proper peak integration can vastly improve the quality of the resulting diffraction data.

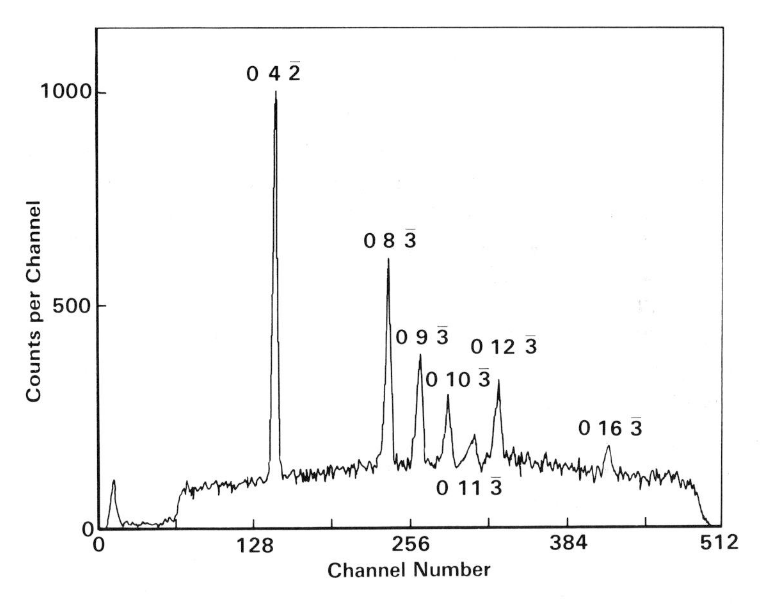

FIG. 3. Example of a frame of output of the linear detector. Data were taken using a crystal of RNase A. Counting time is 120 sec; the beam stop obscures the lower channels. Seven reflections are clearly visible. The high background is due to scattering by H atoms. Resolution for channel 490 is 2 Å. [By permission of *Acta Crystallogr. Sect. B* **B36,** 1826 (1980).]

A number of different integration techniques have been described in the literature. Of particular interest to us was the method of Spencer and Kossiakoff,[14] which was used in processing the diffraction data from trypsin. This technique assumed ellipsoidal shapes of reflections within their boxes and aimed at finding the direction and extent of the half-axes using the pattern recognition technique. A slightly more general technique, which we called "dynamic mask procedure," was derived for the flat-cone diffractometer.[15]

This procedure consists of several separate steps. The first one is to provide the best estimate of the background for each reflection, since in most cases the peaks are only a few percent above background. The lack of computer storage capacity precluded direct saving of all background points, and therefore, the following procedure was implemented. Each frame of data was checked, and all data belonging to reflection boxes were removed and stored on the disk. Remaining points were used to update a "universal background" array computed in a manner suggested by Xuong *et al.*[16] Each universal background point along the detector was recalculated by adding a suitable fraction of the new value (usually $f = 1/16$) to the old value multiplied by $(1 - f)$, and the values for the missing points were obtained by interpolation. When a complete reflection was collected, a polynomial was fitted by least squares to the part of the universal background in the vicinity of the reflection (usually 50 channels each way), and the resulting coefficient were stored together with the reflection. This procedure reduces the amount of background data from several hundred numbers to only three or four per reflection while preserving the information content. Estimated background is available for each reflection even before the data within each reflection box are considered, and this proves to be very useful in detecting the shapes, sizes, and positions of the peaks.

To calculate a mask for each reflection, the data are smoothed and a "statistical filter"[17-19] is applied to distinguish between the data belonging to the peak and the data belonging to the background. These procedures were modified in this case to include the initial estimate of the background based on the "universal background" data. Once the background level and its variance have been established, either on the basis of the points

[14] S. A. Spencer and A. A. Kossiakoff, *J. Appl. Crystallogr.* **13,** 563 (1980).

[15] L. Sjölin and A. Wlodawer, *Acta Crystallogr., Sect. A* **A37,** 594 (1981).

[16] Ng. H. Xuong, S. T. Freer, R. Hamlin, C. Nielsen, and W. Vernon, *Acta Crystallogr., Sect. A* **A34,** 289 (1978).

[17] G. C. Ford, *J. Appl. Crystallogr.* **7,** 555 (1974).

[18] W. Kabsch, *J. Appl. Crystallogr.* **10,** 426 (1977).

[19] M. G. Rossmann, *J. Appl. Crystallogr.* **12,** 225 (1979).

unlikely to contain peak information or on the basis of the universal background, the background is subtracted from each point and a flag is set, depending on whether the net intensity exceeds the background by less than σ (0), between σ and 2σ (1), 2σ and 3σ (2), or over 3σ (3). An example of the so-called "sigma array" can be seen in Fig. 4. It is quite evident that the peak is contiguous, while the noise in the background is random. We can now create a mask for this reflection from the contiguous part of the σ array, and it will include all points flagged 1, 2, or 3. We call this mask "dynamic" since it is calculated independently for each medium or strong reflection and its shape, size, and position of the center are not constrained.

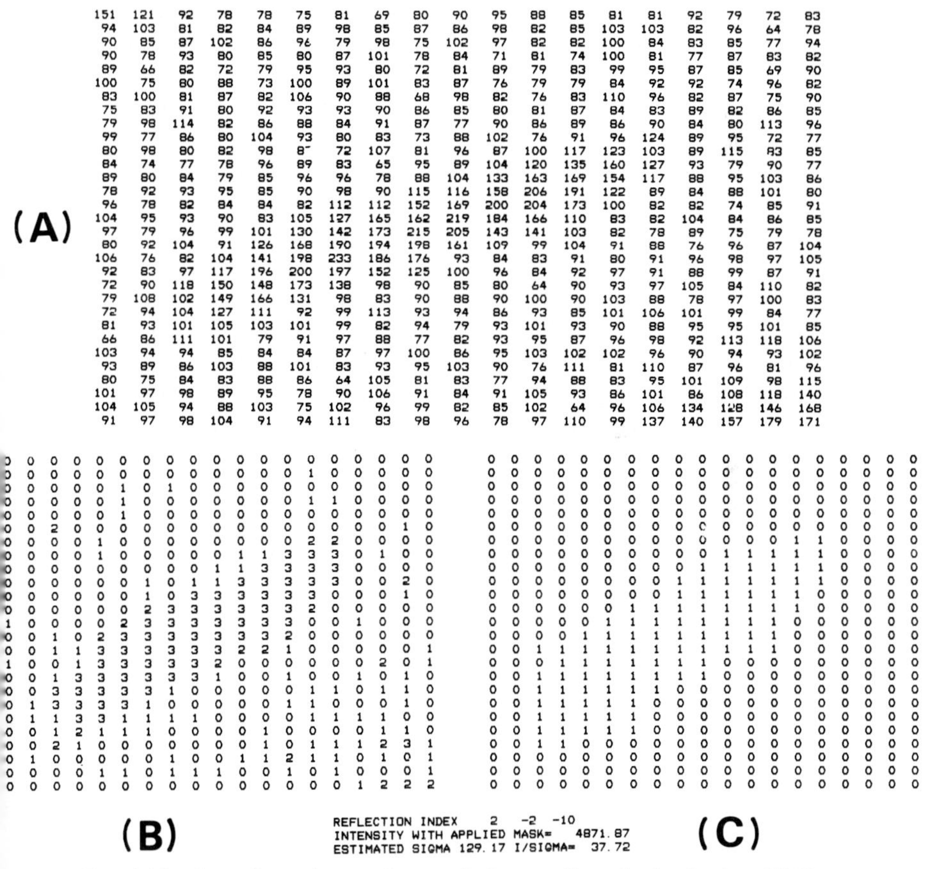

FIG. 4. The dynamic mask procedure applied to a well-resolved reflection. (A) Detector output; (B) σ array; (C) final mask (for details see text.)

The procedure described here will not provide accurate estimates of intensity for weaker reflections [$I < 5\sigma(I)$] which may not create a contiguous mask. In this case we apply a mask calculated for a neighboring medium or strong reflection and include the position of its center of gravity. In this way we can avoid biaising weaker reflections for which the intensities contained in the elements $\sigma = 0$ of the σ array could make a substantial contribution to the integrated intensity. For this purpose we store a number of masks corresponding to different areas on the detector and reciprocal space and update them throughout data processing.

The procedure used to calculate a dynamic mask made the assumption that the box under consideration contains only one reflection. This is not always the case, and while the presence of other reflections can be predicted from the orientation matrix of the crystal, such calculations may be cumbersome in practice, especially for geometries such as normal beam (rotation photography) or flat cone (used in our neutron studies). On the other hand, a comparison of universal and local background can quickly alert us to such a possibility since, if parts of other reflections are present in the area from which local background is computed, the value of local background will be higher than that for universal background.

As a bonus resulting from the dynamic mask procedure, the positions of the centers of gravity for those reflections which were sufficiently strong to describe their own masks became known. Thus the errors in the setting angles ψ and Υ are directly available. These errors are sufficient to provide input data for the orientation matrix refinement procedure of Shoemaker and Bassi,[20] by using formulas derived by Wlodawer *et al.*[21] While the accuracy of the estimate of misalignment of each reflection is not high, this is compensated for by the large number of available corrections. It should also be noticed that this information was strictly a by-product of the normal intensity measurements and was obtained without prolonging the experiment. The orientation matrix was recalculated after each level and provided much better estimates for the next plane to be collected. The absorption correction for the flat-cone diffractometer was described by Santoro and Wlodawer.[22]

The NBS neutron diffraction facility was first used for collecting 2.0 Å data from a large crystal of ribonuclease A. The crystal was mounted with its largest face placed on the flat bottom of a quartz tube and was immobilized with quartz wool. The tube was sealed with silicone grease, dental wax, and epoxy glue. The crystal has not moved and has shown no

[20] D. P. Shoemaker and G. Bassi, *Acta Crystallogr., Sect. A* **A26,** 97 (1970).

[21] A. Wlodawer, L. Sjölin, and A. Santoro, *J. Appl. Crystallogr.* **15,** 79 (1982).

[22] A. Santoro and A. Wlodawer, *Acta Crystallogr., Sect. A* **A36,** 442 (1980).

chemical decomposition in over 2 years.[3] A complete data set to 2.0 Å resolution was obtained by combining the data extending to 2.8 Å collected in the four-circle mode (see above) with the flat-cone data processed by the dynamic mask procedure. Unfortunately, due to the poor signal-to-noise ratio, almost half of the intensities were found to be unobserved [$I < 3\sigma(I)$]. Since the presence of the hydrogen atoms in the model increased the number of parameters, the degree of overdetermination in the neutron refinement would have been very low. As a possible solution, a refinement method specifically designed to increase the ratio of observations to parameters was introduced.[23] In the joint refinement technique, both the X-ray and neutron data were utilized simultaneously in each refinement cycle. This approach was based on an obvious observation that an atomic model for a crystal structure should be consistent with both the X-ray and the neutron diffraction data. Hence more accurate atomic parameters can be expected from a simultaneous refinement with the data from both kinds of radiation than from either separate refinement. Since the degree of overdetermination will be increased in a joint refinement, improved refinement behavior might also be expected. All of the data for such a procedure should, of course, be measured from essentially identical crystals. In particular, for macromolecules, both the X-ray and the neutron data should be measured from identically deuterated crystals in equilibrium with the same mother liquor. Such a suggestion for a joint analysis of X-ray and neutron data from macromolecules was made by Hoppe.[24]

The procedure that was adopted in practice required relatively minor conceptual modifications to the X-ray procedures for stereochemically restrained refinement.[25–27] This procedure introduced stereochemical and other prior knowledge regarding the structure into the least-squares minimization. These geometrical "observations" served as restraints on the atomic parameters.

There may be several qualitatively different kinds of observations. These include the structure factor data, "ideal" bond lengths and angles, planarity of certain groups, chirality at asymmetric centers, nonbonded contacts, restricted torsion angles, noncrystallographic symmetry, and

[23] A. Wlodawer and W. A. Hendrickson, *Acta Crystallogr., Sect. A* **A38,** 239 (1982).

[24] W. Hoppe, *Brookhaven Symp. Biol.* **27,** II-22 (1976).

[25] J. H. Konnert, *Acta Crystallogr., Sect. A* **A32,** 614 (1976).

[26] W. A. Hendrickson and J. H. Konnert, *in* "Computing in Crystallography" (R. Diamond, S. Ramaseshan, and K. Venkatesan, eds.), p. 13.01. Indian Acad. Sci., Bangalore, 1980.

[27] W. A. Hendrickson and J. H. Konnert, *in* "Biomolecular Structure, Conformation, Function and Evolution" (R. Srinivasan, ed.), Vol. 1, p. 43. Pergamon, Oxford, 1981.

limitations on bond and angle fluctuation due to thermal motion.[28] Thus the function to be minimized is in the form of

$$\Phi = \sum \phi_i \tag{1}$$

where each of the separate observational functions ϕ_i is usually (but not always) in the form of

$$\phi_i = \sum \frac{i}{\sigma_j^2} [f_j^{obs} - f_j^{calc}(\{x\})]^2 \tag{2}$$

Here each term relates to a particular observational quantity f^{obs} for which a corresponding theoretical value f^{calc} can be calculated from the set of refinable parameters $\{x\}$ or possibly some subset of these. Each term is weighted by the inverse of the estimated variance for the particular observation or, in the early stages, possibly by some other variance estimate. The joint refinement simply requires adding another term in Eq. (1). Thus we now have

$$\Phi = \phi_{\text{X ray}} + \phi_{\text{neutron}} + \phi_{\text{bonds}} + \phi_{\text{planes}} + \cdots \tag{3}$$

The major tasks involved in implementing the joint refinement procedure were those needed anyway for neutron refinement, namely the incorporation of hydrogen atoms. The changes to the actual refinement program, PROLSQ, to permit simultaneous use of both X-ray and neutron data mainly involved setting appropriate switches for reading the respective data sets, calculating structure factors and derivatives based on the appropriate scattering factors, and including a separate scale factor refinement. Also, a provision for a special class of nonbonded contacts, those involving hydrogen atoms that participate in hydrogen bonds, was included. Changes to PROTIN, the program that prepares the restraint observations for particular protein structures, were more extensive. A variety of program modifications were needed to allow for hydrogen atoms. In addition, new standard groups that included hydrogen positions were compiled. All of the restraint dictionaries were also appropriately upgraded. Distances involving hydrogen positions have been put into special weighting categories.

The results derived from the application of the joint refinement procedure to ribonuclease A[29] show that such a procedure is less likely to lead to serious errors than the separate refinement with neutron data alone (Fig. 5). It can be seen that for these side chains the coordinates corresponding to the joint refinement model are in good agreement with both

[28] J. H. Konnert and W. A. Hendrickson, *Acta Crystallogr., Sect. A* **A36,** 344 (1980).

[29] A. Wlodawer and L. Sjölin, *Biochemistry* **22,** 2720 (1983).

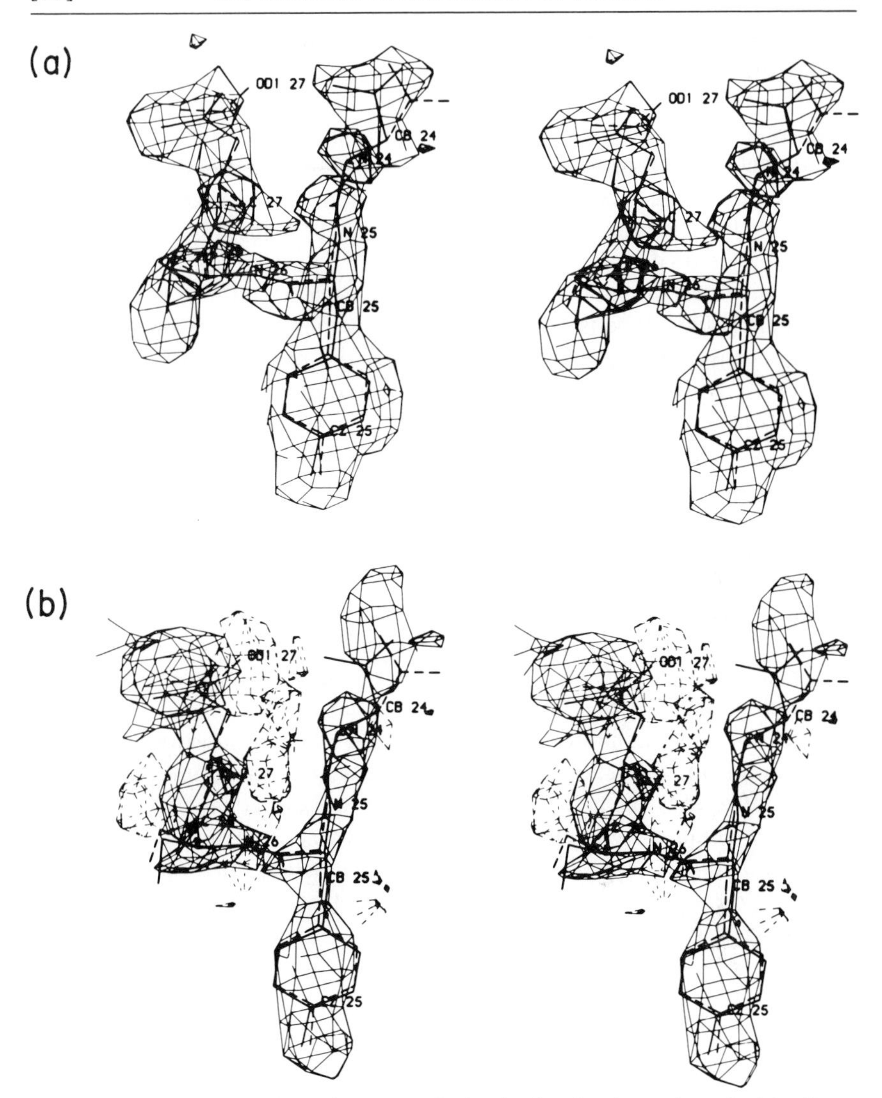

FIG. 5. "Fragment ΔF" Fourier maps calculated using the phases from the joint X-ray and neutron refinement of RNase and subtracting the contribution of residues 24–27. Coordinates after the joint refinement are marked in solid lines; those after the separate neutron refinement are dashed. (a) Electron density contoured at the 5σ level. (b) Nuclear density contoured at the $\pm 5\sigma$ level. Positive contours are solid; negative contours, dashed. [By permission of *Acta Crystallogr. Sect. A* **A38,** 239 (1982).]

maps, while those resulting from a separate neutron refinement are not. Even with an initial model in which many hydrogens were not properly placed, the refinement converged rapidly while the idealized geometry was preserved. The joint refinement, starting from a well-refined X-ray model, achieved a substantial improvement in agreement with the neutron data without appreciable change in agreement with the X-ray data. In contrast, the dramatic decrease in the neutron *R* value during refinement with the neutron data alone was accompanied by great deterioration in the match with the X-ray data and large, meaningless shifts in a number of side chains. This ill behavior of neutron refinement was not simply due to the poor observation-to-parameter ratio of the problem, since an X-ray refinement with a comparable ratio of diffraction data to variables was well behaved.

The neutron data collected using the National Bureau of Standards facility have now been used to refine the atomic parameters in the structure of native ribonuclease A[29,30] and to monitor the amide hydrogen exchange in that protein.[31] A complex of ribonuclease A with uridine vandate, a transition state analog, was studied by Wlodawer *et al.*[32] Neutron data extending to 2.2 Å resolution were collected for insulin (Wlodawer and Savage, unpublished) and have been used together with the X-ray data provided by G. Dodson for structure refinement. Recently neutron data extending to 1.8 Å have been collected from a crystal of bovine pancreatic trypsin inhibitor. This data set had a high proportion of statistically significant reflections (78% of the theoretical maximum, with over half of reflections in the shell at 1.82 Å observed). The structure of this important small protein was refined using these neutron data together with the X-ray data extending to 0.94 Å,[33] resulting in a very accurate model.[34] We consider the National Bureau of Standards data collection facility to be fully operational.

[30] A. Wlodawer and L. Sjölin, *Proc. Natl. Acad. Sci. U.S.A.* **78,** 2853 (1981).
[31] A. Wlodawer and L. Sjölin, *Proc. Natl. Acad. Sci. U.S.A.* **79,** 1418 (1982).
[32] A. Wlodawer, M. Miller, and L. Sjölin, *Proc. Natl. Acad. Sci. U.S.A.* **80,** 3628 (1983).
[33] J. Walter and R. Huber, *J. Mol. Biol.* **167,** 911 (1983).
[34] A. Wlodawer, J. Walter, R. Huber, and L. Sjölin, *J. Mol. Biol.* **180,** 301 (1984).

Author Index

Numbers in parentheses are footnote reference numbers and indicate that an author's work is referred to although the name is not cited in the text.

A

B

C

D

E

F

G

H

I

J

K

L

M

Q

R

S

T

U

V

W

X

Y

Z

Subject Index

A

B

C

D

E

N

O

P

Q

R

S

T

U

V

W

X

Z